AF869485

The Urban Book Series

Editorial Board

Margarita Angelidou, Aristotle University of Thessaloniki, Thessaloniki, Greece

Fatemeh Farnaz Arefian, Silk Cities, The Bartlett Development Planning Unit, UCL, London, UK

Michael Batty, Centre for Advanced Spatial Analysis, UCL, London, UK

Simin Davoudi, Planning & Landscape Department GURU, Newcastle University, Newcastle, UK

Geoffrey DeVerteuil, School of Planning and Geography, Cardiff University, Cardiff, UK

Paul Jones, School of Architecture, Design and Planning, University of Sydney, Sydney, NSW, Australia

Andrew Kirby, New College, Arizona State University, Phoenix, AZ, USA

Karl Kropf, Department of Planning, Headington Campus, Oxford Brookes University, Oxford, UK

Karen Lucas, Institute for Transport Studies, University of Leeds, Leeds, UK

Marco Maretto, DICATeA, Department of Civil and Environmental Engineering, University of Parma, Parma, Italy

Ali Modarres, Tacoma Urban Studies, University of Washington Tacoma, Tacoma, WA, USA

Fabian Neuhaus, Faculty of Environmental Design, University of Calgary, Calgary, AB, Canada

Steffen Nijhuis, Architecture and the Built Environment, Delft University of Technology, Delft, The Netherlands

Vitor Manuel Aráujo de Oliveira, Porto University, Porto, Portugal

Christopher Silver, College of Design, University of Florida, Gainesville, FL, USA

Giuseppe Strappa, Facoltà di Architettura, Sapienza University of Rome, Rome, Roma, Italy

Igor Vojnovic, Department of Geography, Michigan State University, East Lansing, MI, USA

Jeremy W. R. Whitehand, Earth & Environmental Sciences, University of Birmingham, Birmingham, UK

Claudia Yamu, Department of Spatial Planning and Environment, University of Groningen, Groningen, Groningen, The Netherlands

The Urban Book Series is a resource for urban studies and geography research worldwide. It provides a unique and innovative resource for the latest developments in the field, nurturing a comprehensive and encompassing publication venue for urban studies, urban geography, planning and regional development.

The series publishes peer-reviewed volumes related to urbanisation, sustainability, urban environments, sustainable urbanism, governance, globalisation, urban and sustainable development, spatial and area studies, urban management, transport systems, urban infrastructure, urban dynamics, green cities and urban landscapes. It also invites research which documents urbanisation processes and urban dynamics on a national, regional and local level, welcoming case studies, as well as comparative and applied research.

The series will appeal to urbanists, geographers, planners, engineers, architects, policy makers, and to all of those interested in a wide-ranging overview of contemporary urban studies and innovations in the field. It accepts monographs, edited volumes and textbooks.

Now Indexed by Scopus!

More information about this series at http://www.springer.com/series/14773

Fatemeh Farnaz Arefian · Judith Ryser ·
Andrew Hopkins · Jamie Mackee
Editors

Historic Cities in the Face of Disasters

Reconstruction, Recovery and Resilience of Societies

Silk Cities

Springer

Editors
Fatemeh Farnaz Arefian
Silk Cities, Bartlett Development Planning Unit (DPU)
University College London (UCL)
London, UK

Andrew Hopkins
Dipartimento di Scienze Umane
University of L'Aquila
L'Aquila, Italy

Judith Ryser
Urbanist, Silk Cities
London, UK

Jamie Mackee
School of Architecture and Built Environment
University of Newcastle Australia
Callaghan, NSW, Australia

ISSN 2365-757X ISSN 2365-7588 (electronic)
The Urban Book Series
ISBN 978-3-030-77355-7 ISBN 978-3-030-77356-4 (eBook)
https://doi.org/10.1007/978-3-030-77356-4

© The Editor(s) (if applicable) and The Author(s), under exclusive license to Springer Nature Switzerland AG 2021
This work is subject to copyright. All rights are solely and exclusively licensed by the Publisher, whether the whole or part of the material is concerned, specifically the rights of translation, reprinting, reuse of illustrations, recitation, broadcasting, reproduction on microfilms or in any other physical way, and transmission or information storage and retrieval, electronic adaptation, computer software, or by similar or dissimilar methodology now known or hereafter developed.
The use of general descriptive names, registered names, trademarks, service marks, etc. in this publication does not imply, even in the absence of a specific statement, that such names are exempt from the relevant protective laws and regulations and therefore free for general use.
The publisher, the authors and the editors are safe to assume that the advice and information in this book are believed to be true and accurate at the date of publication. Neither the publisher nor the authors or the editors give a warranty, expressed or implied, with respect to the material contained herein or for any errors or omissions that may have been made. The publisher remains neutral with regard to jurisdictional claims in published maps and institutional affiliations.

This Springer imprint is published by the registered company Springer Nature Switzerland AG
The registered company address is: Gewerbestrasse 11, 6330 Cham, Switzerland

Preface

The story of this book is intrinsically linked to the development of Silk Cities. It represents another milestone in its journey. Silk Cities is a growing independent and bottom-up professional and academic initiative for knowledge exchange, research, engagement and advocacy on under-explored contextual and global challenges cities in historic regions face in real-life. Its initial geographic focus was on those countries along the Silk Roads in the Middle East and Central Asia, which suffered from a variety of destructive incidents especially in recent decades, ranging from natural hazards to human induced origins, from earthquakes to wars and conflicts. Therefore, post-disaster reconstruction, disaster management and risk reduction and urban resilience form important themes of Silk Cities activities (silk-cities.org). This is also linked to my personal practice-based experience. My four years of working with reconstruction stakeholders including disaster-affected families during post-disaster urban reconstruction in the context of the historic city of Bam, a world heritage site, triggered my doctoral study and further research and academic activities on the subject matter at international level. This experience was influential in founding Silk Cities.

Post-crisis city recovery is multidimensional and post-disaster reconstruction is a manifestation of physical recovery that must facilitate other kinds of recovery, including psychological, social and economic recovery and enhance future resilience of residents. Cities are complex interconnected social, engineering, economic and administrative systems. Managing urban reconstruction and recovery requires decision makers to understand and deal with layers of complexity, their interactions and their integrating processes. Doing so in an historic urban landscape significantly adds another layer of complexity. This is easier said than done. There is a need for multiperspective and multidisciplinary examinations of cases, listening to different voices and trying new approaches and tools.

The book *Historic Cities in the Face of Disasters* is Silk Cities' fourth collected publication. As acknowledged separately here, chapters of this book are based on reviewed and updated revisions of a selection of papers presented at the 3rd Silk Cities international conference. Silk Cities 2019, titled: Reconstruction, recovery and resilience of historic cities and societies, hosted by University of L'Aquila, 10–12 July 2019. The idea of holding the third Silk Cities international conference around historic cities and destructive disasters emerged from the 2nd Silk Cities conference in 2017, hosted by the Bartlett Development Planning Unit (DPU) at University College London (UCL) in 2017. The focus of Silk Cities 2017 was on reconnecting population with urban heritage in the Middle East and Central Asia that included a theme on post-disaster reconstruction. During and after the related thematic sessions at the second conference, the need for further discussions and more in-depth attention to this urgent matter was highlighted. Soon after, I first visited L'Aquila in 2018, nine years after the 2009 earthquake, and heard the testimony of the city's disaster-affected residents. L'Aquila was a clear case that, when disaster occurs, it is the city and its residents who bear the consequences of insufficient attention paid to complexities of organising urban reconstruction and to their role in the multidisciplinary aspects of city recovery. Post-disaster reconstruction and recovery in historic landscapes is an important yet niche subject matter, and given the fact that L'Aquila, like many disaster-hit cities in the Middle East and Central Asia, enjoyed a rich history of urban life and heritage it made sense to take the conference there.

The large geographical coverage of the papers presented at the conference in fact portrays the subject matter as a global challenge, for which this book and another e-publication, titled *Prerequisites for Post-Disaster Regeneration of Historic Cities* (Silk Cities 2021), serve as frontiers. Aligned with Silk Cities strategy, this book is forward looking and aims to set new directions and to initiate new discussions.

What Next?

Under normal circumstances, we would have been preparing our 4th international conference, but nothing has been predictable nor normal in 2020 and 2021 so far. Confronted by a global health crisis, the challenges the global community had to face because of COVID-19 tested the resilience of us all. With social distancing in place and cautionary measures likely even after successful vaccinations, Silk Cities moved towards digital tools. "Urban Talks" around new directions and critical thinking on our cities in the context of a global pandemic and beyond is one example.

Silk Cities continues to engage with both younger and experienced generations of academics and practitioners and the public who care for and have experience in dealing with real-life urban matters of cities.

Hope you enjoy the book.

London, UK Fatemeh Farnaz Arefian

Geographic coverage of this book—global level (graphic by Maria Diez, Fundacion Metropoli)

Geographic coverage of this book—Italy (graphic by Maria Diez, Fundacion Metropoli)

Acknowledgments

Preparing a peer-reviewed book of this volume during a global pandemic requires collective dedication. The editors therefore are grateful to Nafiseh Irani and Ali Puya Khani, our colleagues at Silk Cities, Maria Diez, and all authors for their commitment to the project, patience and flexibility to pursue it as it was envisioned. Also thanks to Springer colleagues at Urban Book Series who made this book possible in a challenging period of "work-from-home", especially Sanjievkumar Mathiyazhagn, Madanagopal Deenadayalan, Carmen Spelbos and Juliana Pitanguy.

This book has been sponsored by Silk Cities. Chapters of this book are based on reviewed and updated revisions of a selection of papers presented at the third Silk Cities international conference, Silk Cities 2019, entitled: Reconstruction, recovery and resilience of historic cities and societies. It was held at the University of L'Aquila, 10–12 July 2019. Initiated by Silk Cities, the conference was organised by Silk Cities, University of L'Aquila and University College London (UCL). Sponsors of the conference are acknowledged here: Silk Cities; The Bartlett Development Planning Unit (DPU), UCL; The University of L'Aquila; City Part nership Programme, UCL; The DICEAA, Dipartimento di Ingegneria Civile; Edile-Architettura ed Ambientale, Univaq; The DSU Dipartimento di Scienze Umane, Univaq; The Dipartimento di Eccellenza, Univaq, The History Department, UCL. Moreover, the Bartlett DPU at UCL kindly provided travel grants for students from low and middle income countries whose countries encountered destructive disasters. Organising conferences is a collective effort of academic and professional teams. This conference enjoyed support and contribution of the academic team, including strategic advisors, conference conveners and the scientific committee, which reviewed papers for the conference and provided feedback, also as guest speakers. They are acknowledged in alphabetic order: Prof. David Alexander, University College London, UK; Dr. Fatemeh Farnaz Arefian, University of Newcastle, Silk Cities & University College London, Singapore, UK; Prof. Yves Cabannes, Emeritus Professor in Development Planning, Portugal, UK; Prof. Lina Calandra, University of L'Aquila, Italy; Prof. Simonetta Ciranna, University of L'Aquila, Italy; Prof. Julio D Davila, University College London, UK; Dr. Donato Di Ludovico, University of L'Aquila, Italy; Prof. Alireza Fallahi, Shahid Beheshti University (SBU), Iran; Mr. Barnaby Gunning, Independent, UK; Mr. Arif Hassan, Independent, Pakistan; Prof.

Andrew Hopkins, University of L'Aquila, Italy; Prof. Paola Inverardi, University of L'Aquila, Italy; Prof. Cassidy Johnson, University College London, UK; Prof. Hidehiko Kanegae, Ritsumeikan University, Japan; Dr. Alexy Karenowska, University of Oxford, UK; Prof. Ramin Keivani, Oxford Brookes University, UK; Prof. Jamie Mackee, University of Newcastle, Australia; Dr. Roger Michel, The Institute for Digital Archaeology, UK; Dr. Iradj Moeini, Shahid Beheshti University (SBU), Iran; Mr. Babar Mumtaz, DPU Associates, Pakistan; Dr. Florian Mussgnug University College London, UK; Prof. Antonella Nuzzaci, University of L'Aquila, Italy; Dr. Richard Oloruntoba, University of Newcastle, Australia; Dr. Lucia Patrizio Gunning, University College London UK; Prof. Paola Rizzi, University of L'Aquila, Italy; Prof. Salvatore Russo, Iuav University of Venice, Italy; Ms. Judith Ryser, ISOCARP and Fundacion Metropoli, UK; Prof. Antonello Salvatori, University of L'Aquila, Italy; Ms. Anna Soave, DPU Associate, UN-Habitat Iraq Programme, Iraq; Prof. Alessandro Vaccarelli, University of L'Aquila, Italy; Prof. Suzanne Jane Wilkinson, University of Auckland, New Zealand.

Organising the conference to become a success is undoubtedly owed to the support of professional teams at Silk Cities, University of L'Aquila and the Bartlett Development Planning Unit (DPU) at UCL who behind the scene took the responsibility for various stages and tasks during the whole process and made it happen. They are acknowledged here. Professional team at Silk Cities: Maryam Eftekhar Dadkhah, Belgium; Nafiseh Irani, Singapore; Mona Jabbari, Portugal; Ehsan Fatehifar, Iran. Professional team at University of L'Aquila: Carlo Capannolo, Italy; Michela Fazzini, Italy; Sabrina Madia, Italy; Massimo Prosperococco, Italy; Alfonso Pierantonio, Italy. Professional team at the Bartlett DPU at UCL: Hartley, Jacqueline, UK; Alexander, Macfarlane, UK; Ottavia Pasta, UK.

Thank you all!

Contents

About the Editors

Fatemeh Farnaz Arefian is an interdisciplinary expert in disaster management and reconstruction, urban design and development. She is the director of Silk Cities. Her professional life combines industry experience with academic research and education, knowledge exchange and public engagement in Europe, Middle East, Central and Southeast Asia. Her industry experience includes delivering urban development plans and architectural projects, including various post-earthquake reconstruction projects, e.g. participatory housing reconstruction and post-disaster urban design projects. Those first-hand encounter with urban development challenges and disasters, in the context of historic cities motivated her to return to academia and pursue her interdisciplinary doctoral study multi-objective reconstruction programmes in the context of historic cities.

She has authored and edited books and papers, such as her authored book *Organising Post-Disaster Reconstruction Processes* (Springer 2018). She is the lead editor of books *Urban Heritage Along the Silk Roads* (Springer 2019) and *Urban Change in Iran* (Springer 2016). She is the co-editor of eBook *Prerequisites for Post-Disaster Regeneration of Historic Cities* (Silk Cities 2021). Farnaz is also affiliated with the Bartlett Development Planning Unit (DPU) at University College London, UCL.

Judith Ryser qualified architect and urbanist with a social sciences M.Sc., Judith's cosmopolitan professional life in London, Paris, Berlin, Geneva (United Nations), Stockholm and Madrid is focusing on built environment sustainability and researching, reviewing, writing on cities in the knowledge society. She is a life member of ISOCARP (International Society of City and Regional Planners), ex-Vice President, General Rapporteur of the 50th anniversary congress 2015, editor and writer of several publications (e.g. "ISOCARP, 50 Years of Knowledge Creation and Sharing"; with Teresa Franchini 5th and 6th editions of the International Manual of Planning Practice) and member of the Chartered Institute of Journalist. She is senior advisor to Fundacion Metropoli, author and editor of many books and participant in urban projects; senior adviser, book co-editor and co-reviewer of Silk Cities; co-editor and coordinator of CORP (International Conference on Urban Planning and Regional Development in the Information Society); editorial board member, reviewer and topic editor of the Urban Design Group and has written and edited numerous books and articles. She taught at University College London and other universities, is on various scientific committees and mentoring mature students and young planners.

Andrew Hopkins FSA is professor and chair of architectural history at the University of L'Aquila. Architecture and urbanism, together with how people use city spaces in an historical context, is a long-standing interest. His earliest and latest books are *Santa Maria della Salute: architecture and ceremony in baroque Venice*, Cambridge University Press, Cambridge, 2000; and *La città del Seicento*, Laterza, Roma-Bari, 2014. Having taught at L'Aquila since 2004, in the last, post-earthquake, decade he has been involved in several initiatives to do with understanding the issues facing historical cities vis-à-vis disasters, including the exhibition and conference held in the ducal palace of Venice in 2010: L'Aquila 2010, Luogo, Identità, Etica, Ricostruzione, Salone del Piovego, Palazzo Ducale, Venezia 28 Agosto 2010; and recently he was thrilled to be able to welcome the Silk Cities conference of 2019 to the Department of Human Sciences (*Dipartimento di Scienze Umane*), University of L'Aquila.

Jamie Mackee's most recent research has focused on the way in which risk of climate change impacts can be assessed to better protect our cultural heritage. This has resulted in the development of a Cultural Heritage Risk Index (CHRI) that allows stakeholders involved in the management, protection and conservation of cultural heritage to assess risk and plan and manage interventions to mitigate against any potential impacts. The results of this are being published in a forthcoming journal paper.

Jamie's broader interest in old buildings' stems from his eight years teaching in a University and living in Sri Lanka and two years studying in Singapore. His teaching and research work on disasters arises from his work during the Asian Tsunami in Sri Lanka sponsored by the University of Newcastle in 2005. He has two books specifically on cultural heritage and one recent book and book chapter on cultural heritage conservation theory fusing the concepts of Buddhism and Systems theory proposing an alternative to the traditional Eurocentric ideals of conservation theory for dealing with the rich and very unique heritage of the Asian region. Jamie is a co-leader of the School of Architecture and Built Environment's Disaster Research Group, University of Newcastle, Australia.

Chapter 1
Introduction: Towards Multi-perspective and Multidisciplinary Approaches

Fatemeh Farnaz Arefian, Judith Ryser, Andrew Hopkins, and Jamie Mackee

Abstract In a post-disaster situation, whether because of destructive natural or human-induced hazards, the interrelations of historic urban contexts, social, psychological and economic factors present a highly complex scenario for decision makers, practitioners and affected communities and individuals. This chapter makes the case and offers an entry point to this book, historic cities in the face of disasters. Examining historical and contemporary cases of dealing with disasters in historic landscapes, especially the historic built environment reconstruction, provides the opportunity for self-reflection of the part of the global community regarding the past and where theory, policy and practice stand today, to help for directions for a better future. Shockingly, there is extensive evidence that historical experiences have not led to organisational learning in many cases. There is a need to apply a multi-perspective and multidisciplinary examination of this complex subject matter, and this is a phenomenon that occurs as a global challenge which this book aims to explore. Focusing on the bigger picture, the chapter also reflects on views and approaches in different cases discussed in the book and examines their communalities and contrasts to pose a number of key questions and food for thought that can be traced back and in the book. It highlights the everlasting dilemma when thinking about reconstruction in historic landscape, when the obsession with a fixed moment of the past versus updating damaged historic urban fabric; the question of if heritage can be used as

F. F. Arefian (✉)
Silk Cities, Bartlett Development Planning Unit, UCL, London, UK
e-mail: f.arefian@silkcities.org

J. Ryser
Urbanist, Silk Cities, Senior Adviser Fundacion Metropoli Madrid, London, UK

A. Hopkins
Dipartimento di Scienze Umane, University of L'Aquila, L'Aquila, Italy

J. Mackee
School of Architecture and Built Environment, University of Newcastle Australia, Callaghan, NSW, Australia

© The Author(s), under exclusive license to Springer Nature Switzerland AG 2021
F. F. Arefian et al. (eds.), *Historic Cities in the Face of Disasters*, The Urban Book Series,
https://doi.org/10.1007/978-3-030-77356-4_1

an excuse for non-participatory reconstruction approaches and potentials and conditionalities of the application of new and smart technologies for reconstruction of historic cities and their enhanced resilience. Multi-scale examinations offer multidisciplinary, cross-sectoral and complimentary voices from multiple stakeholders. Many chapters adopted different standpoints and tapped on the body of knowledge belonging to a broad range of disciplines and push boundaries of the intersection of urban, disaster and heritage discourses.

Keywords Historic cities · Reconstruction · Disaster management · Built heritage · Social recollection · Multidisciplinary approaches · Multi-perspective approaches · Resilience · Cultural heritage

1.1 Introduction

In a post-disaster situation, whether as a result of destructive natural or human-induced hazards, the interrelations of historic urban contexts, social, psychological and economic factors present a highly complex scenario for decision makers, practitioners and affected communities and individuals. This complexity requires multi-perspective and multidisciplinary approaches with the participation of peoples and communities to ensure that the solutions to these interrelationships are sustainable and resilient. There remains a need for greater reflection, analysis and discussion of the issues as this book begins to explore.

Disasters moved away from being an act of god with the 1990s witnessing the beginning of analysts linking disaster studies intentionally to development, attributing importance to society being considered and establishing an international agenda (Gaillard 2017). International advancements in this field gradually developed thereafter. At the first World Conference on disaster risk reduction held in 1994, the global community devised the first international agenda on disaster management and risk reduction; later, the Hyogo World Conference set the agenda for 2000–2015, followed by the Sendai Framework which has set the agenda for 2015–2030 (UNISDR 2005, 2015). Perceptions of reconstruction have also shifted from replacing lost assets to facilitating the process of producing products. In recent decades, expectations from reconstruction include non-physical opportunities that blend notions of reconstruction and recovery. Urban reconstruction has become the strategic physical agent for broader and multi-dimensional recovery that is linked with future development and resilience, yet the question of "how" in practice this is to be achieved is still underexplored (Arefian 2018). In the meantime, hopes of what can be achieved in reconstruction are being shaped. A key common expectation is to integrate strategies, measures and other tools to reduce future disaster risk in the reconstruction process and to enhance urban resilience. Another aim is to enable active citizen participation in the reconstruction of their physical and social and assets.

In parallel, the domain of cultural heritage studies is also progressing. Specifically, urban heritage gradually became perceived as a dynamic asset for the future instead of a deadweight from the past, moving away from static museum-style conservation of historic buildings and neighbourhoods to the recognition of the many multi-faceted layers of urban life (Ciccozzi 2015; Darlington 2020). Urban heritage in the mid-twentieth century was associated with traditional approaches to conserving historic buildings and town centres. Beginning in European city centres, urban conservation was extended to other parts of the world, driven by organisations such as UNESCO that established a special category of cultural property described as "groups of buildings" in the World Heritage Convention of 1972 and generally associated with towns. Since then, however, overall values resulting from the integration of different urban components gradually came into recognition. This was the result of UNESCO's promotion of an extended approach to urban heritage that goes beyond mere conservation of the built environment and integrates social, economic and functional dimensions. The UNESCO Recommendation on Historic Urban Landscape (HUL) of 2011 provided a significantly more global vision and gave special prominence to the communities that inhabit historic towns and historic centres. This approach also implied a disciplinary expansion of horizons, with an increasing number of inputs being sought from and coming from social sciences (García-Hernández and Calle-Vaquero 2019).

The New Urban Agenda (UN 2017) adopted at the United Nations Conference on Housing and Sustainable Urban Development (Habitat III) in Quito, Ecuador, in 2016 brought together the above and other qualitative aspects envisioned for cities and urban development. Yet, disasters do not wait for fragmented disciplines to come together first. With the increase of disastrous and destructive incidents in historic cities (examples are L'Aquila, Aleppo, Kathmandu, Erbil and Bam), the global challenge remains: what to do when historic cities are hit by destructive incidents? How to balance competing forces and urgent demands in such a context in a highly complex and uncertain post-disaster situation? Acknowledging expectations about dealing with the historical landscape of a city greatly intensifies and amplifies the complexities of post-disaster urban reconstruction, which is already high and requires multi-objective operational mega-systems (Arefian 2018). Even if policy-makers, professionals and indeed stakeholders have not thought about this before, they must deal with it in the field and in practice.

A historical monument or an entire urban fabric defined as "urban heritage" forms part of the same city which also can include contemporary areas; in general, cities mostly have both historic and contemporary areas. Yet the city as a whole witnesses the influence of reconstruction strategies on the way these historic parts are dealt with following major destruction. Large-scale reconstruction of such cities not only has to deal with "heritage conservation projects" but also with urban governance processes and all other complexities of post-disaster reconstruction that go well beyond the scale of a "heritage conservation project" when the city enters the urban reconstruction programme level. This is a combined challenge for the disaster management system and the urban system. There is a need to apply a multi-perspective and multidisciplinary examination of this complex subject matter, and this is a phenomenon that occurs as a global challenge which this book aims to explore. Bringing together the complexities of post-disaster reconstruction and urban development processes of historic cities, something that ICOMOS has acknowledged as needing to be explored (ICOMOS 2014, 2016), its entry point is from the urban disaster side, and it deliberately brings together disaster and urban studies in the context of historic landscapes. Referring to "historic cities" instead of "urban heritage" in the title of this book acknowledges that historic cities are a much broader entity as they deal with various scales of urban development in historic landscapes. Moreover, when tackling disasters in the context of cities, there is a further need to address resilience in the face of disasters as well. The contributions in the book examine the reconstruction and resilience of historic cities and societies throughout the world from multiple disciplinary and complementary perspectives, aspiring to assist researchers and practitioners alike, among them reconstruction managers, urban governance officials and professionals.

This multi-scale and multidisciplinary book offers cross-sectoral and complimentary voices from multiple stakeholders. Although recent disasters of both natural and conflict origins in the Middle East and Central Asia have increased international concern, this issue and awareness of it have existed ever since post World War II reconstruction in other regions, as well as even before. The book, therefore, examines historical approaches to the reconstruction of historical cities and those of more recent times and case studies from Europe, Asia, Africa and Americas that together provide a valuable collection for all those readers who would like to obtain a global overview of the subject matter. This volume enables a deeper understanding of challenges, opportunities and approaches in dealing with historic cities facing disasters at various geographical scales. Bringing together historical and contemporary cases of reconstruction offers opportunities for self-reflection on the past for stakeholders, academics and the global community as well as working towards where theory, policy and practice stand today. Shockingly, there are evidences that historical experiences have not resulted in organisational learning curves or change in many cases.

Disciplines represented in the various chapters address key themes in detail and include disaster management of heritage emergency and reconstruction; urban governance; tactical urbanism and comprise architecture, urban design and planning; management; urban development; architectural history, cultural heritage, the arts, religion; sociology, psychology, behavioural sciences; together with technologies including construction, engineering, information and communication technology (big data, digitalisation, visualisation, social media); tourism and tourism economy; journalism; systems analysis, statistics; economy, political sciences; archaeology; energy consumption, ecology, climatology, meteorology; education; photography.

The examples in the book are chosen from a large geographical range, encompassing developing and developed countries at different stages and holding a broad variety of cultural and political values. They are comprising the following countries in alphabetic order: Chile, Guatemala, Greece, India, Iran, Italy, Japan, Lebanon, Libya, Peru, Poland, Portugal, Spain, Syria, United Kingdom, USA and Vietnam.

1.2 Cultural Heritage, Disasters and Social Recollection

Cultural heritage does not end at monuments and collections of objects (UNESCO, n.d.). It extends to communities' knowledge, skills and social practices that are manifested in various ways and shape relationships between human, natural and the built environments. Cultural heritage is here defined as tangible and intangible culture together with natural heritage and memory. Throughout the world damage to cultural heritage due to the impacts of natural- and human-induced hazards has been enormous. Much of the world's heritage is located in the earthquake-prone zones of Italy, Iran and Japan (the Pacific ring of fire); the conflict-troubled region of Lebanon together with vulnerable and often densely populated areas along coastal belts or rivers is at risk of damage from tidal surges, floods and tsunamis (Jigyasu 2016; Tahboroff 2003). Despite the importance of heritage as a conduit of memory and for propagating collective identity, restoration of damaged heritage is often neglected in post-disaster reconstruction plans and in the development of disaster mitigation strategies. Disaster management agencies often do not distinguish heritage from the general built environment, and there is wide divergence in individual countries' official interest in incorporating cultural heritage into the activities of the disaster management community (e.g. Jigyasu 2016; Tahboroff 2003; Look and Spennemann 2000, 2001).

As a political act, reconstruction is often focusing on the material fabric with little consideration for the pre-disaster context of heritage and population. Bland statements might be invoked, the promise of a modern city with future-proof jobs and real estate fit to compete in the globalised world, but these are rarely delivered and lead invariably to displacement and exclusion. Conversely, in the hands of conservationists, reconstruction consists of rebuilding what in reality was an imaginary historic city with a focus on listed buildings, leading to stagnation and stultifying urban, economic and social development. These deliberations indicate a trend away from concentrating solely on the physical fabric—also for post-disaster reconstruction—and a shift towards criteria related to human needs and the state of the environment.

Many studies in this book stress the need for interdisciplinary considerations in identifying the values to be safeguarded without jeopardising authenticity, development or the organic rebirth of damaged places. In his "Encyclopaedia of Vernacular Architecture of the World", Paul Oliver (1997) demonstrates how vernacular designs and use of materials are not necessarily providing simple climatic functionality but are just as much influenced by a diverse range of cultural motivations. This indicates that prevailing rationales of cultural heritage are subjective and tend to reflect bias towards the physical fabric over social concerns.

Cultural heritage triggers pride in one's own identity. That is why, in wars, one witnesses the destruction of cultural heritage as a tactic aimed at undermining communities and disheartening them. By the same token, shared built cultural heritage has the potential to become a tangible ground for unification and peace making during and after conflicts. It has the healing capacity. However, post-conflict reconstruction is a sensitive matter, and the process of post-conflict reconstruction has the potential either to advance social recovery or to hinder it. Everyone might hope for the sustainability of a fragile peace and of what is to be reconstructed, but nevertheless there remain fears of failure and relapse. Consequently, the inclusion of all the affected communities is crucial as is strong coordination which is even more critical than the issue of post-earthquake reconstruction. Aleppo is a classic case (Chapters 15 and 10).

Extending heritage beyond the built heritage, the importance of cultural heritage is not only the cultural manifestation of monuments and listed buildings or maintaining cultural diversity. It is also about the wealth of knowledge, skills and social traditions that can be a key in preparing for disasters, managing disasters and recovering from it as the case of Otsuchi, Japan (Chapter 6).

Contained in archives, libraries and museums, the existence and survival of documentary memories of those communities whose built fabrics were significantly damaged, or even completely destroyed, is essential for the reconstruction of community identity and building resilience. Hence, the creation of a (Chapter 5) shared project between the relevant territories and stakeholders aimed at preserving the vast documentary heritage that survives, often in highly precarious conditions. This was fundamental for rebuilding the local communities and demonstrated a way of thinking that went beyond merely considering the built fabric.

From a cultural heritage standpoint, the critical analysis of several reconstructed earthquake sites in Italy (Chapter 13) compared their divergent approaches to both the preservation of historic monuments and mundane built fabrics, as well as highlighting the damage done to historic architecture through demolition after seismic events. The rationale attributed to these approaches varied from security and the necessity to minimise displacement, to better functionality and modernisation for economic reasons.

Extending heritage to the relationship between small towns and their rural landscape in Italy is the examination of the rural landscape and Marcite meadows of the Ticino as driver for resilience. The study (Chapter 23) aimed to understand how these landscapes were, and if so to what extent, part of the identity, memory and image of the local inhabitants and what kind of participation from the local community was possible in anticipation of having to face potential disasters.

1.3 Urban Reconstruction, Recovery and Resilience

Resilience in general disaster discourse relates to both pre- and post-disaster situations. On the latter, it strongly underpins post-disaster reconstruction, with notions and aims of building back better and with adaptive capacity driving the process. Resilient thinking in the context of urban settings has advanced from simple engineering perspectives towards socio-ecological perspectives, recognising that the social systems of human settlements are inextricably linked to each other and to the ecosystem they use and depend on (UN-Habitat 2017; Global Facility for Disaster

Risk Reduction-GFDRR 2016). Resilience in the face of disasters is the "ability of a system, community or society exposed to hazards to resist, absorb, accommodate, adapt to, transform and recover from the effects of a hazard in a timely and efficient manner, including through the preservation and restoration of its essential basic structures and functions through risk management" (UNISDR 2017). Thus, the built environment should be designed, located, built, operated and maintained in a way that maximises the ability of built assets, associated physical and institutional systems and the people to withstand, recover from and mitigate for the impacts of extreme natural- and human-induced hazards (Bosher 2008, p.13).

As a strategic physical agent for physical, economic and social recovery, as well as resilience-oriented improvements, large-scale urban reconstruction and retrofit is in itself a complex matter, at present well acknowledged in disaster management, reconstruction and urban development studies and practice. Urban reconstruction connects all aspects of urban life and development before and after the crisis, initial expectations and envisioned improvements from the reconstruction activities and the way they are managed. Following a disaster, the concept of resilience ultimately should be translated into the governance of urban systems in the short-term phase of reconstruction as well as in the learning loop and transformations acting as milestones for any mid- or long-term urban development decisions (Schuetze and Chelleri 2011).

The constant narrative across four nations, Japan, Iran, Lebanon and Italy rich in cultural heritage and at the same time impacted by significant natural- and human-induced hazards, is one of resilience and the awareness that the future can be better than the past. These stories are, firstly, ones that embrace change and see this as an opportunity. Secondly, they appreciate that change is also embedded within the social and natural systems and in this context see multidisciplinarity and community led approaches as a means of achieving resilience, particularly regarding intangible and tangible heritage.

Rising out of the devastating L'Aquila earthquake come accounts of the work of these Italian communities attempting to deal with the inequalities of an education system damaged by the substantial impacts of the earthquake. Education poverty is a term adopted (Chapter 26) to explore the deficiencies that arise when schools are disrupted through hazardous events and the steps that can be taken by the community to overcome these issues. Similar scenarios of resilience when considering the impact of natural hazards upon a hospital and in urban centres are pursued adopting multidisciplinary approaches ensuring the resilience of tangible cultural heritage (Chapter 37).

From Japan (Chapters 6 and 33) come stories of how intangible heritage practices and traditional knowledge both play a role in investigating methods for managing evacuations in the immediate aftermath of significant disaster events. Intangible cultural practices offer an alternative to the resilience management of evacuation centres with demonstrable evidence of success.

Lebanon's story (Chapter 19) is one of many years of corruption, conflict and the misuse of an incredibly rich and diverse cultural heritage, culminating in the recent devastating explosion in Beirut Harbour. This heritage had been appropriated by a particular worldview of a modern city that is at odds with that of the residents and traditional approaches to protecting this heritage. However, ultimately, the story is one of hope and resilience as there is a way of exploring the current situation and providing a way forward in which there will be the opportunity to return the city to its people

1.4 The Dilemma: "Where It Was—How It Was" Versus Updating Damaged Historic Urban Fabric

The book provides evidences that the dichotomy between "where it was and how it was" approach and a broader range of criteria about what to retain, what to repair and what to reconstruct is still prevalent. Therefore, revisiting the understanding of sustainability and resilience is required. First proposed in the Brundtland Report by the World Commission on Environment and Development (1987), the tripartite dimension of sustainability, an overlap of the economic, the social and the environmental signifying the feasible, equitable and liveable, respectively, is still widely applied. More recently though in "Urban Sustainability in Theory and Practice: The Circles of Sustainability" (Paul James et al. 2015) refined the circles of sustainability into economics, politics, ecology and culture. Sylwia Ulicka (2020) in Designing Sustainable Cities went a step further by questioning material culture altogether when designing for sustainability. Contesting the paradigm based on the domination of nature, she considers architecture as alienating humanity from nature by separating outside space from inside space, thereby also increasing alienation between social groups. In her view, this insulation from the processes of nature is at the very heart of unsustainability, and energy-oriented ecological design reflects this contradiction. New ecological thinking led to using principles of life as the basis for key concepts of architectural design. According to Kurokawa (1998), value of life resides in diversity, while a shift to information forms part of a paradigm for improving the coexistence between humans and nature. Life as information, together with life as chemistry, evolution and natural selection, the gene as test of time and the cell as biology's atom are the building blocks of life, elaborated by Paul Nurse in "What is Life, understanding biology in five steps" (Nurse 2020). This evolution of knowing is relevant to this book as it attempts to extend guidance for physical and sociocultural intervention beyond narrow knowledge of specialised disciplines to understanding humans as part of nature and drawing on the symbiosis between nature, scientific technology of reason, feelings and psychological needs.

The prevailing dichotomy can be contested on two accounts, one related to the definition of history and the meaning of heritage, the other to the dynamic of cities, societies and their values which they reflect. Built environments and landscapes alike are constantly undergoing transformations, due to many influences, human-induced disasters being only one of these. Degradation and decay are the most obvious imprints of the passage of time, but also demolition and reconstruction, whether triggered by disruptive natural events or human interventions.

Without descending into random post-modern relativism, both abstract human-made definitions and concrete perceptions and preferences of their time continue to influence decisions about the post-disaster treatment of heritage. It is important therefore to specify the understanding of history, together with prevalent social and political values, on which the latest theories or practices of post-disaster reconstruction rest. Where does history start, where does it end? In determining heritage, does history extend to the present, to times within living memory, or does it refer to certain periods specified by historians and rank ordered into abstract memory by conventional wisdom?

These questions may point to the futility of pursuing any definite or unique answer, thus queering the pursuit of an ideal solution for heritage reconstruction. They also provide the challenging rationale adopted in this book to resort to a pluri-disciplinary, multi-cultural approach of exploring this issue in a wide range of geographies, while cascading across diverse scales and timeframes. The purpose of this book thus reaches beyond case study comparisons. Instead, it establishes synergies between academic disciplines—especially across the natural sciences and humanities—and experience acquired through practice, when trying to understand the material and immaterial impacts of disruptive events from which solutions for reconstruction can be developed. This implies considering a complex set of conventional and subjective perspectives of the main stakeholders involved which, due to the severity of damage, often goes beyond local interests.

From the disaster-development perspective, a disaster—whether of human-induced or of natural hazard origins—disrupts the developmental path of the disaster-hit city and region. Post-disaster reconstruction thus forms part of the new, overall development path of the city and its hinterlands. It ought to link the past and the new adjusted future. It is often the case that reconstruction provides space for new discussions and ideas. Some chapters analyse how historical and contemporary reconstructions led or hindered new discussions and ideas which have been influential in shaping new futures. For example, in Italy, the earthquake and its post-disaster reconstruction in Marsica became a platform for cultivating new discussions about urban expansion and modernisation, as well as restoration and new technologies at that moment in time in the country (Chapter 10).

Yet, again in Italy, coming full circle (Chapter 16), studies on the case of reconstruction in L'Aquila post-2009, revealed, in a relatively shocking way, how obstinate the obsession with rebuilding "where it was and how it was" continues to be. This desire to entirely obliterate any evidence of the earthquake damage to the city and its inhabitants has in large part been embraced by the local *Sopraintendenza* (Cultural Heritage Authority in Italy) and restorers, while local participation was mostly hindered. The case was made that younger generation of architects ought to have a primary right to make a contemporary mark on L'Aquila, but that this has been ignored because it had been decided by the powers that be that the old façades in the city had to be kept intact—even artificially in the cases where newer buildings were constructed behind these (Chapter 9). Interestingly, in the same urban reconstruction, the desire to inject high-technology smart urban infrastructure was apparent. Indeed the case of L'Aquila exemplifies very contradictory approaches within the same reconstruction project: the city wants a rapid transformation in order to become a smart city that can compete with other European cities, even if these projects are not well defined, but it also wants to look precisely old, stuck in a desired past moment (Chapters 29 and 19). An implicit desire to enhancing future local tourism and tourism economy of L'Aquila might offer missing links bringing those contradictions together, although it was never explicitly set out or articulated cogently of by the decision makers.

It seems to be the case that post-disaster reconstruction, wherever in the world it may be, is still an undertaking that, because of its usually local and specific nature, is something that continues to occur on a local, regional or national level and that the specific approach adopted can range from experimental and innovative to traditional and highly conservative.

1.5 Non-Participatory Reconstruction: Heritage as an Excuse?

Within disaster management discourse, practical experience in international cases has showcased the significant advantages stemming from the active participation of people in the reconstruction of their houses in pursuing multi-dimensional recovery, including social and psychological recovery. It is now widely recognised that participatory reconstruction enables people to bounce back from their traumas. Given the above context, the very existence of recent non-participatory approaches is truly surprising. Yet, while contemporary international frameworks such as the Sendai Framework advocate the active engagement of disaster-affected citizens in reconstruction, there are still cases of historic cities that do not engage their local citizenry in reconstruction or limit their decision-making. This phenomenon appears not to be a consequence of either being a developed or a developing country, as it happens in both groups of countries, thus appearing to be much more what could be described as a mindset.

Citizen participation in reconstruction in the face of disaster emerged as being highly significant across many of the studies in different contexts. After an earthquake in Talca Chile (Chapter 20), for example, the local community, emotionally attached to its urban fabric, exercised its capacity of collective resilience by securing the reconstruction of the communal school and the central square connected to it. This was contested by the local authorities claiming that local communities could not deliver comprehensive reconstruction. The same argument was applied even more forcefully to the reconstruction of Benghazi Libya (Chapter 17) where residents cleared post-civil war rubble and rebuilt their homes and public spaces with the support of international NGOs. Nevertheless, the authorities opted for the top-down model of "disaster capitalism", including the intention to hire a multi-national private company with international experts and capital to eradicate the historic fabric and replace it with modernity to turn Benghazi into a world city, despite the fact that the same experience in Beirut, Lebanon, exposed that as a completely false aim (Chapter 19). Many chapters on the context of L'Aquila, Italy also criticised the hinderance of citizen participation by the government, reconstruction manager and restorers.

Pre- and post-disaster planning and reconstruction of historic cities amounts to a very complex system, traditionally dominated by physical and technical considerations, with only partial recognition of material losses on the part of affected communities as well as to the detriment of their emotional needs and how they identify with their city, which is now considered an equally important and inherent part of the cultural heritage of these cities. Not only is citizen participation deemed essential in reconstruction, but that it is also essential to forge the recuperation of social recollection and identity to achieve an integrated knowledge of and understanding of cultural heritage.

The either-or argument for and against civil society, bottom-up versus political top-down approaches is misguided, because local initiatives are not, and do not claim to be, in a position to undertake large-scale reconstruction of infrastructure or whole city quarters. Nor are government institutions advised to going it alone, without involving the affected local population, if their aim is to achieve sustainable solutions for a disaster-affected city and its population. A more integrated approach combining heritage maintenance, urban development and disaster management is more promising.

The meaningful participation of residents and offering them greater choices during the reconstruction of their houses and properties, which may have or not have heritage value, certainly increases operational complexities. Therefore, it is about keeping the balance, and as a consequence, there might not be a perfect museum-style restoration of the built heritage, but there may be more social satisfaction—or at least less social dissatisfaction together with a greater sense of belonging and more authentic urban life, which in turn contributes positively to the economic and social recovery of the city.

1.6 A Web of Disciplines for Analyses

Many authors adopted different standpoints and tapped on the body of knowledge belonging to a broad range of disciplines. From one side, this can be seen a challenge in terms of extracting communalities through comparing in what ways case studies were identifying trends of action within diverging contexts. However, from the other side, it is a welcoming phenomenon, a manifestation of the desire and efforts for pushing boundaries of the intersection of urban, disaster and heritage discourses.

Addressing multiple scales of reconstruction, some authors brought into the equation tourism and the aspect of tourism economy which is inevitably associated with historic cities and their urban heritage. The tourism economy of Hoi An became a driver for establishing heritage maintenance systems at a larger scale in its current pre-disaster state that can be useful in any future disasters (Chapter 21). As part of post-disaster evacuation strategies (Chapter 33), simulations were developed for major tourist sites in Japan, while the unspoken desire to become a World Heritage Site became soon apparent in the city of L'Aquila (Chapter 9). It appears that the contribution of the tourism economy to a city can strengthen the importance of the historic layers of the town and its governance, yet it comes with the intrinsic question of "reconstruction for whom"?

Addressing education and the younger generations, an educationalist (Chapter 28) developed a didactic guide for teachers in order to promote resilience in schools and prepare children and their parents to learn to become more amenable to and accepting of innovative reconstruction techniques in earthquake-prone cities. Another educationalist investigated how an involvement with cultural heritage can assist in reducing the number of drop-outs by school children in places that have faced disaster (Chapter 27).

In L'Aquila, Italy, a photographer (Chapter 4) used pictures that had been taken before disaster struck, including those taken from family albums, to reconstruct a narrative of the ruined places and thereby detect how inhabitants identify with these places emotionally as well as materially.

A long view was taken by an archaeologist (Chapter 24), when examining how ancient civilisations in Guatemala had dealt with the gods to stave off disasters by sacrificing their most precious assets, their children. Another historical approach (Chapter 8) examined how long-range cultural impacts had transformed the tangible and intangible fabric of Shanhjahanabad India and turned it into a productive contributor to its contemporary cultural economy.

The example of a "play street" (Chapter 22) represents the voice of an urban authority which was trying to realise tactical urbanism to enhance social capital, thereby contributing to enhancing the community resilience. By building up trust, the authority was able to organise an informal event and engage with marginalised group of residents of a deprived urban neighbourhood in Iran.

A study (Chapter 3) examined narrative as tool to recover the lost heritage in Aleppo Syria. Narrative has the potential to translate the invisible knowing into telling and through the telling of both experts and non-experts. The healing capacity of heritage is likely to be unleashed through narrative.

In Europe, for some planners and builders, post-disaster reconstruction was seen as an opportunity to experiment with new urban theories (Chapter 2), which were applied with the dual objective of protecting public and private interests and of improving old and vulnerable urban contexts. Such objectives are difficult to pursue, and outcomes do not always meet initial expectations, especially when the difference between functionalist and identity approaches was investigated. As with many other cases, citizen participation in terms of an identity approach to reconstruction emerged clearly as the most successful strategy when adopted.

From an ecological perspective (Chapter 7), the study of memory, agency and resilience was applied to the contradictory role of water in the vulnerable historic landscape on the northern coast of Peru. It addressed the issue of native knowledge in managing environmental phenomena, lost with modern urbanisation. The study of traditional knowledge of the environment and the sustainability of the built environment in Bam, Iran (Chapter 14), examined missing opportunities during reconstruction and provided a way of achieving resilience in the face of significant natural hazards.

These many deliberations indicate a trend away from concentrating solely on the physical fabrics of cities—also in the context of post-disaster reconstruction—and instead shifting towards criteria related to human needs and the state of the environment.

1.7 New Approaches and Technologies

Advancements within the discourses and debates in the fields of disaster management and risk reduction take place in parallel to other advancements, not only in urban theories and heritage conservation but also in technology. Yet these steps forward need to be framed by and aligned with wider discourses about urban reconstruction and development to play an effective role in assisting the city and the community.

Historically, modern technology played a part in the operational choices made between the options of demolition or provisional and emergency restoration of the damaged structures, and these involved not only structural considerations, but also cultural, historical and logistical ones. In some historical cases, due to their preferential treatment, monumental buildings were the main edifices on which restoration experimentation has been conducted, in particular in the early twentieth century, when reinforced concrete systems started to be applied worldwide. Two restoration interventions of the 1930s analysed comparatively are the church of Santa Lucia in Italian Marsica and the Franciscan mission of San Jose y San Miguel Aguayo in San Antonio, Texas, where reinforced concrete assumed an important role in the re-definition of the lost spaces and not only the replacement of structural elements (Chapter 12).

At the material-architectural—urban end of the spectrum, the Coventry case study (Chapter 11) is a relatively rare example of two blatantly modernist approaches to post-disaster reconstruction. After extensive World War II damage, much more was subsequently demolished in the process of redesigning and building the new cathedral, together with a modern city centre whose reconstruction was controversial; as was the second phase, attempting to revive the alienated historic centre with a heterogeneous path from the past to the future.

Although the reliance on new technology has regularly played an important part in reconstruction, reinforced concrete used in the 1930s can hardly be described as friendly in terms of citizens' participation and nowadays also seems unfriendly in terms of materials in a heritage context (Chapter 12). The recent earthquakes in Italy of the last decade or so have also constituted an occasion to set up a sort of "catalogue of damage" for these monumental, highly visible structures, such as bell-towers that play a significant part in the identity of a city and on which a reliable strategy for emergency interventions could be based (Chapter 18).

Focusing on more recent technologies and approaches, some researchers experimented with technical advancements which can open opportunities and help with better decision-making when dealing with heritage monuments including, for example, detailed local soil microzoning (Chapter 31) or enhancing their energy consumption performance of historic buildings during reconstruction and retrofit towards environmental sustainability (Chapter 36). Another example is the virtual reconstruction (Chapter 30) of an ancient site, St. Augustine's Abbey in Canterbury, United Kingdom, which had long since disappeared, leaving mere foundations behind. Focusing on memory and identity, they explored how visitors were evaluating such experiences and analysed them in terms of accuracy, reality and authenticity. New technology was found extremely useful in some instances but in other cases presented problems and issues of cost–benefit over the long term did not always come out in favour of the new technologies (Chapter 18). Despite this, new technology of a twenty-first century type also appears useful today, with proposals of protocols for the typological and parametric representation of townscapes, introducing the application of a building information modelling (BIM) tool for planned reconstructions (Chapter 32). This process aims to promote in-depth knowledge of the site conditions: open data collection, typological survey and visualisation of the city's evolution over the longue durée prior to the earthquake. Not only has significantly better analysis of the historical past of a town been proposed, but a further step is the application of information technology (IT) models to combine adequate technological retrofitting with resilience, while considering the character of the places being reconstructed. In the future, it will be interesting to look back to see how successful such cutting-edge technologies of today will have been to bridge the gap between traditional reconstruction and preserving historical memory.

Importantly, the twenty-first century has been progressively witnessing the emergence and use of digital technologies. Big data, artificial intelligence, social networks and digital mass communications have become indispensable elements of our daily lives in a very short time and are shaping our societies. The ever-growing new ways of direct communications such as Twitter, Facebook, Instagram, Telegram, WhatsApp, Viber, TikToc and so on, cross-cut hierarchies making for real-time updates and news broadcasting (Chapters 34, 29 and 25). The examination and analysis of whether the use of social networks (Chapter 34), in particular Twitter, has led to improvements in the management of emergencies and post-emergency situations within public institutions not only took into account post-realities, but also interrogated the use of social media by public institutions as potentially preventively building resilience among its citizens. While the examples investigated largely demonstrated that there was significant failure to engage, avenues were opened up for future best practice in resilience building. Resorting to social media to obtain immediate and up to date post-disaster information and to mobilise human resources for the management of relief operations has become increasingly widespread. The floods in Lorestan Iran (Chapter 35) are an example of how social media were able to assist, even in a country with political resistance to them.

1.8 From Shaping the Agenda to Implementation in the Field

Reconstruction happens in a highly politicised context, at both national and international levels when dealing with donor countries, funding, humanitarian organisations and international agencies. It takes different forms at different levels. At international level, the expected principles of the donors and funding agencies influence the vision for a disaster-hit city that will then be translated into how the 'new developmental path of the city is envisioned. In other words, the international agenda sets the direction for recovery and reconstruction, although it will be the responsibility of the national or local authorities to implement these in the field. At the national level, the political agenda will also set the direction, or green light, for reconstruction strategies and approaches such as being participatory and engaging people, or of being developer-driven, market-oriented or state-driven; it will also determine the issue of centralisation or decentralisation in reconstruction.

Some chapters (e.g. Chapters 10 and 15) provided evidence of such connections between international and/or national political relations and influence over many aspects of recovery and reconstruction. Additionally, such international influence increased in post-conflict reconstruction and recovery cases. Reconstruction of historic cities also attracted the attention of other kinds of international agencies and stakeholders which are not usually involved, such as UNESCO, ICOMOS and ICOP; national and local heritage-related institutions also intensified the politicisation of reconstruction. But intensified politicisation makes decision-making difficult in a context that is already traumatic, chaotic and uncertain and facing gigantic social, economic, technical and administrative demands.

Intergovernmental bodies and NGOs are often involved at the point of initial disaster relief and temporary emergency measures undertaken by the affected local communities. Planning for post-disaster reconstruction in most countries is in the hands of governments, local authorities and business leaders who often ask for the cooperation of international agencies and private companies in reconstruction planning, investment and implementation. These stakeholder groups have their own agendas, including the built environment professionals and their corporate institutions. This is the moment when the local population may lose ground and decisions are taken about their future with little consideration for their cultural and emotional needs.

Despite recognition of the advantages of participatory reconstruction and especially the owner-driven approach, surprisingly, the empirical evidence examined in this book reveals the existence in practice of different political and managerial approaches in the case of historic cities. While the examples of Bam and, potentially, Hoi-An present owner-driven approaches, the more recent cases have a developer-driven, market-oriented approach to reconstruction in Aleppo and a state-driven, heritage-oriented one for rebuilding in L'Aquila and Beirut (e.g. Chapters 21, 34, 10 and 3). It might be argued that a developer-driven or state-driven reconstruction of privately owned properties (houses, shops and so on) will reduce the practical complexities in the field of safeguarding the historic identity of the city. Conversely, giving people a stronger role in decision-making increases complexity because the operations will be more closely related to housing and urban development, leading to more layers of complexity as house owners and residents will be active figures within the system. The key is maintaining a balance between various priorities or, from the managerial perspective, taking account of various objectives. The devil lies in the detail. The case (Chapter 21) was made for the need for pre-disaster policies and planning for post-disaster reconstruction of the heritage city Hoi-An in Vietnam before disaster strikes.

In the field, authorities deal with day-to-day decision-making and the question of how to achieve their stated goals and the objectives of their projects and programmes. While a progressive agenda and technical advancements increase the level of expectations, the authorities in charge of reconstruction management, urban development and heritage conservation have to deal with, and try to solve, practicalities in the field. Their task is precisely how to implement their strategies and how to get things done. Some chapters dealt directly with various voices speaking about managerial aspects, practicalities and procedures in the field that are often neglected. This is the case with handling the emergency management dealing with heritage of monuments in Emilia-Romagna, Italy (Chapter 25). Even in the period of pre-disaster and resilience building, the "play street" initiative may sound simple (Chapter 22), but it was a challenge and required various approvals within a specific sociopolitical and administrative context. Some chapters also addressed managerial aspects indirectly, when the discussion of their findings took note of significantly poor coordination, approaches and procedures (e.g. Chapter 15). These issues are establishing the basis of why management and organisational aspects are important. Even the best ideas will fail in practice if "the way" they are implemented is set to fail.

1.9 Structure of the Book

Reaching across the various epistemologies of established academic disciplines and the ontologies of practice in a wide variety of geographies and cultures, the book establishes communication between various physical scales of the built environment. The structure of the book consists of four complementary and overlapping parts which connect related chapters that address the past, present and, potentially, the future.

Part I on heritage and collective memory discusses the tangible and intangible dimensions of collective memory and how they can be put to use in rethinking the reconstruction of cities and regions devastated by natural- or human-induced disasters.

Part II contrasts historic and contemporary reconstruction approaches. This part starts with narratives on historic reconstruction approaches, starting with Western countries after the World War II and natural hazards. It then moves towards the most recent reconstruction cases and approaches. Chapters provide narratives about resorting to modern technology, contemporary architecture and a widespread albeit contested view that cities are constantly changing and thus do not need to mimic a vanished historic past.

Part III, on society, governance and collective resilience, focuses on the essential support from both governance and citizenry to achieve greater future resilience, including engagement with younger generations and how to achieve that.

Part IV on bringing the twenty-first century into reconstruction is the final part and is future-oriented. Chapters investigate new and emerging concepts, technologies and communication instruments to create not only more integrated, future-proof resilient cities but, most of all, people-focused post-disaster environments.

1.10 Concluding Remarks

Bringing together scholars from different disciplines to study the complex issue of dealing with disasters in the context of historic cities is ambitious and challenging, yet worthwhile. Being a frontier book, it offers an entry point for shaping an integrative approach to the reconstruction of historic cities as well as understanding the challenges and opportunities of layers of history and social recollection, as well as new ideas and technologies on offer. In essence, this book advocates a multidisciplinary approach at different scales to deal with the subject matter. It explores a web of relevant and complementary disciplines which ought to work strategically together in a balanced way to achieve a better reconstruction of historic cities. Bringing together historical and contemporary cases of reconstruction provides the opportunity for self-reflection of the part of stakeholders, academics and the global community regarding the past and where theory, policy and practice stand today. Shockingly, there is extensive evidence that historical experiences have not led to organisational learning in many cases.

Evident is the book's emphasis of the importance of people's participation in heritage maintenance and reconstruction that links with social and urban recovery and resilience. The integration of people's participation and heritage has a facilitative role for social recollection and future resilience, and compromising on one of them damages reconstruction and recovery. Yet, despite recognition of the participatory reconstruction approach, surprisingly, empirical evidence examined here reveals different existing political and managerial approaches being adopted in practice for cases of historic cities.

Political context and agenda will support or hinder advancements for a better and more inclusive recovery and reconstruction. At both strategic and operational levels, managerial aspects condition implementation strategies in the field. Even in the case of people's participation, people will be just one element of the bigger reconstruction and recovery system, and even the best ideas can fail in practice if they are not thoughtfully organised.

New ideas, tools and technologies, from educational tools to photography and digital communication, potentially offer new opportunities to take the discourse further. There is a need for better communication and integration between advanced technology, new ideas and reconstruction. It appears that reconstruction at present is missing out on the benefits of the rapid advancements that are taking place in many other fields. They could be useful to bring the voices of people and authorities together and help achieve better recovery. Empirical evidence from both historical reconstruction and contemporary cases reveals that there is huge potential for including smart technology in reconstruction. However, advances such as augmented reality and smart grids cannot achieve their full potential if they are not framed within the broader discourse of reconstruction and development that includes community, and potentials of social media need to be further adapted to reconstruction processes. Technology alone cannot tell the story.

The book offers stakeholders, whether policymakers, practitioners or academics in multiple disciplines and those working in practical fields as well as the wider interested public, opportunities for self-reflection. At the same time, by taking a forward-looking direction, the book is an invitation to think ahead and come together pre-disaster.

References

Antonello Ciccozzi A (2015). Where-it-was-as-it-was. The protection of cultural heritage and the seismic safety of buildings at L'Aquila. In: Etnografia e ricerca qualitativa, pp 259–276

Arefian FF (2018) Organising post-disaster reconstruction processes. Springer, Cham, Switzerland

Bosher L (2008) Introduction: the need for built-in resilience. In: Bosher L (ed) Hazards and the built environment: attaining built-in resilience. Routledge, London, pp 3–20

Darlington J (2020) Fake heritage: why we rebuild monuments. Yale University Press, New Haven and London

Gaillard JC (2017) Natural hazards and disasters. In: Richardson D, Castree N, Goodchild MF, Kobayashi A, Liu W, Marston RA (eds) International encyclopedia of geography: people, the earth, environment and technology. Wiley, Oxford, UK, pp 1–15

García-Hernández M, Calle-Vaquero M (2019) Urban heritage in Oxford bibliographies. https://doi.org/10.1093/obo/9780199874002-0208

Global Facility for Disaster Reduction and Recovery, GFDRR (2016) What did we learn? the shelter response and housing recovery in the first two years after the 2010 Haiti earthquake. Washington DC, USA. Available from: http://www.recoveryplatform.org/assets/publication/GFDRR/haiti-what-did-we-learn_GFDRR.pdf. Accessed 15 February 2021

ICOMOS (2014) Tangible risks, intangible opportunities: long-term risk preparedness and responses for threats to cultural heritage. Proceedings of the ICOMOS Scientific Symposium, 31 October 2012, Beijing, China. Available from: http://openarchive.icomos.org/1509/1/ICOMOS_2012_Proceedings_final.pdf. Accessed 9 March 2016

ICOMOS (2016) Post-trauma reconstruction colloquium proceedings, 4 March 2016, Charenton-le-Pont—France, Available from: http://openarchive.icomos.org/1707/1/ICOMOS%2DPost%2DTrauma_Reconstruction_Proceedings%2DVOL1%2DENGok.pdf. Accessed 9 April 2016

James R, Magee L, Scerri A, Steger MB (2015) Urban sustainability in theory and practice: the circles of sustainability. Routledge, London

Jigyasu R (2016) Cause culture cannot wait! Post -earthquake recovery of Nepalese cultural heritages. In: Post-trauma reconstruction. Proceedings of the 1-Day Colloquium at ICOMOS Headquarters, 4 March 2016. Presented at the Post-Trauma Reconstruction., France, pp 7–8

Kurokawa K (1998) From the eco-city to the eco-media-city, A+U Architecture and Urbanism 333

Look DW, Spennemann DHR (2000) Disaster management for cultural properties. CRM 23(6):3–5

Look DW, Spennemann DHR (2001) Disaster preparedness, planning and mitigation. CRM 24(8):3–4

Nurse P (2020). What is life, understanding biology in five steps. David Hickling Books,

Oliver P (1997) Encyclopaedia of vernacular architecture of the world. Cambridge University Press, Cambridge

Schuetze T, Chelleri L (2011) Climate adaptive urban planning and design with water in Dutch polders. Water Sci Tech 64(3):722–730. https://doi.org/10.2166/wst.2011.688

Sylwia U (2020) Designing for Sustainability: questioning our material culture. In: Bürstmayr A, Stocker K (eds) Designing sustainable cities. Birkhäuser, Basel

Tahboroff J (2003) Natural disasters and urban cultural heritage: a reassessment. In: Kreimer A, Arnold M, Carlin A (eds) Building safer cities: the future of disaster risk. World Bank, Washington, DC, pp 233–240

UNESCO (n.d.) UNESCO—what is intangible cultural heritage? [website]. Available at: https://ich.unesco.org/en/what-is-intangible-heritage-00003. Accessed 18 February 2021

UN-Habitat (2017). The new urban agenda. Available from: https://uploads.habitat3.org/hb3/NUA-English.pdf. Accessed 15 February 2021

UNISDR (2005). World Conference on Disaster Reduction, Kobe, Hyogo, Japan Hyogo Framework for Action 2005–2015: Building the Resilience of Nations and Communities to Disasters (No. A/CONF.206/6), World Conference on Disaster Reduction. Kobe, Hyogo, Japan. Available from: http://www.unisdr.org/files/1037_hyogoframeworkforactionenglish.pdf. Accessed 15 February 2021

UNISDR (2015) Sendai framework for disaster risk reduction 2015–2030 (No. UNISDR/GE/2015-ICLUX EN5000 1st edition). Geneva, Switzerland. Available from: http://www.preventionweb.net/files/43291_sendaiframeworkfordrren.pdf. Accessed 15 February 2021

UNISDR (2017) Terminology (Website). Available from https://www.unisdr.org/we/inform/terminology. Accessed 6 Aug 2020

World Commission on Environment and Development (1987) "Brundtland report": our common future. Available from: https://sustainabledevelopment.un.org/content/documents/5987our-common-future.pdf. Accessed 15 February 2021

Part I
Heritage and Collective Memory

Chapter 2
Thinking About Post-disaster Reconstruction in Europe: Functionalist and Identity Approaches

Francesca Fiaschi

Abstract This chapter analyses reconstructions of historic centres in selected European cities damaged by earthquakes and wars. The latest Italian reconstructions have sparked renewed interest within the scientific community of the issue of the preservation of historic centres. Beginning briefly with significant examples of post-disaster reconstruction in Europe spanning a period of almost three centuries, the focus then shifts to emblematic cases of reconstruction that have been significant for the development of urban approaches. Although important examples of reconstructions such as Lisbon, Thessaloniki and Messina are mentioned, the main analysis concentrates on reconstruction post-World War II emphasising recent Italian reconstructions following earthquakes during the last decade. A review of the literature on the history of reconstruction allows the key concepts to be traced: those that emerged from urban planning treaties, charters and international conventions, and how they have influenced patterns of reconstruction up to the present day.

Keywords Post-disaster reconstruction · Rebuilding · Identity · Functionalism · Urban planning · Europe

2.1 Introduction

Cities are created, built, transformed, destroyed and rebuilt. They are sometimes abandoned in ruin and often transformed into sites of memory (Nora 1992). Experience shows that rebuilding a city, especially a historical one, is not a quick and easy task. Indeed, if the reconstructions that have taken place since the World War II are considered, it is clear that the theoretical predictions of an average duration of 10–20 years for rebuilding are not far from reality (Kates et al. 1977). There are also cases, such as that of Dresden, destroyed by Allied bombing in 1945, that demonstrate that the process of reconstruction in ongoing and does not really come to a conclusion (Voisin 2007): the reconstruction becomes part of the ordinary evolutionary

F. Fiaschi (✉)
University of Montreal, Montréal, QC, Canada
e-mail: francesca.fiaschi@umontreal.ca

© The Author(s), under exclusive license to Springer Nature Switzerland AG 2021
F. F. Arefian et al. (eds.), *Historic Cities in the Face of Disasters*, The Urban Book Series,
https://doi.org/10.1007/978-3-030-77356-4_2

process of the city. Reconstruction must be understood as an extended process that inevitably generates a new city (Corboz 2001). Through the choices that are made in the reconstruction process, these offer an opportunity to better understand urban reality and the possibility to improve the built environment.

Destruction takes lives and spaces and architecture. It interrupts daily functions, relationships and lifestyles. The analysis here shows that if reconstruction is as much as 75–90% of the built frame of a city, inevitably it is a large-scale urban project. The inevitable question is what were and what are the urban principles that guide the reconstruction? In particular, reconstructing a severely damaged historic city forces a focus on its history and a comparison of its diverse strata (Corboz 2001). The multidimensional reality (Garbini 2008) of the historic city centre forces one to think about the reconstruction of this artefact (Rossi 1978) and the evolution of its urban form, meaning and function. How, in other words, can this artefact be re-created?

Even though the specificity of any reconstruction relies primarily on the political, economic and social conditions of the destroyed city and its territory (Edgington 2009; Neal 1997), the role of contemporary knowledge and its influence on the models used for reconstruction is analysed. Indeed, this chapter reads post-disaster reconstruction as an opportunity for planners and builders to experiment with new urban theories. The latter are applied with the dual objective of protecting public and private interests and of improving old and vulnerable urban contexts. Such objectives are difficult to pursue and do not always meet initial expectations.[1] Voldman has shown that in the case of the reconstruction of French cities (Gien, Dunkerque, Havre) in the 1945–1954, town planners came to regret a reconstruction that focused too much on architecture and not enough on the city. The inhabitants, in turn, were disappointed with the fact that the reconstruction failed to take account of both the past and their present wishes (Voldman 1997).

Whether following war or a natural hazard, reconstruction always has to confront the same question: should the city be reconstructed identically, or rather according to a functionalist perspective based on a contemporary urbanistic and architectural language? This is the vexed question (*vexata quaestio*) that has yet to find any widely accepted solution. The most recent Italian reconstructions of the beginning of the twenty-first century drew yet again the attention of the scientific community to the issue of the preservation of historic centres, initially raised during the great post-World War II reconstructions. The literature review on the history of the reconstruction shows that the questions "how, when, where, for whom and why to reconstruct?", obviously, represent an unchanging historical set of issues and, as such, they show an absolutely topical character, even today, in relation to the post-earthquake reconstructions in Italy.

Since World War II, the theme of reconstruction has been associated with a deep reflection on modern architecture, both in terms of the relationship between the

---

[1] In the sense that even if the goal is achieved, the result may not necessarily be the expected one. An example is the CASE project in L'Aquila city in Italy, that was born to respond to: give work to private entrepreneurs; the need to experiment with new anti-seismic technologies; and, finally, a home for the displaced before winter, which is very harsh in this part of Italy. They have generated a series of other social problems that were not foreseen in the propaedeutics evaluations.

periphery and the city centre, the expansion of the city itself, and in relation to the foundation of new cities. In addition, the question of modernity has consistently raised further questions dealing with the relationship between the old and the new.

Every reconstruction builds on prior construction in terms of legislation, technology, skills, knowledge and practices. It also builds on recognised mistakes to be avoided and as such becomes a laboratory for testing theories and influencing new ones. Beginning with significant examples of post-disaster reconstruction in Europe over a period of almost three centuries, although important examples of reconstruction such as Lisbon, Thessaloniki, Messina are mentioned, the analysis focuses on reconstruction from the post-World War II period. Through a review of the literature on the history of reconstruction, the key concepts that emerge are traced, and their influence on the patterns of reconstruction up to the first decade of the 2000s is investigated.

2.2 The Birth and Evolution of Post-disaster Reconstruction Patterns for Historic Cities

The question "How should we rebuild?" lies at the roots of each reconstruction. The choices that follow depend on the historical, socio-economic and political factors of the moment. In Europe, reconstructions carried out between 1700 and 1900 sparked little debate: cities were simply rebuilt. The reconstructions answered the need to return to normality and fulfilled the functional and rational exigencies to rebuild the destroyed urban fabric and to improve infrastructure and construction. In this scenario, authenticity was not part of the reconstruction priorities. It was a reconstruction imposed from above, not the result of a consensus. Participation of the population in reconstruction was not considered. In fact, until the post-World War II period, there was no real debate about "historic city reconstruction" as the concepts of urban context, urban heritage and historic centres of the city were still germinating. Experts, town planners, hygienists and architects began to question the fate of old city centres only after demolition policies had been implemented in major European capitals during the second half of the 1800s to clean up what were considered unhealthy urban environments.

From the Portuguese of Lisbon (1755) in the eighteenth century, to the Greeks of Thessaloniki (1917) in the early twentieth century, the sole voice in the choice of reconstruction strategies was essentially institutionally linked to the political power that governed at the time. This entailed a reconstruction of the city through global urban planning projects (Rossa 2012) that gave architecture within the programme the role of representing power, safeguarding memory and, in cases such as Thessaloniki, changing the image of the city by affirming a new national identity (Mazower 2007) or Northern Italy after World War I (Treccani 2014).

The preservation of memory had a fundamental role in maintaining continuity with the past and was carried out through the insertion and reconstruction of destroyed monuments within the new city plan. These types of scheme, understanding and fulfilling the needs of the population were clearly not a priority. The city was viewed as a single unit, and decision-making was centralised. Hence, the choice of a clean slate in leading to a reconstruction of the city based on rational urbanism through new formal and typological parameters, on site or on another site. This was the most widely adopted solution.

The reconstructions of Lisbon and Thessaloniki are clear examples of this. The nineteenth century marked an important moment in reflections about the city in both town planning and urban conservation. The principal interventions in the great European capitals (Paris, Madrid, Barcelona for example) were marked by the search for an image of the city corresponding to the emerging new bourgeois class that envisioned economic, military, social and political power through the evisceration of the earlier fabric of the city centres.

In this period, town planners like Camillo Sitte and Charles Bulls began to fight against what is now considered the barbaric practice of gutting that erased entire sections of the historic urban fabric (Smets 1992). Through their writings, they influenced other intellectuals like Patrick Geddes and Gustavo Giovannoni who, in the second decade of the twentieth century, began to promote their theories about the development/preservation of old city centre. Geddes advanced the notion of "conservative surgery" (Geddes 1915) , and Giovannoni promoted the idea of "selective clearing of the built fabric" ("*diradamento edilizio*") (Giovannoni 1913a, b) . Between 1913 and 1931, Giovannoni pursued his inquiry into the relationship between the "old" and the "new", the results of which continue to represent an important line of inquiry into how to rebuild historic centres. These two theories, "conservative surgery" and "selective clearing of the built fabric" represented alternative responses and are certainly less invasive to that offered by the dominant demolition policy, popular in Europe in the second half of the 1800s (Choay 1992).

2.2.1 Early Reconstruction Patterns: Lisbon and Functionalist Reconstruction

One of the earliest recognised plans for reconstruction that inaugurated what would become a recognisable pattern is that of the Portuguese capital of Lisbon, destroyed in 1755 by an earthquake, followed by a tsunami and a fire. The violence of the earthquake was felt in half of Europe and especially in North Africa. The theoretical basis of the Lisbon reconstruction plan stemmed from the treatise on town planning, "Dissertaçao", by Manuel de Maia, an urban planner appointed by the Marquis de Pombal to draft the reconstruction plan for the Portuguese capital (Rossa 2012). In the treatise, the concepts of renewing, rebuilding and restoring, while not explicitly laid out, tend to describe the philosophy of the programme of reconstruction. De Maia planned the rebuilding of Baixa down to the water, the rational renewal of the urban fabric and the restoration of damaged monuments. For the first time in urban history, there was a plan that integrated all the components necessary for reconstruction: urban design, legislation, financial engineering and programme architecture (Rossa 2012). The novelty of the plan lay in the fact that it reconciled urban rationalism and the preservation of memory (post 1755: Igreja do Carmo, Convento da Ordem do Carmo, Chiado) and as symbols of the old town. It also introduced important innovations for infrastructure and construction: anti-seismic technology, the first step towards a policy of prevention.[2]

Two centuries later, in 1917, the ancient centre of the Greek city of Thessaloniki was destroyed by fire (Petropoulos 1980). Reconstruction in Thessaloniki was here, however, aimed at the affirmation of a new national identity (Mazower 2007). The reconstruction marked the transition from an Ottoman to a European city. The plan not only provided for the reconstruction of the area destroyed by fire, but it also targeted the entire urban area by giving precise indications for the city's extra-urban development. The historical centre was completely redesigned, well beyond the destroyed area, deploying an orthogonal grid in which the distinctive element of the new system was a monumental civic axis: a great path crossing the city and connecting the coast with the hillside zone. The memory of the city was partially renewed through the preservation only of the monumental buildings considered to be representative of the historical continuity between the modern city and the ancient city of the Roman-Byzantine era. This process of de-Ottomanization demonstrated both the will of the Greek ruling class to redefine the image of the city by erasing previous Ottoman influence and the contribution of the architect-urbanist French Hebrard who revisited the vestiges of the past by reassembling them in a new architectural-urban system (Lonero 2001). The Hébrard plan aimed to recover pre-existing patterns and to contextualise them structurally, as well as to impose the typical urban patterns of the Beaux-Arts School. The central area was completely redesigned even beyond the destroyed area with the help of an orthogonal grid: paths parallel to the coast with transverse crossings which define the new blocks with the fully fragmented urban front. This distinctive element of the new urban system was the new monumental civic axis of the city, in a central position and perpendicular to the coast: this great path joins the coast with the hilly area crossing it, in sequence, spaces important audiences (Fig. 2.1).

This new way of understanding the urban artefact, introduced by European urban planners, had two important consequences: the first was that the planning of a city that was considered as a single organism on which planners intervened on all scales and took into account its future development; secondly, the plan involved the elaboration of a body of legislation approved along with the project.

At the beginning of the twentieth century, widespread experimentation with new architectural principles makes it possible to identify urban renewal strategies: many cities at the time structured their new extensions according to the "rules of modernity" wherein neoclassical urban principles followed the rationalist principles enunciated

[2] "La gaiola" was a wooden cage, which improved a building's anti-seismic response. The infrastructure and the sewage and road systems were also improved, with differentiated platforms (Rossa 2012).

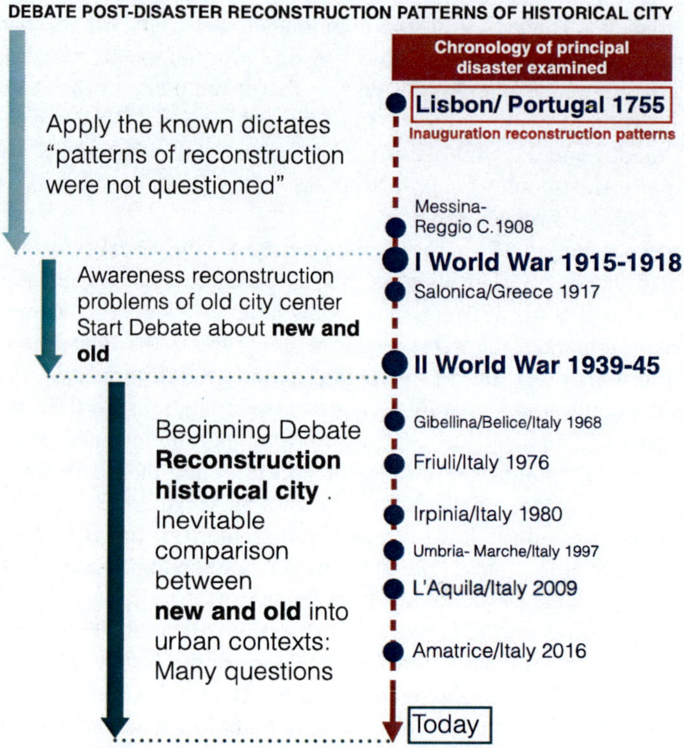

Fig. 2.1 Evolution of the debate over post-disaster reconstruction of the historical city in Europe

in the CIAM Urbanism Charter (1933). The Borzì plan for Messina was based on these principles.

In 1908, Messina was destroyed by a violent combination of an earthquake and tsunami. The reconstruction process, approved by the Borzì plan, did not begin until the 1930s due to delays caused by World War I. The decision regarding how to rebuild the old city was not well-received by the city's inhabitants. The original characteristics of the city, always appreciated by the community, became unrecognisable in the course of reconstruction and generated a sense of dissatisfaction throughout the process and after its completion (Farina and Melluso 2012). Criticism of the plan adopted for the reconstruction of the city highlights what was significantly different from the usual process of city formation. In general, cities are created around a historical core and are composed of parts that are either integrated or juxtaposed. In Messina, however, where the historical centre was largely deleted and preserved only occasionally and just segments of districts or isolated buildings were restored. As a consequence, the city is primarily founded on the morphological characteristics of its original urban plan (Farina and Melluso 2012), yet looks nothing like it and lacks its historical atmosphere despite respecting the pre-Earthquake groundplan.

Messina can be compared with Lisbon. Both are port cities that suffered destruction and chose similar reconstruction strategies: taking into account the typo-morphological principle of the original city in order to reconstruct the new one. Of particular importance was the shape of the urban plan that was generated from the typological definition of a building unit, which became the generating unit of each urban area. Zone and type were the starting points of the project from which the morphological order and the organisation of infrastructure services were deduced (De Solà Morales 1978).[3]

2.2.2 *Reconstruction: "Where It Was, as It Was"*

Reconstruction after the First World War introduced the model of "where it was, as it was" ("*dov'era, com'era*").[4] It consisted essentially of rebuilding everything according to how it was on the original site with variations on the theme.[5] Due to financial constraints, reconstruction deployed older technology using recovered debris and local materials, mostly in a vernacular style. The cities were rebuilt in situ by repairing the destroyed buildings without paying too much attention to the philological approach (Treccani 2014). The reconstruction of the bell tower of San Marco in Venice, important in the context of the debate on the conservative restoration of monuments, will introduce the method of intervention "where it was, as it was".

In 1902, the debate over how the reconstruction of the bell tower should have been undertaken after its collapse culminated in the Venetians unanimously accepting "where it was, as it was", thus rejecting the opinion of the famous architect Austrian Otto Wagner who promoted the reconstruction of the bell tower where it was but not how it was.

The proposal made by Luca Beltrami for the reconstruction of the bell tower consisted of maintaining its original shape but with a different structure of reinforced concrete covered in bricks, in other words a compromise that safeguarded the shape

---

[3] M. De Solà-Morales, *Verso una definizione. Analisi delle espansioni urbane dell'800*, in "Lotus", 19, 1978, p. 27, "Forma di progettazione urbana che, partendo dalla definizione tipologica di un'unità edilizia, rende sistematica la sua generalizzazione per tutta un'area come meccanismo di ordinamento complessivo. L'area ed il tipo sono i punti di partenza del progetto dai quali deriveranno deduttivamente l'ordinamento morfologico e l'organizzazione dei servizi infrastrutturali".

[4] The expression was introduced by the Venetian playwright Luigi Sugana in the aftermath of the collapse of the bell tower of San Marco. Luigi Sugana says: "*Dopo la morte di una persona cara quale sarebbe il desiderio che più impetuoso sorgerebbe nel nostro cuore? Poterne, se il miracolo fosse possibile, rifarne l'immagine, infondere in essa novamente la vita. Perciò è nostra certezza che il campanile debba risorgere com'era e dov'era*", "After the death of a loved one what would be the most impetuous desire that would arise in our heart? To be able, if the miracle were possible, to remake its image, to infuse life into it anew. Therefore, it is our certainty that the bell tower must rise again as it was and where it was". Archivio storico "*La Stampa*" articolo "Gli artisti di Venezia per la riedificazione del campanile", 22 luglio 1902, p. 2.

[5] There were structural and stylistic variations: correction of the street layout, redevelopment of places, construction of public facilities and remanent of architectural styles.

and aesthetics by bringing back the image of the square before collapse. After all, this variant of the "as it was" (structural type) can be read today as a beginning of prevention where the reconstruction of buildings is done in their original volumes but integrated with new technologies.

The example of the bell tower of San Marco remains significant for the issue of conservative restoration of a single building, which collapsed due to structural problems mainly due to age in a vulnerable historical and artistic context (Plant 2002). This example is less significant when talking about the reconstruction of entire urban contexts (entire cities) with historical and artistic value, destroyed due to natural or anthropogenic disasters.

It is only with Gustavo Giovannoni (1917) that a leap in scale occurs and includes urban restoration, therefore no longer just architecture. His speech was fundamental for the reconstruction of entire historic centres destroyed by the Second World War, as in the case of Warsaw and Dresden. These cities have had their historic centre rebuilt "*à l'identique*", based on the historical, iconographic and literary documentation available. However, the application of a reconstruction of "where it was, as it was", with the aim of recovering lost urban symbolism produced many false histories. These operations marked the intention to want to erase war from historical memory and to want to reestablish continuity with the past interrupted by the war. Such initiatives highlighted the inability of modern architecture to express the city's past symbolism, let alone arouse emotions linked to memories of a collective memory. This political strategy turned the cities of Dresden and Warsaw into open-air museums and triggered gentrification (Voisin 2007).

Giovannoni observed that post-World War I reconstruction in Italy was not scientifically based, nor capable of valuing monuments in their entirety, but rather considered as opportunities to rethink the destroyed parts from scratch (Giovannoni 1917). In Italy, the primary intent was to redefine the image of the city by eliminating symbols and meanings associated with their enemy, Austria. For this purpose, the construction of memory was based on the reconstruction process which demolished parts of the ancient city and built ossuaries, squares and memorial boulevards.[6] The restoration architecture played a role in the reaffirmation of national identity (Ojetti 1920). Such reconstruction in these cases used the restoration of monuments as an instrument of liberation, to affirm Italian nationality. It was embedded in a nationalist political project aimed at the Italianisation of their reclaimed land (Treccani 2014).[7]

In France, reconstruction consisted essentially of repairing and rebuilding everything "on the spot" and "identically", even though there were structural and stylistic variations: correcting street layouts, redesigning squares, building public facilities and reworking architectural styles. According to Ford, the reconstruction plans for

[6] Redipuglia ossuarie 1935, work of Giovanni Greppi; Tempio votivo di Venezia lido 1935, work of Giuseppe Torres; Tempio della Pace of Padova 1934, work of Antonio Zanivan. We also have the triumphal arch of Bolzano 1928, work of Marcello Piacentini and Sacrario Militare dello Stelvio 1932, work of Pietro del Fabro.

[7] Trentino Alto Adige, the provinces of Trento et Bolzano, Fiume and Zara and bordering areas. The provinces of Veneto et Friuli occupied by Austrians during the war.

French cities "are not bad", especially from the point of view of contemporary engineering practices, but they lacked fundamental elements: a global vision and a long-term forecast of the community's economic and social development. This was particularly true of the plans for the cities of Arras, Reims, Liévin and Verdun (Ford 1920). The aim was to apply modernist solutions that had been proposed without success from the end of the nineteenth century.[8] However, these reformist solutions were not deployed because of political instability, economic difficulties and an individualistic mind-set of practitioners, both urbanists and architects, that rejected the application of all-encompassing plans (Vayssière 2009). Reconstruction was carried out using contemporary skills and technologies and local materials in an essentially vernacular language. There were no principal architects who were able to impose an overall vision, as would be the case for much post-1945 reconstruction. During post-World War I, a variety of regional styles were proposed resulting in much historical pastiche, including "medieval" houses, Haussmann-like buildings, Louis XIII, Rococo buildings and Art Deco styles. The recovery of used materials became a veritable industry.

2.2.3 Reasonable Reconstruction in Situ: Between Functionalism and the Vernacular

The two world wars resulted in the most radical socio-political changes in history. World War I had upended the old imperialist political systems, aligning the institutional political reality with the changes engendered by the industrial revolution. World War II radically transformed the European way of life by dividing the world into two spheres: capitalist and communist. In the post-1945 period, the east and the west had both suffered considerable damage and had the same problem to solve: how to rebuild?

There were essentially two problems that confronted planners. Within historic centres, it was necessary to repair damaged buildings and desirable to rebuild those that had been bombed. The second problem was how to take advantage of such a widespread reconstruction to redevelop the city, improve residential standards and build new urban agglomerations to decongest old cities (Raja, 1986). In addition, there was the need for rapid economic development to allow Europe to return to some sort of normality. To achieve these objectives, modern or rational architecture, without ornamentation or structural and formal complications, seemed ideal for confronting problems of time and money because it was based on industrial production and standardised components.

From the outset of World War II, historic cities had become a strategic military bombing objective: a clear sign of recognition of the exceptional symbolic, economic and social values attributed to historical cities. Strategic bombardments

[8] Improve traffic, hygiene, road front alignment and the appearance of the entire city (Ford 1920).

caused serious damage to the functional and social structure of such urban agglomerations, as well as to their physiognomy. What resulted influenced architects and town planners (Mougel 2014) because the theory of strategic bombing studied not only how to destroy a city, but also how to think of ways to make the city more resilient (Vauthier 1930). The new model envisaged a "structured and airy" city divided into zones, a concept cherished by the representatives of functionalism in the 1950s. It also promoted the use of new materials: reinforced concrete and steel.

Beginning in 1942–1943, planners set out precepts that would dominate the urban planning scene in the second half of the twentieth century. These ran the gamut from preventative measures against bombing to reconstruction plans: aeration and the reduction of density in city centres, open construction of new housing estates, widening of streets, functional separation of space, development of green belts (Konitz 1989). At the same time, rationalisation and standardisation of construction developed. This would later lead to the industrialisation of housing production. Military town planning functioned as a legacy model for the type of planning for contemporary urban structures that was advocated by the modern movement. The analogy between the work of Le Corbusier and Le Vauthier is an obvious example (Konvitz 1989). This version of contemporary urban history challenged the conventional approach according to which war interrupts the normal development of cities and can therefore be ignored. Vauthier showed how war may have changed and influenced the conception of urban planning of a city in the long term, and at the same time, how it highlighted the strengths and weaknesses of contemporary urban contexts.

Post-war reconstruction was thus an important stage in urban history given the extent of the damage inflicted on European territory and elsewhere. It opened up an entire field of experimentation for modernist urban theories and new construction technologies. In the face of such destruction, even the debate on historic centres was transformed as the problem of reconstruction came to be viewed through a new lens. Indeed, the lines separating tradition and modernity became more difficult to draw from a panorama of ruins.

The solutions adopted in the reconstruction of major European capitals introduced new, hybrid patterns where rebuilding the old city employed architectural languages that oscillated between the modern language of rational urbanism and vernacular historicist languages that retained little authenticity either in terms of construction techniques or in terms of urban form and architectural typology. These solutions synthesised the current architectural debate about the relationship between the old and the new and the need for functional and rational cities. They privileged the in situ restoration of the historic city through a mixed architectural language: historical and new. Modern cityscapes emerged in cities such as Coventry (Bullock 2002), Le Havre and Dunkirk (Voldman 1997). Even in cities that sought to rebuild everything "where it was, as it was", a similar process of reconstruction resulted in urban landscapes that were far from identical to their predecessors. What were considered improvements were made wherever possible. Three models of reconstruction emerged: reconstruction in situ "where it was, as it was"; reconstruction on another site and reconstruction on the same site, both using the principles of modern architecture (Fig. 2.2).

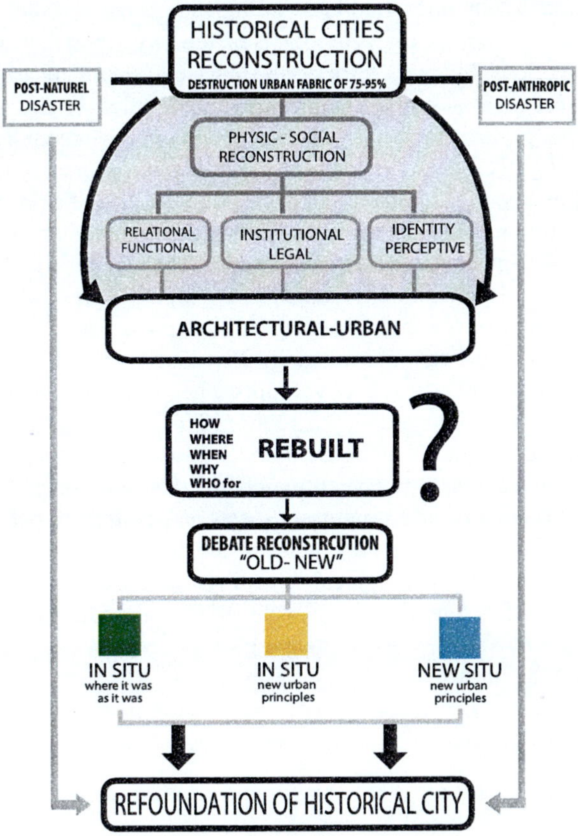

Fig. 2.2 Models for the reconstruction of historical cities following natural or anthropic disaster. The reconstruction of historical cities with an urban fabric destruction of 75–90% amount to rebuilding the city. The diagram shows the physical and social reconstruction analysis following an architectural and urban point of view and offers three interpretative models: In situ "where it was as it was" (Green Square); In situ, new urban principles (Yellow Square); New Situ and new urban principles (Turquoise Square)

Interest in the reconstruction of historic city centres was manifested not only by architects, engineers and town planners, but also by civil society: inhabitants of the city itself. However, planners and architects saw the destruction of the historical context as a chance to modernise urban structure, civil society showed a strong attachment to the lost ancient nuclei (Cutolo 2016). Throughout the second half of the twentieth century, this diversity of approach led historic centres down two different paths. As Diefendorf has shown, some cities chose modernisation in tandem with the safeguarding of some significant buildings or isolated blocks. Others made major efforts to preserve the historical character of the central nucleus. Hannover, Hamburg, Frankfurt and Stuttgart followed the path of modernity; Münster, Freiburg and Dresden rebuilt their old centres. Cities like Lubeck, Cologne, and Munich adopted a mixed approach (Diefendorf 1993) .

The reconstruction of countries in Eastern Europe proceeded in a different direction. In Poland, because of strong nationalist feeling following German occupation, Warsaw's ancient centre was rebuilt "where it was, as it was", in a late-Gothic Renaissance style, typical of Eastern European cities. This operation expressed the intention to erase the historical memory of war to restore continuity with the past. Such an emotionally understandable initiative underscored the inability of modern architecture to express the symbolism of the city's past, let alone evoke memory-related emotions. Planners resorted to superficial, cosmetic architectural forms wherein the facade was a reproduction of the traditional style and hence immediately decipherable by the inhabitants as part of their collective memory. Reconstruction proceeded along two lines: the vernacular and popular aspect and the more official, historicist monumental architecture representing the idea of a political system with a totalising ambition, which became the dominant urban and architectural feature.

The modernist movement almost never operated in the old centres of cities. Their ideas were much better suited to the *tabula rasa* of the periphery, where Corbusian ratios of pure volumes were used to express a new language for new working-class quarters and rational housing. The need to quickly overcome the wounds of war prevented overly reflective approaches to architectural and urban design, as they would have delayed reconstruction. This is why the aesthetic problem, that is to say, the formal quality of the architecture, was unhesitatingly relegated to the background. In this regard, it must be remembered that the Modernist Movement had focused on the functional and structural aspects of architecture but underestimated its symbolic and formal content, and even less the psychological and perceptive aspects. Indeed, the physico-psychological sensations of constructed space which form the "experience" of architecture, were excluded from the game of design (Raja 1986). From this, it follows that even the meaning of modernism has been reduced to the functional and the linear.

2.2.4 Towards a Conservative Reconstruction

The destruction in 1968 of Belice, in southern Italy, following an earthquake, offers an example of the reconstruction of a historic centre in which the urban and architectural spatial organisation was completely new, including its localisation and name. The new Gibellina presents itself as "neither where it was nor how it was" (Fabbro 2013). Gibellina also demonstrated the limits of the generalisation of the principles of modernism. These principles were applied to the reconstruction of an historic centre without taking into account the identity of the place. It was a centralised "reconstruction by the State" (Cannarozzo 2009).

The failure of Gibellina's reconstruction, together with the concepts emerging from the cultural debate on historic centres, such as integrated conservation, the safeguarding of cultural heritage, authenticity and memory, led to more conservative reconstructions post-earthquake from the 1970s to the present day. Clear evidence of this trend were the reconstructions of Friuli, Umbria-Marche and L'Aquila. While the first two reconstructions were characterised by a process of decentralisation, because of the different political conditions, the severity of the damage, the city's typology and its status as a regional centre, the reconstruction of L'Aquila was undertaken through centralised management.

The reconstruction of Friuli begun in 1976 was carried out without the intervention of central government and was self-managed by the institutions of the Region that managed each stage from the legislative phase to implementation. It was a good example of subsidiarity from the outset to its conclusion. The concept of urban and social continuity lay at the base of this reconstruction, which both respected the destroyed urban and socio-economic structure and yet, at the same time, aimed at improving the anti-seismic aspects of the built environment as well as the capacity to develop the local economy through urban planning instruments. The particular historical conditions of institutional, political and social unity together with the status of being an autonomous region, contributed to the success of this reconstruction. Indeed, in the philosophy that guided it, there was a mix of autonomy, regionalism and democracy from below as well as a profound technical, architectonic and engineering culture, a combination of factors considered essential to the success of this reconstruction (Fabbro 1986).

The response to the earthquake of 1997 in Umbria drew significantly on the Friuli experience. Its reconstruction deployed a decentralised system of governance that gave decision-making powers to local municipalities and demanded direct accountability from private individuals. This reconstruction philosophy aimed at preserving history, rebuilding the integrity of social life as well as the urban dimension. The goal was to keep the original urban fabric and its vocation intact.[9]

The model of reconstruction for the historic centre of L'Aquila, still ongoing, stands apart because the state re-appropriated its central role and power. In the first three years of emergency and initial reconstruction, it imposed the model of reconstruction to be followed, also after management would be handed over to the region. This return to a form of centralisation had, however, notably different, and negative results compared to those obtained by the Marquis of Pombal. The reconstruction of the historic centre of L'Aquila follows the philosophy of previous reconstructions that sought to preserve the original urban spatial organisation but not the social, relational and functional past. The decision to build residential neighbourhoods, the C.A.S.E. project, aimed at giving the victims shelter in the aftermath of the disaster, was a drastic choice made without the participation of the inhabitants.[10] This altered the habits of the population of the highly consolidated historical centre and in certain ways the future of the city. Over ten years after the earthquake, the reconstruction of the historic centre continues to follow "where it was, as it was", but it is estimated that it will still take another decade, if not longer, to complete it. Even in the case of a successful reconstruction, where the minimum time for completion remains up to another decade, nothing will be as before because the process of social transformation continues apace.[11] After 10–20 years of reconstruction, could the population that will live in such a reconstructed historical centres ever be the same? The answer is obviously no. For whom is the historical centre being rebuilt? Can the preservation of urban identity help to maintain social unity?

[9] Declaration of Split, 1971.

[10] Ecologically Sustainable Anti-seismic Complexes, for about 14,000 earthquake victims of L'Aquila.

[11] In reference to the reconstruction of the Friuli (Fabbro 1986).

2.3 The Basic Concepts

By tracing over the *longue durée* the links between post-disaster reconstruction models and concepts emerging from urban theory, it is clear that the period from 1945 to 1968 marked a significant turning point (Fig. 2.3). From the reconstruction of Lisbon to World War II, reconstruction entailed the application of new urban theories related to the concepts of a global urban plan and a rational city. Planners sought a typology that represented the first functional city model and incorporated the principles of nineteenth-century hygienist theories. These principles worked on a *tabula rasa* and entailed the demolition of preceding urban fabrics, the isolation of monuments, the construction of major arteries and an orthogonal urban mesh.

In 1876, Baumeinster wrote the first urban planning manual in which he described the importance of the isolation of the monument as a method of regenerating an historical urban fabric (Giambruno 2012). By contrast Camillo Sitte, in his *The Art of Building the City* (1890), introduced the concept of the ancient urban environment by relating it to the concept of conservation that, until then, had referred only to monuments (Sitte 1980). Soon after Sitte, Charle Buls, in *The Aesthetics of Cities* (1894), advanced the theory of liberation and "*diradamento*" (Smets 1992), later taken over

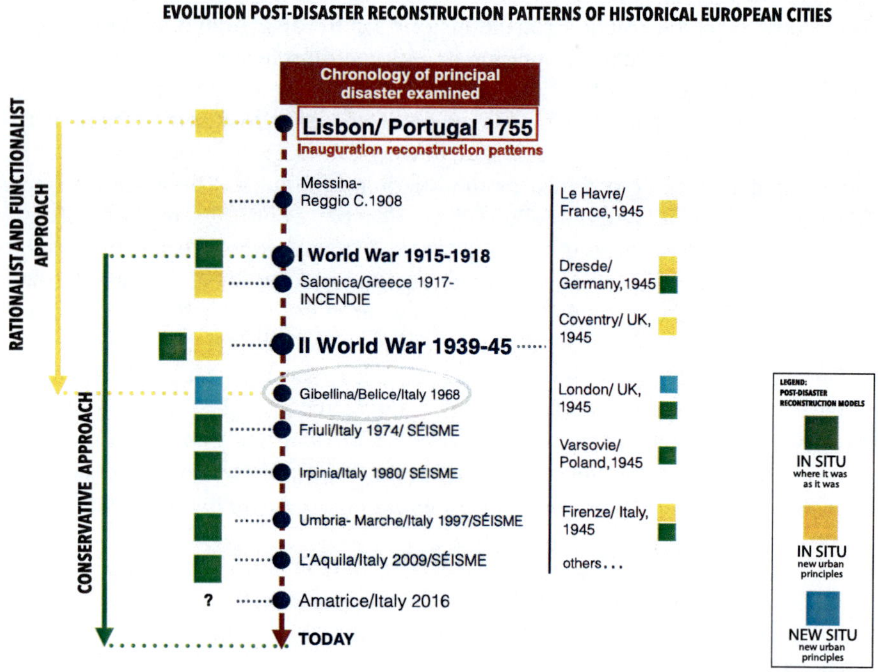

Fig. 2.3 Chronology of post-disaster reconstruction models of historical European cities. In situ "where it was as it was", green square; In situ, new urban principles, yellow square; New Situ and new urban principles, turquoise square

and enriched by the Italian architect and engineer, Giovannoni. Buls recognised the aesthetic value of minor architecture. He laid the foundation for tracking the fibres of Giovannoni's old district and linked urban planning with the problems of safeguarding and transforming historic environments.

In response to devastating urban practices, the new generation of intellectuals, led by Sitte and then represented by Giovannoni, Geddes and others, promoted the surgical, conservative regeneration of historical quarters. They recognised the unique link between monuments and minor buildings in the urban environment, a move that ultimately led to the birth of the concept of urban context. Despite the affirmation of the notion of preservation of the old centres and the urban context, the reconstructions conducted after the World War I were influenced by other priorities, dictated, above all, by the wish for national reaffirmation through a new image of the city, as well as by the meagre economic resources available. In reality, reconstruction was replica pastiche rather than identical, following the model "where it was, as it was".

Giovannoni influenced the restoration charters of 1931 and 1932.[12] He introduced the notion of urban historic heritage into a more general vision of spatial planning. In parallel with the affirmation of the concept of urban heritage, the International Congress of CIAM, held in Athens in 1933, introduced the concepts of the *modern city* and the *functional city* by establishing a link between architecture and urbanism (Le Corbusier 1971).[13] This cultural landscape opened up the possibility of experimenting with two approaches during post-war reconstruction: the modernist and functionalist approach following the principles dictated by the CIAM and a more conservative approach dedicated to preserving the old centre "where it was, as it was". Both approaches were used and were chosen depending on the political message the authorities wanted to transmit. Striking examples are the reconstructions in Eastern Europe of Dresden and Warsaw, compared to those carried out in the North-West, in France at Le Havre and in England at Coventry. The former emphasised a strong nationalist tendency and the latter articulated what is considered a reasonable modernisation.

In 1951, the eighth congress of CIAM, entitled *On the heart of the city*, affirmed the importance of the city centre as a bearer of historical, urban and social continuity. Continuity was to be respected, yet without the new buildings copying the old ones. The role of the city centre was defined as being a container for the different trends of modern and historic architecture. The issue of including the new in the old centre was now the fulcrum of debate about the reconstruction of old city centres.

---

[12] Athens Charter of Restauration 1931; Charter of the Italian Restoration 1932 and the following instructions from 1938.

[13] CIAM-International Congress of Modern Architecture was founded in 1928 in La Sarraz in Switzerland. Until 1959, they were the most valuable and effective tool for the dissemination and discussion of architectural and urban ideas that characterised the development of the so-called modern movement. The main animators were Le Corbusier and S. Giedion, to which were added the contributions of the architects most involved in the treatment of extremely current and urgent operational issues such as minimum housing (2nd CIAM, Frankfurt 1929), of the rational district. (3rd CIAM, Brussels 1930), the functional city (4th CIAM, Athens 1933, Athens Charter), housing and leisure (5th CIAM, Paris 1937).

The problem of pastiche and falsification had already been taken into account in the 1932 *Italian Charter of Restoration* and the subsequent *Instructions* of 1938. For the preservation of all that history has given to buildings in terms of meanings and functions, or historical dignity (*dignità storica*), the construction of buildings in older styles was prohibited, ruling out historical falsification and promoting the insertion of new buildings that respected the previous urban context. In this regard, the typo-morphological studies conducted by the Italian, Saverio Muratori, and those that followed, were propitious.[14] They provided tools for practitioners to learn how to decode the city and to understand it prior to designing their interventions.

In the 1960s, with the *Gubbio Charter*, the terms "old centre", "old environment" and "downtown" became more clearly articulated as the *centro storico* ("historic centre"). Gubbio's *Charter* (Carta di Gubbio 1960) defined the historic centre and gave directions on how to and how not to work within it: thus, no outdated "clearance" and no isolation of monuments.[15] Restoration was the sole instrument for interventions in historic urban fabrics, now viewed as unique organisms. This declaration was reaffirmed by the Venice Restoration Charter of 1964, even though this was essentially dedicated to the restoration of unique monuments, and thus similar to the Athens Charter of 1931.

Gibellina represented the last concrete example of the re-foundation of a historic city in another place, in accordance with the modern urban principles of CIAM, which would then go into crisis due to the failure of the theories of the modern movement as regards modern urban planning (Alexander 1964; Rossi 1978), the planning and reconstruction of the historic centre of cities, the aesthetics of the city's architecture (Mumford 2011) and finally, the absence in planning of a strong social component (Jacobs 1961). They go so far as to challenge all the conceptions of the Athens Charter. The centralised model, designed by the government, was imposed on a functional city without taking into account the real places and needs of the population. This failure was identifiable from the first moments of reconstruction through to the evident dissatisfaction on the part of the inhabitants still present in the city.

The end of CIAM can be linked to the failure of Gibellina's reconstruction, which was based on its principles. It serves as a warning which ought to guide successive post-disaster reconstructions towards a more conservative approach. The "where it was as it was" model characterised the reconstructions of Friuli (1976), Irpinia (1980), Umbria (1997) and L'Aquila from 2009 to present. This model found its theoretical principles in cultural thought developed from the 1970s to today.

The 1970s were the years of heritage, development and reuse of historic centres. International organisations such as the United Nations Educational, Scientific and Cultural Organisation (UNESCO), the Council of Europe, International Centre for the Study of the Preservation and Restoration of Cultural Property (ICCROM) and the International Council on Monuments and Sites (ICOMOS) which, with the publication of charters, declarations and recommendations (Fig. 2.4), helped to define the

[14] Muratori 1950, Caniggia and Maffei 2008; Aymonnino 1977; Rossi 1978.

[15] In English this is "selective clearing of the built fabric".

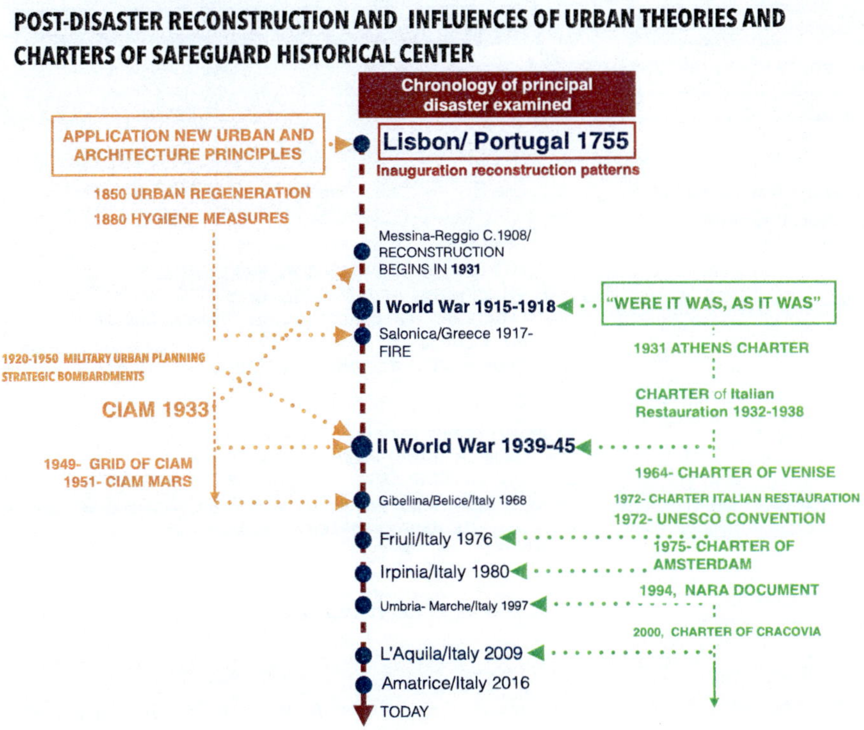

Fig. 2.4 Influences of Charters and urban theories on reconstruction models

concept of the historic centre, its heritage character, economic value (Siena 1973) and its vocation (ICOMOS 1976a, b).[16]

Renewing the concept that the urban context is also characterised by its inhabitants helped to orient the debate about historic centres towards the needs of society. The 1971 Split Declaration is the first European document to discuss the protection of the original social context of historical centres together with its original vocation. Subsequently, with the 1972 Italian Restoration Charter, the historic centre was henceforth considered an urban project to be included within plans for city reconstruction through integrated conservation (Amsterdam Charter 1975) (Fig. 2.5).

The cultural value of historic centres was recognised (ICOMOS 1981) and so the academic world (ICOMOS 1983) became involved in the pursuit and acquisition of multidisciplinary knowledge in order to understand monuments and the urban contexts in which they operated (ICOMOS 1987a, b). Currently, historic sites are considered as places where urban culture is produced and where the main purpose of conservation is believed to be to preserve the physical elements through which

[16] UNESCO: United Nations Educational, Scientific and Cultural Organization, www.unesco.org; ICCROM:The International Centre for Conservation" et The Roma Center, www.iccrom.org; ICOMOS: International Council On Monuments and Sites, www.icomos.org.

Fig. 2.5 Chronology of the basic concepts of urban theories and International Charters and their impact on the reconstruction patterns

urban identity manifests itself (ICOMOS, 1987a). Nara's 1994 document endowed cultural heritage with the role of preserving community memory (Nara, 1994). On this occasion (Krakow Charter 2000), the concept of the authenticity (ICOMOS, ICCROM, UNESCO, JAPAN, CANADA 1994) of cultural heritage was introduced and, in a world characterised by globalisation and standardisation, the protection of authenticity has come to be seen as a means to understand the collective memory of humanity. Contrary to the Venice Charter, the concept of authenticity refers not only to values, based on fixed criteria, but also to cultures and the diversity of their heritage. In this way, the importance of preserving cultural diversity related to different contexts as part of human rights became prominent (Stockholm 1998). Subsequent statements and documents up to the Krakow Charter of 2000 build on this concept and continue to affirm the importance of the diversity and specificity of urban heritage and its identity (Krakow Charter 2000). These are the elements now to be taken into account in the regeneration plans of historic centres and, consequently, in reconstruction plans after disasters.

2.4 Conclusion

Thinking is the necessary first step before taking action. Thinking about the reconstruction of a city that has been destroyed prompts a series of questions, which, 75 years after the greatest reconstruction of cities in the history of humanity, have yet to find an unequivocal answer. Even today, doubts about "how to rebuild" lie at the root of every city reconstruction.

Reflecting on the evolution of the historic city demonstrates that building and rebuilding its urban fabric as if it was a palimpsest has been the rule rather than the exception. In different situations, sudden destruction or planned *tabula rasa* have allowed new values and meanings to be attributed to places, transforming them into monuments or sites of memory. Many specific places have become a testament that can be handed down to posterity as an expression of collective memory, a condition that has been difficult to attribute to an entire historic urban centre. Instead, since their foundations, historic city centres are the result of reconstructions, modifications and partial and gradual adjustments of their minor and vernacular architecture. Traditionally, the transformation of the great capitals of Europe did not provoke debate: cities have been rebuilt. Reconstructions responded to the desire to return to normality and the wish to rebuild the destroyed urban fabric and improve it in terms of infrastructure and construction, without asking the question of the danger of creating false histories or raising the problem of the introduction of the new into the old.

The debate resulting from World War II led to consideration of the reconstruction of historic centres as a problem to solve, where the solutions discussed and explored created further problems. Hence, the choice of reconstruction is functionalist identity-based, or a combination of the two to achieve modernisation as in the case of French cities. The functionalist solution meant putting into practice most of the urban planning principles conceived by the strategic bombing theory of the 1920s, which deeply influenced the modern architectural movement. Essentially, it was an opportunity to boost city development, improve residential standards and build new urban agglomerates to decongest old cities and make them more resistant to future potential devastation in wartime, using materials such as cement and steel. On the other hand, the "identity" solution, which can be identified in more conservative reconstructions, sought to safeguard the historical character of central nuclei, preserving the historical urban settlement and its monuments. The result is a replica of the holographic images of the old city. It is, in conclusion, a copy.

What emerges from this reading of past reconstructions is that when there was no participation by the local population in the reconstruction process, dissatisfaction ensued. From this point of view, the question can be asked: are the figurative and perceptive expedients that determine the pastiche reproduction of the original façades along the streets truly sufficient to satisfy the needs of the inhabitants? The 1970s represent the years of heritage, development and reuse of the historic centre, placing society at the centre of the process.[17] International organisations joined the

[17] Rehabilitation, restructuring.

debate on "the historic centre and its values" and produced a series of declarations, conventions and conference proceedings dedicated to the safeguarding and rebuilding the heritage of historic cities.[18] These contributed to the evolution of thinking towards the affirmation of concepts such as the authenticity, diversity and specificity of urban heritage, the vocation (ICOMOS Charter of cultural tourism, 1976) of urban context and urban identity, all of which pushed the reconstruction of historic cities post-Gibellina towards a more conservative approach.

But questions remain. Why does the reconstruction of a historic city centre currently raise so much concern? Surely this part of the city plays an important role in the life of its inhabitants. It is the source of the identity of the community that lives there; it is a catalyst of social functions and relations in the city and between the city and its territories. Once the city is damaged, the historic centre loses its role. The decommissioning of the historic nuclei of Gibellina, for the more functionalist choice of a new city, or the isolation of L'Aquila in a red zone, have had negative repercussions on the psychological and economic well-being of the affected population.

During the long process of reconstruction, cities undergo an inevitable physical and social transformation. In this scenario, the above-mentioned concepts are destined to change, and therefore, urban identity also changes. The preservation of the identity of places should guarantee spatial, relational, emotional and functional continuity. Is it possible to preserve identity, which by definition evolves (Vinsonneau 2002), fuelled by the social and urban changes of the places it represents? If so, how can urban identity be defined and, especially given its intangibility, how can it be found in the tangible world?

References

Aymonino C (1977) Lo studio dei fenomeni urbani. Officina Edizioni, Roma

Alexander C (1964) Notes on the syntesis of form. Harvard University Press, Mass. Cambridge

Bullock N (2002) Building the post-war world. Modern architecture and reconstruction in Britain. Routledge, New York

Caniggia G, Maffei L (2008) Lettura dell'edilizia di base. *Collana Saggi e Documenti*. Alinea Editrice, Firenze

Cannarozzo T (2009) Rapporto da una periferia territoriale: la Valle del Belice 1968-2008. *Giornale di critica dellarchitettura*. http://www.antithesi.info/0newf/leggitxt.asp?id=569. Accessed 20 Sept 2017

[18] Charters of Athene 1931, 1933; Charter of the Italian Restoration, 1932 and its applications 1938; CIAM, The heart of the city 1951; Charter of Venice 1964.

Charter of Amsterdam (1975) Congress on the European Architectural Heritage. European Council. https://www.icomos.org/en/and/169-the-declaration-of-amsterdam. Accessed 1 June 2018
Charter of Krakow (2000) Principles of conservation and restoration of build heritage. http://smartheritage.com/wp-content/uploads/2015/03/KRAKOV-CHARTER-2000.pdf. Accessed 1 June 2018
Charter of Gubbio (1960) Actes du Colloque Nationale pour la Sauvegarde et la régénération des centres historiques. Italia Nostra. https://www.italianostra.org/wp-content/uploads/2010/04/Carta-di-Gubbio.pdf. Accessed 1 June 2018
Choay F (1992) L'allegorie du patrimoine. Paris: Seuil. trad. it. 1995. L'allegoria del patrimonio. Officina, Roma
Choay F (2013) La Conférence d'Athènes sur la conservation artistique et historique des monuments. Editions du Linteau, Fermanville
Corboz A (2001) Le territoire comme palimpseste et autres essais. Les Editions de l'Imprimeur, Besançon
Cutolo D, Pane S (2016) Il centro storico: una questione aperta, in "Alla scoperta della città antica. Esperienza e conoscenza del centro storico nell'Europa del Novecento", Macerata: Quoblibet Studio, pp 13–68
De Solà Morales M (1978) Verso una definizione. Analisi delle espansioni urbane dell'800, Lotus, 19.27
Diefendorf D (1993) In the wake of war. The reconstruction of German cities after world war II. Oxford University Press, New York, pp 74–75
Edgington DW (2009) Recostructing Kobe. University of British Columbia Press, Vancouver
Farina G, Melluso V (2012) Messina, 'L'architettura della ricostruzione. Metodi, processi e modelli di riferimento della città nuova', in *Storia e progetto* nell'architettura, pp 35–60
Fabbro S (1986) La ricostruzione del Friuli. Udine: Il Campo-Ires
Ford GB (1920) Town planning in the devastated regions of France. In: The American City, XXII/3
Garbini R (2008) Il Linguaggio delle città. Cittalia Anci Ricerche. https://www.cittalia.it/wp-content/uploads/2020/01/città-P-2008-04-Linguaggio-delle-citta.pdf
Geddes P (1915) Cities in evolution: an introduction to the town planning movement and to the study of civics. London Williams & Norgate. trad. it. 1970. Città in Evoluzione. Milano: Il Saggiatore
Giambruno M (2012) Verso la dimensione urbana della conservazione. Alinea, Firenze
Giovannoni G (1913a) 'Vecchie città ed edilizia nuova' in Nuova Antologia, XLVIII/995/I, giugno, pp 449–472
Giovannoni G (1913b) Il "diradamento" edilizio dei vecchi centri. Il quartiere della Rinascenza in Roma, in Nuova Antologia 48(997):53–76
Giovannoni G (1917) 'Per la ricostruzione di città e borgate distrutte', in *Nuova Antologia* LII/1084, pp 156–165
Jacobs J (1961) The death and life of great American Cities. Random House, New York
Kates R, Haas J, Bowden M (1977) Reconstruction following disaster. MIT Press, Cambridge
Konvitz JW (1989) Représentations urbaines et bombardements stratégiques, 1914-1945, in: Annales. Economies, sociétés, civilisations, 44/4, pp 823–847
ICOMOS (1976) Charter of cultural Tourism. https://www.google.com/url?sa=t&rct=j&q=&esrc=s&source=web&cd=6&ved=2ahUKEwjT95iX6LfkAhULh-AKHTFZBWsQFjAFegQIAhAC&url=http%3A%2F%2Fwww.lacult.unesco.org%2Flacult_en%2Fdocc%2FChartCultu rTour1976.doc&usg=AOvVaw1cgNMUaNoz8JDqWMi8d1YB. Accessed 1 June 2018
ICOMOS (1981) Charter of Burra. Charter of Australia on the conservation of places of cultural significance. https://australia.icomos.org/wp-content/uploads/Burra-Charter_1981.pdf. Accessed 1 June 2018
ICOMOS (1983) Declaration of Rome. https://www.icomos.org/en/charters-and-texts/179-articles-en-francais/ressources/charters-and-standards/185-declaration-of-rome. Accessed 1 June 2018
ICOMOS (1987a) Charter of the safeguard of historical cities Washington. https://www.icomos.org/charters/towns_e.pdf. Accessed 1 June 2018

ICOMOS (1987b) First Brazilian Seminar about the Preservation and Revitalization of Historic Centers, Itaipava. https://www.icomos.org/en/charters-and-other-doctrinal-texts/179-articles-en-francais/ressources/charters-and-standards/194-first-brazilian-seminar-about-the-preservation-and-revitalization-of-historic-centers-itaipava. Accessed 1 June 2018

ICOMOS, ICCROM, UNESCO, JAPAN, CANADA (1994) The NARA document on authenticity. https://www.icomos.org/en/charters-and-texts/179-articles-en-francais/ressources/charters-and-standards/386-the-nara-document-on-authenticity-1994. Accessed 1 June 2018

ICOMOS (1998) The Stockholm declaration: declaration of ICOMOS marking the 50th anniversary of the Universal Declaration of Human Rights. https://www.icomos.org/en/what-we-do/focus/human-rights-and-world-heritage/179-articles-en-francais/ressources/charters-and-standards/372-the-stockholm-declaration. Accessed 1 June 2018

Italian restauration charter (1972) http://www.bollettinodarte.beniculturali.it/opencms/multimedia/BollettinoArteIt/documents/1527078909279_10_Carta_del_Restauro_1972_122.pdf. Accessed 1 June 2018. Venice Charter (1964). https://www.icomos.org/centre_documentation/bib/2012_charte%20de%20venise.pdf. Accessed 1 June 2018

Fabbro S (2013) 'Visioni e realizzazioni moderne e postmoderne, di ricostruzione post-catastrofe. Quali lezioni per l'urbanistica?', in Ricostruzione post-terremoto e post-catastrofe: https://www.google.com/search?

Le Corbusier PJ (1971) La charte d'Athènes. Editions Points, Paris

Lonero G (2001) 'Gli envois de Rome di Eugène Beauduin: lo studio dell'antichità come lettura della composizione urbana', in Annali di architettura. Rivista del centro internazionale di Studi di Architettura Andrea Palladio di Vicenza, 13

Mazower M (2007) Salonicco, città di fantasmi: Cristiani, musulmani ed ebrei tra il 1430 e il 1950. Garzanti, Milano

Mougel G (2014) 'Des représentations urbaines à la destruction des villes. Les bombardements stratégiques sur l'Allemagne (1940-1945)', in: La ville en ébullition: Sociétés urbaines à l'épreuve [online]

Mumford L (2011) Il mito della macchina. Il Saggiatore, Milano

Muratori S (1950) Vita e storia delle città. Rassegna critica di architettura 11–12:3–52

Neal DM (1997) Reconsidering the phases of disasters. Int J Mass Disasters 15. http://www.ijmed.org/articles/335/download/

Nora P (1992) Les lieux de mémoire: Les France. Gallimard, Paris

Ojetti U (1920) I nani tra le colonne. Fratelli Treves, Milano

Raja R (1986) Architettura post-Industriale. Editori Riuniti, Roma

Rossa W (2012) Il Piano per Lisbona dopo il terremoto del 1755. Atti dei seminari internazionali dei terremoti di Noto 24 ottobre 2008 e Lisbona 10 ottobre 2008. Terremoti e ricostruzioni tra il XVII ed il XVIII secolo. Palermo: Edibook Giada, pp 87–94

Rossi A (1978) Architettura della città. Clup, Milano

Plant M (2002) Venice: fragile city 1797-1997. Yale Univ. Press, New Haven, Conn.

Petropoulos E (1980) Salonique. L'incendie de 1917. Edizioni Barbounakis, Salonicco

Siena Conference, L'Unità (1973) Nuova vita per i centri storici. https://archivio.unita.news/assets/main/1973/10/01/page_003.pdf. Accessed 28 Apr 2019

Sitte C (1980) L'arte di costruire le città. L'urbanistica secondo i suoi fondamenti artistici. JacaBook, Milano

Smets M (1992) 'Sitte et Buls: la nozione di contesto', in: Zucconi, G., *Camillo Sitte e i suoi interpreti.* Milano: Franco Angeli, pp. 57–60

Treccani GP (2014) Tracce della grande guerra. Architettura e restauri nella ricorrenza del centenario', in *Harchistor*. 1/I, pp 134–179

Vayssière B (2009) Relever la France dans les après guerres: Reconstruction ou réaménagement? Guerres mondiales et conflits contemporains. Presses Universitaires de France, Paris, pp 45–60

Vauthier P (1930) Le danger aérien et l'avenir du pays. Berger-Levrault, Paris

Vinsonneau G (2002) 'Le développement des notions de culture et d'identité: un itinéraire ambigu', in *Carrefours de l'éducation*, 14/2, pp 2–20

Voisin C (2007) 'Le centre, la mémoire, l'identité. Des usages de l'histoire dans la (re) -construction du Nouveau marché de Dresde', in Espaces et sociétés, 2007/3, p 87

Voldman D (1997) La reconstruction des villes françaises de 1940 à 1954: Histoire d'une politique. 1997. L'Harmattan, Paris

Chapter 3
Old Souks of Aleppo: A Narrative Approach to Post-conflict Heritage Reconstruction

Judy Mahfouz

Abstract The ongoing conflict in Syria has caused major destruction to its cities especially in Aleppo and its historical urban fabric. Moreover, the physical destruction of the heritage has caused the disruption of Aleppo's cultural and social practices embodied in its historical city. The current heritage management approach is a top-down practice reinforced by the Syrian law excluding the engagement of the public in the decision-making process. This research focuses on the post-conflict reconstruction of the destroyed heritage sites and the inclusion of the local community and their cultural and social practices. The research analyses *narratives* (past, present and future) related to the lost heritage in Aleppo from both the experts and non-expert groups. Such research helps to raise awareness of heritage's healing capacity, and it is an integral role for recovery in the aftermath of a crisis. The data collection methods incorporated semi-structured interviews and site visits of the destroyed heritage in the Old Souks of Aleppo.The results of the research first identified historical, economical and social drivers that underlie the *past* connections between the heritage and its people. Second, it diagnosed the *present* trauma of losing identities and other problems restricting the process of reconstruction. Third, the research assimilated a collection of *future* suggestions for the recovery of both the built and cultural heritage provided by the participants. The chapter shows that the interrelations of the built and cultural heritage, and associated meanings and values, played a dominant role in the continuation and progress of Aleppo. Therefore, it reinforces the leading role heritage can play in rebuilding the city and healing society's conflict wounds. The chapter advocates for an inclusive approach to ultimately guide the recovery process in a post-conflict society.

Keywords Cultural heritage · Post-conflict reconstruction · Recovery · Narratives · Healing capacity · Souk · Aleppo · Al Medineh · Syria

J. Mahfouz (✉)
Universitat Internacional de Catalunya (UIC), Barcelona, Spain
e-mail: judymahfouz@gmail.com

© The Author(s), under exclusive license to Springer Nature Switzerland AG 2021
F. F. Arefian et al. (eds.), *Historic Cities in the Face of Disasters*, The Urban Book Series,
https://doi.org/10.1007/978-3-030-77356-4_3

3.1 Introduction

The consequences of the Syrian conflict on most of its cities are numerous, especially the destruction of their heritage. An example is Aleppo, one of the oldest continuously inhabited cities in the world (Mansel 2016) that has witnessed devastating destruction to its historical sites. The conflict in Aleppo, more popularly known as "the Battle of Aleppo", has caused severe damages to both the built and cultural heritage. The city was divided into two sides, the western region controlled by the government forces and the eastern region controlled by the opposition forces. The eastern region also included the historical city of Aleppo known as the "Old City of Aleppo" or Al Medineh in Arabic. When the eastern region was reclaimed by the government at the end of December 2016, the Battle of Aleppo had finally ended leaving behind most of the Old City in ruins (Fig. 3.1). In addition to the destruction of its built heritage, the evacuation of the Old City's inhabitants has interrupted the social and cultural processes that embodied this historical site.

The UNESCO Beirut office, the Syrian Ministry of Culture Directorate of Antiquities and Museums (DGAM), heritage experts and local architects/engineers, in the beginning of 2017, quickly started a preliminary heritage reconstruction meeting to organise activities such as damage assessments to be used to identify strategies and to coordinate action for recovery. The rehabilitation process started with rebuilding a few iconic monuments such as the Grand Mosque and its destroyed Minaret. As of 2019, the reconstruction of Al Sakkateyyeh Souk was completed as a Pilot project for the old souk's rehabilitation with the help of Agha Khan foundation.

The question, however, is whose heritage they are planning to rebuild? At this very critical time, it is crucial to combine both the national and cultural process of

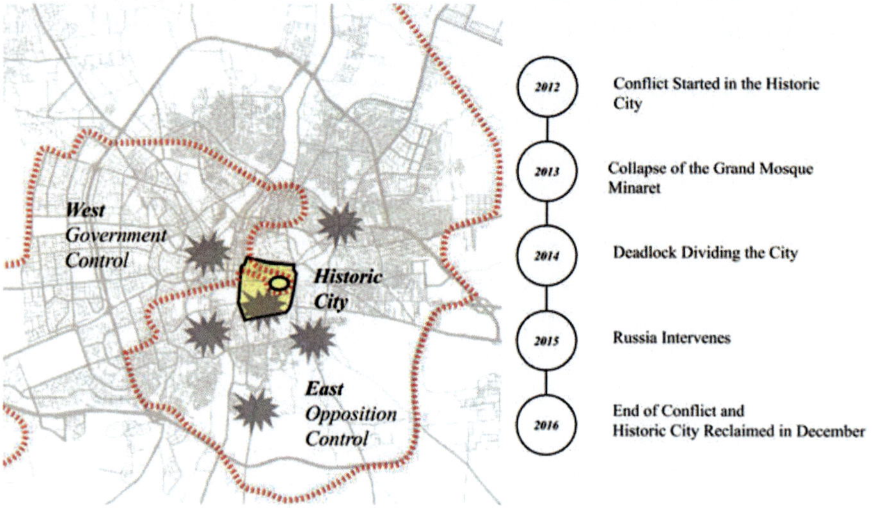

Fig. 3.1 Battle of Aleppo, the timeline and map showing conflict zones (author)

heritage in Aleppo's heritage reconstruction activities. To optimise the early recovery process, it is important to encourage cultural and heritage resilience by managing local capacities, alleviating sufferings and mitigating problems and adapting to the change inflicted by conflict. This research argues that heritage has a healing capacity and can be used as a stimulus for recovery in post-conflict societies.

In Aleppo, the dominant approach towards heritage has been the top-down one where the elite heritage experts make most of the decisions on heritage conservation. There is a need to integrate an alternative and more inclusive approach for post-conflict reconstruction that also frames heritage as a cultural process through the collection of experiences, identities, intangible heritage, memories and performances from the people. To respond to this need, this chapter analyses the narratives of the past, the present, the future (heritage life cycle) of the dominant heritage (experts) and alternative heritage (non-experts) and integrate their voices into the heritage reconstruction process. The research used semi-structured interviews[1] with both groups and field visits in the old souks as the focus of the research and a narrative analysis of the interviews to find common patterns. The expert group or *managers* represented the dominant heritage framework as they are the key decision makers for heritage associated work. For the non-expert group, interviews were conducted with *producers* or the shop owner and the *users* who visit the place for different purposes.

3.2 Old Souks of Aleppo

3.2.1 Historic Urban Fabric

Aleppo's historic city is situated on top of an ancient civilisation dated back to 5000 BC. Because of its strategic location, it was ruled by numerous empires, and it is evident in its overlay of its architectural styles that include: Roman, Byzantine, Seljuk, Mamluk and Ottoman. At the heart of the ancient city sits the Aleppo Citadel. The old part of the city is surrounded by 5 km thick walls with nine historical gates of the ancient city, and many of them are well-preserved (Fig. 3.2). The ancient city includes souks (markets), khans (inns), hammams (public baths), qeysariyes (manufacturing units), madrasahs (schools), as well as places of worship including both mosques and churches (UNESCO 1980). Aleppo "mixed East and West, Islam and Christianity. Until 2012, Aleppo was distinguished by its peaceful character" (Mansel 2016). Throughout the years, Aleppo witnessed dramatic destruction as it "has been won and lost by a succession of empires, sacked by myriad invaders and reduced to rubble by epic earthquakes" (Time 2012).

[1] The study included a total of 16 semi-structured interviews, selected to be representative of different backgrounds. The interviews were divided into two groups: experts (managers) and non-experts (producers and users). All interviews were conducted 2019, in Arabic and were translated by the author.

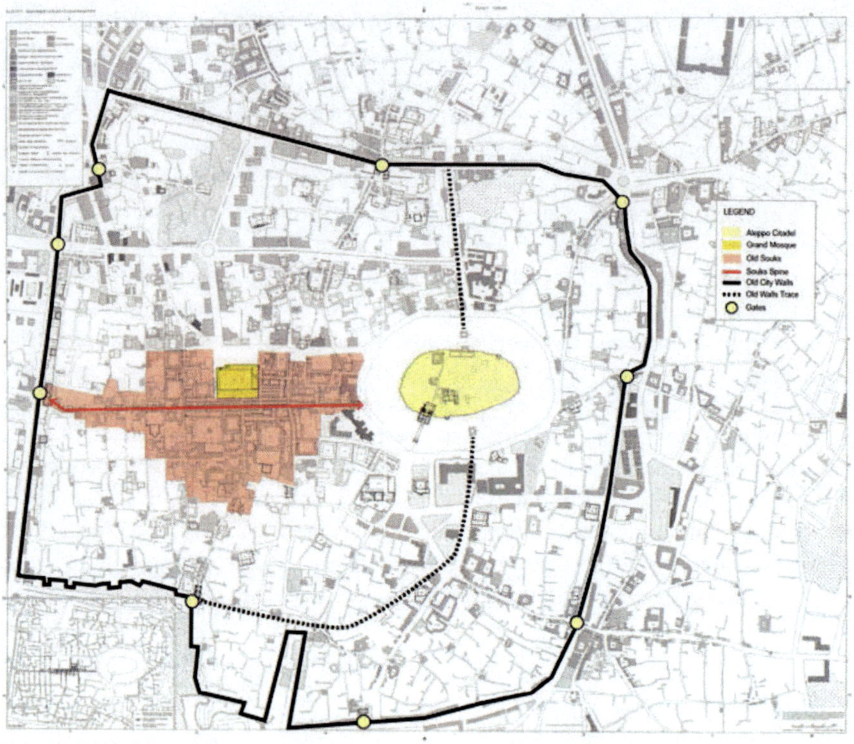

Fig. 3.2 Map of the Historic City of Aleppo (illustration by author, base map courtesy of University of Aleppo)

3.2.2 The Old Souks Before and After

The ancient city has been listed as a UNESCO World Heritage site in 1986 and this is a powerful validation of its historical value. "Commercial activities reached their peak in the fifteenth and sixteenth centuries, when Aleppo was the third most important city in the Ottoman Empire after Istanbul and Cairo" (UNESCO 1980). The heritage site, that housed commercial activities known as "Al Medineh Souks" or old souks of Aleppo, is located between Aleppo Citadel and Bab Antakya (Gate). The old souks of Aleppo were among the severely destroyed heritage sites and will be the focus of this study. For centuries, the old souks of Aleppo were the beating heart of this great ancient city. The name Aleppo paints an image of the rich marketplace of trade and commerce from traditional textiles, carpets, furniture, food to the very famous Aleppo soap. The city Aleppo was located strategically along the Silk Roads acting as a threshold connecting the East to the West. Ibn Jubayr, a traveller from Andalusia, praised Aleppo's old souks and narrated, "large markets arranged in long adjacent rows so that you pass from a row of shops of one craft into that of another until you have gone through all the urban industries" (Mansel 2016, p. 3).

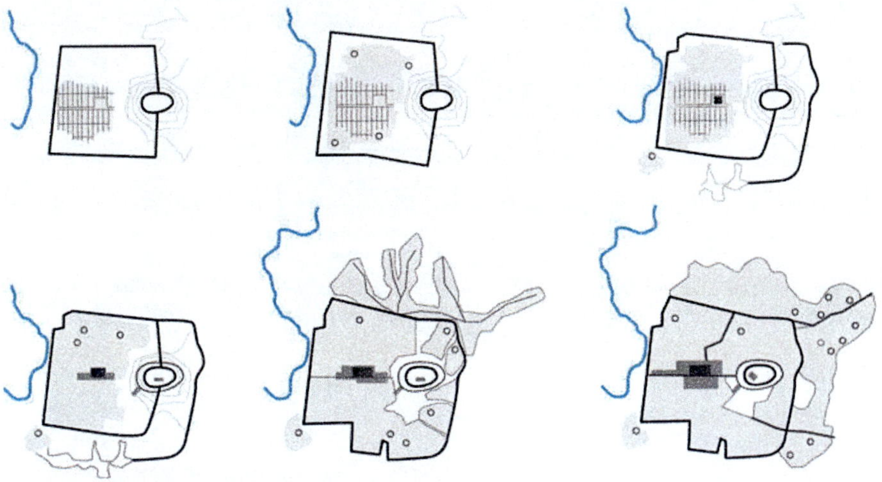

Fig. 3.3 Historic Evolution of Aleppo from the Hellenistic period up to the nineteenth century according to Sauvaget (1941) (illustration by author)

Initially, the wide streets of the Hellenistic city established the souks by subdividing the markets along its main axis and slowly it grew into a network of long and narrow corridors covered by tall vaults. Figure 3.3 presents historic evolution of Aleppo from the Hellenistic period up to the nineteenth century as described by Sauvaget (1941). The storefronts were covered by beautifully decorated wooden doors, and some doors acted as entrances to larger public spaces known as Khans (WMF 2018). Although because of the introduction of the sea connections, the city's commercial activities declined, the souks managed to maintain their original features and activities up until the Battle of Aleppo in 2012 (Fig. 3.4).

The situation rapidly changed after the start of the conflict. According to the reports published in February 2017 by the Syrian Ministry of Culture Directorate General of Antiquities and Museums, only 5.9% of the ancient city is in good condition, and the rest is categorised as lightly damaged (26.5%), heavily damaged (34.7%), almost destroyed (24.1%) and completely destroyed (8.8%). As the numbers show, unfortunately, today most of the ancient city sits in ruins. Old souks of Aleppo have suffered one of the worst damages, and some sections of the old souks are reduced to rubble and are completely on the ground. (Fig. 3.5).

The resilience of the people in Aleppo is worth mentioning. It is incredible to see the number of Aleppines who are currently engaging in the reconstruction of their historic city. Some shop owners who remained in Aleppo are gathering for clean-ups and fundraising. Many are repairing their shops and its surrounding fabric on their own expense. There are shop owners in the middle of the rubble and the darkness opening their shops to display their products. Moreover, most engineers are actively volunteering and offering their expertise for the recovery of the heritage through

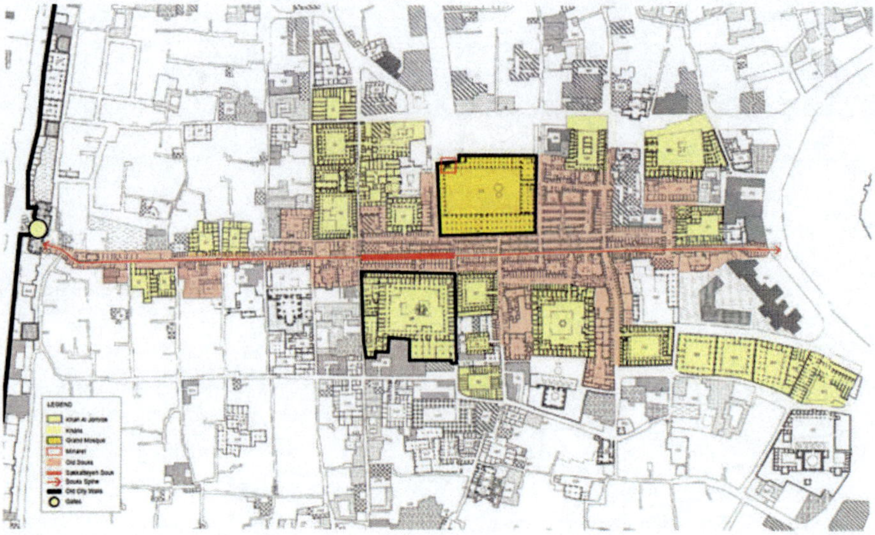

Fig. 3.4 Old souks of Aleppo Map (illustration by author, base map courtesy of University of Aleppo)

Fig. 3.5 Old souks after the conflict (author 2019)

attending meetings with the municipalities and conducting field work damage assessments. Educators and heritage experts are educating the public and re-igniting the connection with the lost heritage through passionately transferring their knowledge. For example, one leading organisation taking on this educating role is the Archaeological Society of Aleppo which is offering heritage visits and tours amidst the destruction to encourage public re-engagement with their heritage. In addition, schools are planning day visits with students to the Aleppo citadel to fill the historical connection gap found in the younger generation.

3.3 Heritage as a Healing Capacity

3.3.1 What Is Heritage?

To understand the consequences of the destroyed heritage at the national and community level in a post-conflict setting and specifically the destruction of the old city of Aleppo, it is crucial to understand how heritage is defined and treated in literature. There are two main definitions of heritage in the literature. One defines heritage as a national identity represented by material objects or sites such as mosques, cathedrals, temples, museums, or libraries managed to conserve the original physical image. The other defines heritage as cultural and social practice embedded in the social relations, traditions or rituals passed on from one generation to the other open to change and transformation. For Laurajane Smith (2006, p. 1) "heritage was not only about the past – though it was that too – it also was not just about material things – though it was that as well – heritage was a process of engagement, an act of communication and an act of making meaning in and for the present". Both definitions emphasise the past of a context, and it is the past where the nations or cultures drive their sense of identity either from the material form or cultural practice. In that sense, the destruction of the heritage in Aleppo is contributing to the destruction of both the nation's and people's identities.

- Dominant Heritage

The national definition of heritage first emerged in the early nineteenth century in the West promoting new movements that romanticised structures and preserved their original physical state. Later, it was developed to support the national symbols and its authority. Smith (2006) explains that there is a dominant discourse regarding heritage that influences the way heritage is perceived, discussed and implemented. She calls this discourse "authorised heritage discourse" (Smith 2006), which she defines as discursive practice led by experts or elites that decide what objects/sites to be conserved and passed onto future generations" (Smith 2006). It undermines the social and cultural practices and limits who has the authority and voice over heritage and who does not. Also, it highlights certain cultural values and meanings on one hand and omits others on the other hand.

In the context of Syria, Munawar (2018) has demonstrated that the heritage practice in Aleppo is a top-down approach by showing evidence from the Syrian Antiquities Law (1963), "wherein the property of archaeology only belongs to the government. This has generated a top-down approach which eventually caused archaeology to become a topic dominated by society's elite" (Munawar 2018, p. 3). Therefore, the dominant heritage practice in Aleppo is like the "authorised heritage discourse"; consequently, the current approach is outdated and needs to incorporate a practice that engages the public and their cultural heritage.

- Alternative Heritage

According to Smith (2006), the authorised heritage discourse "does not tell the whole story". Her main argument is that heritage is not just an object or site, "rather, heritage is what goes on at these sites" (Smith 2006, p. 44). She suggests that heritage is a process where users actively engage in the present through acts of remembrance using the sites as "cultural tools".

She proposes six concepts that engender heritage as a process rather than a thing. First, heritage is the act of *experiencing* the past in the present, situated in its original context. The second component of the cultural heritage is its connection to creating an *identity* and sense of belonging. She makes a clear distinction between the national identity represented by grand monuments or objects, and the collective identity that is constructed by the continuation of the inherited daily practices or traditionsthat leads to her third concept, and the *intangible* heritage where heritage is identified by the cultural expressions such as music, oral histories, language and folklore.[2]

This encourages the idea of *memory* and act of remembering, the fourth concept of heritage as a cultural process. Smith (2006) explains how memory is a concept that is in tension with the national ideologies since they are not inclusive to other pasts or histories that might not align with the past or history the nation aims to conserve for future generations. Maurice Halbwachs explores first the idea of collective memory as a cultural or social practice by arguing that "it is in society that people normally acquire memory. It is also in society that they recall, recognise and localise their memories" (Halbwachs 1992 translation, p. 38). Memories are embedded within the daily practice of a community, while history is selected by elites and professionals. Pierra Nore (1989, p. 13) best explains the difference between memory and history as he argues: "we should be aware of the difference between true memory, which has taken refuge in gestures and habits, in unstudied reflexes and ingrained memories, and memory transformed by its passage through history, which is nearly the opposite: voluntarily and deliberate, experienced as a duty, no longer spontaneous; psychological, individual and subjective; but never social, collective or all encompassing". Memory is an important factor to retrieve the past to reinforce identities especially in the context of Aleppo when the built and cultural heritage are both severely destroyed. *Reconstructing individual and collective memories* has the potential to encourage physical heritage reconstruction in Aleppo.

The fifth concept outlined by Smith is heritage as *performance*, which is the performance of active remembering situated in a heritage site which inspires "the exchange of meaning and memory" (Smith 2006, p. 67). Therefore, she reinforces by stating: "The engagement of emotion and the sharing of this emotive experience or performance, together with sharing of acts of remembering and memory making, are vital elements of the glue that creates and binds collective identities" (Smith 2006, p. 70). The final idea of heritage is the *sense of the place*. She argues "Heritage, particularly in its material representations, provides not only a physical anchor or geographical sense of belonging, but also allows us to negotiate a sense of social

[2] Smith (2006) describes folklore as it the intangible heritage embodied within the oral histories and knowledge.

'place' or class/community identity, and a cultural place or sense of belonging" (Smith 2006, p. 75).

Both definitions of heritage are very crucial in the context of Aleppo because the material form or the place represents the cultural and social practices. Without the built form or place, the cultural and social practices will not be able to manifest itself into existence and continue for future generations. In Aleppo, there is an alignment of identity and meaning making at the national and community level to a certain extent. However, it is very important to emphasise and to note, and Aleppo is a unique case when it comes to post-conflict reconstruction because the state contributed partially to its destruction and currently is responsible for its reconstruction.

3.3.2 Post-conflict Heritage Reconstruction

Two important points exist in discussions about post-conflict reconstruction of cultural heritage. Firstly, there is an intentional and deliberate destruction of heritage as "heritage is one of the first weapons drawn in a conflict often for fighting propaganda battles" (Munawar 2018, p. 7). "The story that really concerns us here—the story of how such objects have been used in times of war either to create and enhance bonds of collective identity or to dissolve those bonds of identity through their destruction or removal—is a very old one indeed" (Ascherson 2005, p. 18). This was very evident during the attacks on Sarajevo's national library and bridge during the Balkan wars where the destruction of the national identity through the destruction of its heritage was used as a strategy to undermine the opposing forces. However, the victims of this destruction are mainly the people, and through this disruption, the identity was at risk of being forgotten and changed.

Secondly, there are several debates on when the heritage reconstruction process begins. In the aftermath of a disaster or conflict, the phases of recovery are the following: humanitarian relief, early recovery, long-term recovery, development and risk reduction. The reconstruction of cultural heritage has been in most cases integrated in the long-term recovery plan because protecting people's lives and addressing their humanitarian needs are the main priorities. However, Barakat (2005, p. 29) has outlined why it is very crucial to recover heritage early on by stating: "Just as destruction of cultural property may be the vehicle of conflict, so it is the continuing thread of cultural behaviour and traditions that provides the basis for reconciliation and recovery (Warren 2005). This is why the recovery of cultural heritage should be central and not peripheral to post-war reconstruction". In that sense, communities have the right to have their cultural heritage protected and safeguarded at an earlier stage of recovery.

In a post-conflict recovery, it is crucial to reconstruct and reclaim the lost cultural heritage and identities at an early stage to enhance the healing process for the community who suffered from the trauma of the conflict and deliberate destruction. Also, it is important to actively engage the community in the recovery process. Yet, in the case of Aleppo, there is a clear gap in the process of the built heritage reconstruction. In March 2017, a UNESCO action plan was drafted by the Beirut office that encouraged the inclusion of the social and cultural practices. However, there is no clear evidence of the implementation of this action plan or follow ups and monitoring-related activities incorporated in the reconstruction process.

3.4 The Quest for Hidden Narratives for Aleppo's Lost Heritage

Narrative is known as the act of telling real events, characters, or settings embedded with emotional expressions producing meaningful content. Hayden White (1980, p. 5) raises the question of the nature of narrative and reflects on how it is inspired by "the very nature of culture and, possibly, even on the nature of humanity itself". The author also argues that narrative can be the solution of how to "translate knowing into telling" (White 1980). Narrative is an important tool to recover the lost heritage in Aleppo because it has the potential to translate the invisible knowing into telling, and through the telling of both experts and non-experts, this research explores various meanings and values of heritage.

The research adopted field observations and research interviews, which addressed three main themes that reflect the three dimensions of the life cycle of heritage: the past, the present, the future (Munawar 2018). The first theme explored the past, and its main objective was to better understand the old souks of Aleppo before the conflict, in addition, to collect memories from the participants through the act of remembering. This is a crucial theme because it helps participants to engage in the reconstruction of memories of their heritage. The second theme explored the present situation, and its main objective was to dive into the current emotions towards the destroyed heritage. This theme also explored the first visits, or recent memories, after the end of the conflict and their first encounters with the destruction of their heritage. In addition, the theme aims to better understand the current problems and challenges that are complicating the reconstruction process of the old souks of Aleppo. The aim of future theme was to discover what kind of image the people of Aleppo envision for the old souks, the original image of the past, the commemoration of the conflict or a completely new image reflecting the contemporary aesthetics. It also shed some light on the perspectives on the role of public administration and shop owners. In summary, it reinforced the idea of community resilience, "a process linking a set of adaptive capacities to a positive trajectory of functioning and adaptation after a disturbance" (Norris et al. 2008, p. 130). Together, they identified common connections in the past, diagnosed the different trauma and their effects in the present and finally grouped future suggestions.

Field observations included visiting the old souks and its surrounding (Fig. 3.6), as well as specific site visits to three important built heritages. The old souks of

Fig. 3.6 Entrance from Aleppo Citadel and one open shop among the ruins (author 2019)

Aleppo including its neighbouring heritage landmarks and major heritage sites are gradually undergoing reconstruction and rehabilitation work.

Three heritage sites visited were Souk Al Sakkateyyeh (Fig. 3.7), Grand Mosque of Aleppo (Fig. 3.8) and Khan Al Jomrok (Fig. 3.9). The Al Sakkateyyeh Souk

Fig. 3.7 Souk Al Sakkateyyeh during reconstruction (author 2019)

Fig. 3.8 Grand mosque of Aleppo reconstruction work (author 2019)

Fig. 3.9 Khan Al Jomrok Courtyard and entrance wall (author 2019)

is a market corridor that houses 60 shops and is located at the centre of the Al Medineh Souks axis. This project has been completed as a pilot project funded and managed by Agha Khan Foundation. The Grand Mosque of Aleppo is one of the most important landmarks of Aleppo. Its Minaret that stood for 950 years has been destroyed completely, and most of the mosque walls, columns and ceilings have traces of conflict. The project is managed by the government and is strictly secured. The minaret stones are organised and numbered across the mosque's central courtyard. Khan Al Jomrok is one of the most important and prestigious caravanserais in the ancient city of Aleppo. On the lower floors, Khan Al Jomrok contains mainly shops, small workshop spaces and one central mosque. And on the first floor, there are offices and stores (previously the inn's bedrooms for travellers). The shop owners of this Khan collaborated to rebuild the destroyed parts of this khan.

3.4.1 *Narratives of the Past: Drivers of Connections*

Heritage is a process to actively engage with the past (Smith 2006), but when the physical engagement has been disrupted because of a conflict, narratives or remembering offer an alternative to engage with the past. It also facilitates the revealing of the invisible past engagement with heritage. The main drivers that guide the connections, meanings and values between the people of Aleppo and their heritage in particular the old souks of Aleppo are (a) history observation, (b) economic activities and (c) social interactions.

As the narrative analysis showed, Aleppines visit the old souks to experience history through observing the historical building styles and materials, the intangible heritage and traditional activities that ultimately connect them with their ancestors and the inheritance they have left behind. This historical connection enhances the sense of heritage where one's emotions and sense of self are truly engaged (Smith 2006). Visiting built heritage creates a sense of belonging and the sense of place because the historical experience enhances the creation of sentimental meaning, attachment and identity. One narrated the memory of historical connection with the built form:

> It transfers you back to the rich eras, always the old heritage is rich with all its landmarks, its archaeological sites, its architecture, and you see how it was built 100 years ago without cranes, and you look at the ceiling to see large stones [blocks], and they found a way to build it. They had great architectural and building techniques.[3]

In addition, the narratives of the past also included an economic driver that is fundamental for the connection between the old souks of Aleppo and its people. There is a commercial or economic hierarchy that influenced the historic urban fabric and its planning. The historic morphology consists of the various types of retail shops, mosques, khans and small factories where the economic activities transition from one space to the other to create one strong connected loop of economic production. The merchants of Aleppo were praised and were generally considered one of the most respected members and of highest status among the Aleppine society, they were known for their unique craftsmanship. The souks housed an assemblage of various traditional products and daily household essentials in one location, therefore, many of the narratives described the high density of activity that dominated the old souks. To describe the economic high activity in the corridors of the markets, one narrated: "*shoulders touching shoulders from how crowded it used to be.*"[4] As a result, it can be observed from the collective narratives that the inherited economic activities are another integral driver for the connection with the heritage site.

The old souks of Aleppo bustled with life and activity and nurtured the most important ingredient for the well-being and the resilience of a city, its social interactions. The beauty of the old souks is that it acted as a community hub for all levels of society gathering to engage socially through economic or leisure activities. Professionals from all types of fields would meet in the khans and enhance their professional networks. Shop owners narrated their daily connections with each other and the support they received from one another. In that sense, one can recognise the connection with the historical site because of its authentic social structure and bonds. One of the shop owners narrated:

> In my earlier days, we had a system, not every person who comes here can open his shop. No, it was a must to know who his father was, and there should be an origin.[5]

In conclusion, the past narrative findings shed some light on the influential role the old souks of Aleppo have historically played in the lives of its culture and society. Crucially, it accentuates heritage healing capacity and its leading potential to recover a post-conflict society through the restoration of its devastated heritage and cultural processes. The historic economic activity of the old souks of Aleppo has the potential to activate the city's economic status. Restoring the souk by deeper understanding of the dynamics of the producers and users of the space will provide better guidelines to strategically recover both the built and cultural heritage.

---

[3] Anonymous declaration, man.

[4] Anonymous declaration, shop owner 03.

[5] Anonymous declaration, Shop Owner 01.

3.4.2 Narratives of the Present: Effects and Problems

"Cultural heritage must be recognised as a crucial element of the recovery process immediately following the end of an armed conflict and not be considered a luxury to await attention later" (Stanley-Price 2005, p. 1). This statement can be evidently observed from the present theme narratives of the research participants. The narratives described the trauma of social destruction, and that the society has been subjected to significant adjustments.

The present narratives clearly showed that the people of Aleppo are suffering from the trauma of losing their ancient identity and the emotional crisis of the disruption of their socio-cultural activities. There is a social trauma associated with losing the normalcy of life and the cultural practices engendering the heritage. The shop owners' narratives described the trauma of losing the long-life work inherited from their fathers and grandfathers. Furthermore, it was shocking to learn that most of the narratives described the belief of intentional destruction aimed to remove and erase the Aleppine cultural identity. One narrated his first encounter with destruction of his shop in Khan Al Jomrok:

> The most painful moment is when I returned to Khan Al Jomrok. Basically I feel [pausing to hold tears] that they destroyed part of my life, so it is not an issue of rebuilding, it is an issue of lost memories, our mosque [in the khan Al Jomrok], and this type of building style is Ottoman, so I rebuilt this mosque with the help of the merchants [shop owners] here, and we collected an amount, the mosque's crown [on top of the dome], was covered in silver; so when I returned, its precious silver [pointing to the mosque] was removed I could not find it here, and it was removed and taken. It seems that they already knew exactly what it was, it was not an accidental destruction, it was deliberate, studied, organised, and they had plans![6]

The present narratives also exposed overall problems restricting the reconstruction of the old souks as well as restricting the continuation of cultural heritage. People identified key problems, which need immediate intervention and action in the post-conflict response. Among them are: the political instability and international sanctions, the diaspora and uncertainty about their return, the disorder of the heritage management and the lack of local skills, the improper selection of pilot projects and the lack of localising the construction methods, complex and time consuming public administration procedures and lastly the technical problems and hazards occurring on the sites such the landslides of debris and unstable damaged buildings during heavy rainfall. One narrated the problem of the diaspora and its consequences by expressing:

> The problem is, they [shop owners] have left 7 years ago, and it would be very hard to re-shift their business to the old location. Half of the city is literally empty; the people are gone. With the remaining current half, only 30% would return to work in the old city. There are very few people who returned and opened their shops in the old souks. Therefore, only 10% of the old souks would return until the war officially ends.[7]

---

[6] Anonymous declaration, Shop Owner 01, declaration.

[7] Anonymous declaration, Engineer 01.

3.4.3 Narratives of the Future: Suggestions

Its is crucial to integrate local capacities or narratives to the post-conflict reconstruction. As Barakat (2005) argues, too often, post-conflict societies are viewed as passive recipients of help rather than active assets of reconstruction. In addition, their inclusion could reinforce the recovery of the two most important basic requirements after conflict: "to reaffirm a sense of identity and to regain control over one's life" (Barakat 2005, p. 6). The narratives of the future explored participants' creative and pragmatic suggestions for the reconstruction of Aleppo's lost heritage. There is an opportunity to integrate their voices in the decision-making process.

Through the future narratives, the research participants shared their aesthetic ideas and vision for the reconstruction of their heritage. They acknowledged the adjustments that have been imposed because of the conflict and proposed ideas to embrace and adapt to the change, although all agreed and emphasised the importance of maintaining the original features. One expressed:

> [...] to rebuild the old city like it was, it will not return, it will be a new city, I think they will rebuild it with the old image but with new materials.[8]

One of the experts (an engineer) perceived the destruction as an opportunity to repair the pre-conflict damages. He mentioned that this will help adjust informal infrastructure systems that would otherwise have been impossible to repair when the souks were continuously active and fully occupied. For example, the facades of the old souks had a network of hanging electrical lines deforming the image and historical urban fabric of the old souks, and the engineers of Al Sakkateyyeh souks have introduced an underground infrastructure incorporating newer electrical and water systems and planning. Preciously the roofs did not include suitable water drainage systems, and the reconstruction provided the opportunity to introduce them.

The future narratives included conversations on the responsibilities of the main stakeholders, public administration and shop owners, towards the reconstruction of the old souks. It emerged there is a need for a mediation strategy to reduce tensions between the two stakeholders. Participants suggested the public administration should better define roles of shop owners and community-based initiatives and their scope of work. Some also stressed the importance of participation of the shop owners and using the local capacities in the reconstruction. In terms of protecting the historical urban fabric, some urged the public administration to enforce policies to avoid illegal construction.

The narratives supported the notion of regaining some control over their lives through actively engaging and proposing solutions to their problems. For the abandoned souks, participants proposed empty shops to be temporarily used by the government. They suggested for the government to support those struggling shop owners with loans and other financial support to encourage them to rebuild their shops. In parallel, they urged the public administration to circulate warnings and penalties for

[8] Anonymous declaration, women focus group.

the shop owners who can rebuild their shops but have not yet started. Some participants urged the return of the shop owners if the shop has minor damages and to rebuild it by using their own resources to set a leading example for others.

Shop owner participants proposed to form committees and elect a representative for each souk to coordinate between them and the public administration. One encouraged:

> One should pursue to rebuild his shop and do his best to rebuild his city, one shop after another, as everyone pursues to build their shops, the old souks of Aleppo will be recovered.[9]

3.5 Conclusion: Rethinking the Built Heritage Reconstruction

The old souks of Aleppo are composed of two complementing components. First is the cultural processes that are activated by shop owners and people that create meaning through the experience of the past and the performance of the social dynamics within the historical urban fabric. Second is the built form that is preserved and managed by experts who deeply understand the traditional building techniques needed to recover the original image of the old souks. A *narrative approach* can integrate the dominant heritage (experts) and alternative heritage (non-experts). Narratives integrating the community's voice can better influence more inclusive guidelines and future policies. Ultimately, the heritage recovery has the capacity to heal war wounds through the collective remembrance accessing the emotional connections of the lost heritage. The analysis of past, present and future narratives can present a guidance for achieving a shared vision for the heritage reconstruction process to realise that healing capacity. Since the state has contributed partially to the destruction, it is important to play its role by strategic and early international agreement and supervision to safeguard the built, and cultural heritage to ensure all stakeholders has a role in the heritage reconstruction process.

Remembering those Aleppines who actively engage in the reconstruction of their historic city, one can learn immensely from the resilience of these champions of Aleppo, despite all odds, they are actively engaging to reconstruct memories and rebuild their heritage, and to end with a hopeful narrative expressed by one of the participants:

> Aleppo, how many times did it **fall** and how many times did it **rise**, so in God's will, it will be rebuilt back better.[10]

[9] Anonymous declaration, women focus group.

[10] Anonymous declaration, women focus group.

References

Anderlini SN, El-Bushra J (2004) Post-conflict reconstruction. In: International Alert and Women Waging Peace,. Inclusive security, sustainable peace: a toolkit for advocacy and action (pp 61–68). International Alert and Women Waging Peace, London and Washington DC. Available at: http://www.inclusivesecurity.org/wp-content/uploads/2012/04/39_post_conflict.pdf

Arnstein SR (1969) A ladder of citizen participation. J Ame Inst Plan 35(4), 216–224. American Institute of Planners, Boston. Available at: http://lithgow-schmidt.dk/sherry-arnstein/ladder-of-citizen-participation.pdf

Ascherson N (2005) Cultural destruction of war and its impact on group identities. Cultural heritage on post-war recovery. ICCROM Conserv Stu 6:17–25. Available at: https://www.iccrom.org/sites/default/files/ICCROM_ICS06_CulturalHeritagePostwar_en_0.pdf

Barakat S (2005) Post-war reconstruction and the recovery of cultural heritage: Critical lessons from the past fifteen years. Cultural heritage on post-war recovery. ICCROM Conserv Stu 6:17–25. Available at: https://www.iccrom.org/sites/default/files/ICCROM_ICS06_CulturalHeritagePostwar_en_0.pdf

Barakat S, Zyck S, Studies H (2009) The evolution of post-conflict Recovery. Third World Quart 30(6):1069–1086. Available at: https://www.tandfonline.com/doi/abs/10.1080/01436590903037333

BBC (2016) Aleppo: key dates in battle for strategic Syrian City [online]. Available at: https://www.bbc.com/news/world-middle-east-38294488

Halbwachs M (1992) On collective memory. University of Chicago Press, Chicago

Jigyasu R, Murthy M, Boccardi G, Marrion C, Douglas D, King J, O'Brien G, Dolcemascolo G, Kim Y, Albrito P, Osihn M (2013) Heritage and resilience: issues and opportunities for reducing disaster risks. United Nations, India. Available at: http://nrl.northumbria.ac.uk/17231/

Mansel P (2016) Aleppo: the rise and fall of Syria's great merchant city. I.B.Tauris, London

Munawar NA (2017) Reconstructing cultural heritage in conflict zones: should Palmyra be Rebuilt. EX NOVO J Archaeology 2:33–48

Munawar NA (2018) Rebuilding Aleppo: public engagement in post-conflict reconstruction. ICOMOS University Forum 1:1–18

Nora P (1989) Between memory and history: les lieux de mémoire. In: representations. 26, pp 7–25. [pdf]. Available at: https://eclass.uoa.gr/modules/document/file.php/ARCH230/PierreNora.pdf

Norris FH, Stevens SP, Pfefferbaum B, Wyche KF, Pfefferbaum RL (2008) Community resilience as a metaphor, theory, set of capacities, and strategy for disaster readiness. American J Comm Psycho 41(1–2):127–150. Available at: https://doi.org/10.1007/s10464-007-9156-6

Reuters (2016) Timeline: the battle for Syria's Aleppo City. Available at: https://www.reuters.com/article/us-mideast-crisis-syria-aleppo-timeline-idUSKBN1412BO

Reuters (2018) Long reach of U.S. sanctions hits Syria Reconstruction [Online]. Available at: https://www.reuters.com/article/us-mideast-crisis-syria-sanctions/long-reach-of-u-s-sanctions-hits-syria-reconstruction-idUSKCN1LI06Z

Sauvaget J (1941) ALEP Essai sur le développement d'une grande ville syrienne des origines au milieu du XIXe siècle. Texte et album, Librairie orientaliste Paul Geuthner, Paris

Smith L (2006) Uses of heritage. Routledge, London

Stanley-Price N (2005) The thread of continuity: cultural heritage in post-war recovery. Cultural heritage on post-war recovery. ICCROM Conserv Stu 6: 1–16). Available at: https://www.iccrom.org/sites/default/files/ICCROM_ICS06_CulturalHeritagePostwar_en_0.pdf

Stephens J (2014) Fifty-two doors: Identifying cultural significance through narrative and nostalgia in Lakhnau village. Int J Heritage Stu 20(4):415–431. Available at: https://doi.org/10.1080/13527258.2012.758651

Syria DGAM (2018) State party report on the state of conservation of the Syrian cultural heritage sites. Syria Arab Republic: Directorate General of Antiquities and Museums. Available at: https://whc.unesco.org/en/soc/3795

Time (2012) Brief history of Aleppo: a great world city now in the grip of war [Online]. Available at: http://world.time.com/2012/07/27/brief-history-of-aleppo-a-great-world-city-now-in-the-grip-of-war/

UNESCO (1980) The conservation of the old city of Aleppo: Syrian Arab republic—(mission). Available at: https://unesdoc.unesco.org/ark:/48223/pf0000042161

UNHCR (2018) Comprehensive protection and solution strategy: protection thresholds and parameters for refugee return to Syria. Available at: https://data2.unhcr.org/en/documents/download/63223

Vecco M (2010) A definition of cultural heritage: From the tangible to the intangible. J Cultural Heritage 11(3):321–324. Available at: https://www.sciencedirect.com/science/article/abs/pii/S1296207410000361

Warren J (2005) War and the cultural heritage of Iraq: a sadly mismanaged affair. In: Barakat S (ed) Third World Quarterly 26(4/5):815–830

White H (1980) The value of narrativity in the representation of reality. In: Mitchell WJT (ed) On narrative. University of Chicago Press, Chicago, IL, pp 1–23

World Monuments Fund (2018) Souks of Aleppo [online]. Available at: https://www.wmf.org/project/souk-aleppo

Chapter 4
Photography for the City, Between the Need for Protection, Conservation and Civic Identity

Simona Manzoli

Abstract This chapter analyses the potential of photography, an extraordinary tool to explore city recovery based on social, psychological, economic and cultural heritage and to link urban resilience with cultural heritage. Photography reproduces a variety of contexts and events: daily intimate ones such as family photos and public identity-making ones, such as celebrations and religious rites that take place in strategic places within the city, such as squares and churches. With reference to the latter, photography stands as a direct witness to unknown settings and transformations, as memory of the lost heritage during restorations and catastrophes, thus assuming great documentary value. The earthquake which struck L'Aquila in 2009 gave the opportunity to highlight the documentary value of photos. The aim of this chapter is to draw attention to the importance of both analogue and digital photos, by using the example of the earthquake of 2009. The project of a large historical photographic collection about the cities which were struck by geophysical disasters represents the source and the underlying theme for reconstruction and also the occasion for citizens to regain their collective consciousness. It requires civic participation and public commitment since everyone has got photos and can effectively contribute to the project. This leads to a real civic recovery of the heritage and therefore shortens distances between the old and the new generations on the one hand and the cultural heritage on the other hand. The awareness of the importance of heritage and of the broader concept of cultural landscape are the basis for the rediscovery of a city's civic identity, for the material reconstruction of the city and are also a stimulus to recovery, in terms of social cohesion, psychological well-being and resilience.

Keywords Photography · Cultural heritage · Earthquake · Cultural landscape · Civic identity · L'Aquila · Italy

S. Manzoli (✉)
University Gabriele d'Annunzio, Chieti, Italy
e-mail: simona.manzoli@unich.it

© The Author(s), under exclusive license to Springer Nature Switzerland AG 2021

F. F. Arefian et al. (eds.), *Historic Cities in the Face of Disasters*, The Urban Book Series,
https://doi.org/10.1007/978-3-030-77356-4_4

4.1 Introduction

Photography, an extraordinary and versatile instrument of knowledge and investigation for different approaches, has been experiencing, during recent times not only an interesting and diversified methodological enlargement, but also an important critical and normative approval by the Ministry of Cultural Heritage and the Ministry of Cultural Assets, Attractions and Tourism (MIBACT).[1] This was possible thanks to the drafting and the implementation of a strategic plan for developing photography in Italy (2018–2022). The three areas for the direct programming of the Plan—heritage, contemporary creation and education and training—outline more clearly the cultural field of reference and the value of photography in a broader sense. It is now recognised as "historical heritage and a contemporary language, an instrument of memory, of expression and understanding of the real, useful for the inclusion and for the growth of an autonomous critical sensitivity that is coming from the citizens".

The extraordinary ability of photography to produce a visual memory (Adacher 2010), linked to the increasing spread of the instrument and the impact of digital technologies, leads to think about its use, especially in contexts affected by natural hazards such as floods and earthquakes.

The shocking psychological and emotional impacts that these tragedies have on the population can be contained by constant work of reconstruction, thereby strengthening the historical memory of the city (Guidoboni 2017; Schenk 2017; Mantini 2018). This could be realised in synergy with citizenship, as a bridge between the old and the new generations, in terms of continuity and of civic re-appropriation of the historical and monumental fabric of our centres.

Therefore, photography stands as the ideal instrument to implement, in a real methodological practice, the need to involve the population who live in the centres inflicted by disastrous incidents, a need which has always been hoped for but never realised.

4.1.1 Photography in the Historical-Artistic and Monumental Field

Photography is not just a medium for conveying images of daily life and intimate contexts and events such as family photos, or public and identity ones such as parties and religious rites that take place in strategic places of the city like churches and squares. It is also an instrument of extraordinary importance for the study of architecture, of monuments and, in a broader sense, of the entire historical and artistic heritage. Here it is used as a first form of protection of a cultural asset and online resources available to all.

---

[1] Ministero per i Beni e le Attività Culturali e per il Turismo.

A photograph tells and describes and "photographs" the context during a unique and extraordinarily valuable "here and now" (*hit et nunc*). It is an instrument that helps to reconstruct monuments and contexts since, as amply emphasised by the German photographer Max Hutzel: "A good picture always tells more than a thousand words"[2] (Schallert 2011).

The first form of protection of a cultural asset is, in fact, its photographic reproduction. The photographic campaigns, created by the bodies responsible for the protection of works of art and kept at the National Photographic Cabinet,[3] established in 1895 and merged in 1975 into the current Central Institute for the catalogue and documentation of Rome[4] (Callegari 2005) are inspired by this concept—which was a milestone in the historical-artistic field at the beginning of the twentieth century.

The systematic study of institutional photographic collections permitted to probe and reconstruct certain contexts that were not known until now, as well as others that were profoundly modified by earthquakes and restoration interventions. It also helped to proceed with the recovery of illicitly stolen works of art.

This is the case of the complex framework of the measures of protection and restoration in Abruzzo from the Unification of Italy to the Second World War (1860–1940) with related methodological guidelines (Pezzi 2005). It has been made possible by a comparative study of archival documents and historical photo-documentation relevant to a specific ministerial body such as the Superintendency.[5]

More structured and research-based surveys in public and private photographic archives have enabled us to understand the damages caused by the earthquake and the restoration choices which affected the Collegiate Church of Saints Cesidio and Rufino in Trasacco in the Marsica area (Manzoli 2015).

In the early twentieth century, photographic campaigns became part of protection operations. Thanks to such a campaign it was possible to recover the liturgical codes of the Collegiate Church of Santa Maria Maggiore in Guardiagrele in the antiquarian market, a large part of this heritage (Manzari 2005, 2012; Corso 2006, 2010) which was stolen in 1979.

Finally, the research cases presented here show how photography is a highly multidisciplinary instrument of crucial importance in the field of historical artistic studies. It has become irreplaceable for the study of conservation needs which have to be encouraged in terms of civil commitment and continuity. With reference to photography, the web is a privileged space of knowledge accumulation, dissemination, sharing and bringing systematic application into play. Gradually, small and big photographic collections, not just those linked to history of art, have been digitised, classified and entered online (Atlante 2016), an irreplaceable prerequisite for wider accessibility of photographic material as an open data source.

The analysis of the photographic resources available on the web started from the macro containers—international or national consortia and censuses, characterised

[2] "Ein gutes Foto sagt immer mehr als tausend Worte".

[3] Gabinetto Fotografico Nazionale.

[4] Istituto Centrale per il Catalogo e la Documentazione di Roma.

[5] Soprintendenze.

by strong centralisation and an interactive approach. This is the case of "Pharos: the international consortium of photo archives", an international consortium of fourteen photographic archives of European and North American art history. The task is to consolidate an accessible digital platform of 25 million images with the multidisciplinary project "Census of photographic collections in Italy. Memory, identity, future" the beating heart of the strategic plan for the development of photography in Italy (2018–2022). The aim of this census is to make available a systematic knowledge of this collection, by means of a network project open to all photo-collection holders, as an essential prerequisite of setting upon policies for its protection and enhancement.

The large foreign photo collections which are available online—like Photo Archive. The Getty Research Institute of Los Angeles and the Image archive, photo Marburg. German Centre of Documentation for History of Art of the Philipps University of Marburg[6]—together with a considerable amount of informative material on documented realities are similar, in terms of form and structure, to the photo libraries of German research institutes of art history in Italy (the Bibliotheca Hertziana; the Max-Planck-Institute for History of Art[7] of Rome and the Max-Planck-Institute for History of Art in Florence[8]) . These are physical places in which photography is complementary to bibliographic material for a broader methodological approach to the research and historical-artistic disciplines.

The photo libraries of Federico Zeri[9], Roberto Longhi[10] and Bernard Berenson[11] are more detailed and inseparably linked to the most illustrious exponents among the international experts. The photograph, as well as the material artefact itself become the object of the connoisseur's reflection, who affixes handwritten notes on the back of the photo. These notes are faithfully included in the complex operations of digitalisation and online entry.

The irreplaceable appeal of the web has also involved the oldest company in the world, which is still operating in the field of photography and image communication, the Florentine Alinary brothers[12] founded in 1852, a leading exponent of the generation of great photographers.

Several and very rich photographic collections[13] make up the stratified photographic heritage of the Central Institute for the Catalogue and the Documentation of

[6] Bildarchiv Photo Marburg, Deutsches Dokumentationszentrum für Kunstgeschichte of the Philipps Universität Marburg.

[7] Institut für Kunstgeschichte.

[8] Kunsthistorisches Institut in Florenz.

[9] Federico Zeri Foundation, Alma Mater Studiorum University of Bologna.

[10] Foundation of History of Art Roberto Longhi, Villa "Il Tasso", Florence.

[11] Villa I Tatti, The Harvard University Centre for Italian Renaissance Studies, Florence.

[12] Fratelli Alinari.

[13] Archive National Photographic Cabinet, Archive of the Photographic Department, Antiquities and Fine Arts Fund of the Ministry of Public Education, National Aerial Photography Library and many others belonging to photographers and/or photographic companies, collections.

Rome,[14] one of the most important public collections of national historical photography. The Abruzzo realities are shown in several archives.[15] All these archives comprise historical and contemporary documentation of the rich historical, artistic, regional, including monumental ethnographic, pastoral and anthropological heritage.

In addition, many other institutions are holding art collections. They range from the photographic archives of the various Superintendencies and universities, to the photo libraries of various research institutions,[16] museums,[17] archives of national tourist associations,[18] as well as banking foundations. To summarise, a partial overview of a digital world is constantly in progress.

4.1.2 The Documentary Value of Photography in Cases of Emergency and Post-emergency

The case studies presented above are linked to the great contribution of photography to the historical-artistic fields of research and studies of monuments. They introduce an element of great depth related to the nature of photography. Photography is a real document, regardless of factors such as the subject represented, the cultural, temporal and geographical context in which it is located, or the author of the shot and whether he/she is a professional photographer or anyone else. Photography is therefore of particular value in case of an emergency and post-emergency as it can assist with disorientation, anxiety and resilience.

A paradigmatic example is given by an aspect of photography that has perhaps not yet been fully explored: the wedding photo album. It may seem improbable that a photograph created to record a private event like a marriage may actually be configured as a double document, but this is the reality. The wedding picture, which is born essentially as a remembrance of a rite and a unique moment, can also be considered as an important and irreplaceable document by the student, whether he/she is an historian or an anthropologist, who is reflecting on the evolution of the marriage rite over time, or by an art historian who is studying the transformations of monumental and historical-artistic contexts over the centuries, as in my case.

I have searched and analysed historical pictures to reconstruct the internal and external layout of the Collegiate Church of Saints Cesidio and Rufino in Trasacco in the Marsica area (Manzoli 2015) after several earthquakes and restorations. Among

---

14 Istituto Centrale per il Catalogo e la Documentazione di Roma.

15 The Medieval Abruzzo photographic archives of the Department of Literature, Arts and Social Sciences of the University G. d'Annunzio of Chieti-Pescara (the first photographic reality in Abruzzo relating to the Census of Photography in Italy), the photographic archive "Domenico Nardini" of the Library "Melchiorre Delfico" of Teramo and the photographic archive of the Museum of the People of Abruzzo (Museo delle Genti d'Abruzzo) "Claudio Leno de Pompeis" of Pescara.

16 The Institute of Art History of the Giorgio Cini Foundation in Venice, the Department of Photography of the Central Institute for Graphics in Rome.

17 The Photo Library of the Vatican Museums.

18 Documentation Centre and Archive of the Italian Touring Club—TCI, Milan.

them, I found a photo of a marriage celebrated before a controversial restoration project carried out in the nineteen-sixties. It has shown me with evocative simplicity all the infinite possibilities of photography and its innate ability to tell, create and—how could I forget—surprise. This practical example highlights the great inner potential of photography which is endowed with "an added value with respect to a text" (Schallert 2011). The example shows that by establishing a link between a context (struck by catastrophes) and the population (contribution of the marriage pictures), photography has a dual documental and commentary value for the reconstruction of earthquake struck Trasacco.

The very high documentary value of photography, of great visual impact, has already strongly emerged since the great earthquakes at the beginning of the century in Messina and Reggio Calabria in 1908 and the Marsica area in 1915. "The need to document has always been the first goal of the State, in order to offer an exhaustive frame of reference to memorise the overall extent of the damages and to act as a starting point for the continuous and dynamic process of knowledge which is essential to schedule the activities aimed at the definitive restoration and full recovery of use" (Callegari 2009). This statement finds its validity in the reconstruction of a few monuments. The fundamental role of photography became known during the reconstruction of the Church of San Pietro in Alba Fucens, destroyed by the earthquake of 1915 (Valerio 2014), but not in the case of entire centres, although they have been photographically documented.

"The cities that have been destroyed by the earthquake were rebuilt elsewhere, no longer in their roots. The relationship between people and the places has been abruptly interrupted, broken off" (Callegari 2009). This is the case of the city of Avezzano, flattened by the earthquake in 1915 and rebuilt in another place in order to complete the master plan project of the historic centre of Avezzano and its enlargement. The project which was strongly characterised by the concepts of "expansion" and "rehabilitation" was developed before the earthquake, revised and then applied in two different phases between 1916 and the 1920s (Ciranna 2015). By contrast, the history of the city of L'Aquila shows an uninterrupted settlement (Boero 2018). In a note following the earthquake in 1703 in L'Aquila, the historian Anton Ludovico Antinori recalls how the nobles and the princes of the Kingdom worked to rebuild the historical centre, since the city risked abandonment due to the devastation.

The concepts of disorientation and resilience play a key role in anthropological and psychological aspects. The disorientation concept is a dangerous condition in which everyone is afraid to lose their own domestic references (acting as "indices of direction"). A good illustration is provided by the episode of the "Bell tower of Marcellinara" told by Ernesto de Martino. For the elderly local shepherd, the bell tower represented the point of reference of his circumscribed domestic space, without which he felt effectively lost and in a state of anguish. In this case, the bell tower (a monumental element, part of the architectural heritage) plays a central value for the elderly shepherd, without which he falls into a deep distressing condition of existential crisis, the same crisis that the citizens of Avezzano have lived during the devastation in 1915. Their city and its monumental elements had disappeared. Writings such as "Avezzano sparita", the "disappeared Avezzano" (Lustri 1988) and

"L'altra Avezzano", the "other Avezzano" (Mastroddi 2005) document the loss with the original context of its roots. In psychology, resilience is the ability of an individual to face and overcome a traumatic event or a period of difficulty. Therefore, the identity factors, social cohesion, the community of intents and essential values represent the essential foundation of the "resilient community" (Resilienza delle città 2016).

4.2 The Experience of the 2009 Earthquake in L'Aquila and Projects to Recover the City Memory

An earthquake struck L'Aquila at 03:32 a.m. on 6 April 2009. The photographic depiction of the catastrophe proved to be the most effective and direct means of communicating the situation in one of the most important art cities in the world (Pace 2010) on a global level. The earthquake catalysed attention from the media due to the implications behind the material and cultural reconstruction of its historical centre (Fig. 4.1).

It was depicted as an irreparably compromised yet still throbbing heart and "irreproducible identity nucleus on which to base the material and socio-cultural reconstruction of the city" (Bartolomucci 2014). Valorising and passing on L'Aquila's

Fig. 4.1 L'Aquila (AQ), The castle (author 2015)

strong roots of the architectural, monumental and artistic heritage merged with planning into a single entity were considered to require everyone's participation for its existence and survival. Yet it was widely reported that in L'Aquila, the earthquake "took on the connotations of a destruction of the historical and artistic identity of a territory" (Varagnoli 2010). This was not a new event for the city, as it occurred after previous earthquakes in 1349, 1461 and 1703 the latter known as the "Great earthquake" due to its devastating effects (Boero 2018). Observing continuous reconstructions following all these major seismic events, continuity in terms of rebuilding the historical centre of the city of L'Aquila appears to have been disrupted only in the case of this latest earthquake. This was due to a plan elaborated to isolate the historical centre from its nature as an inhabited centre and to dislocate the population towards the countryside, causing a fracture in the century-long relationship between local people and their places. However, this relationship is the true lifeblood of these locations and requires constant mending rather than tearing.

"Earthquakes can erase a city only if they erase the inhabitants' desire to live it; to fix it, to redeem themselves by rebuilding it" (Ciranna 2015); "Cities must be rebuilt or restored where they are; it makes no sense to do otherwise" (Marconi 2010). These statements by architecture star Renzo Piano, so fitting for the case of L'Aquila, fall within an ongoing debate and within the conflict surrounding the reconstruction, whereby "how it was and where it was" have become milestones and guides in the long and complex process to rebuilt both monuments and contexts.

After the earthquake, civic redemption and a strong attachment to the memory of the city were promoted by many initiatives organised, from 2009 onwards, by citizen associations, committees, research institutes, universities and individual citizens.

These initiatives are still part of an ongoing process today, assuming the role of models and reference points for a critical and functional debate on pre- and post-earthquake operations.

An analysis of the various projects developed in L'Aquila reveals how photography was assigned the mission of communicating—as immediately as possible—the current state of affairs, the progress made, the slow passage between destruction and reconstruction.

An example of an approach based totally on photographic impact is the social and documentary photography project "The state of affairs, geographies and stories from the post-earthquake" (notes for a virtual museum of documentary photography in L'Aquila after the 6 April 2009 earthquake).[19] It was promoted and developed by the cultural association "The camera of its time"[20] as well as thanks to the actual collaboration of more than 100 Italian photographers. This ambitious project defines itself as "a permanent observatory to tell the story of L'Aquila and Italy at the time of earthquakes". It collects documentation concerning not only L'Aquila and the vast seismic crater formed in 2009, and later earthquakes in Central Italy in 2016 and 2017. It also includes areas such as Belice or Irpinia which despite being ravaged

---

[19] Lo stato delle cose. Geografie e storie del doposisma (Appunti per un museo virtuale della fotografia documentaria a L'Aquila dopo il sima del 6 aprile 2009).

[20] La camera del Tempo.

Fig. 4.2 L'Aquila (AQ), Piazza del Duomo (author 2015)

by seismic events were soon forgotten. The project follows a social point of view focusing on the document—the photograph—as an object for socially sharing a condition of marginalisation and destruction, the loss of one's cultural references. Once again, the web is the virtual space where this sharing becomes concrete, where everyone can see and participate in the collective drama, given that the definition of a red area within a town destroyed by an earthquake is more than merely a local or national issue—it is a question that knows no borders (Fig. 4.2).

Therefore, post-earthquake photography becomes truly a collective narration, reaching the awareness that "if documenting via photography corresponds to taking care of the most fragile heart of Italy, it is fundamental that this keeps taking place". Indeed, the will to document and share is what led photographer Giovanni Lattanzi to donate no less than 2000 pictures to the Regional Direction for Cultural Assets of the Abruzzo Region, showing the first chaotic reactions following the earthquake, the initial revelation of the damage and early safety operations. The generosity and intuition of a professional are often revealed in post-emergency contexts, when individual interests and benefits are set aside in favour of approaches that prioritise a general sense of sharing.

The Deputation for Local History in Abruzzi has carried out a project clearly devoted to the need to generate memory, both visual and textual.[21] The project aims to build an overview of memories of the earthquake in L'Aquila through a constantly evolving database updated via a progressive acquisition of texts, images, photographs, videos, bibliographic information and links.

The seismic events of 2009, 2012 and 2016–2017 touched the heart of the scientific community, leading to a debate on the need to create a multidisciplinary system and to set up a joint effort to plan interventions and restoration methodologies. The research project "Art History and Catastrophes (I): Seismic Italy (sta-sis)", curated by the Max-Planck Institute of History of Art in Florence, inaugurated a scholarly approach extending to the full environmental, social, cultural and identity-related context of L'Aquila. It involved young scholars from several disciplines to build a multidisciplinary and group-based study of the post-earthquake context. For this project, L'Aquila once again became a shared workshop and the right city for experimenting new approaches. The institution of a Summer School and a connected research group dedicated to the Abruzzo capital paved the way for a national and global perspective: History of Art and Catastrophe 2019.[22]

Much focus was placed on the relationship between photography and catastrophes, especially in the comparison between a contemporary vision and historical photos from the photo library of the Institute of History of Art in Florence. In 2018, the L'Aquila-born photographer Antonio Di Cecco curated the online exhibition "Photography and catastrophe, Antonio Di Cecco in dialogue with collections from the Photo Library".[23] He used his particular visual narration methodology in photographs that depict the alterations caused by a natural event in a specific territory (what he called urban contrasts), a formula with which he had previously experimented for his book "In full emptiness, a look at the territory of L'Aquila"[24] published in 2013 (Di Cecco 2013). In 2015, the photographs taken for that project became part of the Archive of the Central Institute for Catalogue and Documentation of Rome, projecting an ideal abridgement between historical and contemporary images.

The meeting organised in September 2017 by the National University Council for Art History (CUNSTA) entitled "Heritage and calamities, the role of art historians"[25] presents another opportunity for debate and reflection. It addressed multiple topics concerning the management of cultural heritage in situations of emergency and the need to create an open model, a multidisciplinary system that can be implemented in disastrous circumstances. There was also an online exhibition, "Amatrice in Focus, earthquakes and photography—recording the past, planning the future", designed to offer an opportunity to reconnect the surviving artworks and monuments to their lacerated contexts.

---

[21] Entitled S.I.S.M.AQ., an acronym standing for L'Aquila Memory Information Service on the Earthquake.

[22] Storia dell'arte e catastrofi.

[23] Fotografia e catastrofe. Antonio Di Cecco in dialogo con le collezioni della Fototeca.

[24] In pieno vuoto, uno sguardo sul territorio aquilano.

[25] Patrimonio e calamità. La parte degli storici dell'arte.

Architect Barnaby Gunning set up a creative project bearing the direct and realistic title "How we do things".[26] It was expressing a general dilemma, the need to do something but the lack of knowledge about how to do it. It gathered professional contributions around a cartographic and geo-referenced model and focused on a 3D reconstruction of the city (both pre- and post-earthquake) via a massive collection of photographs. A similar structure and function—also based on geo-referenced cartography and conceived as an open data system—were shared by the project "SismAbruzzo 2009: ABC—Atlas of Cultural Assets" of the Regional Direction for Cultural and Landscape Assets of the Abruzzo Region, in collaboration with the local Superintendencies. Differing from Gunning's "social" project, this initiative was created to satisfy the need to supply an operational tool for ministerial technicians involved in post-earthquake operations. In other words, this is a tool for "professionals of the sector".

The photographic exhibition and correspondent scientific publication "Before and after the earthquake, medieval art conservation affairs in the Abruzzo Region"[27] (Prima e dopo il sisma 2011)—are based on the effective communication methodology experimented on the occasion of the 1909 publication of the illustrated volume by the Italian Photographic Society entitled "Messina and Reggio before and after" (*Messina e Reggio* 1909) which was depicting the disastrous effects of the 1908 seismic event on the cities of Messina and Reggio Calabria.

The exhibition and publication "Before and after the earthquake, conserving medieval art in the Abruzzo"[28] (Prima e dopo il sisma 2011) created a photographic dialogue between historical photographs from the Photographic Archive of the Department of Literature, Arts and Social Sciences of the University "Gabriele d'Annunzio" of Chieti-Pescara and new photographic campaigns shot ad hoc in the post-quake phase, mainly focusing on churches in L'Aquila and on the crater. This clear and direct methodology based on the "before and after" allows the project to focus on the catastrophic effects of the earthquake on the historical-artistic and monumental heritage. Here, photographs become the source of studies on the effect of the earthquake, while the associated texts are documenting and contextualising the cultural horizons in question, as well as the historical and artistic characteristics (Reggiani 2012).

In yet another case, the web allowed a project to go beyond the extemporaneous and fleeting features of an exhibition.

The website "Abruzzo Medievale" features a specific "Post earthquake" collection (Figs. 4.3, 4.4, 4.5, 4.6, 4.7 and 4.8)—currently in the filing phase—and truly enables anyone (as in the case of the aforementioned projects) to confront their knowledge and prejudice and to experience a dialogue of genuine truth with the photographs. Perhaps, such a dialogue is the highest and extraordinarily winning potential of photography, what makes it the ideal medium for a shared approach to memory, one

---

26 Come Facciamo.

27 Prima e dopo il sisma. Vicende conservative dell'arte medievale in Abruzzo.

28 Prima e dopo il sisma, vicende conservative dell'arte medievale in Abruzzo.

Fig. 4.3 L'Aquila (AQ), Church of San Pietro di Coppito (Photographic Archive of the Department of Literature, Arts and Social Sciences of the University "Gabriele d'Annunzio" of Chieti-Pescara, "Post-earthquake" collection, 2011)

that is truly open to contributions from everybody, instead of remaining reserved exclusively to professionals of the sector.

The guidelines traced by the photographic projects described invite the public to reflect on concepts of memory, prevention and citizenship. Every type of preventive action needs strengthening, especially in terms of sharing memory with citizens. They must not be excluded from the material, cultural and social reconstruction of a city destroyed by an earthquake, but should be encouraged to participate instead. This is the only way for citizens to put in practice the civic re-appropriation process concerning the heritage and history of their hometown. Photography in an integral part of this process: it is the key to the system.

4.2.1 Historical Photographical Collection as a Premise for Re-appropriation of Local Memory

As previously argued, photography assumes a strong significance in the case of L'Aquila and other cities struck by calamities, as the testimony of a city, both in terms of transmitting cultural information for the future and in the civic reclamation of the local heritage—two aspects that are deeply connected. Guidelines for a shared path are the next step.

Fig. 4.4 L'Aquila (AQ), Church of San Pietro di Coppito (Photographic Archive of the Department of Literature, Arts and Social Sciences of the University "Gabriele d'Annunzio" of Chieti-Pescara, "Post-earthquake" collection, 2011)

Starting from a methodological and critical consideration as an art historian and addressing the contribution, photography has made to my field of studies. The study of photography could be expanded further in the cases of L'Aquila, and all other cities exposed to high seismic risks.

Based on the aforementioned projects focusing on cartographic and localised management and on "work in progress and open access data implementation" (Gunning, How we do things; ABC—Atlas of Cultural Assets[29]), the active involvement of citizens in the systemic gathering of historical city photographs is valuable. This is especially the case for the ones that precede the cataclysmic event and therefore enable recovery operations to be enhanced by these sources of absolute value. The challenge is to move towards modernity while relying on the active

[29] Atlante dei Beni Culturali.

Fig. 4.5 L'Aquila (AQ), Church of Santa Maria di Paganica (Photographic Archive of the Department of Literature, Arts and Social Sciences of the University "Gabriele d'Annunzio" of Chieti-Pescara, "Post-earthquake" collection, 2011)

Fig. 4.6 L'Aquila (AQ), Church of San Silvestro (Photographic Archive of the Department of Literature, Arts and Social Sciences of the University "Gabriele d'Annunzio" of Chieti-Pescara, "Post-earthquake" collection, 2011)

Fig. 4.7 Bazzano de L'Aquila (AQ), Church of Santa Giusta (Photographic Archive of the Department of Literature, Arts and Social Sciences of the University "Gabriele d'Annunzio" of Chieti-Pescara, "Post-earthquake" collection, 2011)

Fig. 4.8 Prata d'Ansidonia (AQ), Church of San Paolo di *Peltuinum* (Photographic Archive of the Department of Literature, Arts and Social Sciences of the University "Gabriele d'Annunzio" of Chieti-Pescara, "Post-earthquake" collection, 2011)

participation by anyone. The goal is to incentivise the gathering and conservation of private photographs, an aspect also reflected in education, a fundamental item for the "Strategic Plan for the development of photography in Italy", so that the entire territory may be mapped. This implies that the owners of photographs can move independently across a pre-defined data grid (consisting of cartography for identifying places, toponymy for the "Road" tool set up by the Statistics and Toponymy Office of the Municipality of L'Aquila and urban planning) and to upload their photographs and related data (year of the shot, town, occasion) within an open collective online archive.

The idea is to prepare collective and systemic projects and actions which may find practical application in anyone's daily life. The previously listed initiatives appear to travel on parallel orbits without finding a shared outcome, as is also the case of the many collective photographic collection initiatives. We need to create an integrated system not only for L'Aquila, but to concretise overall idea-gathering channels and source containers.

The project for a great historical photographic collection should constitute both the source and the outline for reconstruction and, above all, the moment in which citizens can reclaim their local history. We are speaking of a shared project which implies civic participation and public commitment, given that everyone owns photographs and can potentially offer a contribution. The recent call for tenders "The path of history"[30] issued by the Municipality of Celano (AQ) is relevant here. It is open to all and aimed at collecting photographs of the city. The key goals of the project are memory, awareness and the valorisation of a city that was strongly altered by the 1915 earthquake. Civic reclamation of local heritage assists in shortening distances between older and newer generations and the cultural heritage, which should truly be everyone's heritage.

4.3 Conclusion

The course of events demonstrates how we lose sight of characteristics and notions that are decisive for sustainable life development. L'Aquila is a city with a well-known historical seismicity, but the seismic history of its territories has been moved totally out of sight, with all the resulting consequences on issues such as the prevention and proper construction of buildings following earthquake laws.

The historical earthquakes of L'Aquila and other centres must be kept in mind, because "the earthquake cannot be predicted but must be expected": we must be ready and teach also the new generations to be ready. Paradoxically, since 2009, we have begun to gather information, memories and details of a city and a context that has been destroyed, just to stem the spectre of loss and destruction. However, thanks to the contribution of memory such a process has to begin earlier not later. The numerous reconstructions of our centres following the earthquake devastations

[30] Il percorso della storia.

must teach us that there is continuity even after the tragedy, and this must be based on the relationship established between identity and its heritage. "The urban historical heritage is a place of cultural identity, the only one that can guarantee the psychological, cultural and economic survival of the city for its material and immaterial values" (Bartolomucci 2014). Photography is key to this relationship. It is an instrument of the past that can strengthen the link with historical centres hit by calamities while taking on the challenge of modernity also in a social aspect.

The hope is that the openly multidisciplinary approach of this tour of photography can teach us all to open our specific perspectives to wider sharing, a necessary precondition for true modernity and contemporaneity. The next step is to suggest future government actions, using the vast amount of data, models and debates which has emanated from the tragic events of the last few years, to realise the initiatives proposed and to give priority to guidelines for education in artistic heritage, sustainability and citizenship.

Acknowledgements Special thanks to: Maria Giulia Aurigemma, Manuela Di Miero, Sandra Mammarella, Daniela Manzoli, Verdiana Visco, Alessandro Tomei.

References

Adacher D (2010) Didattica della memoria. L'uso della fotografia come fonte storica. Istituto Abruzzese per la Storia della Resistenza e dell'Italia Contemporanea, Teramo

Archivio Fotografico Abruzzo Medievale, Dipartimento di Lettere, Arti e Scienze Sociali dell'Università degli Studi "G. d'Annunzio" di Chieti-Pescara (website). Available from http://www.abruzzomedievale.it

Archivio Fotografico "Domenico Nardini", Biblioteca "Melchiorre Delfico" di Teramo (website). Available from http://www.censimento.fotografia.italia.it/archivi/archivio-fotografico-domenico-nardini/#archivio

Archivio Fotografico Museo delle Genti d'Abruzzo "Claudio Leno de Pompeis" di Pescara (website). Available from http://www.gentidabruzzo.com/?page_id=696

Atlante (2016) Atlante degli archivi fotografici e audiovisivi italiani digitalizzati, a cura di G. Sergio. Venezia, Marsilio

Bartolomucci C (2014) La conservazione del centro storico di L'Aquila come paesaggio urbano storico. Available from http://www.academia.edu/9481965/La_conservazione_del_centro_storico_di_LAquila_come_paesaggio_urbano_storico

Bibliotheca Hertziana, Max-Planck-Institut für Kunstgeschichte, Rome (website). Available from http://www.biblhertz.it/en/photographic-collection

Bildarchiv Foto Marburg. Deutsches Dokumentationszentrum für Kunstgeschichte, Philipps Universität Marburg (website). Available from http://www.uni-marburg.de/de/fotomarburg

Boero S (2018). La storia e i terremoti: lo studio del passato per le prospettive del presente e del futuro. In: Mariantoni S, Vaccarelli A (a cura di) Individui, comunità e istituzioni in emergenza. Intervento psico-socio-pedagogico e lavoro di rete nelle situazioni di catastrofe. Angeli, Milano, pp 260–270

Callegari P (2005) Il Gabinetto Fotografico Nazionale: storia di un'istituzione tra esigenze conservative e promozione del patrimonio culturale italiano. In Callegari P, Curzi V (a cura di) Venezia: la tutela per immagini. Un caso esemplare dagli archivi della Fototeca Nazionale (cat. della mostra, Roma, 2005–2006). Bononia University Press, Bologna, 55–68

Callegari P (2009) Fotografare le calamità. La fotografia, il territorio e i disastri naturali. In: Puglisi G, Callegari P (a cura di) La furia di Poseidon. II. 1908 e 1968: I grandi terremoti di Sicilia. Silvana Editoriale, Milano, pp 9–17

Censimento delle raccolte fotografiche in Italia. Memoria, identità, futuro (website). Available from http://www.censimento.fotografia.italia.it/

Centro di Documentazione e Archivio del Touring Club Italiano—TCI, Milano (website). Available from http://www.digitouring.it

Ciranna S (2015) Avezzano tra metamorfosi e cancellazione di una storia urbana. In: Ciranna S, Montuori P (a cura di) Avezzano, la Marsica e il circondario a cento anni dal sisma del 1915. Città e territori tra cancellazione e reinvenzione. Regione Abruzzo, Consiglio Regionale dell'Abruzzo, Ufficio Stampa, L'Aquila, pp. 27–38

Come facciamo, Barnaby Gunning (website). Available from http://www.comefacciamo.com

Contrasti urbani (website). Available from http://www.contrastiurbani.it

Corso G (2006) I manoscritti miniati di Santa Maria Maggiore a Guardiagrele. Edizioni ZiP, Pescara

Corso G (2010) Miniature per una collegiata abruzzese. I corali medievali di Guardiagrele alla luce dei recenti ritrovamenti. Edizioni ZiP, Pescara

Di Cecco A (2013) In pieno vuoto. Uno sguardo sul territorio aquilano. Peliti Associati, Roma

Dipartimento di Fotografia dell'Istituto Centrale per la Grafica di Roma (website). Available from http://www.grafica.beniculturali.it/category/4_fotografie

Donazione Lattanzi al Segretariato regionale (website). Available from http://www.abruzzo.beniculturali.it/donazionelattanzi/

Fondazione di Studi Storia dell'Arte Roberto Longhi, Villa "Il Tasso", Firenze (website). Available from http://fondazionelonghi.it/wordpress/en/eng_photo-library/

Fondo post sisma Abruzzo Medievale (website). Available from http://www.abruzzomedievale.it/archivio-fotografico.php?archivio=postsisma

Fototeca dei Musei Vaticani, Città del Vaticano (website). Available from http://www.museivaticani.va/content/museivaticani/en/collezioni/ricerca-e- restauro/laboratori/fototeca.html

Fondazione Federico Zeri, Alma Mater Studiorum Università di Bologna (website). Available from http://www.fondazionezeri.unibo.it/en/photo-archive/zeri-photo-archive

Fratelli Alinari, Florence (website). Available from http://www.alinari.it/en/77/images

Guidoboni E (2017) Il valore della memoria. Terremoti e ricostruzioni in Italia nel lungo periodo. Quellen und Forschungen aus Italienischen Archiven und Bibliotheken 96:415–444

Il percorso della storia, Comune di Celano (website). Available from http://www.comune.celano.aq.it/articoli/1512/bando-concorso-fotografico-il-percorso-della-storia.html

Istituto di Storia dell'Arte della Fondazione Giorgio Cini di Venezia (website). Available from http://www.cini.it/en/photolibraries

Istituto Centrale per il Catalogo e la Documentazione, Roma (website). Available from http://www.iccd.beniculturali.it/it/fotografia

Kunsthistorisches Institut in Florenz, Max-Planck-Institut (website). Available from http://www.khi.fi.it/en/photothek/index.php

Lo stato delle cose. Geografie e storie del doposisma (website). Available from http://www.lostatodellecose.com/laquila/

Lustri S (1988) Avezzano sparita. 91 fotocartoline dopo il prosciugamento del lago Fucino, prima del terremoto del 1915. Legraf, Roma

Mantini S (2018) Coltivare memorie: la Storia dopo il terremoto. In: Mariantoni S, Vaccarelli A (a cura di) Individui, comunità e istituzioni in emergenza. Intervento psico-socio-pedagogico e lavoro di rete nelle situazioni di catastrofe. Angeli, Milano, pp. 249–259

Manzari F (2005) Il ritrovamento del foglio miniato con la Pentecoste dal perduto Antifonario di Santa Maria Maggiore a Guardiagrele. Arte Medievale III 2:139–141

Manzari F (2012) Un felice ritrovamento. Ritorna a Guardiagrele un foglio rubato. Alumina 10(39):42–47

Manzoli S (2015) Fonti fotografiche per lo studio della collegiata. In: G. Curzi (a cura di): La collegiata dei Santi Cesidio e Rufino a Trasacco. Un santuario nella Marsica. Gangemi Editore, Roma, pp 179–188

Marconi P (2010) Cosa fare nelle città distrutte dalla guerra, dai terremoti o dall'uomo? Via Giulia a Roma nel 1939, L'Aquila dopo settant'anni. In: Marconi P, Pallottino E (a cura di) Com'era, dov'era. Dopo il terremoto, o la guerra. Ricerche di storia dell'arte pp 99, 77–101

Mastroddi M (2005) L'altra Avezzano (cat. della mostra, Avezzano, 1998). Di Censo, Avezzano

Messina e Reggio (1909) Messina e Reggio. Prima e dopo il terremoto del 28 dicembre 1908. Società Fotografica Italiana, Firenze

Online exhibition Amatrice in Focus. Earthquakes and photography—recording the past, planning the future (website). Available from https://galerie.biblhertz.it/en/amatrice/

Online exhibition "Photography and Catastrophe. Antonio Di Cecco in dialogue with the collections of the Photo Library" (website). Available from http://www.khi.fi.it/en/aktuelles/ausstellungen/fotografie-und-katastrophe.php

Patrimonio e calamità. La parte degli storici dell'arte (website). Available from http://cunsta.it/eventi-segnalati-da-cunsta/233-patrimonio-e-calamita-la-parte-degli-storici-dell-arte.html

Pace V (2010) Il terremoto del 6 aprile 2009 in Abruzzo: danni, interventi, iniziative e schede. La terra trema: storia e attualità dei sismi in Abruzzo. Kunstchronik 63(2):45–51

Pezzi AG (2005) Tutela e restauro in Abruzzo. Dall'Unità alla seconda guerra mondiale (1860–1940). Gangemi Editore, Roma

Pharos. The International Consortium of Photo Archives (website). Available from http://pharosartresearch.org

Photo Archive. The Getty Research Institute of Los Angeles (website). Available from http://www.getty.edu/research/tools/photo/

Piano Strategico di sviluppo della fotografia in Italia. 2018–2022 (website). Available from http://www.beniculturali.it/mibac/multimedia/MiBAC/documents/1525355716796_Piano_Strategico_di_Sviluppo_della_Fotografia_in_Italia-1.pdf

Prima e dopo il sisma (2011) Prima e dopo il sisma. Vicende conservative dell'arte medievale in Abruzzo (cat. della mostra, Chieti, 2011), a cura di C. D'Alberto. D'Errico Edizioni, Teramo

Reggiani AM (2012) L'Aquila. Una storia interrotta. Fragilità delle architetture e rimozione del sisma. CISU, Roma

Resilienza delle città (2016) Resilienza delle città d'arte ai terremoti. Enhancing resilience of historic cities to earthquakes, "Atti dei convegni lincei, Roma, 2015". Bardi edizioni, Roma

Schallert R (2011) I tesori nascosti d'Italia: l'archivio dello studio Foto Arte Minore di Max Hutzel. In: Caraffa C (ed) Photo Archives and the Photographic Memory of Art History, "Akten des Konferenz, London-Florenz, 2009". Deutscher Kunstverlag, Berlin-München, pp 335–345

Schenk GJ (2017) Sulla necessità di rendere proficua la memoria dei terremoti e sulla problematicità di trarre insegnamenti dalla loro storia. Quellen und Forschungen aus Italienischen Archiven und Bibliotheken 96:445–452

SismAbruzzo 2009 ABC Atlante dei Beni Culturali (website). Available from http://www.abruzzo.beniculturali.it/index.php?it/142/progetti-e-iniziative&pag=0

S.I.S.M.AQ. Deputazione Abruzzese di Storia Patria (website). Available from http://sismaq.org

Storia dell'arte e catastrofi (I): L'Italia sismica (sta-sis) (website). Available from http://www.khi.fi.it/en/forschung/abteilung-wolf/storia-dellarte-e-catastrofi-i-litalia-sismica-stasis.php

Storia dell'arte e catastrofi (2019) Storia dell'arte e catastrofi. Spazio, tempi, società, a cura di C. Belmonte, E. Scirocco, G. Wolf. Marsilio, Venezia

Tatti VI, The Harvard University Center for Italian Renaissance Studies, Florence (website). Available from http://itatti.harvard.edu

Valerio V., 2014. *Gli scatti del tempo puro. Il ruolo della fotografia nella ricostruzione della chiesa di San Pietro ad Alba Fucens danneggiata dal terremoto del 1915*. In: A. Cipriani, V. Curzi and P. Picardi (a cura di): *Storia dell'arte come impegno civile. Scritti in onore di Marisa Dalai Emiliani* (pp. 267–274) Roma: Campisano Editore

Varagnoli C (2010) Tecniche costruttive tradizionali e terremoto. Riflessioni per la ricostruzione in Abruzzo. In: Marconi P, Pallottino E (a cura di), Com'era, dov'era. Dopo il terremoto, o la guerra. Ricerche di storia dell'arte 99, 65–75

Chapter 5
Cultural Heritage as Stones of Memory: The Recovery of Archives in the Marche Crater Area

Giorgia Di Marcantonio, Pamela Galeazzi, and Caterina Paparello

Abstract The seismic events of Central Italy in 2016 and 2017 interrupted the lifeblood of many territories between the Marche, Umbria, Lazio and Abruzzo. The most affected zone included the area of the Sibillini Mountains and the upper Tronto valley. Even if the emergency state is now over, there are still many interventions to be undertaken. Unsurprisingly, the first allocation of funds was directed to the physical reconstruction of the territories, but it is also essential to deal with the recuperation of social recollection and identity. For this reason, an integrated reading of cultural heritage (archives, libraries and museums) was essential to guarantee the survival of the memories of those communities that were significantly damaged or even completely destroyed. Hence, the need to create a scientific project shared between territories and stakeholders aimed at preserving the vast documentary heritage that remains in highly precarious conditions. A first phase of the project was dedicated to the recovery, description, reorganisation and cataloguing of selected archives; subsequently, a second phase of the project will be aimed at reconstructing the identity of the communities damaged by seismic events through a narrative path and an educational strand, so that citizens can fully understand how the possible loss of archives is also the equivalent of a loss of memory itself. Stones of memory is a project dedicated to cultural heritage so as to avoid losing the material and immaterial value that the institutes and places of culture preserve as the principal aim of their mission. This chapter examines the methodological principles and the actions related to this project dedicated to the recovery of archives in the area of the Marche crater. This contribution presents the planning phases of the first two projects begun in the municipalities of Urbisaglia and Ussita. Furthermore, it analyses a case of musealization (SI) and the recovery of local memory.

G. Di Marcantonio
University of Rome 'La Sapienza', Rome, Italy
e-mail: giorgia.dimarcantonio@uniroma1.it

P. Galeazzi · C. Paparello (✉)
University of Macerata, Macerata, Italy
e-mail: c.paparello@unimc.it

P. Galeazzi
e-mail: pamela_galeazzi@msn.com

© The Author(s), under exclusive license to Springer Nature Switzerland AG 2021
F. F. Arefian et al. (eds.), *Historic Cities in the Face of Disasters*, The Urban Book Series,
https://doi.org/10.1007/978-3-030-77356-4_5

Keywords Archives · Earthquakes · Cultural heritage · Memory · Community value · Marche · Italy

5.1 Introduction

The seismic swarm that regularly struck Central Italy in 2016 and 2017 revealed organisational asymmetries, weakness and fragility of the sorts that had already presented themselves forcefully in other recent seismic episodes: Marche and Umbria 1997–1998, L'Aquila 2009, Emilia Romagna 2012. In all these cases, there have been organisational asymmetries, which is a concept of the economy of culture borrowed from the economy and management of businesses. They include the absence of preventive risk management plans determined weaknesses regarding the coordination between central and peripheral bodies, regarding the availability of safe and well-equipped deposits for the shelter of assets and the availability of means and people called to intervene, overall fragility and an absence of a culture of risk prevention, together with weaknesses in the management of cultural assets in a state of emergency. Not even the establishment of the Blue Helmets of Culture (promoted by Italy, they are currently a UNESCO cultural peacekeeping corps) has served to provoke a serious reflection capable of offering the country an effective and efficient operational plan that would be applicable to the fragile territory of Italy, and which would be a constant dialogue between the centre and the periphery. The destruction of entire peripheral communities belonging to the hinterland and territory of Macerata University and how they live and work, has led to methodological reflections in this contribution that applied scientific research that can offer support to the territory, in terms of third and fourth missions (respectively, relations with the territory and the multidisciplinary approach to the problems of the great variety within society: Missione e impatto sociale delle Università italiane, Linee guida dell'Agenzia di valutazione del Sistema Universitario e della ricerca). Support is also envisaged for the reconstruction of identity, community values and the impact of the development of internal areas: mainly mountainous areas at risk of depopulation, identified in various European countries (SNAI, so-called Barca strategy: National strategy for internal areas) (Padfield et al. 2012). Beyond the technical, scientific and methodological components, the intent of this research is to ensure that the various disciplines involved can affect the public's perception of the value of community and the reconstruction of the intangible connections that have historically linked smaller communities to the larger territory, especially the Apennines (Nishikawa 2017, 2018).

5.2 In the Heart of the Crater

With law number 189 of 17th October 2016 and subsequent additions, issued by the Italian government in order to ensure homogeneous reconstruction, a list of municipalities affected by the seismic events of 24th, 26th and 30th August 2016. A total of 140 municipalities were included in the list and 86 are in the Marche region, where the earthquake affected, in particular, the inner areas of the provinces of Macerata, Fermo and Ascoli Piceno, in the south of the region.

Three years after the first seismic swarms, the state of emergency is considered closed, but many interventions still need to be undertaken and many of these concern cultural heritage, including the numerous municipal and diocesan archives preserved across the territory devastated by the earthquake. The first interventions in fact were in favour of the inhabitants of the affected territories because it was necessary to take care of primary needs such as housing, schools and other services in order to return to a certain degree of normal daily life. In other words, it was plausibly argued that the first allocation of funds should be directed to the physical reconstruction of communities, but the appropriateness of dealing with the restoration of collective memory and the recovery of a sense of identity and belonging, the connective tissue of the communities living in the areas hit by the earthquake, were also recognised. For this reason, preservation of cultural heritage, here specifically archival and other documentary records, is considered essential to guarantee the survival of that history and of those life stories. Recovering this heritage is, also, a necessity dictated by practical needs because, in order to ensure the support of the municipal authorities for the reconstruction of both private and public buildings, it is essential for the current documentation of municipal bodies to be accessible. In the case of the community of Ussita, accessing the documentation has been very difficult because the archives were partly located within the municipal building that was condemned immediately after the earthquake. It must be emphasised that the intense activity of recovering cultural heritage that was carried out by the authorities immediately after the seismic events was extensive. In the final report issued by the General Secretariat of the Ministry of Cultural Heritage, Activities and Tourism (hereafter MIBACT), with data updated to 24th July 2017, it was noted that in the Marche Region alone 2073 linear metres were recovered (*Segretariato generale* MIBACT 2017). In the latest estimate of cultural assets recovered, dated November 20th 2017 (MIBACT 2017) and because of the collaboration of the Civil Protection (*Protezione Civile*), the Army, the Carabinieri of the Command for the Protection of Cultural Heritage and the Fire Department, the MIBACT was able to report the following rates of recovery:

4,623 linear metres of archival documentation (110 for Abruzzo, 530 for Lazio, 2,283 for the Marche and 1,700 for Umbria) ;

20,254 historical-artistic and archaeological heritage (265 for Abruzzo, 2,992 for Lazio, 11,386 for the Marche and 5,611 for Umbria) ;

9,780 volumes of books (671 for Lazio, 4,109 for Marche and 5,000 for Umbria) .

The work done by the central authorities has been commendable, but to date there are still many interventions to be carried out aimed at returning and relocating cultural assets to their communities of origin. Specifically, for archives, it is necessary to assess their condition after the state of emergency because the risk of dispersion or dismemberment is very high. Furthermore, the situation of small municipal archives is often quite distant from solid archival theory. In a nutshell, current archival methodology and practice prefers current, working archives to be held distinctly from the repository for historical documents no longer in use except to researchers who access them in a supervised reading room. While in general, documentation from historical archives is being selected and reordered adequately for historical research purposes, in the case of Ussita, the working group found within the same building and storage current papers, archival filings and historical documentation, all mixed up together, and archivists and researchers were equipped only with approximate search tools for the material. In the case of the archives involved in this project, the documentation also had suffered various manipulations due to administrative requirements, such as requests by citizens to obtain access to old building permits which often refer to previous earthquakes and are therefore important for current reconstruction. The aim is to recover these archives and make them accessible, in order to return them to their proper functions and their communities, taking into account the constraints imposed by the methodology of archival science: the threats to and weaknesses of a wounded territory as well as the context of local stakeholders together with the internal and external conditions: i.e. the internal conditions refer to the state of conservation in which the archive was located at the time of recovery (if in a state of serious disorder or the deterioration of the documentation), the external conditions are to be ascribed to the possibility that the municipality has to properly preserve the archive and provide for an assessment of the existence of suitable storage rooms in which the archive can be safely relocated.

5.3 The Recovery of Archives in the Marche Crater Area

"The recovery of archives in the area of the Marche crater" is a project born specifically from the intention to intervene on the most damaged public archives found in the area of the crater, with the aim of creating a methodology of intervention and at the same time establishing narrative paths that ought to be useful to giving back to the communities some sense of belonging to these wounded territories through rediscovering history, stories and memories. Hence, the need to create a project shared between the territories and the stakeholders that was aimed at preserving and enhancing the immense documentary heritage that remains in highly precarious conditions.

The project is coordinated by the University of Macerata and methodological guidelines were outlined, a network of Municipalities was chosen among those affected by the seismic events, local companies specialising in archiving digitization projects were engaged, and the co-financers of the project are the Carima Foundation and the Marche Region (decree of the Office of the Cultural Heritage number 187 of 17th June 2019).

The project is based on the assumption that archival science, as a multidimensional discipline, can penetrate the territories and be both a tool for creating efficiency and transparency as well as an instrument for the construction of memories (Zanni Rosiello 1987; Valacchi 2018). Archival science, therefore, is here presented as one possible aim for reading a territory and its people as well as to understand the value that the archives have within these communities (Giuva et al. 2007). This analysis involves, finally, the measurement and evaluation of the perception that citizens in the affected territories have of the archives through the administration of questionnaires. This is a critical opportunity to evaluate the effectiveness, incidence and impact of the methodologies deployed in active archival science. Furthermore, the project aims to analyse the relationships between the population and the archives to try to understand what the archives represent in the present time and what they might represent not only in the reconstruction and future development of these territories, but also and above all in the reconstruction of the communities that live there (Da Nova 2016; Dorsi 2016). Archival science, as an active discipline in the territory, is considered to be able to provide citizens with useful tools to reconstruct the connective tissue that makes them communities and allows them to stand out in terms of identity values (Valacchi 2018). By fully accepting what is established by the Faro Convention (Council of Europe 2005, art. 2, art. 12) in relation to access to cultural heritage and democratic participation, the project develops within the citizenry a sense of identification with the archives, persuading communities to perceive the cultural heritage they contain as their own. This also helps them to fully understand the need to preserve this documentary heritage and pass it on to future generations. No less an important aim is to guarantee archival access even to the youngest citizens, so that they too can understand the value these historical documents hold both as a community binder and as identity builders. It is believed, in fact, that to develop a real sense of belonging to the territory it is important to recognise oneself in one's own cultural heritage. In other words, the methodological approach on which the project is based tends to go beyond the simple physical reconstitution of the archives to aim at reconstructing the memory and history of the communities. Furthermore, it is hoped that the project will contribute to the debate regarding the evolution of the archival discipline and on the ethical evolution of the profession (Di Marcantonio 2019; Feliciati 2018).

The first phase of the project is dedicated to the description, reorganisation and inventory of 13 archives located in locations along the Apennine ridge, mostly public, but also of religious institutions. The second step is focused on the creation of a narrative thread aimed at reconstructing profiles of community identities, so that citizens can fully understand how the possible loss of archival material amounts to a loss of memory itself. Each and every archive is assigned a strategy in line with the needs represented at the time of their inspection. Despite having different technical and scientific requirements, the projects share a common thread linked to memory, both individual and collective. These projects, in fact, aim at bringing forward the notion of the archive as having community value, one of the pivots on which the community ought to be founded. That is to say, archives as a fundamental part of the cultural heritage of the community that chooses to recognise itself in this cultural heritage and on the basis of this memory chooses to rebuild the links between communities (Council of Europe 2005, art. 2; Porter and Kramer 2018).

5.3.1 The Historical Archive of the Municipality of Urbisaglia

The reorganisation of the historical archive of the Municipality of Urbisaglia was the starting point of this project. Urbisaglia is located in the outermost part of the seismic crater and the minor impact that the earthquake had on its territory permitted a more immediate and safe action for the archive. Fortunately, it was kept in a building that suffered only a limited amount of damage.

The scientific criteria at the base of every activity of public archival science are historical method and the respect for the temporal modality with which the collection of documents was formed. This is based on respect for the principle of origin, which consists in maintaining or returning to the original order series of documents, taking into account the institutional context in which they were produced (Cencetti 1939; Valenti 1969, 1981; Pavone 1970; Carucci 1983).

This specific project began with a census of the papers that permitted a first summary view of the documentation preserved in the archive, such as the Council Books of income and expenses and other related documents and correspondence relevant to the juridical state of Italian *anciene regime* systems (*Consigli, Libri delle entrate e delle uscite, Lettere dei Superiori, Lettere delle Comunità, Tabelle delle tasse* tutti atti e carteggi pertinenti stati giuridici propri degli ordinamenti italiani di Antico regime). The investigation also analysed documentation relating to various aggregate archives that, together with the official papers of the Municipality, tell the history of Urbisaglia through the actions of the various charities including the *Congregazione della carità* (Congregation of Charity) that operated there. The archives preserve their documentation, which was merged in 1937 into the *Ente comunale di assistenza – ECA*, when they managed the civic hospital while the *Opera Pia Buccolini* managed the *Monte di Pietà* and responsibility for the housing fund for the poor and elderly. In addition, there are the papers of the *Asilo Giannelli* and the *Pio Sodalizio Petrini* (Fig. 5.1).

This documentation, quantifiable in 50 linear metres and ranging from the mid-nineteenth century to the second half of the twentieth century, is of great interest for the reconstruction of the social history of Urbisaglia through the correspondence, repertoires, registers, reports and various other documentation produced by the several pious institutions which were founded by the legacies of some local benefactors during the nineteenth century. These papers speak of the importance of charitable work for the development of Urbisaglia. They gave a boost to the growth of the community, promoting the construction of hospitals, developing a better hygienic

Fig. 5.1 Historical archive of the Municipality of Urbisaglia (authors)

system and a modern health system. These charities also ensured that the education system involved the poorest sections of the population with the opening and management of kindergartens and schools.

The project continued by checking the documentation on the basis of a previous summary list produced by the archivist Pio Cartechini in the 1960s. Once the documentation had been checked, cataloguing began. A title (original, if present, or assigned by the archivist) was given to each archival unit. The other information provided were the chronological extremes, a brief description of the content and salient features of the piece described, as well as information relating to medium, language, state of preservation (with indication of any type of damage) and external characteristics such as binding, dimensions and pagination, the progressive numbering of all the pages of a manuscript and other archival documentation (1r-1v; 2r-2v). This cataloguing permitted the recognition of each documentary series and enabled them to be brought back to the original described units. Thus, the archive was reconstructed to a state as close as possible to its original.

The time partition of the documentation was established as "Antico regime" (Old Regime XVI century—1797); "Municipalità" (Municiple 1797–1798); "Prima Restaurazione" (First Restoration 1800–1805); "Regno d'Italia" (Kingdom of Italy 1805–1814); "Governo provvisorio napoletano" (Provisional Neapolitan Government or Murattiano, named after Gioacchino Murat who was at its head (1814); "Governo provvisorio austriaco" (Provisional Austrian Government 1815); "Seconda Restaurazione" (Second Restoration 1815–1861), "Regno d'Italia" (Kingdom of Italy 1861–1900).

This cataloguing led to the creation of two research instruments: the former is an analytical inventory of the documentation of the Municipality before Italian Unification; the latter is a list related to post-unification documentation up to 1900. Currently, it has not been possible to carry out a proper archival arrangement on the post-unification documentation, because only a small part of it is available. being stored in various parts of the municipal areas. It will be the subject of a second intervention.

The historical archive of the Municipality of Urbisaglia is now available to scholars and to the community; the many stories it contains are waiting to be brought to light. The studies and activities (exhibitions, workshops, readings) that can now be done based on the archive's papers that will permit deeper knowledge of the territory and of the people who have lived there. They will help to develop in the community a sense of belonging to the places where they live, thus making the archive alive, vital and above all contemporary.

5.3.2 *The Historical Archive of the Municipality of Ussita*

The intervention on the documentation of the Municipality of Ussita is still ongoing and provides for the recovery, description and reorganisation of it all, divided into current, semi-current and non-current records. The documentation covers a chronological period from 1910 to today, except for a small section that the municipality received from Cardinal Gasparri (1852–1934), concerning deeds and documents produced by the prelate, born in Ussita on 25th July 1871. Over the years, the Municipality of Ussita, from the date of its establishment, has never carried out the authorised selection procedure of the documentation produced, which normally involves shredding what according to the law can be destroyed, such as invoices and tax returns after ten years, such documentation in this instance was never subjected to scrap operations or regularly inventoried and hence survives in full.

Most of the municipal archival material was located in the Town Hall. The building, constructed between 1927 and 1929, has a basement and three floors. Seismic events heavily damaged the entire structure which is condemned and, for security reasons, completely propped up with scaffolding. The situation inside the building appeared to require immediate critical work, so that the research group considered it impossible to manage the documentation on site. For this reason, the University of Macerata decided to organise the transfer of the entire archive, in accordance with the Archival and Bibliographic Superintendency of Umbria and the Marche, to the Franciscan Library "San Giacomo della Marca", a private ecclesiastical institution located in Falconara Marittima which was considered suitable to host the documentation. In fact, in the seismic crater there was no suitable space available to safely host the archive, and because of this the University has become a hub of local strength and resources (Fig. 5.2).

Inside the Town Hall, the archive was distributed over several levels: on the upper floors, the work group found documentation relating to the current archive which, after the earthquake, largely had lost the folders and lay on the floor in the centre of the individual rooms of the building in a complete state of disorder. The basement, where the semi-active archive was kept, presented a less complex situation, but many shelves had collapsed due to seismic shocks. Despite the state of disorder of the documentation, in order to ensure that the transfer did not further disrupt the archival series, the following sections were preventively identified:

Fig. 5.2 Town Hall of Ussita (authors)

Administration: protocols; administrative correspondence; resolutions; municipal administrative decisions; records; accounting (final balances, estimates, treasury); registry office; electoral office; municipal police.

Aggregate archives: archives of the charitable organisations (Congregation of Charity and Municipal assistance body); papers related to the Sanctuary of Macereto; archive of Cardinal Piero Gasparri.

For administrative reasons, related to the difficult situation created after the earthquake, some parts of the archives of the municipality of Ussita have been found in other municipal buildings which are also condemned (Fig. 5.3).

As can be imagined, the area in which the working group would have had to act to recover the documentation and prepare it for the transfer, was, and still is, a red zone that was accessible only by authorised operators. This involved a long bureaucratic procedure, conducted by the team of the University of Macerata, the Municipality of Ussita and the Superintendency in order to make use of the support of the Fire Department for the recovery of documentation and the Carabinieri Unit for the protection of cultural heritage.

From the outset, recovery operations have been extremely difficult. The Fire Department has guaranteed safety and security by providing men, skills and mechanical means to transfer the documentation from the Town Hall to the external area,

Fig. 5.3 Public school of Ussita in which part of the archive was stored (authors)

where the operators provided for the positioning of the folders in boxes, indicating the type and number of documentary pieces inserted. The use of mechanical means was not sufficient given the state of collapse of the building, and therefore, the firefighters have created an extraordinary operational access plan (manual movements between the floors, passages through windows and the like) (Fig. 5.4).

Once the archive was transferred to the Franciscan library, work on the documentation proceeded with the scrapping operations approved by the archival and bibliographic superintendence for Umbria and the Marche. Considering the fundamental importance of the papers related to the technical office and their usefulness for the reconstruction that has just begun within the municipality, it was decided to digitise the whole series to make it usable as soon as possible. At the same time, the entire archive was catalogued. The description of the archive is aimed at creating an analytical inventory for filing documentation and a summary of the documents relating to the current archive.

Fig. 5.4 Project group at work (authors)

5.4 The Glory of a Small Homeland - Between Institutions and Museums and Collective Identity: The Case of the Gasparri Museum in Ussita

The municipality of Ussita was established in 1913 (Law number 660 of 29th June 1913), achieving its independence from that of Visso, on which it was dependent historically. Its establishment was the result of a wider programme of administrative reorganisation that came about as a consequence on arrangements that the Extraordinary General Commissioner Lorenzo Valerio issued on the occasion of the annexation of the Marche (Santoncini 2008; Dragoni 2012; Paparello 2016). Ussita's independence had already been fiercely supported in the years of the Roman Republic and the Restoration (Pirri 1920; Fiorelli 1962). In the post-unification period, the Visso "*mandamento*" (administrative district under the government of a magistrate), like that of Camerino, was annexed to Macerata Province through the "*Circoscrizione territoriale*" promoted by Minister Marco Minghetti and by Lorenzo Valerio (December 1860). The administrative geography—as it was redrawn—did not correspond to the diocesan division that existed until the 1980s and had the effect of increasing internal conflicts within the local population, reinforcing moves towards separatism that was characteristic of the microhistory of the Apennine lands. The centralisation of the administration in Visso established in the Napoleonic era was perceived as foreign and as having little basis in the identity of the population, in which the memory of the autonomous *inter pares* administration that had for centuries governed the five medieval "*guaite*" of the free municipality of Visso: Guaita Plebis (Visso), Guaita Uxitae, Guaita Montana (Castelsantangelo sul Nera), Guaita Villae (Villa Sant'Antonio) and Guaita Paggese (Cupi, Macereto and Aschio; Cardona, Chierici, 1986) still lived on. The division into "*guaite*" can still be seen and interpreted today and makes the modern-day municipalities of Castelsantangelo sul Nera, Ussita and Visso inextricably linked by a continuity of landscape and identity values, in particular from the agricultural and farming point of view and in their historical and artistic nature (Capriotti and Vitiello 2017; Capriotti 2019).

Moreover, the fear of losing the independence it had acquired is, even today, an enduring sentiment within the social organisations of Ussita and can be recognised in the actions of political decision-makers, in their resistance to reticular forms of managing both cultural and natural heritage, which would instead be the most suitable for promoting the territory as a *museo diffuso* (Montella 2009, 2016). Indeed, while in Europe since the 1970s, the *nouvelle muséologie* was moving towards forms of integrated management of the historical-artistic and naturalistic heritage between the territories; in Italy, despite authoritative rumours, the border and peripheral territories have historically shown opposition to giving up part of their autonomy in favour of administrative and managerial forms of cooperation which would, on the other hand, be the best form of conservation and enhancement, promoting the territory as a diffuse museum (the concept is similar to that of open air museum, although not identical).

Nevertheless, the achievement of Ussita's municipal autonomy was not a simple process and, for the local community, was prolonged and spread out over seventy years starting around 1859, nourishing the desire for municipal independence consistent with the historical processes of Risorgimento and post-Risorgimento united Italy, in particular during the Giolitti era. With the Royal Decree number 5242 of 11th January 1880 (published in the Gazzetta Ufficiale No. 41 of 12th February of the same year, which came into force on 4th March, but was applied administratively only from 1906 onwards), the districts of Ussita, Castel S. Angelo and Croce were granted financial autonomy: separate revenue and expenditure. This began a progressive drive towards full autonomy; a petition for such was supported in the Chamber by deputies Cesare Mattia Giuseppe Silj and Anselmo Ciappi in 1908, but it was not taken up and debated in the Senate at the same time; such a move was repeated in 1912, again without result. A "*pro autonomia*" Committee was set up on 1st January 1913: the collective body managed to obtain approval for separation from Visso municipal council into three autonomous municipalities (13th March 1913), from the Royal Chamber of Deputies (assembly of 6th June) and from the Senate (agenda of 18th June). The onset of World War I, however, rendered unenforceable the provisions which the appropriate local commissions were producing regarding the territorial demarcations of the Ussita and Castelsantangelo sul Nera municipalities that were being created and also regarding distributions of income and expenditure (Pirri 1920, p. 236). During the 1920s, the whole process of administrative definition was fully complete and the building of Ussita Town Hall was symbolically granted. It was built through public subscription, largely subsidised by Cardinal Pietro Gasparri and other high-ranking clergy, prominent figures and local landowners; the architect Filippo Sneider, of the well-known family of architects of the Sacred Apostolic Palaces, was appointed to draw up the plans. The building is a small but prestigious neo-Gothic palazzo, renovated several times, which was seriously damaged by the earthquake that struck central Italy in 2016. While this would not be the place to discuss the reconstruction, it is nevertheless appropriate to reflect on the symbolic components comprising the civic building and on the community purposes that the premises have historically fulfilled. This community, strongly impacted by the earthquake, broken apart for a long time and where the places of memory and of daily life have been disrupted, still needs to restore the links that have been severed, to trace and retrace identity lines that cannot exclude the Town Hall even more so because, in this case, it has fulfilled other community uses, both as a war memorial and as a temple to civic pride.

5.4.1 The Glory of a Small Homeland: The Mythography of a Community

The Town Hall was opened on 8th September 1929, having been constructed between 1927 and 1929 also with the aim of fulfilling the function of a monument to the dead of the World War I and a lapidary shrine, as we see from the inscription "*Ussita ai suoi figli caduti nella grande Guerra*". This stands out on the façade of the building, among the commemorative plaques set into it, but is now partly destroyed. This unusual architectural combination of civic hall and community memorial takes on distinctive features in a country in which each individual parish was erecting separate monuments to its own soldiers. The case of Ussita is an example of the cultural and educational programme of the twenty-year Fascist period, aimed at commemorating the Great War in the Italian Risorgimento tradition, by interpreting the conflict as the last war of national unification. Significantly, when the Lateran Pacts were first drawn up, Giovanni Gentile and Gioacchino Volpe, two major figures in the national cultural debate who came from very different theoretical and educational positions, found points of convergence in the narrative of fascism as the culmination of the process of national unity and identity. The Concordat had made possible a new narrative regarding the relationship between Catholics and Risorgimento secularism, which had previously stood as the religion of the homeland (Baioni 1994, 2006). For example, a few months after the signing of the "*Conciliazione*", the review «*Critica fascista*» reinterpreted the Concordat in a neo-Guelph key "as the true crowning of the Risorgimento" «*come vero coronamento del Risorgimento*» also ascribing to fascism the merit of having restored to the nation its religious faith, which was effectively popolare and national: «effettivamente popolare e nazionale» (Baioni 2006, p. 158).

This interpretation certainly did not escape the notice of the cardinal who signed the Lateran Pacts. He was Pietro Gasparri, an enlightened expert in canon law and a seasoned diplomat, who had been the Vatican's Secretary of State since 1914 and hailed from Capovalazza, a district of Ussita, where he maintained constant relations with his native town (Vercesi 1932; Fiorelli 1960; Fiorelli 1962, in particular pp. 85–113; Spadolini 1972; Moroni 1977; Corsetti 1999; Dalla Torre 2007; Paolini 2012; Pettinaroli and Valente 2020).

He was responsible for donations, works of restoration, road links and a strenuous defence of the autonomy of Ussita, even to the extent that he obtained written assurances from Benito Mussolini during the signing of the Pacts (cf. minutes of the Lateran Treaty and Concordat of 11th February 1929, document belonging to the Gasparri collection). This document, together with a collection of letters, rough drafts, telegrams, certificates of merit, signed photographs and photograph-albums, forms part of the heritage that the cardinal himself wanted to display in Ussita Town Hall from the time it was inaugurated. The collection, also known as the "Gasparri museum", had all the features of a period room (for a study referring to the territory of the Marche: Dragoni and Paparello 2020), to which were added the mythographic aspects of local pride and the rules of exhibiting that belong to Risorgimento period educational museums. This typology, largely borrowed from the models of Italian regional exhibitions of the early twentieth century, were multiform in nature: conservative, expository, celebratory and documentary. As in the case of the Gasparri museum, the monumentalisation of the document, already intelligently theorised by Jacques Le Goff (1978), was the positivist aesthetic canon on which the displaying of documents framed in a row and preserved in treasure chests was based. Together with a wide selection of documents attesting the cardinal's scientific and diplomatic endeavours, five other pictures of a celebratory type and of a more properly historical-artistic nature were exhibited in the Gasparri room. These included a set of decorations for merit, diplomatic honours and distinction medals, of high value both symbolically and as hoarded wealth. In 1929, when the Town Hall was opened, goods and documents were permanently deposited by the cardinal himself to the municipal authority, subject to a clause countersigned for acceptance by the Podestà, as follows:

> [...] all the pictures that I gave back to the Council and that are found in the main room of the Council Headquarters have not been given to the Council of Ussita as their property, but rather they have been given on loan in this sense, that they must remain in the Council Headquarters, until the Council of Ussita acquires its council autonomy and hence without a temporal limit. But if the day would come thaat the Council was merged with another Council, they must be restituted without doubt to my nephew Pietro Gasparri or his descendants to be preserved as a record of the Gasparri family (cf. Gasparri heirs, private archive, collection of letters dated 9th and 10th September).[1]

[1] «tutti i quadri che io ho rimesso al Comune e che trovansi nell'aula principale del Palazzo Comunale non sono stati dati al Comune di Ussita in proprietà, ma gli sono stati piuttosto affidati in deposito in questo senso cioè che essi debbono rimanere nel Palazzo Comunale, finché il Comune di Ussita conserverà la sua autonomia comunale e ciò senza limite di tempo. Ma se giungesse il giorno in cui il Comune venisse aggregato ad altro Comune, essi debbono essere restituiti senz'altro al mio nipote Pietro Gasparri o ai suoi discendenti per essere conservati come ricordi nella famiglia Gasparri»

This kind of museum-temple to the memory of the community, glorifying this illustrious figure, also built up the myth on which the long-desired municipal autonomy was founded, creating a space for a rediscovered continuity between secular and religious loyalty.

Evidence of the union of sacred and civic loyalty was also seen in the processional-type canopy, or "*baldacchino*", honouring the bronze bust of the cardinal, which is still to be found in the building, although in a different position; arrangements are to be made shortly for moving it to make the area safe after the earthquake. The bust is the work of the sculptor Pietro Canonica (1869–1959), a scholar in Venice and Rome and an artist of international renown who also became established in the Vatican (Talalaj 2004; Forti 2015; Santese 2017). The work was commissioned through public subscription and is accompanied by the inscription «*Ussita al suo figlio Pietro Gasparri – In memoria – 11 febbraio e 7 giugno 1929*»—these being, respectively, the dates of the signing of the Concordat and its ratification. In the same year, a further bust was commissioned from the artist by the *Ecole Catholique* in Paris, the institution in which the cardinal was appointed to teach canon law from 1880 to 1898 (Delpal 2004; Fantappie 2004; Nacci 2012).

A replica of the sculpture, unsigned but taken from the cast, is kept in the Pontificia Università Lateranense, where it was displayed in 1959 on the 25th anniversary of the cardinal's demise. In addition, there are reports that Canonica's cast (plaster, 40 × 16 × 20 cm) is preserved in Stresa, together with a collection of 23 other items that the artist donated to this Piedmontese municipality in 1955. Further investigations aimed at outlining the network of relationships on which the quoted commissions were based are postponed to a later time, for reasons of expediency (Fig. 5.5).

The Gasparri collection was dismantled sometime during the 1970s, coinciding with some works to upgrade the Town Hall promoted by the mayor, Nicola Rinaldi, a former deputy. In these circumstances, the Gasparri collection was packed away and not on exposition to the public and was the subject of thefts and dispersal, currently a theme being researched by this author. Remarkably, there was nevertheless no desire to remove or conceal the cardinal's likeness: it was re-sited at the base of the monumental staircase and placed, in the ancient manner, on a reclaimed column originating from the *Pollentia-Urbs Salvia* archaeological area, thus contributing, with this symbological act, to building up myths about him.

Another stage in the plan that Pietro Gasparri applied to creating his identity in his native Ussita can be traced in the building of the town's cemetery. Together with his cousin, the above-mentioned Cesare Silj, Pietro Gasparri obtained in emphyteusis from the Municipality of Visso the land on which the medieval defensive *castrum* stood and took charge of all the expenses for restoration, renovation and urbanisation. Re-purposing, for the benefit of the community, the greatest medieval archaeological site and a symbol of the Ussita's "guaita" bears witness to the deep, historicized Lateran culture of which the cardinal was the greatest exponent in those years. The monumental cemetery, rendered gravely inaccessible by the earthquake, was constructed to a project designed by the engineer Aristide Leonori (Nuzzo 2005, 2016), although only a part of the original design was carried out. It is hoped to be able to retrace a copy of this project during reorganisation: the architects Filippo

Fig. 5.5 Raccolta Gasparri, Town Hall, 1924 circa, archival fund "Balelli" (57, 24), Macerata (photo courtesy of the Biblioteca "Mozzi Borgetti")

Sneider, quoted above, and Nazareno Biscarini, a leading figure in post-unification Eclecticism (Menchetelli 2013), also took part in the enterprise, which involved many other noble families too. Biscarini was asked to create the tomb of Francesco Arsini, an entrepreneur and local benefactor who was related to the Gasparri family. The decorative details on this sepulchre, although ruined today, exemplify the taste for celebratory ornamentation typical of memorial tombs. Cardinal Gasparri, however, was buried inside the cemetery chapel built on the remains of the ancient church *intra muros*, at the foot of the watchtower, in perpetual glory and public memory (Fig. 5.6).

5.4.2 *The Gasparri Museum and an Opportunity to Rebuild Collective Memory*

The value of territorial cultural heritage, if one goes beyond idealistic value-related and aesthetic categories, has progressively assumed a richer and more up-to-date anthropological significance (Montella 2015). Concepts of historical-cultural value, collective memory and identity enliven a wide-ranging interdisciplinary debate that has also had an impact on museum studies, starting with Henri Rivière's *Nouvelle*

Fig. 5.6 Funeral of Cardinal Pietro Gasparri, in the background the Monumental Cemetery of Ussita,1934, archival fund "Balelli" (57, 32), (Courtesy of Biblioteca "Mozzi Borgetti", Macerata)

Muséologie (Dragoni 2018). Taking up the title of a well-known conference of the International Council of Museums, *Le Musée au service des hommes, aujourd'hui et demain: Le rôle éducatif et culturel du musée* (ICOM 1972), we hope that a careful scientific investigation can contribute to making the Gasparri Museum come alive again, interpreting it also as a community museum, a stronghold of memory of what is preserved and what is lost and a space for social and collective living. The neologism *glocalism* is a term for the value that the identity component takes on in the global race: a theme on which to reflect seriously, imagining needs, *aujourd'hui et demain.*

5.5 Conclusion

In the current state of research, it is difficult to draw conclusions based on scientifically measurable data about the impact of actions on the territory and the communities. However, there is a clear intention to involve other actors of scientific research pertaining to the historical, human and economic sciences in order to produce a measurement of results, accountability documents and other pillars for territorial development strategies (Cerquetti 2017, 2019). Undoubtedly a first outcome is given by the expressed interest of other stakeholders; in fact, during the writing of this chapter, further working tables were activated with other actors of the territory (e.g., the involvement of the Municipality of Castelsantangelo sul Nera) . This data strengthens our assumption about the role of the humanities in the reconstruction of the territory after the earthquake. These, in fact, are called to give meaning to both urban and collective identity rebuilding. In other words, the results of scientific research should provide tools for territorial planning and public value management, as well as educational and cultural accessibility guidelines. The actions related to public archival science are therefore pursued here in order to enhance audience engagement and translate documentary value into shared memory. Not just papers but stones.

References

Baioni M (2006) Risorgimento in camicia nera. Studi, istituzioni, musei nell'Italia fascista. Carocci, Roma.

Carucci P (1983) Le fonti archivistiche: ordinamento e conservazione. Carocci, Roma

Cencetti G (1939) Il fondamento teorico della dottrina archivistica. Archivi VI 7–13

Capriotti G, Vitiello R (2017) La storia dell'arte ai tempi del terremoto (Marche 2016–2017). Annali di critica d'arte. Nuova serie 1, 11–31

Capriotti G (2019) Il territorio ferito: il patrimonio storico-artistico delle Marche dopo il terremoto. In: Smuraglia C (eds) La Costituzione, 70 anni dopo, pp 101–106. Viella, Roma

Cerquetti M (2017) From value assessment to public value creation and measurement in the museum sector. Theoretical approaches and critical issues in a changing world. In: Cerquetti M (ed) Bridging theories, strategies and practices in valuing cultural heritage. EUM, Macerata, pp 121–147

Cerquetti M (2019) What gets measured gets done. Misurare e valutare l'archeologia pubblica. Il capitale culturale. Stud Value Cult Heritage Suppl 9:553–569

Corsetti A (1999) Le "memorie" del Cardinal Gasparri. Osservazioni e congetture. In: Corsetti A (ed.) Scritti, pp. 157–174. Le lettere, Firenze

Council of Europe (2005) Council of Europe framework convention on the value of cultural heritage for society, Faro 2005. Available from https://rm.coe.int/CoERMPublicCommonSearchServices/DisplayDCTMContent?documentId=0900001680083746 Accessed 05 July 2021

Da Nova R (2016) Alberi della memoria. Modalità ed esempi degli interventi di ripristino degli archivi friulani dopo il sisma del 1976. In: Azzolini C, Carbonara G (eds) Ricostruire la memoria. Il patrimonio culturale del Friuli a quarant'anni dal terremoto. Forum Edizioni, Udine, pp 237–245

Dalla Torre G (2007) La vicenda poco nota delle Memorie del cardinale Gasparri. Aracne, Roma

Delpal B (2004) Pietro Gasparri, professeur à l'Institut catholique de Paris. Mélanges de l'école française de Rome 116(1):91–107

Di Marcantonio G (2019) L'archivista: una figura in costante trasformazione. *Officina della storia*. https://www.officinadellastoria.eu/it/2019/01/09/larchivista-una-figura-in-costante-trasformazione/. Accessed 7 Aug 2019

Dorsi P (2016) Memorie di carta: gli archivi, In: Azzolini C, Carbonara G (eds) Ricostruire la memoria. Il patrimonio culturale del Friuli a quarant'anni dal terremoto. Forum Edizioni, Udine, pp 231–235

Dragoni P (2012) La memoria sul territorio: i musei civici delle Marche. In: Severini M (ed) Memoria, memorie. 150 anni di storia nelle Marche, pp 38–65. Il lavoro editoriale, Ancona

Dragoni P (2018) Germain Bazin, il museo e la museologia. In: Bazin G (ed) Les temps des musées, preface and translation of P. Dragoni, pp 7–27. Edifir, Firenze

Dragoni P, Paparello C (2020) Guglielmo Pacchioni a Pesaro: l'allestimento come atto critico. In: Dellapiana E, Failla MB, Varallo F (eds) La conferenza di Madrid del 1934: un dibattito internazionale (Università degli Studi di Torino—Politecnico di Torino, 26–27 febbraio 2018), pp 103–120. Sagep Editori, Genova

Fantappie C (2004) La formazione teologica e giuridica di Pietro Gasparri a Roma nel Seminario dell'Apollinare. Mélanges de l'école française de Rome 116(1):115–140

Feliciati P (2018) Per una qualità ed etica della mediazione archivistica. In: Di Marcantonio G, Valacchi F (eds) Descrivere gli archivi al tempo di RIC-CM. EUM, Macerata, pp 21–29

Fiorelli L (ed) (1960) Il cardinale Pietro Gasparri, Roma: Pontificia Università Lateranense

Fiorelli L (1962) Ussita, terra di uomini illustri. Lineamenti storici, profili biografici, itinerari turistici con illustrazioni. Tip. Armellini, Roma

Forti M (2015) Testimonianze artistiche del Novecento nella basilica di San Pietro. In: Iacobone P (ed) La basilica di San Pietro, pp 145–155. FMR art'é, Bologna

Giuva L, Vitali S, Zanni Rosiello I (2007) Il potere degli archivi. Usi del passato e difesa dei diritti nella società contemporanea. Bruno Mondadori editore, Milano

ICOM (eds) (1972) Le musée au service des hommes aujourd'hui et demain: le rôle éducatif et culturel du musée. Parallel title. The museum in the service of man, today and tomorrow: the museum's educational and cultural role. The Papers from the ninth general conference of ICOM. Grenoble, 1971, Paris: International Council of Museums

Le Goff J (1978) Documento/Monumento. Enciclopedia Einaudi 5:38–43

Menchetelli V (2013) Nazareno Biscarini (1835–1907). In: Belardi P, Bori S (eds) 1861–1939: l'architettura della Perugia postunitaria, pp 171–183. Fabrizio Fabbri, Perugia

MIBACT (2017) Sisma centro Italia: recuperate oltre 30 mila opere d'arte, oggetti archeologici e libri, circa 5 chilometri di archivi e oltre 1000 interventi di messa in sicurezza. http://www.beniculturali.it/mibac/export/MiBAC/sitoMiBAC/Contenuti/MibacUnif/Comunicati/visualizza_asset.html_933113829.html. Accessed 9 Sept 2019

Montella M (2009) Valore e valorizzazione del patrimonio culturale storico. Mondadori Electa, Milano

Montella M (2015) Cultural value. In: Golinelli GM (ed) Cultural heritage and value creation. Towards new pathways, pp 1–51. Springer, Cham

Montella M (2016) Economia e gestione dell'eredità culturale. Dizionario metodico essenziale. Wolters Kluwer, Assago; CEDAM, Padova

Moroni A (1977) Atti della tavola rotonda su la figura storica del card. Pietro Gasparri di Ussita. Macerata, 17 maggio 1973. Giuffrè, Milano

Nacci M (2012) L'anima giuspubblicistica di Pietro Gasparri e le Institutiones Iuris publici. Apollinaris 85:603–622

Nishikawa E (2017) Development in earthquake countermeasures for heritage buildings in Japan. Archeomatica 8/1. https://www.mediageo.it/ojs/index.php/archeomatica/article/view/1371. Accessed 9 June 2020

Nishikawa E (2018) Earthquake as Heritage—examples from Japan. https://www.iccrom.org/news/earthquake-heritage-examples-japan. Accessed 9 June 2020

Nuzzo M (2005) Lo spazio sacro della Terza Roma: dinamiche di insediamento, aspetti devozionali e caratteri formali degli edifici religiosi dei nuovi quartieri della capitale tra Leone XIII e Pio X (1878–1914). Mélanges de l'école française de Rome 117(2):497–518

Nuzzo M (2016) Roma 1913. Artisti e architetti per il centenario costantiniano: Biagio Biagetti, Eugenio Cisterna, Giovanni Battista Conti, Aristide Leonori e Guglielmo Palombi. In: Canella T (ed) L' impero costantiniano e i luoghi sacri, pp 637–670. Il Mulino, Bologna

Padfield C et al (2012) Green paper fostering and measuring ´third mission´ in higher education institutions, Valencia

Paolini G (2012) Un «pontiere» fra le due rive Tevere: Cesare Silj e la Questione romana. *Nuova Antologia* 147 (fasc.2263), offprint, 10–56

Paparello C (2016) Alienazioni eccellenti e museo pubblico: un riesame documentario. In: Paparello C (ed) La Storia e il Museo. Documenti e proposte per la valorizzazione del patrimonio museale, pp 51–111. Il Formichiere, Foligno

Pavone C (1970) Ma è poi tanto pacifico che l'archivio rispecchi l'istituto? Rassegna degli Archivi di Stato XXX 1:145–149

Pettinaroli L, Valente M (2020) Il cardinale Pietro Gasparri, segretario di Stato (1914–1930). Heidelberg University Publishing, Heidelberg

Pirri P (1920) Ussita. Notizie storiche con illustrazioni e documenti. Tipografia poliglotta vaticana, Roma

Porter ME, Kramer MR (2018) Creating shared value. How to reinvent capitalism—and unleash a wave of innovation and growth. In: Lenssen GG, Smith NC (eds) Managing sustainable business. An executive education case and textbook, pp 323–346. Springer, Dordrecht

Santese BM (ed) (2017) Museo pietro canonica a villa borghese: museo e casa d'artista. Palombi editore, Modena

Santoncini G (2008) L'unificazione nazionale nelle Marche. L'attività del Regio commissario generale straordinario Lorenzo Valerio (12 settembre 1860–18 gennaio 1861). Giuffrè, Milano

Segretariato generale, MIBACT (2017) Relazione conclusiva attivita' coordinate dalla Di.Coma.C a seguito degli eventi sismici 2016-2017 del Centro Italia 24 agosto 2016 – 6 aprile 2017. http://www.beniculturali.it/mibac/multimedia/MiBAC/documents/1502211250548_Relazione_del_Segretariato_Generale_MiBACT.pdf. Accessed 9 Sept 2019

Spadolini G (ed) (1972) Il cardinale Gasparri e la questione romana con brani delle memorie inedite. Le Monnier, Firenze

Talalaj MG (2004) Pietro canonica: lo scultore prediletto dell'ultimo zar. Studi piemontesi 33:93–101

Valacchi F (2018) Archivista, parola plurale. Archivi XIII(1):7–13

Valenti F (1969) A proposito della traduzione italiana dell'"archivistica" di A. Brenneke. Rassegna degli Archivi di Stato XXIX, pp 441–455 (now in Grana, D. (ed) 2000. Filippo Valenti. Scritti e lezioni di archivistica, diplomatica e storia istituzionale, Roma: Ministero per i beni e le attività culturali)

Valenti F (1981) Riflessioni sulla natura e struttura degli archivi. *Rassegna degli Archivi di Stato* XLI, 9–37 (now in Grana, D. (ed) 2000. Filippo Valenti. Scritti e lezioni di archivistica, diplomatica e storia istituzionale. Ministero per i beni e le attività culturali, Roma

Vercesi E (1932) Tre segretari di Stato: Consalvi, Rampolla, Gasparri. Libreria editrice emiliana, Venezia

Zanni Rosiello I (1987) Archivi e memoria storica. Il Mulino, Bologna

Chapter 6
Intangible Heritage and Resilience in Managing Disaster Shelters: Case Study in Japan

Miwako Kitamura

Abstract The Great East Japan Earthquake has taught Japan many valuable lessons, one of which is the importance of preparation for disasters that can occur unpredictably and overwhelm all defences built within the limits of an anticipated threat. In the case of an excessive challenge, unexpected resources might be required. This chapter investigates one such type of resource in the context of a small town in Iwate Prefecture. Intangible heritage encompasses several types of traditional culture, including speech, rituals and movements, such as forms of interaction and festivals, beliefs and ceremonies and dance. These forms of intangible heritage are greater than just an inheritance, similarly, functioning as the foundation of a sense of community. Focusing on three disaster shelters in three districts of the coastal town of Otsuchi, the study examined links between establishing and managing disaster shelters and intangible cultural heritage resources in communities. It was found that of the three, one lacked a connection to the intangible cultural heritage related to the given district, whereas the other two had connections. Managing the shelter in the district without heritage binding the community did not go well, with an absence of desire to take responsibility or make decisions. Contrastingly, in the other shelters the interviewees often mentioned their surprise at how well everything had gone in the wake of the disaster. As the chapter argues, the continuation of traditional culture, such as local dances and festivals, may well be the key to preparing for disasters. The case studies in this study can adapt to communities around the world—cultural preservation and sustainable community disaster management plans realised by carrying on the culture and disaster management practices.

Keywords The Great East Japan Earthquake · Intangible cultural heritage · Folk culture · Evacuation shelter · Otsuchi · Iwate Prefecture · Japan

M. Kitamura (✉)
Graduate School of Engineering, Tohoku University, Sendai, Japan
e-mail: kitamura.miwako.s5@dc.tohoku.ac.jp

© The Author(s), under exclusive license to Springer Nature Switzerland AG 2021

F. F. Arefian et al. (eds.), *Historic Cities in the Face of Disasters*, The Urban Book Series,
https://doi.org/10.1007/978-3-030-77356-4_6

6.1 Introduction

This chapter explores how festivals and deer dancing, an intangible cultural heritage, practiced in the aftermath of the Great East Japan Earthquake, demonstrated the power of resilience in rebuilding communities that were devastated by the disaster.

This study investigates the role intangible cultural properties played in the successful operation of disaster shelters in the areas of Eastern Japan in the immediate aftermath of the Great East Japan Earthquake (referred to as the earthquake) in 2011. The operation of disaster shelters in the town of Otsuchi on the Sanriku coast in Iwate Prefecture was investigated following the earthquake. The data were collected through a survey of the literature and interviews, and the relationship between the operation of the evacuation shelters for the benefit of the victims and the intangible cultural properties in the affected areas immediately after the earthquake were examined.

The earthquake occurred around 14:46 Japan Standard Time, on 11 March 2011. The epicentre was approximately 130 km east-southeast off the Oshika Peninsula, Miyagi Prefecture, off Sanriku, about 24 km deep (Jst.go.jp 2011). More than three prefectures suffered damages.

In Fukushima Prefecture, a nuclear power plant was damaged by the tsunami that ensued, resulting in a Level 7 nuclear event. In many affected areas during the immediate aftermath of this complex series of disasters, support from the central government was not forthcoming making it necessary to conduct initial activities such as the rescue and evacuation of those impacted by the impacted communities. Even in areas that suffered similar damage, a number of communities successfully managed their shelters, while others were unable to manage them successfully (Bousai.go.jp 2016).

In this study, possible disaster prevention approaches are researched by illuminating the relationship between intangible cultural properties and the operation of disaster shelters in communities before and during the earthquake. In particular, the state of affairs immediately after the disaster in Otsuchi, Iwate Prefecture (Fig. 6.1), in a rural area of the Sanriku coast, is examined to determine the importance of intangible cultural assets and successful management of disaster shelters for the community.

6.2 Background

Immediately after the devastation caused by the earthquake, a number of coastal areas were hit by the tsunami and inundated to an extent that exceeded expectation, and as a result many sites lost power. People in the affected areas survived only through mutual aid. This study investigated how shelters in the impacted areas were operated immediately after the earthquake, while critical infrastructure was interrupted. It built on literature review on the historical background of Otsuchi, Iwate Prefecture and

Fig. 6.1 Map of Otuchi

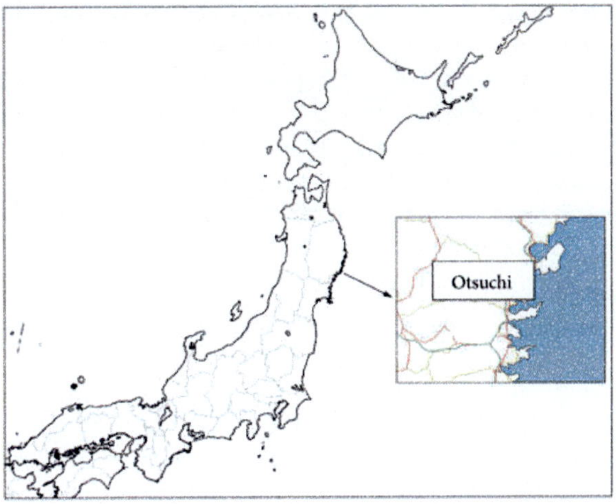

interviews. The town of Otsuchi was divided into three distinct areas: the town centre, the fishing village and the agricultural district. Interviews were conducted with three representative refuge leaders in each of the three distinct areas immediately after the earthquake.

The investigation was undertaken into the way in which the wisdom of the ancestors of local communities was cultivated in the form of intangible cultural assets and further explored as to whether these assets formed part of evacuation centre. management. The faith, festivals and dance, which form part of the intangible cultural properties that are active in the area, were a significant influence on the success of the shelter operations. In these interviews, therefore, the focus was on the management of the evacuation centres immediately after the earthquake and the maintenance of the intangible cultural assets in the community before the earthquake.

Interviews were conducted using ethnographic methods of participant observation, involving a prolonged stay in the disaster area as part of the context of the interview.

6.3 Intangible Cultural Properties in North Eastern Japan

The Great East Japan Earthquake severely damaged the Sanriku region, which is located in the northern part of Japan. Moreover, the cold weather worsened the condition of agricultural activities of the region. Such events have resulted in repetitive periods of poverty and famine due to damages to the sea and seawater. Nevertheless, the Sanriku region is sufficiently fortunate to have a coastline, which enables it to support people fishing in the fertile waters.

However, historically, the district of Otsuchi has faced severe famine and heavy taxes from the central government, and the Otsuchi area was frequently disadvantaged with poverty. Therefore, local people started relying on the natural resources available in Sanriku, especially the sea, to survive the harsh daily life. A unique culture of songs and dance was created to express gratitude for the understanding of the historical background of these communities. Considering the results of the improvement and reconstruction of Sanriku region after the disaster it is necessary to elucidate its history and culture before the disaster.

The importance of intangible cultural heritage is not only a legacy, but also as a contribution to social cohesion (UNESCO 2019). For example, in community festivals, where an individual or people from different groups and backgrounds can feel like members of a larger community. It fosters a sense of identity and responsibility within a larger society. The essence of intangible cultural properties is the rich cultural knowledge, traditions and skills that are passed on from generation to generation (UNESCO 2019). Hashimoto (2016) reported that the intangible culture and folk performing arts preserved on the coast of Iwate Prefecture are a spiritual pillar of the local community and are not maintained by limited, local resources alone but are enjoying new sources of support.

Historically, local people of the Sanriku region have lived with nature and protected their culture at the same time. However, after World War II, the urban development was implemented with a focus on regional economic growth rather than safeguarding the social networks and characteristics of the region. As previously mentioned, the Sanriku region was repeatedly hit by massive tsunamis. Although many stone monuments and documents related to the tsunamis remain, the history of these tsunamis is a negative factor for regional economic development. Hence, the existence of these monuments is generally ignored.

Intangible cultural heritage thrives on a foundation of the community, and knowledge, traditions, skills and customs provide support for and are inherited from elders, being passed on from generation to generation, or to other communities. According to Hyoki, folklore-related aspects of culture have begun in recent years to develop intangible cultural properties, such as festivals, folk beliefs and dances. Deer dancing in the Otsuchi district in the Sanriku region is an example that is based on the legend of the ancient deer, in which a grown deer willingly hits its horn against a solid surface and folds it from the root to encourage the regeneration of a new horn (Fig. 6.2). Deer dance, which symbolises rebirth, was performed after the earthquake not only for the tsunami evacuees but also other people with the hope of rebirth from the earthquake. These are developing new value as an economic resource. They incorporate aspects such as the ability to gather financial support for areas affected by the disaster where the subsequent recovery was sluggish.

However, due to rapid economic development, more people in the Sanriku region are looking for workplaces in cities rather than fishing. The outflow of young people in Otsuchi caused a decline in birth rate and an increase in the ageing population. Against this background, the number of deer dancers in Otsuchi has dropped dramatically. Therefore, they have relaxed the strict rules for becoming a deer dancer, expanded the region and attributes where deer dancing can be taught and continued to feature deer dance at festivals. In this manner, the culture was maintained. Despite changes in the form, the intangible cultural properties of deer dancing have been inherited. They

Fig. 6.2 Traditional deer dance performed within the community a deer can be seen in the foreground (author)

are expanding the areas and attributes for teaching the tradition, which is related to the fact that many tsunami evacuees naturally gather despite traditional deer dance halls being unregistered as emergency shelters.

Oshima (2019) focused on the economic effects of this development and on the way intangible culture proceeded in local culture. Other studies also found that community resilience was strengthened through intangible heritage. Beel et al. (2017) studied how community heritage built bottom-up in a rural area of Scotland supported community resilience. Lahournat surveyed the state of Kagura, an intangible cultural asset in the Ogatsu area of Miyagi Prefecture following the earthquake, and folk performing arts there strengthened the connection among people in the affected areas through a reflection of shared identity that may likely lead to the recovery of the affected community (Lahournat 2016). The intangible cultural activities of communities are an object of study beyond Japan, although not many works have explored the relationship between intangible cultural properties and shelter management that is explored in this study.

In the research area of Otsuchi, this study reviews the example of leaders of folk dance groups who are also community leaders and played a major role in developing community resilience after the earthquake. This study investigated the case where the relationship between people through folk performing arts played a significant role in the operation of shelters in the study area, namely the town of Otsuchi. It helps to understand the resilience of local communities after the disaster that is linked with successful evacuation centre management as a good starting point for subsequent community rebuilding.

6.4 Intangible Cultural Property and Identity

Traditional cultural expressions, including festivals and the accompanying songs and dances, may seem to have lost importance, as the lack of successors in Otsuchi, where the falling birth rate and ageing of the population were increasing before the earthquake (Figs. 6.3 and 6.4).

However, in some communities traditional cultural expressions remained, especially those communities which continued their livelihood through fishing. Such livelihoods, where cooperation within the community is essential, can continue the culture and traditions of the community. For example, Kirikiri district communities whose fishing industry is the centre of livelihood. Thus, the community frequently loses fishermen in marine accidents. In such unfortunate cases, a widow's life is supported by the community. Often, a new husband is brought from other fishery areas to continue life in the dead fisherman's community. Kirikiri district communities have continued to help one another because of the risks of disasters, such as tsunami and marine accidents. The social connections of the Kirikiri community,

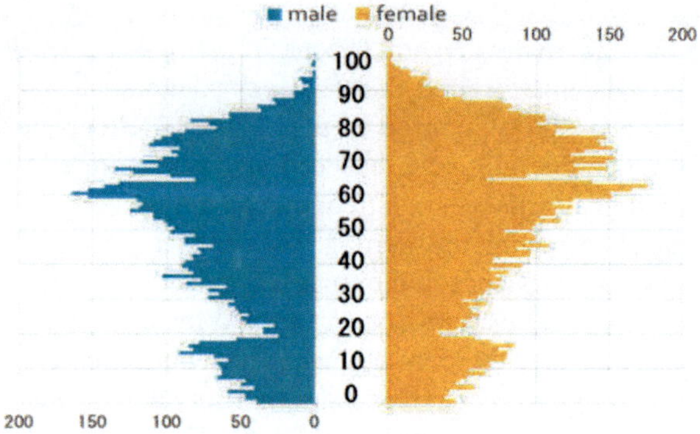

Fig. 6.3 Demographic pyramid of Otsuchi (Japanese National Census 2010)

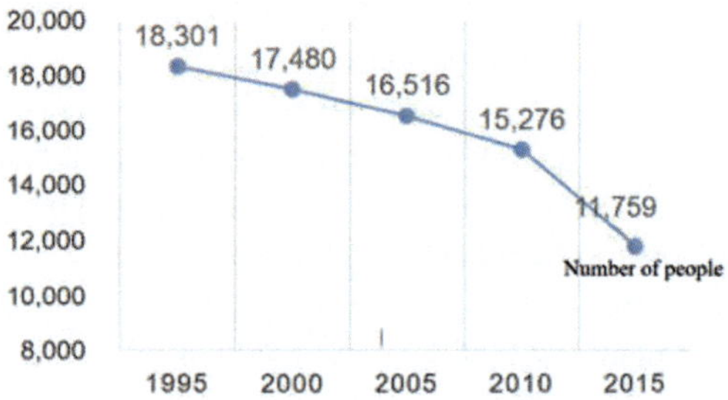

Fig. 6.4 Estimate decreasing the number of populations of Otuchi, 1995–2015 (Japanese Statistics Bureau)

have also accumulated through cultural and religious events. For example, Kirikiri Tiger Dance paid tribute to the crew of the Sengoku Ship who saw the national battle in Edo when the wealthy merchant Kirikiri Zenbei, who built huge wealth through trade during the Edo period, was prosperous (Fig. 6.5). Historically, the community of Kirikiri has been taxed heavily by the central government and harshly treated. Thus, the local people have joined forces to overcome various difficulties.

Another example is Kagura, which is the tradition of the performance of Shinto music and dance. The Mawari Kagura has been performing for more than 340 years, travelling along a 150-kilometre stretch of Iwate's Sanriku coastline from Kuji in the north to Kamaishi in the south. On New Year's Day, the performers carry a "Gongen-sama" (lion's head), which is a transfer of the spirit of the Kuromori Shrine, around the village and perform prayers for exorcism and fire-bursting at the front yard of the house. At night, they entertain people with a night time Kagura performance at the tatami room of the house where they stay. In response to the wishes of the deceased, they offer Kagura Nembutsu to the souls of the deceased. Since the Great East Japan Earthquake, its role has become even more extraordinary and has been designated as a National Important Intangible Folk Cultural Property.

Many old communities disintegrated as people moved to different places after the earthquake, but their shared identity provided support to people's hearts as they slowly recovered from the disaster. This identity was in part hosted in the intangible cultural heritage of festivals and dances.

Festivals and folk art have also been effective externally and after the earthquake, attracting attention from the broader world as connection or network nodes that

Fig. 6.5 Kirikiri Tiger Dance at local festival (author)

can support affected areas. Local traditional culture can create a dynamic force that widens the circle of sympathy to reach a broader area (Matarasso 2001). Traditional activities of intangible culture are part of the life of the region and must be taken into account in the typical living environment relative to the residents (Watanabe 2019). While there have been many studies of folk performing arts and their relationships to local communities, this research examines the role of intangible heritage both during disaster recovery and in the immediate wake of a disaster. Examining examples of how folk performing arts of affected areas influenced refuge management can demonstrate its relationship to resilience.

6.5 Overview of Otsuchi

Otsuchi is a town on the coast of Iwate Prefecture, with a population of around 12,000 in 2015, although it was 15,000 before the earthquake, in 2010. The size of the population is declining, and its average age is increasing. In 2010, the population had already decreased by about 20% relative to 1995, and 32% of the residents were over 65 years old (Stat.go.jp 2010).

The earthquake caused enormous damage to Otsuchi. There were close to 1300 casualties (Table 6.1), around 8% of the population. The tsunami struck town hall, and many of those who were in town hall at the time of the tsunami died, including the mayor. Automobile fuel washed out of vehicles by the tsunami caught fire, and Otsuchi burned for about three days after the earthquake. Under these circumstances, the victims lived for a long time without the necessary infrastructure to deliver water, gas and electricity (Osaka Medical College 2011). This made the emergency shelters even more important for the victims (Alexander 2000).

As this study found, in some places, the disaster shelter management was confused, and in others, it was smoothly conducted. To better understand how these shelters functioned during the response and recovery phase in the aftermath of the earthquake, interviews were conducted in the three main districts of Otsuchi (Fig. 6.6).

The town centre, Machikata: This area saw significant direct damage from the tsunami and a resulting massive fire. This is the centre of Otsuchi, and the town hall and a commercial area are located here. There are many service industry businesses located here. This area bears little relationship to intangible cultural properties.

Table 6.1 Human damage caused by the earthquake in Otsuchi (Otuchi Town 2013a, b)

Human casualties of the Great East Japan Earthquake	The number of death and missing persons
The number of the dead	870
The number of the missing persons	416
Total	1286

Fig. 6.6 Map of inundation in Otsuchi, 1896–2011 (Author)

The fishing village of Kirikiri: This village received significant tsunami damage. It is the site of a fishing village and is a coastal area. It is the site of intangible cultural assets, such as Sanriku Kagura and Toramai Shrines.

The rural area Kozuchi: This area experienced no flooding due to the tsunami. It is located on a hillside and away from the coast. Many residents here are engaged in agriculture and the service industry. The intangible cultural property of the deer dance is actively pursued.

6.6 Machikata

Machikata, is the centre of Otsuchi Town is the place of the Town Hall and main shopping district. During the earthquake, a massive fire broke out in this area due to the aftereffects of the tsunami.

Previously, there had been evacuation drills for the centrally important functions of the city were hosted in sites that were, like the Town Hall, in an area that was flooded after the tsunami. However, the tsunami following the earthquake was larger than accounted for in evacuation drills (Archive.town.otsuchi.iwate.jp 2013b).

At the Town Hall, during the actual disaster, it proved impossible to follow the example of the evacuation drills. For this reason, the disaster response headquarters were set up in front of the government offices in a lower area of Machikata resulting in administrative organisational structures being significantly damaged (Otsuchi Town 2013a, b).

The earthquake verification report prepared for Otsuchi Town (2013a, b) included interviews with those involved in shelter management. In Machikata, a public hall located on higher ground became a refuge.

Local residents testified that this shelter was non-functional. The following are extracts from an interview with a woman in her fifties who was born and raised in Machikata district. At the government's request, this woman was appointed a leader of the town refuge. She stated:

> All the victims became suspicious of others because of the fear provoked by the tsunami. Many people were employees, so if I told them to do something, it would work, but no one took the lead and tried to do anything. The atmosphere kept me from wanting to be a leader or getting involved in this disturbing situation.

According to the 2013 Earthquake Verification Report from Otsuchi Town, the following problems in the operation of evacuation centres were noted, and the situation in town was considered to have been confusing (Otsuchi Town 2013a, b).

- The disaster was not anticipated, and several evacuation shelters did not host survivors.
- The level of supplies were insufficient.
- Evacuation centre management was not clearly defined.
- The designated evacuation shelter was isolated due to damage or blocked roads.
- Privacy was prioritised and implemented through creating partitions using cardboard, meaning disaster victims could not communicate well.

In Machikata, where festivals and occasions that involve traditional culture were not common evacuation operations were complex, due to the absence of recognisable community leaders and a lack of trust among survivors, which created uncertainty and confusion with evacuation procedures. The lack of faith in human relationships immediately after the earthquake in this area of the city hindered response and recovery.

6.7 Kirikiri

Kirikiri is the site of a village, whose inhabitants live on fishing. It has provided a flourishing trade and since ancient times have often been a port town. In the Kirikiri area, residents express a deep respect for the sea god and the mountain god. Because the sea is the source of their livelihood, fishers often leave their homes for long periods and require help from the community. To live in this harsh natural environment, strict observance of local rules is necessary. Together, these factors have led to the development of Sanriku Kagura shrine, a folk cultural activity unique to the Sanriku coast around Kirikiri. Sanriku Kagura combines the gods of Kagura and folk expressions in dances and songs (Miichi 2016).

During the GEJE, about 100 people (about 4%) of the 2500 residing in Kirikiri experienced human damage (Town.otsuchi.iwate.jp 2013a). In the Kirikiri evacuation shelter, community ties were deeper than before the earthquake, the shelter leadership was improved, and many victims were able to live comfortably within evacuation shelters. The *Guji* of the shrine acquired the responsibility for the shelter. Guji refers to a head priest who is in charge of the shrine. If one compares a shrine to a company, a priest is an employee, and the representative of the company is called Guji. In other words, there is only one Guji in each shrine because the Guji is the representative of the shrine. According to traditional custom, the Guji had the role of managing the community in an emergency, which meant that the selection of the leader could be smooth. The local Guji who was interviewed mentioned that:

> Traditionally, the Guji should be the leader if there something urgent requires leadership, so I became the one in charge of the shelter quite naturally. It was a quake that I had never experienced before. No phone call, no electricity, no information. Anyway, I saw a fire coming from the centre of town, and so no one could come to help, so we had to manage.

As in Machikata, Kirikiri's infrastructure was blocked by the cascading effects of the aftermath of the earthquake, tsunami and fire. However, a significant difference appeared in the state of the Kirikiri evacuation shelters as opposed to those in Machikata. After the earthquake, the surviving villagers gathered at the shrine to discuss what to do and created a disaster response headquarters in the district, as, typically, the residents gather at the festival shrine when there is an emergency. They organised themselves, around their strength within the community, giving each person a role. This experience in this emergency immediately after the earthquake brought the organisation of a festival to mind. The Guji continued to state that:

> Anyway, I wanted to help the inhabitants by removing the debris, so people who could handle bulldozers gathered, and I immediately began working to rescue all the residents I could.
>
> Many other community leaders gathered at the festival shrine, and I was able to move quickly.
>
> Women survivors usually made food during the festival, so those who did gathered immediately. They brought me some food, rice and vegetables. Women are accustomed to making meals (sic), so even in an emergency, they know how to make a fire and where to find water. The women began to prepare food immediately after the earthquake.

It is clear here that if the acquaintances did not evacuate to the evacuation shelters, they rescued themselves in anticipation of possible tsunami damage. This is possible because the residents often meet with each other during festival preparations and dance practice. The rapid reaction of women led to things naturally happening because of their traditional practice of serving a large meal for the festivals.

In Kirikiri, locals immediately took life-saving actions, removed debris, secured roads and installed heliports. At the evacuation centre, the distribution of cooked food and support supplies, the securing of fuel and the maintenance of the communal baths were all carried out successfully (Takezawa 2013).

6.8 Kozuchi

In Kozuchi, some residents evacuated to the deer dance hall, although it was not an official shelter facility. Several years before the earthquake, a group wished to gather many people to learn and perform the deer dance, and the group accepted anyone, not just people from Kozuchi. Those who evacuated here after the earthquake were those who had earlier studied the deer dance in this place.

The deer dance has been passed down for a long time without interruption. It is a traditional dance from the district, but fewer and fewer people were participating due to the declining birth rate and ageing population. For this reason, over the course of several years before the disaster, the traditional rules of the deer dance were changed to match the times, and anyone interested in the dance was invited without any limitation on residential area. Those who evacuated to the traditional deer dance centre on the mountain side were those who had studied the deer dance. During an interview with the leader they said:

> The deer dance plays a vital role in the festival. Many come to see the dance during that time, and we even see traffic jams. During the festival, so as not to bother the locals, we chose deer dancers to direct traffic. We put a yellow ribbon on the arm to indicate the one directing traffic. Even in the evacuation shelter after the earthquake, the evacuation leader put a yellow ribbon on to let everyone know who the leader was. This made it easier for those who were evacuating to know who the leader was.

This technique, which enabled easy identification of the leader in the evacuation centre, protected life in the refuge. The leader further mentioned discussed:

> There are many part-time farmers who live in this area, and we have stocked a great deal of staples like rice and miso for festivals and our other customs. Because much food is distributed to the people who attend the festival, and it is challenging to prepare everything all at once, it is customary for people to always have ingredients ready for the festival.
>
> The people here refer to themselves as mountain people and to the people on the coast as sea people. The groups interacted at the festival, so if they ran into trouble, they could help each other.
>
> I feel that the bond is even stronger than in a normal healthy relationship when people get to know each other through traditional culture.

In Kozuchi, many people practiced agriculture, and they customarily served meals and snacks to those who came for festivals and dance practice. Their stockpiles for that purpose were invaluable as this saved the lives of many after earthquake. Further, during interviews it was mentioned that:

> Local people respected everyone involved in traditional culture, so leadership was natural. The people in the region respect the dance. At the time of the earthquake, it seemed quite reasonable to the residents that the dance leaders would be the shelter leaders.

In the Kozuchi district, residents involved in intangible cultural heritage such as faith and festivals demonstrated leadership. Those who were respected in the local community became shelter leaders in emergencies. Further comments from the locals stated that:

> Recently, my dancing friends and I spoke about the splendour and wisdom of the ancestors. We worked hard on disaster prevention before the earthquake, but what helped in the real emergency was the activities of the intangible cultural property community: rice and miso, which are indispensable for daily life. Traditionally, they are stockpiled for festivals, dance practice and ceremonial occasions. I never thought they would be useful in an earthquake.
>
> The women are accustomed to making large meals for the festivals and dances, and they said that their bodies moved naturally to do the same thing even after the earthquake. I think this preparation is terrific. I did foresee the festival preparations being useful in emergencies or natural disasters.

The deer dance leaders clearly stated that residents' active participation in local intangible cultural property not only increased community cohesion but also supported disaster response.

6.9 Findings

To understand the state of the evacuation shelters immediately after the earthquake, their operations were reviewed in the Machikata district, the commercial centre of Otsuchi; the Kirikiri district, where the fishery industry is centred; and the Kozuchi district, which hosts the agricultural production. Interviews were conducted with the person responsible for each shelter. The findings showed that the involvement of the community with intangible cultural assets was an important factor that affected post-disaster emergency shelter management immediately after the earthquake.

It is clear that participation in collective intangible cultural events whether to be related to faith, dance and festival as a cohesive community was related to community resilience in the wake of the response and recovery activities following the earthquake. The success of the management of evacuation centres was closely related to the collective survival of victims , their ability to self-organise in the absence of basic infrastructure, familiarity to collective efforts, and respect to appointed leaders. These cases made a case that the community's preparation for participation in intangible cultural events prior to the disaster was greatly useful for disaster resilience.

This study highlights that collective community festivals and traditional performing arts dances had great value in disaster response and recovery after the earthquake. The community respects people involved in traditional performing arts. Such community respect is essential for leaders in disaster recovery in times of emergency. Local knowledge of local culture and customs enhances the resilience of the community compared to the outsider knowledge assumed by the elite leaders who are not familiar with the community and who usually sent into assist in the aftermath of a major disaster. It is vital to respect the culture of communities and benefit from the traditions and knowledge of community people, not just during disaster recovery. The above contribute to better operations for disaster management which are essential in our disaster-prone Japan and around the world. For example, it might be difficult for developing countries to afford to maintain large-scale disaster management plans that require significant amounts of money, yet tapping on local knowhow and traditions might save resources. It is imperative to harness the latent culture and knowledge of the region, as identified in this study.

6.10 Conclusion

This chapter investigated the relationship between the management of evacuation centres immediately after the earthquake and intangible cultural assets in Otsuchi.

In the Machikata district, where there is little relationship to intangible cultural properties, evacuation centre management was confused. By contrast, the leaders in intangible cultural property successfully managed evacuation shelters in Kirikiri and Kozuchi.

Communities are threatened by many natural hazards, which come suddenly, and if they are on a large scale, it may take time for official resources to begin rescue. This means that survivors must determine how to survive until rescue operations. The festivals, rituals and dances that appear in the course of life in communities can play the role of disaster management and prevention activities that support communities. Better understanding of such intangible cultural properties, daily invisible local habits, and valuing daily human relationships can strengthen self-help public assistance for disaster resilience. In this way, disaster prevention and mitigation are enhanced for a community.

References

Alexander D (2000) How to write an emergency plan, 1st edn. DUNEDIN, London, pp 65–67

Beel D, Wallace C, Webster G, Nguyen H, Tait E, Macleod M, Mellish C (2017) Cultural resilience: the production of rural community heritage, digital archives and the role of volunteers. J Rural Stud 54:459–468

Disaster Management in Japan (2016) Report on disaster victim support at evacuation centers [online]. Available at: http://www.bousai.go.jp/taisaku/hinanjo/pdf/houkokusyo.pdf. Accessed 26 Sep 2019

Hashimoto (2016) 太朗舘野. 2016. '橋本裕之 [震災と芸能—地域再生の原動力] [芸能的思考]'. 演劇学論集 日本演劇学会紀要 62:128–134. https://doi.org/10.18935/jjstr.62.0_128

Japanese National Census (2010) [online] Available at: http://www.stat.go.jp/data/kokusei/2010/. Accessed 26 Sep 2019

Japanese Science and Technology Agency (2011) The great East Japan earthquake information from official websites [online]. Available at: https://www.jst.go.jp/pr/pdf/great_east_japan_earthquake.pdf. Accessed 26 Sep 2019

Lahournat F (2016) Reviving tradition in disaster-affected communities: adaptation and continuity in the kagura of Ogatsu, Miyagi Prefecture. Contemp Jpn 28(2):185–207

Matarasso F (2001) Recognising culture a series of briefing papers on culture and development [online]. Available at: http://www.culturenet.cz/res/data/004/000567.pdf#search=%27Local+traditional+culture+creates+a+dynamic+force+that+spreads+the+circle+of+empathy+over+a+wider+area.%27. Accessed 26 Sep 2019

Miichi K (2016) Playful relief: folk performing arts in Japan after the 2011 Tsunami. Asian Ethnol 75(1)

Osaka Medical College (2011) Annals of the Osaka Medical College Doctor's Association [online]. Available at: https://www.osaka-med.ac.jp/deps/omcda/report/pdf/report_36.pdf#search=%27津波で流された車に残っていた燃料に引火し、大槌の町は地震後約3日間燃え続けた%E3%80%82%27. Accessed 26 Sep 2019

Oshima A (2019) 無形民俗文化財の「変化」を考える—特に文化財指定との関連で— [online] Tobunken.repo.nii.ac.jp. Available at: https://tobunken.repo.nii.ac.jp/?action=pages_view_main&active_action=repository_view_main_item_detail&item_id=3136&item_no=1&page_id=13&block_id=21. Accessed 26 Sep 2019

Otsuchi Town (2013a) Otsuchi Town Great East Japan earthquake verification report [online]. Available at: https://www.town.otsuchi.iwate.jp/fs/1/9/9/5/9/3/_/kensyo.pdf. Accessed 26 Sep 2019

Otsuchi Town (2013b) 東日本大震災津波[大槌町被災概要]2013_4_1 - 大槌町震災アーカイブ [online]. Available at: https://archive.town.otsuchi.iwate.jp/search/archive/1277.html. Accessed 26 Sep 2019

Takeuchi K (2019) <研究論文> 心理学の課題としての無形文化財保護: 世代間の関係性という時間の枠組みによる発達心理学の新たな可能性 [online] Hdl.handle.net. Available at: http://hdl.handle.net/2433/190356. Accessed 26 Sep 2019

Takezawa K (2013) 津波の破壊に対抗する被災コミュニティー: 大槌町の避難所に見る地域原理と他者との関係性 [online] Minpaku.repo.nii.ac.jp. Available at: https://minpaku.repo.nii.ac.jp/?action=pages_view_main&active_action=repository_view_main_item_detail&item_id=3858&item_no=1&page_id=13&block_id=21. Accessed 26 Sep 2019

UNESCO (2019) UNESCO—what is intangible cultural heritage? [online] Available at: https://ich.unesco.org/en/what-is-intangible-heritage-00003. Accessed 26 Sep 2019

Watanabe Y (2019) 8 「祭り」という文化伝承・継承空間 [online] Hdl.handle.net. Available at: http://hdl.handle.net/2433/176383. Accessed 26 Sep 2019

Chapter 7
Water Gives, Water Takes Away. Memory, Agency and Resilience in ENSO—Vulnerable Historic Landscapes in Peru

Rosabella Alvarez-Calderón and Silva-Santisteban

Abstract The valleys of the north coast of Peru comprise a rich cultural landscape that over many centuries facilitated the development of complex societies, due to a year-round temperate climate, abundant natural resources and fertile soils. This area is also vulnerable to episodic ENSO (El Niño Southern Oscillation) events, in which elevated sea temperatures and changes in wind patterns lead to intense rains, flooding, droughts and disruptions in the coastal food chain. Historically, El Niño events have varied significantly in intensity, frequency and location, so that while some communities experienced devastation, others benefited from an increase in water, the expansion of green areas and vegetation and the creation of new ecosystems. The societies of the pre-Hispanic period like the Moche, Sicán and Chimú were skilled at understanding, adapting and transforming these diverse territories and in adjusting to the ebbs and flows of water through technological, managerial, religious and performance-based approaches. In times of environmental stress, human sacrifices were a way for people, usually leaders and elites, to commune with their deities to give thanks and ask for relief, while also serving to negotiate and consolidate political and symbolic power, prestige and privilege. These narratives and experiences of the past have the potential to offer modern-day rural and urban communities valuable lessons in rethinking how to live with water and increase resilience, especially in the face of more frequent ENSO events, climate change and urban expansion. Learning from the past in order to ensure a better future, however, also requires consideration of failure, fear and perplexing perceptions of risk. Finally, it is valuable to understand how some markers of social complexity, like deep economic and symbolic ties to a place, and the desire for stability, can hinder flexibility and lead to responses that ultimately increase vulnerability to El Niño events.

Keywords El Niño Southern Oscillation · North Coast of Peru · Resilience · Sacrifice · Memory

---

R. Alvarez-Calderón (✉) · Silva-Santisteban
Pontificia Universidad Católica del Perú, Lima, Peru
e-mail: Bellarb@hotmail.com

© The Author(s), under exclusive license to Springer Nature Switzerland AG 2021

F. F. Arefian et al. (eds.), *Historic Cities in the Face of Disasters*, The Urban Book Series,
https://doi.org/10.1007/978-3-030-77356-4_7

7.1 Introduction

What leads societies to sacrifice what is often their most valuable resource, healthy young human life? In the north coast of Peru, archaeologists have found evidence of offering sacrifices, meant as a way to commune, supplicate and negotiate with supernatural forces in order to obtain a benefit, or to ask for relief from a negative condition (Klaus and Toyne 2016). At the Chimú site of Huanchaquito-Las Llamas, located near the present-day city Trujillo, dating between the years AD 1400–1450 (Prieto et al. 2019; Romey and Clark 2019), and in a plaza in the ancient city of Sicán, about 200 km north of Trujillo and dating around the period known as the late Middle Sicán (AD 1050–1100; Klaus and Shimada 2016), archaeologists found the remains of humans, animals and objects of value that display evidence of having been subjected to ritualised violence and killing. These remains were found in layers of dried mud and sediment, and, in the case of Huanchaquito, several sets of adult, child and camelid footprints, implying that the killings were done while the ground was still wet. These events would have been a response to severe El Niño events, which caused flooding and intense rains, often followed or preceded by severe droughts. The same life-giving water that these ancient societies had learned to harness through the construction of elaborate irrigation and agricultural systems was now flowing uncontrollably, damaging their canals, fields and monumental architecture, destroying homes and workshops and killing and injuring people and animals.

The Sican and Chimú societies lasted for hundreds of years and developed a repertoire of responses to cope with these periodic, if often unpredictable, events, involving construction and settlement patterns, irrigation and agriculture infrastructure and flood control. However, these would not have been sufficient to cope with all crises (Schaedel 1992; Vreeland 1992). The timing of the sacrifices, done shortly before the collapse of these societies, suggests that people perceived that their valuable offerings were not well received by the gods and failed to stop the disasters they were facing. Considering the significant political, social and symbolic investment that these sacrifices entailed, it may seem devastating to realise that these communities took a risk that failed to bring relief, especially if the resources, lives, time and energy devoted to preparing and carrying out these offerings ultimately increased their vulnerability. For the Chimú, these disasters would have left them particularly vulnerable to conquest by the Inca empire, an event that took place between AD 1462–1470 (Rowe 1948).

Resorting to sacrifices would also indicate that the strategies that had been successfully used in the past to address similar challenges (technology, management, social organisation) had failed, or that the social and natural environment had undergone a dramatic shift, to which these societies failed to prepare and adapt. For instance, in the case of the Chimú, the period known as the Late Intermediate Period in the north coast (AD 900–1500) roughly corresponds to the Medieval Warm Period/Medieval Climatic Anomaly (AD 950–1250), which led to a drier, warmer, more stable climate. The mass child and camelid sacrifices came at the end of this period, thus implying that the Chimú were unable to successfully adapt to this new normality (Romey and Clark 2019).

When considering the symbolic and social value of these offerings, as well as how and why they failed in their intended objectives, it is useful to use these lessons from the past to examine critically the way modern societies anticipate, react and respond to El Niño and other crises. What needs to be considered is the pre-Hispanic period, which refers to the history of the societies that developed in the area up the arrival of the Spanish conquistadors in AD 1532. While we have a better understanding of ENSO events today and have access to more sophisticated tools and strategies to predict, prepare and mitigate their negative effects, the contemporary north coast is a vastly different scenario than it was during the pre-Hispanic period (Fig. 7.1). Now this area has a significantly larger population, a complex and sprawling network of cities, towns, roads, infrastructure, industry and commerce, much of it built in at-risk areas like floodplains and ravines. In the past hundred years, the north coast has been affected by several severe El Niño events, about once every 10–20 years since 1925. How are we responding to modern El Niño events in ways not unlike the offering of sacrifices in the past? While highly symbolic and performative actions serve an important political, economic and social agenda, they may even increase the vulnerability and reduce the resilience of a community by diverting much needed

Fig. 7.1 A street in the historic centre of the modern city of Trujillo, in the north coast of Peru. During ENSO events, the city centre is vulnerable to flooding, which has happened as recently as in the summer of 2017, during a Coastal El Niño event (Álvarez-Calderón 2017)

resources, expertise, time and energy. Are these disasters and failures prompting us to rethink our relationship with the environment, with water, with sustainability and with the actual carrying capacity of the territories we inhabit and transform? What can we learn from the failures, from the perplexing decisions past societies made that left them ultimately more vulnerable, and the values, agendas and mentalities that we can attempt to understand from the archaeological evidence?

This chapter explores challenges regarding vulnerability and environmental stress, resilience and sustainability and what we can learn from narratives of the pre-Hispanic past of success and failure in the face of El Niño and similar water-related environmental stress and hardships, based mainly on two archaeological case studies from the ancient Chimú and Sicán societies.

7.2 Understanding the North Coast of Peru and the Societies of the Pre-Hispanic Past

7.2.1 The Geographic Setting

The north coast of Peru is defined by the presence of a warm equatorial landmass and cold coastal water temperatures, originating from the nutrient-rich Humboldt cold water current, which impacts water temperature, wind patterns and wildlife. These conditions inhibit significant coastal rainfall and give the area a year-round temperate, mostly sunny and dry climate and diverse ecosystems. Meanwhile, the regions located closest to the border with Ecuador are affected by the El Niño[1] warm-water ocean current. Over the centuries, the north coast has also been affected by larger patterns, such as the Medieval Warm Period and the Little Ice Age (Rick and Sandweiss 2020; Fagan 1999; Sandweiss and Quilter 2008; Billman and Huckleberry 2008). The north coast is particularly susceptible to environmental disruption and stress, brought about every time the warm El Niño ocean current ventures further south, displacing the Humboldt current, lessening and even changing the direction of the northeast trade winds and bringing humid air to the coast.

The author has experienced and studied three recent El Niño events while living in the city of Lima, in the central coast of Peru. In 1982–1983, the normally arid and humid city experienced intense rains which lead to flooding. In 1997–1998, Lima experienced some flooding of the Rímac River. The "year without a winter", followed in 1999 with an event known as "La Niña", since it had the opposite effect. This event led to the formation of a temporary lake in Piura, in northern Peru, and most significantly, to the "greening" of many loma ecosystems, hill environments which remain mostly dry in summer and become green and lush with vegetation in winter due to the intense humidity and fog, which condenses against the hills.

[1] El Niño (the child) is the name which Paita fishermen gave to this phenomenon many decades back, based on the fact that it became noticeable through biological indicators around Christmas. The name thus refers to the Christ child and its association with the Catholic Christmas.

While some lomas become green every year, some only do so during El Niño events. Finally, there was the 2016–2017 episode known as "Niño Costero" (coastal Nino), in which the city of Lima again suffered flooding, though not so much rain. In March 2017, the resilience of the city was tested when the main water treatment plant, La Atarjea, was damaged by flooding and landslides known locally as huaycos. Due to mud and debris clogging the infrastructure, water was cut off to the entire city for several days until the system was repaired, and water service was restored gradually.

Interdisciplinary research has revealed that rather than a phenomenon local to the Peruvian coast, El Niño refers to global events related to shifting ocean currents and wind patterns, which end up affecting worldwide climate patterns, thus explaining the name "*El Niño Southern Oscillation*" (Fagan 1999). Recent research points at the existence of at least four different "flavours" of ENSO events—Eastern Pacific, Coastal, Central Pacific/Modoki, La Niña—(Sandweiss et al. 2020). The effects tend to last 1–2 years, sometimes leading to "years without winter" in the central coast, which happened as recently as 1997. There is evidence of more severe events lasting up to several decades, informally known as Mega-Niños. El Niño events usually bring unpredictable and extreme weather into areas that are normally dry, and where rain is rare. While abundant water is disastrous for many communities, as can be seen in official government risk reports (Villacorta et al. 2008; Medina and Luque 2008), El Niño also brings welcome water to arid areas, which can even lead to the creation of new ecosystems, recharging natural aquifers and groundwater sources. One of the most significant examples of this is La Niña lake, a temporary body of water located in the Sechura desert near Piura, which has formed at least twice in the past few decades (1997–1998 and 2017). This lake has been recorded to reach an area of 2326 km^2 and attracts many species of flora and fauna.

El Niño events tend to place societies under significant environmental, social and economic stress and have often served to test their resilience and capacity to adapt successfully to challenging conditions. The ways in which people understand, address and act in response to El Niño events have shaped the development of complex societies and their relationship with land and water. For example, thousands of years ago, communities spared by the catastrophic effects of El Niño and/or by the loss of coastal land due to rising sea levels at the end of the most recent Ice Age, would have taken advantage of this situation by offering refuge to people dispossessed of resources and land. This situation facilitated the formation of strong leaderships and communities, which explains the rise of monumental architecture in societies that did not yet use pottery, and did not yet rely on irrigation-fed agriculture as their main resource during the period known as the Preceramic (6000–1800 BC; Billman and Huckleberry 2008).

The frequency and intensity of El Niño events also have a significant impact on vulnerability, resilience, and how the memory of water-related events becomes part of the social narratives of the past. For example, frequent El Niño events, even if mild, can be very disruptive if it means that every ten years or so, a community will have to learn to cope with 1–2 years of flooding, increased rainfall and sometimes drought, with significant damage to housing, agriculture and hydraulic infrastructure. While a strong, well-organised society is more likely to have the capacity to repair canals and roads relatively quickly, damaged agricultural fields are more difficult to restore since they involve a double loss: destruction of existing crops, and thus potential disruption of the food supply; and the need to repair the fields to restore fertility (Fagan 1999).

7.2.2 *The Cultural Setting: The Sicán and Chimú Societies*

The Sicán was a local polity that ascended to regional supremacy between AD 850–900 in the area known as the Lambayeque Valley Complex, which includes the valleys of Motupe, La Leche, Lambayeque-Reque and Zaña. Their most important settlement is the Sicán complex, located in the present-day Pomac forest (Fig. 7.2), which consists of a series of large-scale flat-topped platform mounds made of adobe bricks with plazas and open spaces between them, and it had an important ceremonial and symbolic function, being the main place for ancestor veneration and with a tradition of rich burials and accompanying rituals (Fig. 7.2). There were other sites with monumental architecture like Chotuna-Chornancap (Fig. 7.3), the place where the mythical founder of this society, Naylamp, is supposed to have arrived to from the sea, as well as rural communities like Sialupe and Batan Grande, which engaged in the production of craft and utilitarian objects like pottery and metalwork (Klaus and Shimada 2016).

The main site of Sicán experienced significant environmental stress between AD 1050–1100, likely as a result of a particularly strong El Niño event, which was

Fig. 7.2 View of the Sican complex and the Pomac forest, showing the earth pyramids eroded by rainfall (Alvarez-Calderon 2009)

Fig. 7.3 Site of Chotuna-Chornancap, the mythical origin site of Naylamp, founder of the Sican society (Alvarez-Calderon 2009)

itself preceded by a 30-year drought. By the year AD 1050, there is evidence of growing tension between elites and non-elites, which can be seen in the discovery of a multi-stage offering sacrifice known as Matrix 101, located in the plaza between Las Ventanas and Loro pyramids. Matrix 101 contained the remains of more than 170 individuals, mainly healthy and strong adult males, very likely members of the Sican elite, who were ritually killed by being buried alive and thrown into a pit cut deep into the dried mud in at least two different episodes. The sacrifice also included the remains of large ceramic vessels for consuming chicha, a fermented alcoholic drink made of maize; disarticulated crania; dog remains; and at least 1.000 carefully broken ceramic pots, many of which featured the image of the Sican deity and which looked as if they had been symbolically decapitated/sacrificed. Not long after these sacrificial events, the main ceremonial and monumental buildings were deliberately set on fire and abandoned (Klaus and Shimada 2016). It can be argued that the elite and leaders of Sican society made themselves more vulnerable by using luxury objects, in which the commoners had to invest significant time and resources to manufacture, to portray themselves as having a power that derived from a special relationship with the gods, and the capacity to negotiate with supernatural forces on behalf of the people. This power granted the elites great privilege and prestige, but added to their burden of being responsible for the welfare of the people regarding events they had little real control over, which placed them at great risk if they failed to uphold their part of this social contract.

Following the collapse of the Sican society and the abandonment of their main ceremonial sites, the Chimú state flourished for several centuries (eleventh–fifteenth centuries) and built a sprawling city of walled compounds, plaza, cemeteries, water reservoirs and gardens called Chan Chan, though Chimú control and impact was felt over most of the north coast. The Chimú engaged in long distance trade and built an extensive system of agricultural fields and hydraulic works for irrigation. However, the time period over which this society developed was marked by frequent tensions and even warfare between different polities. Notwithstanding, for several centuries the Chimú developed and even thrived as a society, until the mid-fifteenth century, when the mass sacrifice at Huanchaquito-Las Llamas suggests a desperate measure, a gift to the gods asking for relief from an environmental crisis, as well as the impending threat of conquest by the Inca (Prieto et al. 2019).

Chimú society experienced significant landscape stress during the years AD 1400–1450, so leaders chose to resort to an extraordinary sacrifice of over 140 young healthy children from different ethnic and geographic origin, as well as more than 200 camelids, which were violently killed by cutting open their thorax cavity at the site of Huanchaquito-Las Llamas, sometime between AD 1400–1450 (Prieto et al. 2019, Fig. 7.4). After death, the bodies were carefully arranged on the still-wet ground, with the children facing the ocean and the camelids facing the mountains. It is intriguing to consider the significant amount of planning and investment that went into this elaborate ritual, what ultimately was a desperate attempt to bring an end to a negative situation by offering the most valuable resource, essentially the future of their society. Although it is unclear how much longer the rains and flooding lasted

Fig. 7.4 Mass sacrifice of children and young camelids at the Chimu period site of Huanchaquito-Las Llamas (Courtesy of Gabriel Prieto, Huanchaco Archaeological Programme)

after this dramatic event, the city of Chan Chan and Chimú society as a political system collapsed and were taken over by the conquering Inca empire not long after this event.

7.3 Discussion: Crises and Offering Sacrifices

The belief in the need and effectiveness of sacrifice arises from a shared belief that the Andean landscape is filled with *huacas*, a term derived from the Quechua language which refers to any object, place, animal, shrine or person perceived to be imbued with sacredness. In a world where these sacred forces were believed to have power over life and death, people offered sacrifices in order to communicate with, appease and negotiate with these deities, to ask favours, to prevent disaster and to ask for relief. Sacrifice also had an important political, social and symbolic function, especially ritual violence that was performed in front of an audience, as it could serve to manipulate, persuade and intimidate, and both consolidate and challenge hierarchies, power dynamics, social relationships and privilege within a community (Klaus and Toyne 2016).

Sacrifice, and the rituals, processes, and material culture associated with it, became part of the repertoire of responses that ancient societies used to address environmental stress, complementing technological and managerial responses. These rituals had two important components. First, a particularly valuable gift such as healthy human life would please the deities and even be used to atone for transgressions and grievances. Second, they served to appease a frustrated community into believing their leaders were making an effort to do something to ease their suffering, and that they retained some semblance of control over their lives and environment. As researchers and citizens, it is tempting to believe that we can use the memory and experiences of the past in an objective manner to create a narrative intended to increase the resilience of modern cities and communities and thus ensure a safer and better future for the people, by recommending policies, practices and planning tools based on what the archaeological and historical records reveal.

Considering that the cities in the present-day north coast of Peru are rapidly expanding, especially into high-risk areas like floodplains and ravines; the realities of climate change, a growing population and more frequent, yet very irregular El Niño events that tend to vary significantly in their intensity and area affected, the importance of the teachings of the past, and the necessity of communicating these experiences to the public in a compelling manner, cannot be underestimated. This is significant, especially when considering that much of this urban expansion rarely considers the realities of geography, topography and even where rain and water should go, thus leaving the city and many of its people—usually the poorest and the disenfranchised—at great risk. However, it is just as important to understand how people evaluate and perceive risk and vulnerability at the household and community level, what informs their decision-making processes and what are their priorities and values. When people make decisions that seem perplexing to us in the present, such as building and living in areas known to be prone to flooding, does this happen due to a lack of knowledge, a failure of communication, planning and memory, or due to not having a "bigger picture" view of events, actions and their consequences? Or are these decisions the conscious result of a different agenda, and a different set of priorities, values, expectations and beliefs?

7.4 Narrative and Memory to Increase Resiliency of Present Societies

The use of narratives of the past, usually oral histories, regarding social and natural disasters as both warning and recommendation, has a long history both in Peru and in other parts of the world. In the late sixteenth century, the Spanish chronicler Miguel Cabello Balboa recorded the story of Fempellec, a leader of the Lambayeque people and descendant of the mythical founder Naylamp, whose grave misdeeds led to intense rains and drought, and who was ultimately punished by his own people (Cabello Balboa 1951 [1586]: 327–329, in Moore 1996). Similarly, in Japan the tradition of kataribe storytelling aims to communicate the memory of destructive events from one generation to the next, as well as the use of physical markers like tsunami stones, which serve both as memorials and as warnings of which areas have been affected by historical tsunamis (Kohlstedt 2016). A key aspect of these narratives is that they offer agency to ordinary people. The tale of Fempellec implies that disasters are preventable through good behaviour, and even when they happen, they can potentially be stopped through acts of atonement. Regarding kataribe, it communicates an acceptance that while natural phenomena like tsunamis, earthquakes and typhoons are inevitable, people can mitigate their negative impact by applying the lessons of the past.

These narratives, however, can also be used to understand the more perplexing decisions made in anticipation and in response to El Niño, namely how do people evaluate risk, what do they cherish and how do these values conflict with the need to increase resilience and reduce vulnerability to El Niño? These decisions and actions also indicate just how far some people are willing to go to preserve or regain certain power dynamics, inequalities and hierarchies; a sense of normalcy and stability, of control over their lives and their social environment; and a sense of place, belonging and identity. Decisions on what is worth preserving and protecting tend to reflect values, a sense of identity, and what gives stability, meaning and significance to people's lives (Lowenthal 2014).

Finally, we should consider how fear, frustration, and anxiety shape decisions, places and social dynamics, especially post-disaster. What did different people fear more than the possibility of death, how did they evaluate short- and long-term needs, risks and gains? The behaviour of many city dwellers in Peru during the 2020 COVID-19 pandemic offer a valuable case study, especially regarding those who defied the government-imposed lockdown and restrictions. When considering that so many people in Peru already live precariously, lack job security, savings, work in the informal sector and live in substandard housing, the pandemic made these inequalities and vulnerabilities even more critical. It can be argued that many people violated the lockdown because the desperate need to work, to provide for their families and to survive was perceived as something they had control over and was more urgent than the seemingly more remote possibility of acquiring the disease.[2] When looking at the decisions people have made in the past and continue making in the present regarding El Niño and other episodes of environmental stress, it is important to consider what the consequences are of these decisions, and how people would have known if their actions, notably the sacrifices, were successful or not. For instance, did ancient societies truly believe that offering sacrifices would actually stop the devastating effects of too much or too little water, especially considering the significant costs and investment of resources involved? This is a difficult question to answer, since the value of sacrifice as an offering to deities is an important part of many belief systems and organised religions, both ancient and modern. Sacrifice, in a symbolic form in the shape of a consecrated communion wafer, is still perhaps the most significant part of the Roman Catholic mass. Anthropologist Jane Baxter describes the extraordinary cost of sacrificing healthy children, in the case of the Chimú mass sacrifice: "They were sacrificing their future and all its potential. All the energy and effort aimed at preserving the family and leading society towards the future. All that is lost when a child dies" (Romey and Clark 2019: 50).

The failure of past actions to solve the challenges of a crisis should lead us to explore current responses to crises like El Niño in a more critical manner and especially to consider if they are mainly serving the same purpose than sacrifice did in the past. Sacrifice was and maybe remains a powerful, symbolic and highly visible gesture, which serves important sociopolitical and symbolic purposes, such as lifting morale and bringing about social cohesion. It is often done in desperation and even at great cost, but it can also leave communities increasingly vulnerable, as significant investments in resources, time and energy are directed to these modern-day "performances". It is telling that in contemporary Peruvian society, it is common for leaders and especially politicians to be symbolically "sacrificed" (usually removed from office or forced to resign, sometimes even having to face criminal charges) after natural and social disasters, or following political scandals. Similarly, in times of crisis, leaders and especially people in government are expected to "be present" and engage in acts of "helping out" victims, performances which can boost social morale and optimism, especially if they lead to tangible, positive results, and which a savvy leader can use for political gain. This was the case of President Alberto Fujimori, who was often photographed wearing mud-stained jeans and rubber boots in knee-high floodwaters during the 1998 El Niño.[3] Both the local and international press argued that this greatly boosted his popularity and support as well as the people's

[2] Source: https://wapo.st/2RO3t6z.

[3] As was documented by both the local and international press, the 1998–1999 El Niño, and the people's perception of Alberto Fujimori's leadership and performance during this crisis, https://www.economist.com/the-americas/1997/09/25/fujimori-against-el-nino (The Economist, September 1997), https://elpais.com/diario/1998/02/23/internacional/888188412_850215.html (El

morale, especially during his bid for an unprecedented third term. In times of crisis, it becomes important for leaders to show that they are doing something and that the community is not helpless against nature, which also serves to maintain a favourable power dynamic with the rest of society.

The Moche, Sican and Chimú people were skilled at increasing the productivity and carrying capacity of the valleys in the Lambayeque and La Libertad area through the use of an extensive system of agricultural fields and irrigation canals, including intra-valley canals like Ascope and La Cumbre (Huckleberry et al. 2017). They also built societies that thrived for hundreds of years, which indicates an existing capacity to adapt successfully to challenges and possibilities of geography. While a comprehensive exploration of the collapse and abandonment of the main pre-Hispanic cities falls beyond the scope of this chapter, it becomes clear that societies become more vulnerable when their beliefs, their social and political organisation, and especially their perceptions and relationship with the territory and water, fail to adapt successfully to a transforming cultural landscape and to successfully address the social fears, frustration and tensions. This means that when creating narratives associated with memories of the past intended to be used for kataribe-style communication, it is essential to go beyond a simplistic view of focusing on monumental sites, domestic settlements and agricultural and hydraulic infrastructure of the pre-Hispanic past as evidence of societies knowing how to "live with water" successfully. It is just as essential to understand how these ancient societies, from the leaders and elites to the ordinary people, failed and knowingly made decisions that ultimately made them more vulnerable to natural and cultural hazards.

7.5 On Perplexing Decisions and Actions, Values, Fears and Mentalities

When societies invested significant time, energy and some of their most valuable resources on offering sacrifices, what other choices did these people have, and especially what available options did they have that they failed to consider and implement? How strong was their sense of self-preservation, and which people, places and institutions were deemed worthy of being saved, even at very high cost? Who decided what people were to be chosen to be stripped of their individual identities and personhood, to be transformed into offerings to the gods? Looking back, one of the most valuable lessons is just how much some societies were willing to sacrifice to preserve values, a sense of stability and permanence that were considered to be worth more than limited individual lives. David Lowenthal describes this powerful social urge to conserve that which we create as humans and has the quality of transcending the span of single human life; that which gives our lives meaning and significance (Lowenthal 2014).

Pais, February 1998), https://www.clarin.com/ediciones-anteriores/nino-da-mano-reeleccion-fujimori_0_S1YlKOy1U3g.html (Clarin, March 1998).

The complicated relationship with water during the twentieth century of the communities around Ancash located in the northern highlands of Peru serves to illustrate the importance of considering social collective values, fears and agendas, and how these can dampen the perception and evaluation of risk and danger. The area has over 600 glaciers located at over 6000 m above sea level, most of which feed mountain rivers and springs and are essential fresh water sources for human and animal consumption, agriculture and most recently for powering hydroelectric installations that provide electricity to the region. Since the late nineteenth century, climate change and the end of the period known as the Little Ice Age have led these glaciers to retreat and melt, forming glacial lakes perched precariously over populated areas, many with only moraine dams to hold them in place. During the twentieth century, several of these lakes overflowed due to falling rocks and seismic activity, creating large outburst floods and avalanches (1941, 1945 and 1950) that affected nearby towns, cities and industrial facilities, leading to destruction and loss of life. The immediate response of many of the people affected—notably, many high-income urban residents of the city of Huaraz—was to demand that the government intervene to reduce the risk these lakes posed by draining them, at least to lower levels where they were less likely to cause outburst flooding. In other words, these lakes were perceived by many as "controllable threats that State-funded science and technology could contain". However, these same lakes and glaciers were also perceived as sacred,[4] as important sources of freshwater for agriculture and electricity, and with great potential for tourism. While it was important to reduce the risk these lakes posed, it was also necessary to do so in a way that still facilitated the commercial, urban and industrial use of this water. Most importantly, after a devastating 1941 landslide that affected the city of Huaraz, there was also a demand to rebuild in the damaged areas, "as it was, where it was", to restore the city, with its material and urban markers of modernity, while mostly rejecting new zoning designed to prevent rebuilding in areas considered to be high risk. Science historian Mark Carey argues that the 1941 event disrupted established, yet invisible, social boundaries and power dynamics between nature and humans, urban and rural, highlands and lowlands and even in terms of identity, between "Indian" people living in the countryside and off the land and "mestizo" people living and working in the city. Furthermore, the demands to rebuild the city and to rely on science and technology to "fix" the unstable lakes reflect a desire to re-establish order and "social harmony" (the status quo), as well as deeper anxieties regarding identity, race, economic class, the future, the role of government and the relationship with water and nature. In a context where disasters can become opportunities for traditionally disenfranchised people and for different actors, including the State, to implement new political, social, economic and urban agendas, it is important to consider that many who fear change and value the old status will strongly desire a return to "normalcy", even if the measures required to

[4] Glaciers and glacial lakes have been regarded as sacred since pre-Hispanic times, and today the festival of *Qoyllur Rit'I (a Quechua term which can be translated to "snow star" or "bright snow"), which combines pre-Hispanic beliefs with Catholic symbolism and rituals, takes place every year in June near the* Qulqipunku glacier in Cusco which consists of a pilgrimage featuring music, dance and traditional costumes.

achieve this place their community at risk in the long run (Carey 2010). This is because over-reliance on technology to "fix" disasters means avoiding a necessary conversation about how sustainable one's current lifestyle actually is and the need to break a problematic status quo and establish a new relationship with the landscape and especially with water.

7.6 Conclusion: Building Resilience by Learning How to Live with Water

The societies that have developed over the years in the north coast and highlands of Peru were highly skilful and creative at transforming their environment to create complex societies. The same water, however, that gives life, prosperity and makes lands fertile, can also flood, destroy and be scarce when needed. Learning how to rethink and reimagine our relationship with water and land; accepting the inevitability of recurring El Niño events and learning how to prepare for them; recognising the different agendas, values, fears and priorities of the people that come into play when a disaster strikes, are all essential experiences to learn from the past (Gandy 2014). Societies with strong leaders and institutions, and who rely more on internal resources for funding and labour, are more resilient than those plagued by corruption, political turmoil and incompetence, and which heavily rely on external resources like trade (Sandweiss et al. 2020). An uncomfortable lesson is that complexity, stability, sense of place and community, and prosperity, come at a price, which is the loss of flexibility, mobility, and a reduced capacity to adapt to changing conditions, all traits which contribute to a more resilient society. Investments made in monumental architecture, materials goods, large settlements and infrastructure made these societies prosper, but it also made them more vulnerable. Even so, most societies developed a repertoire of responses to cope with challenges, often related to science, technology, leadership and management.

When communities are severely affected by disasters, they typically face a choice: flight, as in abandoning their settlements and often building new ones in different locations; adapting to a new, transformed environment; or restoring/rehabilitating the environment to pre-disaster conditions (Sandweiss and Quilter 2008). In this sense, offering sacrifices was an attempt to adapt to changing conditions. The use of narratives of the past and memory become tools for adapting and even for thriving when used successfully, especially when it comes to learning from the mistakes and problems of the past. Increased vulnerability and loss of resilience are the result of societies being unable to respond properly or even address challenges, not learning from failure, as well as their own internal tensions and conflicts, and unable to successfully adapt to changed circumstances.

Many of the societies of the twenty-first century occupy the same territories as those of the past, but now face modern challenges such as climate change, dwindling resources, a larger population and free-market capitalism operating at both the local and global level. This is why good leadership, comprehensive planning and strong institutions are important to improve resilience and reduce vulnerability. Yet, it is essential to develop individual and collective mental resilience, the capacity to identify internal conflicts, tensions, agendas and forces that may be working towards its own goal, as well as fears and uncertainties, especially fear for loss of status, of choices and liberty, of one's defined place within society.

A significant component that the past can teach us is the importance of agency and the capacity to successfully adapt to changing circumstances. Agency, the sense of having control over one's environment and future, explains the significance of offering sacrifices. They served to persuade a community that the leaders were doing their best to protect them by appealing to the deities.

Following this idea, Mexico City is an illustrative Latin American case study of a city taking steps to address their water-related challenges in a more sustainable manner by working to become a "Water-Sensitive City". The area that covers the present-day city of Mexico was historically a series of interconnected lakes, which eventually became the great Aztec city of Tenochtitlan. When Spanish conquistadors conquered Mexico in the sixteenth century, the Aztec city was mostly destroyed in order to build the Spanish city, and the process to drain the lake's water began. Today, one of the largest cities in the world has developed over "...a closed drainage basin that retains water and allows no outflow to other external bodies of water, such as rivers. It stores the water from rain and streams that run down from the surrounding mountains and then very slowly infiltrate into the ground, due to a thick layer of impervious sediment called the aquitard" (AEP et al. 2015: 37). In order to address the challenges that water poses, the municipal government and the Public Space Authority published "Towards a water-sensitive Mexico City: Public space as a rain management strategy" (2015), the result of an international workshop in which a new paradigm, based not on piecemeal technical-centred "solutions" to individual problems, but on the water-sensitive approach was presented, with the intention of it eventually becoming part of a design and public policy toolkit. According to this document, the way to address water should follow this order: delay, retain, store and reuse, only drain when necessary. Water infrastructure, management and public space design are also integrated into a comprehensive strategy, in order to make water more "visible" and create awareness of the water cycle among the public, while providing necessary and valuable public space.

While it is presently too soon to evaluate the impact of this strategy properly and how it could be adapted to other contexts, this document is a notable case on how to use memory, narratives of the past, public participation and agency as part of an approach to reduce vulnerability and improve the resilience of water-sensitive environments, adapted to the needs of a growing modern city.

Acknowledgements This chapter originated from presentations given in April 2018 as part of the event *Contributions of Historical Heritage to the Design and Management of Sustainable Cities* in Trujillo, Peru, organised by the La Libertad, Peru Architects Association, Chamber of Commerce and Commission for Sustainable Urbanism; in January 2019 at the *First International Workshop on Paleoclimate, Water Use and Environmental Phenomena in Ancient Peru and their Contemporary Impacts*, organised by the National University of Trujillo and East Anglia University (UK) sponsored by the British Council in Peru, the Newton Paulet Foundation, and CONCYTEC-Peru (National Council for Science, Technology and Innovation); and in July 2019 at the *Silk Cities* conference in L'Aquila, Italy. I would like to thank everyone who offered valuable feedback and comments for this chapter, especially Gabriel Prieto, Ana Cecilia Mauricio, Lucia Patrizio Gunning and Anna Porebska.

References

Autoridad del Espacio Público (AEP), Banco de Desarrollo de America Latina (CAF), Embajada de Países Bajos (2015) Hacia una Ciudad de Mexico Sensible al Agua: El espacio público como una estrategia de gestión de agua de lluvia (Towards a water sensitive Mexico City: Public space as a rain management strategy). Ciudad de Mexico, Mexico; Gobierno de la Ciudad de Mexico, Autoridad del Espacio Público

Billman B, Huckleberry G (2008) Deciphering the politics of Prehistoric El Ninno events on the North Coast of Peru. In: Sandweiss D, Quilter J (eds) El Niño, Catastrophism and culture change in Ancient America. Dumbarton Oaks Research Library and Collection, pp 59–76

Carey M (2010) In the shadow of melting glaciers. Climate change and Andean society. Oxford University Press

Fagan B (1999) Floods, famines and emperors: El Niño and the fate of civilizations. Basic Books, New York

Gandy M (2014) The fabric of space. Water, modernity and the urban imagination. The MIT Press, Cambridge, Massachusetts. London, England

Huckleberry G, Caramanica A, Quilter J (2017) Dating the ascope canal system: competition for water during the late intermediate period in the Chicama Valley, North Coast of Peru. J Field Archaeol 43(1):17–30

Klaus H, Toyne M (2016) Ritual violence on the North Coast of Peru: perspectives and prospects in the archaeology of Ancient Andean Sacrifice. In: Klaus H, Toyne M (eds) Ritual violence in the Ancient Andes. University of Texas Press, pp 1–24

Klaus H, Shimada I (2016) Bodies and blood. Middle Sican human sacrifice in the Lambayeque Valley Complex (AD 900–1100). In: Klaus H, Toyne M (eds) Ritual violence in the Ancient Andes. University of Texas Press, pp 120–149

Kohlstedt K (2016) Tsunami stones: ancient Japanese markers warn builders of high water. https://99percentinvisible.org/article/tsunami-stones-ancient-japanese-markers-warn-builders-high-water/. Accessed 12 Sep 2019

Lowenthal D (2014) Sylvester Baxter lecture: conservation past and present. Lecture presented at Harvard University Graduate School of Design. https://www.youtube.com/watch?v=a4u9HWNcpKQ&t=1346s. Accessed 13 Sep 2019

Medina L, Luque G (2008) Zonas críticas en la región de La Libertad, informe técnico preliminar. Lima, Instituto Geológico Minero y Metalúrgico, Perú

Moore J (1996) Architecture and power in the Ancient Andes: the archaeology of public buildings. Cambridge University Press

Prieto G, Verano JW, Goepfert N, Kennett D, Quilter J, LeBlanc S et al (2019) A mass sacrifice of children and camelids at the Huanchaquito-Las Llamas site, Moche Valley, Peru. PLoS ONE 14(3): https://doi.org/10.1371/journal.pone.0211691

Rick T, Sandweiss D (2020) Archaeology, climate, and global change in the Age of Humans. PNAS 117(15):8250–8253

Romey K, Clark R (2019) Sacrificio inconcebible: El misterioso asesinato de niños hace años en lo que hoy es Perú. National Geographic en Español 44(2):28–51

Rowe J (1948) The kingdom of Chimor. Acta Am 6(1–2):26–59

Sandweiss D, Andrus F, Kelley A, Maasch K, Reitz E, Roscoe P (2020) Archaeological climate proxies and the complexities of reconstructing Holocene El Niño in coastal Peru. PNAS 117(15):8271–8279

Sandweiss D, Quilter J (eds) (2008) El Niño, catastrophism and culture change in Ancient America. Dumbarton Oaks Research Library and Collection

Schaedel R (1992) Paleotechnology and flood control on the Peruvian north coast. In: Ortlieb L, Machare J (eds) Paleo ENSO-records international symposium, extended abstracts. ORSTOM-CONCYTEC, Lima, pp 285–286

Villacorta S, Ochoa M, Núñez S (2008) Zonas críticas por peligros geológicos de la región Lambayeque, primer reporte. Instituto Geológico Minero y Metalúrgico, Lima, Perú

Vreeland J (1992) Indigenous response mechanisms to periodic climatic disasters on the Peruvian north coast: Paleotechnological repertory in 1578. In: Ortlieb L, Machare J (eds) Paleo-ENSO records international symposium, extended abstracts. ORSTOM-CONCYTEC, Lima, pp 329–330

Chapter 8
Intangible Cultural Economy, a Mould for Tangible Urban Built Fabric—The Case of Shahjahanabad, India

Abhishek Jain

Abstract Historic Silk Route cities have always been known for their trade culture and lifestyles which contribute to generating tangible urban fabric and built character. Residential quarters of such historical cities are often based on trade communities. Usually, they grew organically with time and led towards interdependency between their tangible urban fabric and intangible sociocultural pattern. Such living pockets are known by traditional different native names like "Mahalla", "Katra", "Bara", "Pol" and "Ahata" in different parts of India. This chapter envisions future urbanism for traditional trade-based residential pockets in old cities of India through a revival of their cultural traits, especially the ones which have a direct connection with the Silk Route crossing through India which were important trade centres from the past to the present. These cities have proved to be interesting networks of such trade-based residential pockets which act as a laboratory for the evolution of cultural features based on communities having a direct relationship with trade and trade routes, focusing on India. The chapter illustrates that these traits, based on cultural economy, generating richness and diversity in old heritage trade-community-based neighbourhoods, are the main key to the cultural heritage in historic old cities which arose as a by-product of trade routes. The research has been restricted to current cultural, social and spatial characteristics, to study the "Puras of Shahjahanabad" and their intangible heritage in the City of Shahjahanabad (Old Delhi). The historical transformation of Mahallas (neighbourhoods) in relation to a city plays an important role in understanding its significance on how an intangible heritage of a trade-based community plays an important role in giving birth to a tangible one in the form of urban built fabric. The research explores the idea of the cultural economy in the old city of Shahjahanabad in India through a study of the trade-based community neighbourhoods having historic morphological structures and communal intangible traits related to the Silk Route. It helps to understand the qualities of the physical character of such trade-community-based neighbourhoods in old cities, especially when they are in their transition period from historic trade towns to contemporary commercial zones of an expanded city.

A. Jain (✉)
Shahjahanabadi Foundation, Delhi, India
e-mail: aj0880886@gmail.com

© The Author(s), under exclusive license to Springer Nature Switzerland AG 2021
F. F. Arefian et al. (eds.), *Historic Cities in the Face of Disasters*, The Urban Book Series,
https://doi.org/10.1007/978-3-030-77356-4_8

Keywords Silk Route Cities · Organic morphological structure · Cultural economy · Architectural heritage · Mahalla · Shahjahanabad · India

8.1 Introduction

The research aims to understand the historic Silk Route as a productive resource for the cultural economy in the old city of Shahjahanabad in India. The Silk Route has led to the formation of various trade-based intangible social communal structures in traditional neighbourhoods situated around the spine trade market in Shahjahanabad. As Silk Route cities were not exclusively about trade but also about cultural lifestyles of various trade-based communities they generated specific building typologies which supported both trade and income from other activities. This connection of the city of Shahjahanabad with the Silk Route over centuries is not only fuelling the cultural economy of the city in present days but has also the potential to act as a heritage resource for the betterment of the residents of the city through heritage tourism. Sadly, these resources are diminishing due to lack of sensitive regeneration policy and methods of resilience which is harming the image of Shahjahanabad as a global Silk Route city.

For that reason, the production of a heritage sensitive urban regeneration policy becomes a primary objective in re-establishing the significance of the Silk Route as an entity which was the birth place of diverse intangible cultural economies connected with communal lifestyles as well as tangible built heritage resources. The research examines the cultural economy, lifestyle and traditional built fabric as a diminishing potential heritage resource of the city. It is based on a series of historic trade-based, gated community neighbourhoods surrounding the main Bazaar streets of the city as case studies.

Dharampura and Vakeelpura are identified as the main neighbourhoods of trade-based communities located in the heart of the city and connected with the major Bazaar streets of the city from the Mughal era. These neighbourhoods (Mahallas) have grown organically over time, generating an interesting morphological structure supporting and framing the built typologies with cultural traits of the community. The field research of the Mahallas is based on two interdependent components, its tangible built fabric and the connection of building typologies with intangible cultural traits. The study of the trade-based communities focuses on their anthropos, demographics, day-to-day activities and built fabric on the one hand and on the transforming indigenous cultural economy on the other hand (Fig. 8.1).

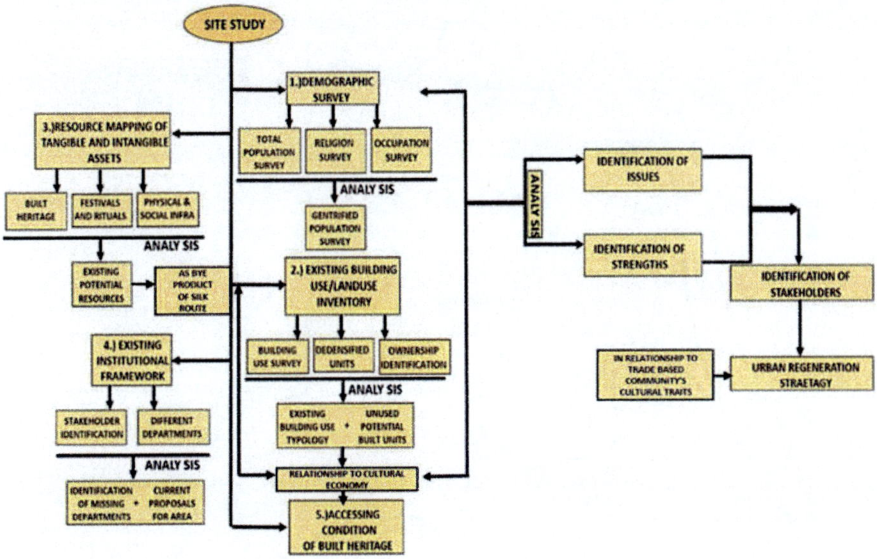

Fig. 8.1 Research methodology

8.2 The Silk Route and the City of Shahjahanabad (Old Delhi)

Delhi forms an integral part of the morphological structure of the Silk Route as part of seven cities. Regardless of whether it was the capital of India, it was always an important city over centuries, together with Mehrauli, Siri, Tughlaqabad, Jahanpanah, Firozabad Shergarh and Shahjahanbad next to the river Yamuna which in medieval times acted as an extension of the Silk Route under a succession of kingdoms and rulers. This led to the addition of new typologies of settlements and built fabric. They are the trade-based gated communities (Mahallas); the "revenue villages" with agriculture based trade markets on the outskirts of cities; and the Serais with settlements around them along the Silk Route as part of its morphological structure. They comprise various monumental Islamic tombs, mosques and other built structures which are acting as a rich built heritage of the present city of Delhi, capital of India. The older seven cities got destroyed and left abundant historic ruins in the form of fortresses and small walled complexes which are currently maintained by the Archaeological Survey of India. Newer developments around them were initiated by the Delhi Development Authority in the 1960s as part of the Greater Delhi Development Plan after independence of India in 1947. However, this also led to the loss of cultural trade along the extended Silk Route. Nevertheless, the last historic city among the seven cities, the walled city of Shahjahanabad is still inhabited by many traditional as well as transformed cultural trade activities and acts as the main inner core city of extended New Delhi (Fig. 8.2).

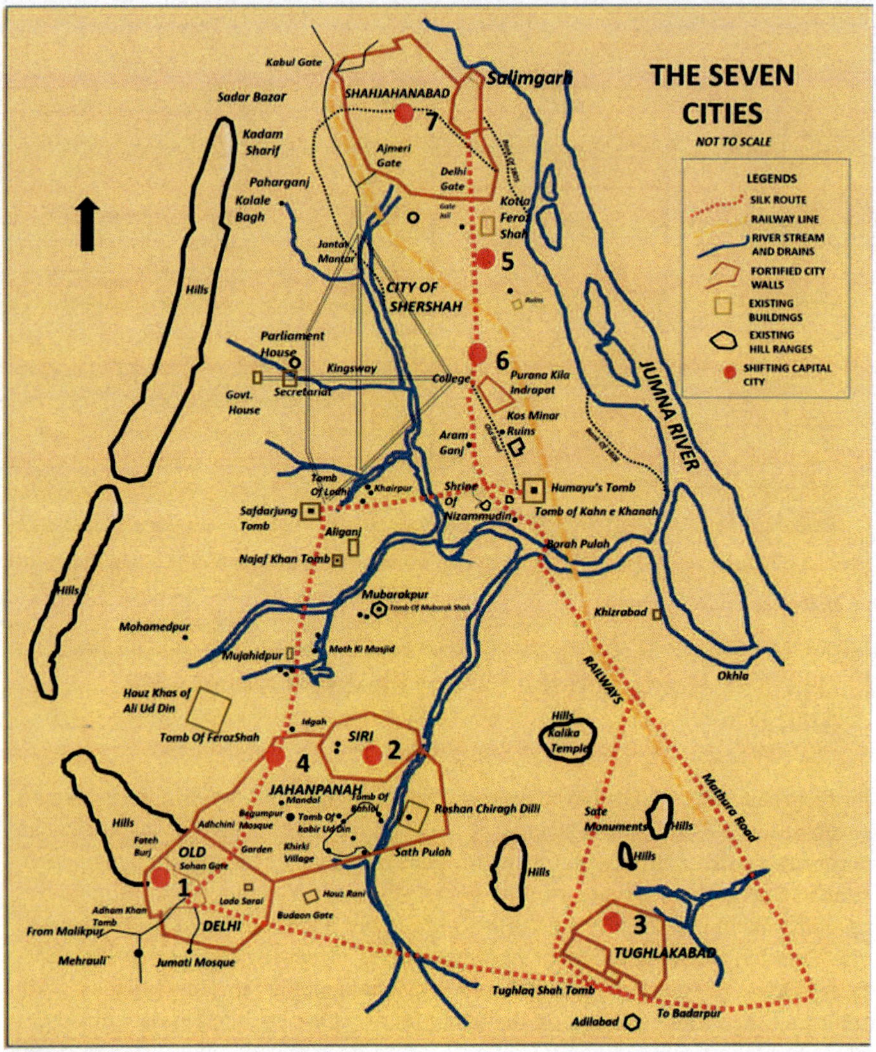

Fig. 8.2 Map of seven cities of Delhi showing the location of the extended Silk trade route through the seven cities of Delhi (author, based on Hearn (1928), The Seven Cities of Delhi)

The walled City of Shahjahanabad was the capital city of the Mughal Empire established in mid-seventeenth century after other six cities in the Delhi triangle had been designated as capital by previous rulers. The city of Shahjahanabad was planned and laid out with the main commercial spine comprising the Bazaar streets for trade and shopping, public squares for resting and other areas for public activities in the city. The organic development of further residential quarters occurred with the extension of the Silk Route. Residential districts were demarcated and grew organically in the form of clusters of residences called "*Havelis*" (mansions). Over

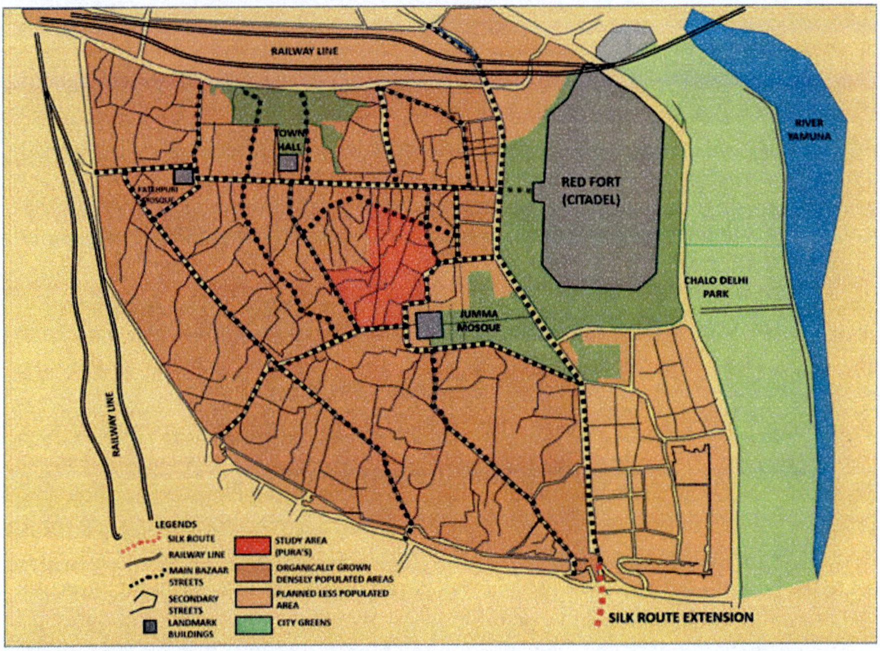

Fig. 8.3 Map of Shahjahanabad showing the location of Pura's of Shahjahanabad (author)

time community-based gated neighbourhoods called "*Mahallas*" formed, based on either religion or occupation. Another typology of residential quarters also came into being called "*Katras*" which were gated neighbourhoods accommodating craftsmen with their commercial workshops, always next to the city's primary Bazaar street.

The Mughal court had nobles of different ethnicities, religions and sects. The people from the Hindu community like Jains, Khatris and Banias were the major trade communities and an important part of the city and the Royal court. Over time the city acquired different layers of urbanisation which led to further gentrification and increase in population in the 1700s. With increased production of traditional craft goods like Mughal jewellery, textiles and gold thread embroidery, the economy grew and the tradesmen groups expanded, leading to an increase in trade-community-based neighbourhoods in the heart of the city (Fig. 8.3).

8.3 Baniya Community Mahallas (Pura's) of Shahjahanabad

"Pur" and "Pura" are suffixes meaning "city" or "settlement", used in several place names across the Indian subcontinent, Southeast Asia, Afghanistan and Iran. The word "Pura" is the oldest Sanskrit language word for "city". It finds frequent mention in the Rigveda, one of the four canonical sacred texts of Hinduism, mostly dating between c. 1500–1200 BCE (Macdonell and Keith 1995).

Dharampura and Vakeelpura are mainly Baniya community-based gated Mahallas. Dharampura accommodates the Jain Baniya community which includes known jewellers of the city and Vakeelpura is occupied by the Non-Jain Baniya community. The Mahallas grew out of each other. In the eighteenth century, Raja Harsukh Rai (Mughal Imperial Treasurer) was one of the important nobles of the Mughal court and therefore given the right by the Mughal king to build a Jain temple in a central location of the city, which later evolved into a Mahalla called Dharampura. When the area started to grow the neighbourhood Vakeelpura came into existence next to it as another gated Mahalla, accommodating a community which took care of the city's law-related activities and housing the Vakeel's (lawyers) of the city (Fig. 8.4).

From the mid-1600s to the late 1700s, these neighbourhoods comprised two important Havelis belonging to the main nobility of the Royal Mughal court, respectively, known as Ustad Hamid (master builder of Red Fort) and Nawab Shahdi Khan. This shows how a strong connection with the Silk Route has transformed the city from a Royal noble Muslim based population to a mixed trade-based city. The entrance street to the Dharampura neighbourhood was known as Kucha Ustad Hamid based on the name of the nobleman. It now accommodates mansions of the jeweller community. The entrance alley of the Vakeel Pura neighbourhood was called Gali Gulian which had a direct link to Jama Masjid (Friday Mosque of the City) known as "Chhatta

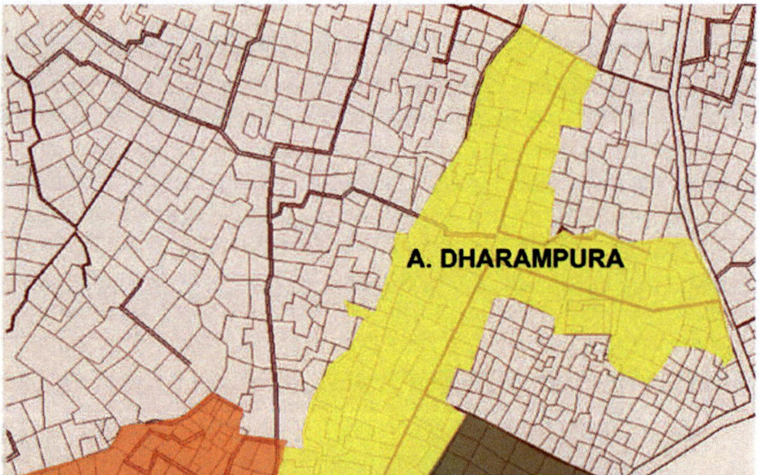

Fig. 8.4 Map showing the location of the study area, the Puras in Shahjahanabad (author)

Shahji" based on the name of Nawab Shahdi. In the early 1800s, the Mughal court gave this area to "Raja Harsukh Rai" (Imperial Treasurer) under the reign of Akbar II (Mughal king), to build the first planned Jain Temple at Shahjahanabad. It was not only destined for the community but also to obtain grand monumental buildings inside the neighbourhoods to enhance the visual experience of the city for traders. Over time the community started to settle around the temple. Further, greatly ornamented Jain temples were built in the area. It got soon converted into a series of gated Mahallas which became a new built typology of the city around these beautiful public buildings as their centres.

The Aggarwal Jain Panchayat (Social Management Group for Community) was established to maintain community rights and customs. A further Baniya community associated with legal trades of the Mughal court settled in an another extended gated Mahalla called Vakeel Pura which housed the famous Chah Rahat well (an another existing heritage resource currently in very bad condition) with a Persian wheel providing water to surrounding Havelis and Jama Masjid in 1700.

8.3.1 Physical Urban Structure of Dharampura

Dharampura belongs to the Baniya community's gated neighbourhood of Shahjahnabad built around Naya Mandir and Bara Mandir. Dharampura consists of a trade-based neighbourhood inhabited by the Baniya community of India, the Vakeelpura neighbourhood of Treasurers, as well as the Vaishnav and Punjabi community and it has a relationship with the Silk Route. It has two major "*chowks*" (public neighbourhood squares) with community buildings accommodating major public activities of the neighbourhood. A series of other gated Mahallas are branching out of this main Mahalla, one of which being Vakeelpura. The neighbourhood of Dharampura is directly connected to two Bazaar streets which are the working place of the trading community residing in these Mahallas. Together they meet and intersect with Chandni Chowk Street, the primary commercial street of Shahjahanabad and spine of the city which has been laid out as the continuation of a trade route from the Silk Route entering Delhi. Thus, the whole morphology is like an extension of a trade route which is a by-product of the existing Silk Route through Delhi, India (Fig. 8.5).

Both Dharampura and Vakeelpura have a very strong social structure due to the morphological structure of the Mahalla which is a by-product of streets branching out of the Bazaar as a result of connecting the Silk Route to the city. Urban elements like chowks have been developed over time in the neighbourhood and are common to both of them. Certain alleys with dead ends contain a cluster of Havelis for extended joint families with small street intersections acting as spaces of interaction for the families themselves.

Organic expansion has generated semipublic zones, fostering further levels of community interaction. In Dharampura, these alleys with cul-de-sacs were named after the number of Havelis they contain, e.g. Chaughara (Street with four houses), Panchghara (Street with five houses), Chhehghara (Street with six houses) and

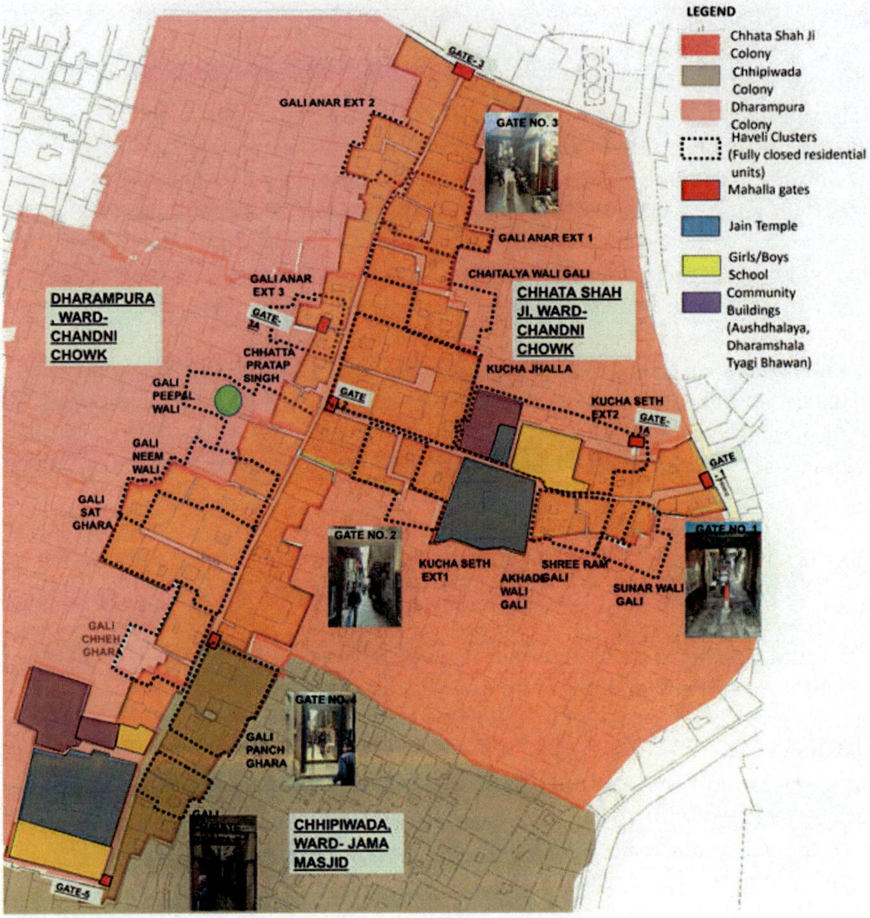

Fig. 8.5 Map of Dharampura with major zones and neighbourhood elements (author)

Satghara (Street with seven houses) and so on. In both neighbourhoods, these clusters of Havelis are directly connected to the main spine street of the neighbourhood where mixed land use is consisting of daily convenience shops attached to the large Havelis. This generated a different typology of Havelis in the neighbourhood, once famous for their grandeur and craftsmanship. They are explained in greater detail below.

In these Mahallas, the Haveli typology which surrounds the main spine Street are huge in scale and have shops attached to them. These Havelis are often connected to the main wide spine street of the city through an alley with a gateway with a canopy for security guards known as "*Chhatta*". The main spine street of the Mahalla has Chowks at every street intersection and contains social amenities. Other ornamented grand public buildings were designated for the community as well as for visual

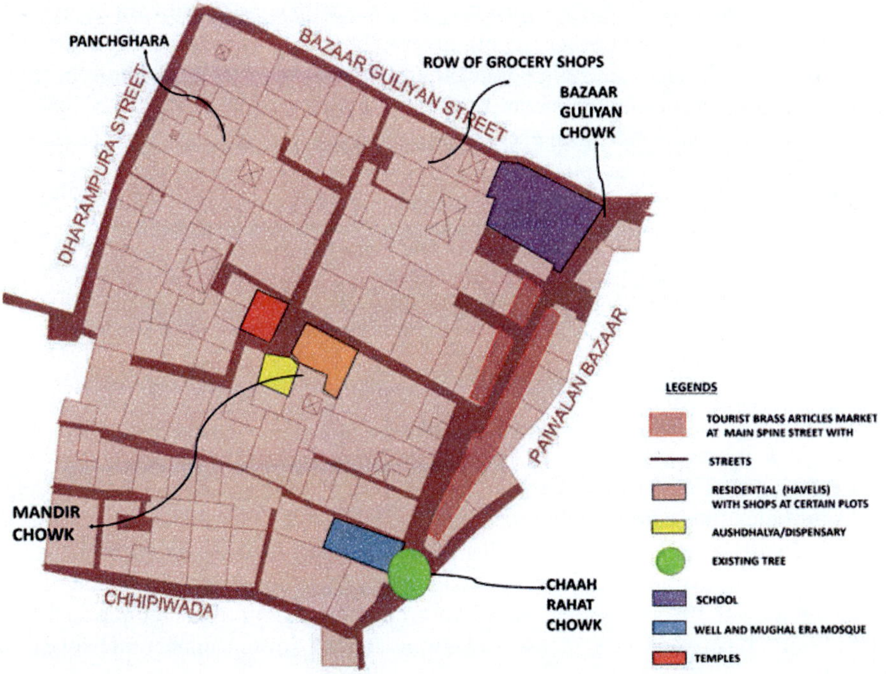

Fig. 8.6 Map of Vakeelpura showing amenities and connection with main commercial spines (author)

experience and use of both domestic and international traders who explored the city for trade in historic times and acted as catalyst of flourishing trade in the city. These buildings are still there in the form of Jain/Vaishnav Temples, Jain/Vaishnav Aushdhalayas (Ayurvedic dispensaries which are defunct now), the Jain Library and Jain Schools which promote religious learning as well with modern education. These Chowks also have some betel leaf shops acting as the main place of interaction for men in the evenings, as well as temples where women congregate in the mornings, leading to a strong social cohesion and constant street life in the Mahallas (Fig. 8.6).

8.4 Potential Intangible Traits of Mahallas in Need of Conservation

Mahallas are rich in intangible heritage in the form of community festivals and processions associated with temples which are generating street activities and enriching the city's intangible heritage. The following are some important festivals contributed by the temples of the Dharampura Mahalla. Paryushan/Anant Chaudash is a festival and procession organised annually during the auspicious month of "Bhadrapad".

"Mahaveer Jayanti" is the birth anniversary of the 24th Jain Tirthankara or monk Mahaveera. Finally, Navmi Mela is a big procession taking place annually at every ninth day of Paryushan. It is a big colourful and historically rich procession for the whole city. It passes through all central neighbourhoods and the city's main Bazaar streets, in which holy water from every temple well has been collected and taken to Lal Mandir, the city's main Jain temple for the coronation and holy bath ceremony of idols of gods. It is carried by men of the community who are bejewelled in their traditional attires with fine jewellery (as this is a neighbourhood of jewellers) and are acting as "*Indras*" (caretakers) of main deities. These festivals and processions are attracting religious tourism and thus contributing to and stimulating the cultural economy of the city. Besides offering local inhabitants more economic opportunities these community traits also enhance social resilience in case of disasters by assisting the city and the community through social and economic resources.

Connected with the Jain temples and initially performed as cultural practice to enhance trade connections between different communities, the Dharampura processions have a lot of associational and cultural value for the local people today. The area contributes four cultural Jain processions to the city's cultural heritage, closely associated with a series of Jain temples, seven in total, in these gated Mahallas. They form a temple trail where the procession proceeds from one temple to another, carrying holy water from each temple's holy well and taking it to the temple next to the Red Fort, the city's central Citadel. This constitutes a strong connection between intangible and tangible heritage, adding to the cultural image of the city which is a by-product of the Silk Route and needs to be identified as a resource for heritage tourism. These long processions connecting different landmarks of the city generate a great sense of communal heritage and social cohesion in the City (Fig. 8.7).

The direct connectivity with the extension of the Silk Route and the related development of trade settlements in the city fostered the exchange of lifestyles between the various resident communities, as well as with traders coming from different parts of the world and developed all the local tangible and intangible resources. The resulting mixed cultures became the identity of these trade-based communities and are constituting their current heritage resources. This enabled them to turn the organically grown residential quarters into a laboratory for new cultural traits.

The people of the Vakeelpura neighbourhood are particularly associated with the city's Royal money due to their association with imperial treasurers who formed part of the royal nobility and are now acting as intangible heritage. However, the neighbourhood started to lose its identify when the famous old families moved out which caused major physical transformations. Another important intangible heritage is associated with the intangible activities which took place around the now totally lost functions of the Chah Rahat well.

Fig. 8.7 Map of Dharampura and attached neighbourhoods, showing festival procession routes connecting heritage Jain Temple buildings and main commercial spines (author)

8.5 Physical Transformation of Mahallas Affecting the Image of the City as a Silk Route City

The gated neighbourhoods of Dharampura and Vakeelpura underwent major physical changes historically which had affected the current physical structure of the Mahallas. Major changes occurred in the physical morphological structure of the Mahalla after 1856, mainly due to the British invasion of the city of Shahjahanabad. It led to more migration of tradesmen to the Mahallas as the Baniya community played a major role in terms of investments for the East India Company government.

This expansion of economic uses of the Mahallas led to the de-densification of the residential population, especially in the large Havelis of primary streets. This resulted in the loss of community features which could have contributed to the cultural economy of the city through heritage tourism and were a potential source of social and economic resilience of the Mahallas at trying times.

8.5.1 *Transformation of Building Use in Mahallas and Its Impacts*

The field survey shows that in the Dharampura Mahalla, 56% of buildings out of 119 plots, including plots with only residential activities, got converted into mixed-use buildings with incompatible uses in a residential neighbourhood, this included certain jewellery making workshops and "*godowns*" (warehouses) the proximity of the city's two commercial Bazaar streets, called Dariba Kalan and Kinari Bazaar, acting as important jewellery and lace craft-making clusters, as well as markets and workshop zones for citywide trade. The old connection of residential and commercial uses was once attracting cross cultural experiences for traders, but moving commerce into residential zones is now a major problem. Due to lack of special building use guidelines as part of a sensitive urban regeneration policy, the commercial use of residential buildings displaced 18% of residents from the neighbourhood. Alternatively, increasing the labour market community led to gentrification. This rapid change has also led to the disappearance of the network of "*Chabutra*", the platforms for sitting outside the buildings, as well as the public use of Havelis. The majority of Chabutras in the Mahalla was covered over owing to an increasing number of shops and is badly affecting public interaction on streets (Fig. 8.8).

Similarly, a detailed built use and plot use survey has been undertaken for Vakeelpura to understand the current situation. Approximately, 30% of the buildings have totally converted into commercial use, an alarming situation. Here de-densification of the residential population is due more to its close connectivity to the city's central Friday mosque Jama Masjid which is a grade I listed monument of Delhi and a centre of tourism. A whole-sale market of metal sculptures is emerging there which bears no relation to either the Chah-Rahat or the Vakeel community residing in the Mahalla. Preventing this could increase tourism in the Mahalla and leads to the preservation of both its tangible and intangible heritage resources.

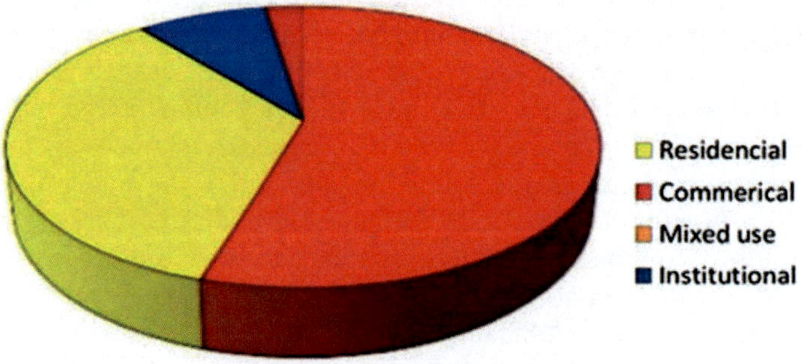

Fig. 8.8 Pie chart showing the building use distribution in Dharampura due to commercial infiltration and reduction of resident numbers, leading to loss of cultural economy traits of the Mahalla (author)

A sequence of physical mapping of the ground floor illustrates the observations from secondary and primary research, including residents' interviews. It shows the physical transformation of the Mahalla from the 1800s till now (Fig. 8.9).

Infiltration of solely trade-based commercial activities in residential neighbourhoods was not part of the city in historic times and was confined to the Commercial Bazaar streets, maintaining them as laboratory of trade-based communal practices to attract traders from the outside. Commercial use infiltration in recent times is leading to loss of sanctity of tangible urban built fabric and intangible community traits in these neighbourhoods. It is the main reason of decay of these heritage resources and harming the image of the city as a Silk Route city. The city is going through a slow constructed disaster, as most of the buildings in these neighbourhoods are getting converted into either commercial or industrial use. This is leading to the depletion of traditional built patterns, as well as of communal heritage without which any Silk Route city is incomplete. Converting the existing cultural economy into an economy without any association with city's culture is also generating a loss of sense of belonging in residents due to the decayed physical built environment. Recently, one of the important Heritage Haveli in Dharampura went through an insensitive adaptive reuse as a Heritage hotel, without involving the local community or respecting traditional features and the character of the Mahalla. This kind of economic development with lack of culture could lead to the loss of important heritage values of the area, the main contributor to the cultural economy.

8.6 Historical Built Typologies and Trade-Based Community Semiotics in Need of Conservation

Different typologies of communal buildings, such as temples, Dharamshala and school buildings have large courtyards in specific locations in the Mahallas. These types of architecture or elements of architectural unity are commonly used throughout the city. Other building typologies also feature these architectural details, such as the multifoliate Mughal arched gateways with platforms and niches on both sides, the stone bracketed sunshades and other features. Likewise, these elements have a direct relationship with the regional style of architecture and its main Islamic influence due to Mughal rule. Meanwhile, these Mahallas have added their own style of motifs as ornamentation of these architectural elements. This not only reflects their communal believes but also cross-cultural influences of ideas and beliefs exchanged by the local trading communities with traders travelling to the city through the Silk Route. Today, these features are acting as the main semiotics to identify the type of Mahalla in a particular community. This is not just reflecting their own ethnicity but the specificity of other international traders they have dealt with in the past. These characteristics are now constituting the heritage value to the Mahalla, as well as the city itself, due to both their craftsmanship and symbolic communal value.

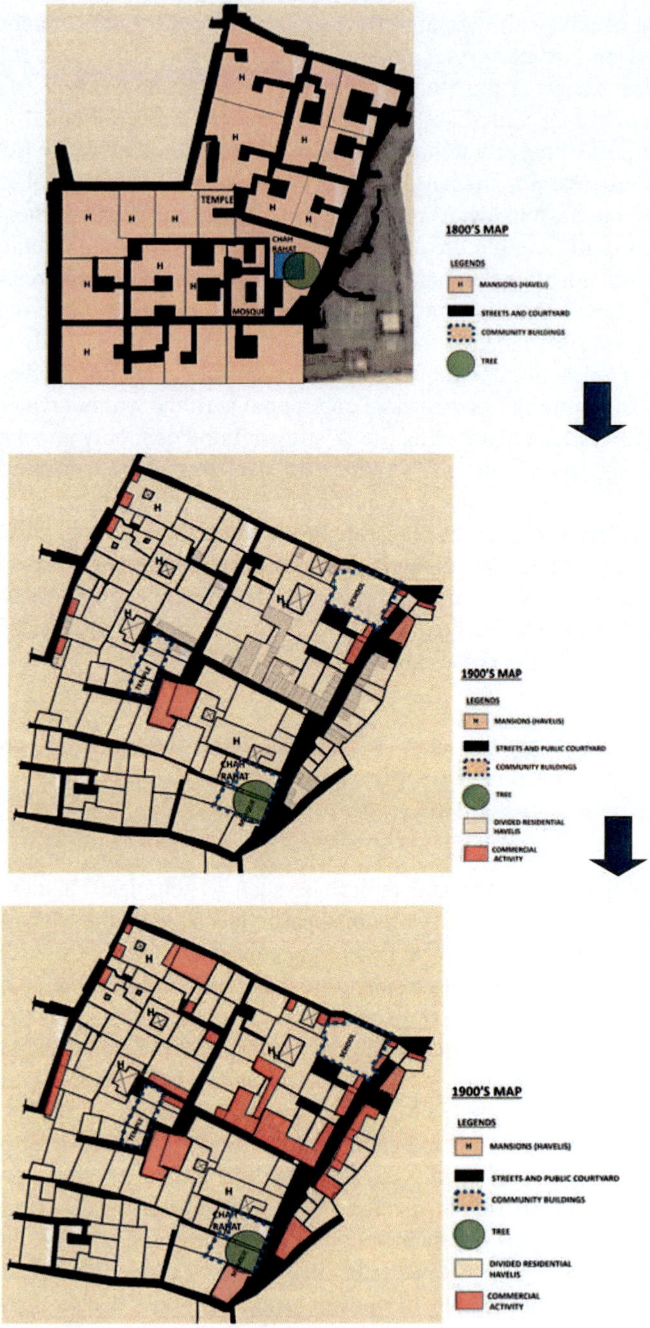

Fig. 8.9 Maps showing chronological urban transformation of the neighbourhood, its plot uses and its urban fabric (author)

Fig. 8.10 Pictures showing the typology of residences on the secondary spine street of the neighbourhood with shops on both sides at lower level, a by-product of the Silk Route and the mahalla based cultural economy (author)

8.6.1 *Haveli (Housing) Typology*

Above are shown a few typologies of Havelis from the spine street of the Mahalla. Large entrance gateways with common motifs are representing the prevailing religion as well as the impact of cross-cultural interaction with different traders who passed through the Silk Route (Fig. 8.10).

8.6.2 *Transformation of Haveli and Its Impacts*

These neighbourhoods have a typical Indian—Mughal housing typology, the Havelis. They are big mansions with huge courtyards typical for the Mughal era. In Dharampura and Vakeelpura, this particular typology was divided further into sub-typologies, which are dependent on several factors, such as land topography, surrounding built typology and proximity to the main spine. Generally, Havelis on the main spine of the Mahalla have raised plinths and shops for day to day needs next to them, some of them have verandas and a third type has its own dedicated sub-alley forming a cluster of houses for extended family living in and sharing common courtyards. These different typologies of mansions support the social and physical needs of the community. Current urban transformations have provoked a chain of issues affecting the social life of the inhabitants and the cultural image of the area. Due to an increase in trade activity, built floorspace has increased by 25% for buildings, converting them into G+3 and making already narrow streets even narrower. In this process, 18% of buildings got partially de-densified from residential population which led to partial or full dilapidation of them, as well as of surrounding buildings which, in turn, are in a constant state of residential population de-densification.

The majority of Havelis were G+1, and these large Havelis used to have bigger courtyards for private and semiprivate activities. However, soon a single unit got converted into a block of multiple residential units leading to the conversion of the semiprivate courtyard into smaller Mahalla Chowks.

8.6.3 Semiotics in the Form of Ornamental Features

The images below demonstrate architectural features as an important part of community identity in the Mahallas. They symbolise certain beliefs, class distinction or auspicious character while creating a landmark quality with associational value for people living there (Fig. 8.11).

These features contributed to a visual delight for traders in historic times and gave status to the city. These physical communal aspects generated the diversity in the city and attracted traders due to specific commodities as well as the identity associated with the community.

Similar kind of patterns used in ornamentation of buildings could be found in jewellery designed by the Dharampura jeweller community based on a mixture of Hindu and Mughal influence (Fig. 8.12).

Fig. 8.11 Pictures showing communal semiotics in the form of architectural features in the neighbourhood (author)

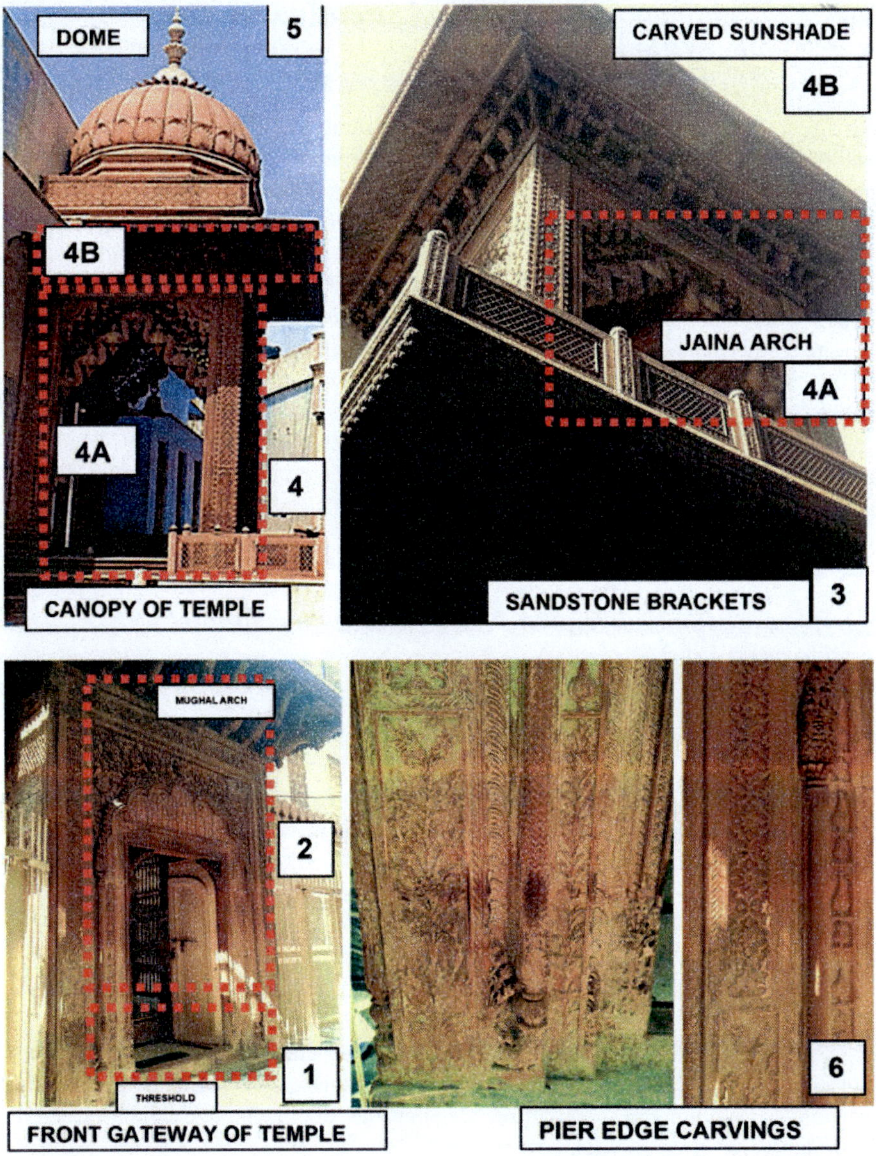

Fig. 8.12 Pictures showing the building typology of public buildings with communal semiotics in the form of architectural details in the neighbourhood (author)

8.6.4 Transformation of Streets and Street Intersection Squares (Chowks)

Inner streets with cul-de-sac of Mahallas were undergoing more vertical expansion of buildings due to expanding families and lack of specific development guidelines. This was creating an imbalance between building height and the width of secondary chowks and streets, leading to further loss of public character of chowks and streets due to less natural light and street widths, which is one of the reasons for decay in urban heritage built fabric. On primary streets of Mahallas temple sunshades and canopies were expanded due to changing needs of the community. This added further coverage of the main chowks and converted them from public to semipublic areas, thereby obstructing visual access to these heritage buildings which, in turn, was affecting the heritage value of the built heritage. All this chaos was a by-product of the lack of Mahalla and community-specific development controls. Nevertheless, social interaction in Mahallas is still active due to the community structure of the area and active social community groups.

8.6.5 Social Impacts Due to Physical Transformation of Neighbourhoods

Physical transformations have led to significant social impacts in the Mahalla mainly affecting the social cohesion of the community. Together with the change of the built activities and gentrification, this has led to the mix of many incompatible communities. Another example of these transformations is the interruption of the continuity of the Chabutra network outside the Havelis which had been generating a lot of visual and verbal interaction in the streets. The decrease of residential population is reducing the usage of communal buildings which, in turn, is making them eligible for uses other than social activities, such as rental storage of goods for industries. Another physical reason is the partitioning of plots due to increasing real estate value and lack of special area norms which was leading to infiltration of commercial activities. This led to the division of joint families and affected the community's social integration. For the same reason, the courtyards which once acted as main interaction zones for neighbours were getting smaller and smaller in size which affected the terrace connections as well.

8.6.6 Synthesis of Social and Physical Issues

These findings are derived from analyses and interviews in both neighbourhoods. In residential quarters of the city of Shahjahanabad, people have a very strong social lifestyle due to the typical compact physical urban growth in the Mahallas with their traditions and special built typologies compatible with their surrounding Bazaar streets. Together they were an integrated by-product of the Silk Route. Modern times brought about their eradication as an important source of resilience for the city which lacks special development controls and regeneration policies to maintain its heritage resources to preserve its image of Shahjahanbad as a Silk Route city. De-densification of residential population due to incompatible transformation of existing building is an another important issue which has led to an increase in the number of dilapidated buildings and thus increasing urban decay in city. Change in built use is eroding the network of connected terraces leading to further loss of interaction spaces in the Mahalla which is adversely affecting its social structure. With increase in trade-based and decrease in traditional craft-based economic activities, the commercial real estate value of the area is increasing day by day leading to vertical expansion. For that reason, old buildings with heritage value are getting torn down. New commercial buildings are being put up in their place without communal architectural features acting as semiotics. This is badly affecting the communal heritage identity of the area and thus cultural tourism through the loss of its Silk Route city image. With unplanned vertical expansions of buildings, the streets are getting less habitable and reduction of street life is adversely affecting social cohesion in the community, leading in turn to loss of cultural traits associated with the economy.

Due to this, increase in land value reverse gentrification is also taking place in Mahallas. In Dharampura, 20% of buildings are accommodating commercial goldsmith workshops. Moreover, they are being transformed partially into cohabited residential space for economically weaker labourers of varied communities with poor sanitation. They are degraded living conditions and are increasing the chances of epidemics in the city as settlements are physically compact in nature and getting overcrowded. For example, in Dharampura in many of the commercially transformed Havelis, one single $10' \times 10'$ room is occupied by 15 to 20 people with a makeshift cooking space and lack of ventilation with small shafts as the only source of ventilation because traditional courtyards are converted into expansion of commercial space.

Buildings in Mahallas are physically connected and have common foundations as well, which can come down in a stroke in case of disasters like earthquakes. For the people of the Mahalla, the decaying urban built fabric is an important issue due to lack of technical assistance programmes from the government for residents and lack of traditional "*Rajmistris*" (masons) in the city who have moved to modern construction techniques, which are incompatible with traditional structures in terms of materials used and maintenance. All these transformations have led to loss of cultural aspects of the community and cultural economy as a traditional source of resilience.

8.7 Recommendations for a Heritage-Based Urban Regeneration of Neighbourhoods of Shahjahanabad

The neighbourhood Mahallas are the main strength of the Silk Route city which enables continuous sustenance of the traditional trade community in association with Bazaar streets. Hence, they are the main heritage resources for urban regeneration and an inseparable part of the urban fabric. An integrated approach is needed when proposing a better urban future, keeping the community traits intact as source of resilience in times of disaster while understanding their relationship with the economy and community identity. An appropriate urban regeneration strategy for such Mahallas should involve the community in the regeneration process, because physical interventions have to be done in relation with intangible traits of the community, such as religion, trade activities and levels of social interaction for different genders at different times of the day. It is important to consider the importance of the community as the living resource of the built urban fabric to maintain the image of the city as a Silk Route city. It is also necessary to resolve the practical problems like lack of Mahalla-specific development norms, while keeping intangible community traits as the main framework into which other regeneration frameworks should fit.

8.8 Conclusion

Making communities aware of their own heritage resources will keep their identity alive in a Silk Route city through their own cultural economy with the help of community participatory programmes. Existing layers of transformation as a product of cultural and physical transformations in cities along the Silk Route are an important focus for an urban regeneration strategy and building process as it leads to a more viable regeneration plan with positive answers for the relationship between tangible and intangible issues.

In India, Islamic Silk Route cities had diverse trade-based communities supporting the cultural economy and the trade network of Silk Routes in the past. Now that the traditional Silk Routes are no longer active for trade they have kept their character as trade centres for domestic trade and in some cases international trade and contemporary Silk Route city communities are still contributing to the city for its image and running the economy.

Developing relationship-based resource mapping of built and cultural traits related to trade evoked by the Silk Route can act as a key entry point for heritage-based urban regeneration of Silk Route cities. The Silk Route is an important resource to identify such communities and their cultural economy in Silk Route cities while assessing current layers of transformation. The revival of this existing physical network at city level can lead to a better cultural and economic network between the extended and the old city.

Acknowledgements I should like to express my gratitude to the advanced architectural theory studio students of MBS (Mind Body Soul) School of Planning and Architecture, years 2013 and 2014, for helping me in conducting surveys and making base maps. I also thank the residents of Dharampura and Vakeelpura for their support during interviews.

References

Akhtar S (2015) The challenges of housing. My liveable city, pp 31–32, 7 Sep 2015

Amos R (1977) Human aspects of urban form—towards a man—environment approach to urban form and design, 1st edn. Pergamon Press, Oxford

Blake P (1991) Shahjahanabad The Sovereign City in Mughal India 1639–1739, 1st edn. Cambridge University Press, Cambridge

Farida N (2015) Hidden morphological order in an organic city. Protibesh 9:31–32

Hearn G (1928) The seven cities of Delhi, 2nd edn. Thacker, Spint & Co, Calcutta & Simla

Hutton T (2015) Routledge, 1st edn. Routledge, New York

Jain J (1975) Pramukh Jain Etihasik Purush aur Mahilayen, 1st edn. Bharatiya Jnanapith, India

Jain A (2018) Intangible communal traits: a mould for Tangible Urban landscape—case of Dharampura, Shahjahanabad. In Urban Futures 3. U.K, 14.06.2018. University of Gloucestershire, Gloucestershire, pp 7–16

Jain A (2019) Futuristic urbanism for organically grown neighbourhoods of Shahjahanabad: a case of Pura's of Shahjahanabad. In: National Seminar, Architecture for Masses—Ekistics. India, 22.2.2019. Jamia Milia Islamia University, Delhi

Jain K (2020) Morphostructure of an organic town: Ahmedabad. Environ Design J Islamic Environ Design Res Centre XI:32–28

Lawrence R (2012) Requalifying the built environment: challenges and responses, 1st edn. Hogrefe publishing, US

Macdonell A, Keith A (1995) Vedic index of names and subjects, 5th edn. Motilal Banarsidass Publishers, Delhi

Nilufar F (1997) The spatial and social structuring of local area in Dhaka City—a morphological study of the urban grid with reference to neighbourhood character within naturally grown areas. The Bartlett School of Graduate Studies, University College London, University of London

Wolfrum S, Brandis N (2015) Performative urbanism generating and designing urban space, 1st edn. Jovis Publishers, Berlin

Chapter 9
The Tree: The Concept of Place After the Earthquake, L'Aquila

Arianna Tanfoni

Abstract A tree is a vital source for the inhabitants of the forest: it offers a safe home. Then the tree gets sick and the animals run away, abandoning it. A place must be *place* before it is attractive and comfortable. The city of L'Aquila is like that tree. Due to the reconstruction as it was and where it was the inhabitants of L'Aquila stopped right before the tragedy and they found themselves in front of fake buildings just like the castles of Disneyland. The few examples of contemporary architecture in the city have been denied by a large portion of the citizens. The city fails to grow in parallel with the reconstruction of its historic centre. Over a decade after the earthquake the inhabitants are tired of waiting. They complain about missed opportunities and delays and no longer believe in life inside the city walls. For many years, they have been living an internal conflict suspended between despair and joy. Tourists, university students and foreigners who move to L'Aquila are affected by the social fervour coexisting with construction sites, the dust and the trucks, which remind locals of the strong identity that this city communicates. The reconstruction is having unexpected implications: it is directed not only to the inhabitants of the place, but to the world. If it is true that a city is made by its people it is also true that without the city, what would people do? A loud voice asks and therefore awaits the return of the historical heritage to give its life back. L'Aquila, unlike the tree has not lost its inhabitants, nor its identity because the concept of place has never disappeared. The resilience of L'Aquila is evident in its motto "immota manet" which means to remain deeply rooted in the earth.

Keywords Resilience · Identity · Concept of place · Cultural heritage · Contemporary architecture · Historic centre · L'Aquila · Italy

---

A. Tanfoni (✉)
Building Engineer and Architect, L'Aquila, Italy
e-mail: arianna.tanfoni@gmail.com

© The Author(s), under exclusive license to Springer Nature Switzerland AG 2021
F. F. Arefian et al. (eds.), *Historic Cities in the Face of Disasters*, The Urban Book Series,
https://doi.org/10.1007/978-3-030-77356-4_9

9.1 Introduction

> If a place can be defined as identity, relational, historical, a space that cannot be defined as identity, relational and historical will define a non-place.
>
> Augé (1993, p. 77)

What is a place and when does it stop being? A tree is a vital source for the inhabitants of the forest: it shelters from the sun and the rain, hides the nests among the leaves, offers a safe home. One day the tree gets sick and the animals run away, abandoning it to its fate. A place must be *place* before it is attractive and comfortable. The city of L'Aquila can be compared to a tree. L'Aquila can be compared to a tree not from urban structure perspective, instead from the organic point of view, with reference to the roots, the local identity, the inhabitants. As a tree offers a home to the animals, the city of L'Aquila was cross the centuries and still is the home of many inhabitants.

This chapter advocates the concept of place and therefore the concept of city as a community of inhabitants even before buildings has never failed in L'Aquila after the 2009 earthquake. The chapter starts with the historical review of the city foundation and its numerous rebirths following the devastating earthquakes. It then provides a local resident's observations, as an architect and a disaster-affected citizen, of attitudes the inhabitants had towards the 2009 disaster and its subsequent reconstruction. The chapter explores issues related to the concept of place, identity, resilience, rejection of modernity and internal conflict that the inhabitants have lived for many years, suspended between moments of hope and moments of disillusionment and abandonment.

9.2 The Tree

9.2.1 Short History of L'Aquila, the City of 99

L'Aquila in central Italy is a city very close to the Gran Sasso, the highest mountain in the Apennines. The city foundation was decided and strongly desired, according to legend, by 99 neighbouring villages—*castles*—that did not want other domination if not that of the crown. King Federico II allowed the inhabitants to build a new city, which would later become one of the most important ones in the kingdom. According to legend, L'Aquila was built in the shape of an anatomical heart, on the reverse design of the city of Jerusalem. The city was founded in 1254 and the design formed four quarters—*quarti*—and each of them included a number of localities—*locali* (Fig. 9.1). Each castle had the right to have a large *locale* (a single portion of the many locali) in proportion to the size of the castle itself. Part of the inhabitants of villages moved to their *locale* and built, in addition to their houses, also a square, a church and a fountain (Spagnesi and Properzi 1972). L'Aquila became the city of 99 with 99

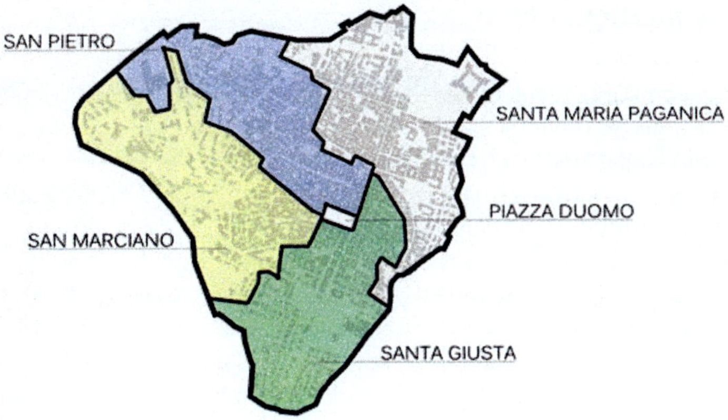

Fig. 9.1 Division of L'Aquila into quarters and the common square Piazza Duomo in the centre (author)

squares, 99 churches and 99 fountains with, in the heart of the city, Piazza Duomo: a large common square, a place for everyone. Since its first foundation, therefore, there has been a strong desire to aggregate and a bond with the territory, although history shows many episodes of problems and conflicts between the inhabitants of different villages within the city walls (Clementi 2011; Lopez 1988). With the construction of the Basilica of Collemaggio, outside the historic walls, and with the papal crowning of the hermit Celestino V in 1294, L'Aquila celebrated the first jubilee of history with the name of Perdonanza Celestiniana (Celestino V's forgiveness).

Destructive earthquakes over the centuries severely undermined the importance that the city reached during the Middle Ages in terms of prestige, culture and trade. The most significant catastrophes were recorded in 1315, 1461, 1703 and recently in 2009. The population responded to these events with attitudes of terror, of abandonment, but also of strong resilience and identity. In the case of the 1703 earthquake, for example, the marquis Marco Garofalo wrote: "*The city of L'Aquila was, it is not; the houses are joined in piles of stone, the buildings not fallen are falling. I don't know anything else that I can say more to consider a ruined city*", considering therefore the city as dead (AA.VV., 1999, Sulle ali dell'Aquila, p. 78). L'Aquila instead wanted to rise again: the destruction led to changing the medieval colours of the city from white and red to black for mourning and green for hope of rebirth, indicating a strong sense of internalisation of the tragedy and resilience. Although devastated by earthquakes over the centuries, the city has always been rebuilt on its own ruins and has never been abandoned or rebuilt elsewhere. The highly seismic territory and the cold climate have never hindered the city to live and have constituted its own strength instead. This has been also the case after the 2009 earthquake.

9.3 The Abandonment

The tree that metaphorically identifies the city of L'Aquila gets sick one day—on the day of the earthquake, 6/04/2009—the branches dry up and the animals run away abandoning it to its destiny.

Abandonment was one of the risks to the life of the city following the earthquake. In a world of possibilities—certainly very different from the ancient world—it is easy to move elsewhere and travel cheaply and in a short time, with the same currency in Europe and with a language common to the whole world, the risk for L'Aquila was depopulation, the death of the city and therefore a useless reconstruction, addressed to anyone. Indeed, data from the National Institute of Statistics of Italy (ISTAT) shows that population suffered a decline in 2011 (66,964 inhabitants: −7.50% compared to the number of inhabitants in 2001, that was 68,642). Recently, however, the data is recovering and increasing (+0.06% in 2018: 69,478 inhabitants; +0.33% in 2019: 69,710 inhabitants) (ISTAT 2020; Tuttitalia 2020). As in the previous earthquakes, at first the city was depopulated, but then it has been enriched by new inhabitants who arrive for study, work or other reasons. The demographic decline therefore was not as massive as feared, although the city suffered moments of strong abandonment, of university crisis and is still experiencing the issue of lengthy restoration of buildings in the historic centre which, however, remain uninhabited. It is also difficult to open or reopen commercial activity in a city subject to an intermittent recovery; therefore, the economy of the city is still struggling to restart.

Arguably, the abandonment that L'Aquila has suffered or still suffers from since the earthquake is of a different nature. It is an identity abandonment for some people as they no longer recognise the city; they have broken the bond with it. Some want to leave, others live in the suburbs, others do not want to go back to the historic city centre because they cannot accept that the city is no longer the same as they remember, and that the city has been completely destroyed not only in its buildings and public spaces, but also in its soul. By contrast, as it will be discussed later, the city also saw those who have kept the bond or those who are rebuilding the bond as the reconstruction progresses.

Another type of abandonment is from a civic point of view. Today L'Aquila suffers a loss of civic sense. In the historic centre, cars are parked everywhere, trash is abandoned in the streets and fountains, recycling is not done and the newly restored monuments are smeared (Fig. 9.2). There appears to be a total indifference to rules, to respect other citizens and to the city itself. Since the centre is constantly invaded by construction site trucks, everyone feels free to arrive anywhere by car, creating discomfort to pedestrians, businesses and residents who often find cars parked in front of their doors (Fig. 9.3). The complex question of the compatibility between the speed of cars and the calm of the historic centres is amplified in L'Aquila at a special time when the historic centre is starting to recover both from the point of view of trade and of identity, which over the years has undergone strong contractions.

The reason for these types of abandonment could be due to the inhabitants no longer recognise the city as their own, but as an ordinary place to which they do not

Fig. 9.2 An example of waste abandoned in the historic centre, L'Aquila (author 2019)

Fig. 9.3 Cars parked in the middle of a square in the historic centre (author 2019)

belong or as a place that they are tired of waiting for. On the one hand, this may be due to the trauma of the earthquake that has abruptly changed the city in road mobility in neighbourhoods and in its heart; on the other, perhaps to the people's non-participation in the reconstruction process which with its large scale as a regional capital.

9.4 The Reconstruction and Rejection of Contemporary Architecture in Historic Centre

The reconstruction of the historic city centre is necessary to reaffirm membership in the city and identity, not in the sense of equality, but of recognition as a separable, unique, individual entity (Lynch 1960). As many other historic cities, in L'Aquila urban life took place and takes place in the old town. The historic centre was the fulcrum of work and leisure activities; therefore, for the rebirth of the city, it was necessary to restore the historic centre and its enormous architectural heritage. Following the earthquake, a great interest in the heritage and history of the city developed. If for some citizens the link with the city was broken, for others it was strengthened. In fact, starting from the loss of the certainty of the city as their home, young people and adults began to spread the news about the history, legends, art and architectural heritage of the destroyed city, especially remembering the value that L'Aquila had during the Middle Ages and its division into four *quarti*. Many began to describe places in the city according to the *quarto* (quarti singular) they belong to. The recent recognition of Perdonanza Celestiniana—the first jubilee in the world celebrated in L'Aquila every year from 1294 in August 28 and 29—as an intangible heritage of humanity by United Nations Organization for Education, Science and Culture (UNESCO) in 2019 will increase this feeling of belonging and connection with the historic city.

The post-disaster priority was clearly the reconstruction of houses and restoration of the architectural heritage recognised as the identity of the city. Reconstruction policies and regulations aimed to restore heritage to the state of before the earthquake by building "*as it was and where it was*" with a strategy to restore buildings using the techniques of the past, trying to hide from the exterior the modern technologies of consolidation and strengthening structures of historic buildings. L'Aquila became the largest construction site in Europe that applies restoration techniques to its entire heritage, sometimes leading to cases of anastylosis. Figure 9.4 shows the typical skyline of the city during the current reconstruction activities that will predictably stay for many more years to come. Since the heritage of the city has been its identity, the psychology of the inhabitants after the earthquake led to the acceptance of a plaster with the colours of the earth hiding the thermal coat behind or stuccoes made with molds and not by hand as in the past (Fig. 9.5).

Yet, the risk of a such strategy is uniformity and spectacularisation, operations that stiffen the city landscape as such historic buildings become virtual objects for

Fig. 9.4 Characteristic skyline of L'Aquila with the construction cranes (author 2019)

Fig. 9.5 Construction site of a building in the historic centre (author 2019)

tourists who look at them only for a few moments and instead do not awaken their awareness of time, of history. Like the Parthenon as an unfaithful memory in which multiple confused past plunges, always new, condemned to survive the projections it arouses, an intimate obsession and heritage of humanity (Augé 2004). History shows instead that great tragedies have also created a field of innovation and redesign of cities with the current styles of the time. If after the earthquake of 1703 they had rebuilt the city "*as it was and where it was*" and if, before that, after the earthquake of 1461 they had done the same and so on back in time today L'Aquila, like other cities devastated by earthquakes, would perhaps be a city of caves and prehistoric houses. The fury reconstruction can be a form of violence, while construction sites represent uncertainty, not the present and not even the past, but a poetic incompleteness (Augé 2004).

The approach to reconstruct "*as it was and where it was*" led to highlighting a common problem throughout Italy that other countries might not have: the rejection of contemporary architecture in historic centres. Addressing this national tendency, the architecture critic Luigi Prestinenza Puglisi states that "*we are not educated in the contemporary world*" (interview for the Italian student magazine "124"). On one hand, the mistake was perhaps to give enormous economic importance to tourism, so each city aims to keep its historical heritage intact in order to be included in the UNESCO World Heritage List. On the other hand, many other authors continue to argue that the historical city is clearly distinct from the new or contemporary city, in the same way as the ancient thought considering it as a unitary organism and enclosed it within a very precise and distinct design of the walls (De Fusco 1999). To them, not only heritage is untouchable as a historical memory and identity, but even the 70s condominiums are considered normal in the common sense and reassuring. Yet on the contrary, a building constructed today, with the architectural language of the current century and generation of architects, is rejected and refused. In the historic centre, there is no space left for contemporary architecture; even a fake or anachronistic style is preferred.

In L'Aquila probably the traumatic consequence of the earthquake has amplified this behaviour. All that is not heritage and therefore not recognised as an identity is refused. The historic centre is therefore stopped at the moment just before the tragedy and tries to restore the condition to the state of many years ago; it wants back what was violently taken away, leading to a paradoxical reconstruction with very little new in an ever-changing society.

For example, great criticisms have been directed towards two buildings of contemporary architecture built in the historic centre after the earthquake. One is the Auditorium del Parco completed in 2012 in a park at the edge of the historic centre (Fig. 9.6). It has been criticised for its building material, shape and contest, the removal of some trees for its construction and finally for its little capacity. Some years after its construction some people still cannot accept it. The other one goes back to the pre-disaster time. It was a 70s reinforced concrete building in Piazza Santa Maria Paganica, almost completely for residential use, and therefore not generally considered as architectural heritage. It was demolished and rebuilt in complete adjacency to the ancient Palazzo Ardinghelli (Fig. 9.7). In this case also the criticisms

Fig. 9.6 Auditorium del Parco in the historic centre (author 2016)

Fig. 9.7 Residential building near to Palazzo Ardinghelli in Piazza Santa Maria Paganica (author 2019)

were very negative: some said that it required putting cantonal stones, others that it required to be painted as the ancient building, others that it would have been better to build it in suburbs.

In normal situations and perhaps in other non-Italian cities, the process of modernisation is gradual, and the introduction of modern contemporary architecture is not subjected to such severe criticisms. Citizens can slowly internalise change and perceive it in parallel with other changes in their lives and the passage of time. In fact, even in L'Aquila, before the earthquake, there was already a modern building

Fig. 9.8 (Left) Modern building close to Piazza Duomo (author 2019). (Right) University of Human Science in Piazza San Basilio (author 2019)

with a glass facade, a few steps far from Piazza Duomo (Fig. 9.8, left). Also, the University of Human Sciences in Piazza San Basilio (Fig. 9.8, right) was also under construction with seismic isolators, an irregular shape and a facade very far from the classic plaster. In these cases, the criticisms were not particularly severe, perhaps because the transformation process was progressing slowly.

The present generation must be able to design cities with pride and not with fear of the past. This generation has the duty to reconstruct and protect historic heritage as its memory and identity but can also build something new and of its own. The reconstruction of L'Aquila must allow the present to manifest itself and to leave a sign of this era in the urban landscape, and as a sign of the earthquake instead of deleting it. It is necessary to have the courage to face the theme of the recovery of some buildings that cannot be restored due to the great extent of the damage. Architects, urbanists and academics have the role of educating people about heritage and providing tools for them to recognise and appreciate heritage and to understand that ancient does not necessarily mean heritage.

Some other opportunities also have certainly been lost in the reconstruction of L'Aquila. For example, there has not been urban plans to regulate reconstruction in parallel with the growth of the city and its urban development. For over a decade, L'Aquila has not been a competitive city not in Europe nor in Italy because it is completely occupied with reconstruction. The city did not find the time or the way, to grow and focus on its development opportunities that could also have led to redesigning some urban aspects for the city of the future. In order to be liveable, a city, as a *place*, must be attractive too.

Today, L'Aquila is perhaps the safest seismically and most energy-efficient city in Italy. Its enormous heritage is being restored and the city is revealing its wonders to the world and to itself, enriching them with new technologies. An example is the Smart Tunnel, an underground structure in the historic centre that houses the gas, electricity and water services, communications and internet infrastructures that are examined in other chapters of this book. The reconstruction and such projects may have unexpected effects with a wide resonance because they are directed not only to the inhabitants, but to the world. L'Aquila may become again a city able to attract tourists, students, professions, new activities and new inhabitants, perhaps not as before the earthquake but with a greater force.

Over a decade after the earthquake there recently started a new fervour in the city and an air of change also in the mentality and in the approach to modernity. As the reconstruction progresses, the city has begun to have other needs and starts asking for more. Finally, there is a small portion of a completely rebuilt area, ready for normal life and the future prosperity and growth, and as one can observe the needs are changing: from the absolute and primary necessity of houses, the priority is now societal needs and social life outside of buildings.

If, on the one hand, the reconstruction is trying to cancel the pedagogical vocation of the ruins and the passage of time (Augé 2004), on the other hand, it is necessary to settle again membership to the city, to restore the link with it and identity and it is essential to restore life. Many years after the earthquake and with the reconstruction process at a good point L'Aquila perhaps feels safer, quicker and stronger: it is time to project the city to the future, to design the city for the next generations because it has great potential for economic and social revival.

9.5 The Concept of Place

Here L'Aquila is metaphorically identified as a tree not in the sense of the abstract structure of the city which, also thanks to the designers and planners, merely imitate the appearance of the ancient and which tries to simplify, for the desire for neatness, just as human thought tries to schematise complex concepts. With regard to that abstract structure, the city of L'Aquila is in fact a semilattice, a natural city—even if its formal structure was partially designed at the time of its foundation—in which each node is linked to the other through multiple and non-univocal and very complex structures (Alexander 2015).

As mentioned, L'Aquila is comparable to a tree in the sense of its roots and relationship with its inhabitants. After the 2009 earthquake, the city is facing a double approach among its population: there are those who have broken ties with the city and those who have kept them or are rebuilding them, even with difficulty and slowly. L'Aquila experienced some sad moments due to the breakdown of the social tissue and dislocation of houses in the news town in the suburbs called progetto C.A.S.E.,[1] offices, shops and strategic points, all that makes a regional capital. The first years after the earthquake have strongly marked the city from a social and relational point of view. This condition may be considered as normal to some extent following a disaster. The state of emergency in the night of the earthquake was total. Probably, as happened after the earthquake of 1703, on the morning of 6 April 2009 someone could easily have said that L'Aquila was a dead city.

[1] Complessi Antisismici Sostenibili Ecocompatibili (C.A.S.E.) means Eco-friendly Sustainable Seismic Complexes.

L'Aquila has been destroyed by earthquakes and has experienced episodes of depopulation several times. However, this phenomenon has always been short-lived, ending with an increase in the population, strengthening the resilience of the city as a *place*, identity, relationship and historical, like the metaphor of the tree. A decade after the 2009 earthquake also the ISTAT data demonstrates an increase in population. As Augè (1993) states *place* is what can be recognised as unique, born from a specific context and with historical–sociological characteristics strongly rooted in the territory. On the contrary, it is a *non-place*. In L'Aquila, the concept of *place* has never failed, despite several destructive disasters that have marked its history. The city has never become a *non-place*. It has retained its identity, relational and historical characters. When allowed, some people in fact tried to continue living in a few accessible places in the historic centre, making space between the construction sites and the dust, studying the history and rediscovering forgotten points of the historical centre. An important example is the reopening of a simple commercial activity, a wine bar, few months after the 2009 earthquake, which had the participation of many citizens who wanted to take back the historic centre, albeit in a very small part. At that time only a small square was accessible, while the entire historic centre was closed and fenced. The reopening of that wine bar played an important role in social reconnection because for the first time after the earthquake people have returned to meet and to a have a place for doing it.

Despite the abandonment of the city by some people as discussed before, the great strength of the city lies in the concept of *place*, which is deeply connected not only to the geographic space and the territory, but to the communal identity and strength of those inhabitants who keep it, passing it down over time, like the sap of the tree. The reconstruction process has been acting as a backdrop to the lives of the inhabitants (Fig. 9.9). Those who arrive in L'Aquila, a decade after the earthquake, perceive the strong identity that embraces the city. The story of a young man, during an interview in 2018 for example, who recently moved to L'Aquila for work reasons, explains it. Although he had lived in Rome and abroad, he had never been to a city where he

Fig. 9.9 A sunday within construction sites in the historic centre (author 2016)

could go to a pub and start to talk to strangers with extreme normality. The concept of resilience is strictly connected to the concept of place. As a place, a resilient city is a city that wants to live despite abandonment and catastrophes because of its inhabitants and the strong concept of identity.

L'Aquila is an example of a city made up of people even before buildings although buildings have been the priority in the reconstruction. Here the concept of place and the concept of city are much stronger than buildings, almost as if space and sociality existed even without architecture or in any case in the expectation of architecture.

9.6 Immota Manet as a Motto of Resilience

In the coat of arms of the city of L'Aquila—which name means *water*, from the original ancient settlement of Aquili, but also the bird *eagle*—two words appear that explain much of the history and strength of this territory. The motto that the city has chosen is taken from the Georgics of the Latin poet Virgil by the humanist Salvatore Massonio and is "*immota manet*".

In the most common and best known sense in the city "*immota manet*" means "*that remains still, that nothing new can happen*" and it refers to the attitude of the inhabitants who not accept the new, showing a typical character of mountain cities and of cold climates.

But perhaps not many know the original meaning of the motto. "*Immota manet*" essentially means—in Massonio's interpretation—that remains deeply anchored to the ground, to identity, to origins, to history, like the roots of a tree.

The city of L'Aquila has always been rebuilt after the many earthquakes because the community felt the need to continue living here. Over the centuries, population did not want to rebuild their homes and lives elsewhere, but wanted to remain in this rough territory. This is the resilience of L'Aquila, a city made up of people even before buildings. L'Aquila is a tree with very deep roots: a *place*.

9.7 Conclusion

It is particularly complex to investigate the response of a city to a disaster a few years after it happened from the perspective of a disaster-affected inhabitant and practitioner architect concerned about the city. There is ambivalence in the thought of the inhabitants who have lived and still live an inner conflict and mixed feelings towards their city. The slow reconstruction process, although it may seem tortuous to the inhabitants, provokes a rapid and constantly changing from social, urban design and urban transformation points of view. From the time of the tragedy to the beginning of the reconstruction until the tenth anniversary of the earthquake, the city has greatly changed.

The case of L'Aquila after the 2009 earthquake offers the possibility of investigating, perhaps with a greater distance in time, the psychological and social aspects of the tragedy, the beginning of reconstruction and the advancement of it, highlighting the theme of resilience and of the concept of place that characterises the identity of peoples. Despite the hard times during the last decade, it's evident how in the last decade population shown great resilience and attachment to the city, as to their *tree*, demonstrating that, centuries after its foundation, the resilient city exists because of its strong identity.

Immota manet, *despite the many earthquakes that the city has suffered.*

References

Alexander C (2015) A city is not a tree, 50th anniv. Sustais Press, Portland
Augé M (1993) Non-places. Elèuthera, Milano
Augé M (2004) Rovine e macerie. Il senso del tempo. Bollati Boringhieri, Torino
Clementi A (2011) Storia dell'Aquila dalle origini alla prima Guerra mondiale. Laterza, Bari
Dati ISTAT (2020) Available at: www.dati.istat.it. Accessed: 2 Aug 2019
De Fusco R (1999) Dov'era ma non com'era. Il patrimonio architettonico e l'occupazione. Alinea, Firenze
Lopez L (1988) L'Aquila—le memorie i monumenti il dialetto. G. Tazzi, L'Aquila
Lynch K (1960) The image of the city. MIT Press, Cambridge
Spagnesi G, Properzi PL (1972) L'Aquila—problemi di forma e storia della città. Dedalo, Roma
Tuttitalia.it (2020) Available at: www.tuttitalia.it. Accessed: 2 Aug 2019

Part II
Historic and Contemporary Reconstructions of Historic Cities

Chapter 10
Marsica: One Hundred Years on

Simonetta Ciranna

Abstract This chapter revisits reconstruction after the disastrous earthquake in 1915 that devastated many small and medium towns in Marsica, Italy. The historical, political and geographical context of the Marsica reconstruction, the premises and diverse ways in which it was tackled, and the outcome of the reclamation, extension or requalification of the urban structure and building typology of its "resurrected" town centres, constitute a reference point in the history of earthquakes and reconstruction policies in Italy. They also portray the background of present-day situations and problems. Such questions concern to the fullest extent the resistance and resilience of the urban centre and its social and historic collective heritage, particularly where the consistency and identity of buildings are also the outcome of reconstruction processes after repeated destructive seismic events, as in the case of Marsica, many other areas of Italy and other countries.

Keywords Marsica · Earthquake · Reconstruction · Avezzano · Italy

10.1 Introduction

The 1915 earthquake struck Marsica, and in particular—but not solely—Avezzano, its largest town, causing over 30,000 victims and practically razing entire inhabited centres, coming just a few years after the one that destroyed Messina and Reggio Calabria in 1908. The destruction of residential buildings and monumental heritage as well as the economic and social system took place at a time of economic and social growth and of debate about town planning and urban building development. A debate that arose in the context of the most important and celebrated nineteenth-century transformations of major cities and capitals of Europe and, in Italy, starting from the unification and with the renewed energy of the first decade of the twentieth century. Those discussions concerned the ability of ancient town centres to adapt to

S. Ciranna (✉)
University of L'Aquila, L'Aquila, Italy
e-mail: simonetta.ciranna@univaq.it

© The Author(s), under exclusive license to Springer Nature Switzerland AG 2021

F. F. Arefian et al. (eds.), *Historic Cities in the Face of Disasters*, The Urban Book Series,
https://doi.org/10.1007/978-3-030-77356-4_10

the requirements and aspirations of the modern age, of economic and social progress imposing a rapid upgrading of urban infrastructure and fabric.

Such ambitions and choices were hindered by the scarce economic resources of local townships and, with the earthquake, by the drama of the emergency and the urgency of responding to the primary needs of survivors. In May of the same year, the emergency was aggravated by Italy's participation in the World War I.

Reconstruction consequently proceeded slowly, accelerating only during the late 1920s under the drive of the Fascist government. It was marked by many different solutions to urban reconstruction and to the preservation of historical heritage. With regard to the former, it ranged from the total demolition of the ancient centre of Avezzano and its reconstruction on the same site with the implementation of a vast adjacent expansion, to total or partial abandoning of urban centres. In this last case, as, e.g., in Aielli, their reconstruction happens in safer locations, even involving expansion plans in line with the landscaping ideas and perceptions of Camillo Sitte, like the architect Gustavo Giovanni's proposals for Celano (Ciranna and Montuori 2017). With regard to the latter, the architectural heritage was either abandoned or sacrificed, with the moving and recomposing of architectural elements deemed visually important, or else "restored".

The Marsica earthquake also played a major role in accelerating the formulation of seismic regulations suited to the related territories. To enhance constructional safety, new materials and building methods were adopted—in particular the use of reinforced concrete—by both private and public sectors, whether residential or not, and for works on the historical and monumental heritage (Montuori 2019; Fiorani 2004–2007).

The questions faced and solutions adopted in the post-1915 reconstruction continually reoccur, albeit in different historical, geographical and cultural contexts, whether to be after a war or natural hazard. Damage and the need for recovery inevitably give rise to debates about the role and relevance of expertise on historical buildings, the preservation of their importance and tangible and intangible value and about the sometimes radical desire to replace them with contemporary models.

10.2 Rebuilding: Where and How?

Given the above historical background of the 1915 earthquake, urban reconstruction was oriented according to the extent of the areas destroyed and the economic importance of the small and medium towns of the Marsica region.

For the Avezzano City, however, decisions were made to completely demolish remainders of its original centre and to rebuild it with a radical change of road network and buildings. This choice involved the obliteration of the central location of the church of San Bartolomeo that previously had morphological, civic and religious importance. Recent studies showed that such decisions were influenced by the nineteenth-century urban layout created by the engineers Sebastiano Bultrini and

Loreto Orlandi prior to the seism, which tragic event in some ways encouraged a more radical application (Ciranna and Montuori 2015).

The original general plan, finally drawn up between 1906 and 1912, proposed a considerable expansion and major interventions on the old centre towards its "modernisation". The expansion, adjacent to the old centre, was planned in a north-north-west direction, towards the railway station, as being a commercially and hygienically more suitable area, where the designers created a grid system cut from some diagonal roads, low council-house type dwellings, apartment blocks and small houses. The "modernisation" of the old centre's road network and original buildings also involved the opening of new main roads on a "trident" pattern focusing on the Orsini-Colonna castle. The plan was slow in being adopted and became enforceable only in October 1916 after the earthquake and applied starting from the expansion. The expansion, in fact, in the post-earthquake situation was indeed less complicated in attributing land ownership, which had become dramatically unreadable in the original old centre. The expansion's infrastructure swallowed the old centre, rendering it totally unrecognisable (Fig. 10.1).

Inside the expansion area, immediately after the World War I, a considerable complex of economic and council housing was built between 1920 and 1924 in the area of the former estate named "Chiusa Mazzenga" by the Nationale Building Union (Unione Edilizia Nazionale, U.E.N.). It was an intermediate institution created by decree in 1917 as the germination of the Unione Messinese (1910), and later, the Unione Edilizia Messinese (1914) which, after the Messina earthquake in 1908 had, by law, assumed a considerable role in the reconstruction of economic and council housing in stricken areas (Unione Edilizia Nazionale 1921). In Avezzano and Marsica the U.E.N. also followed the realisation of much private residential property, consisting of apartment blocks and small houses in the historicistic or liberty style. Only later were sites identified for the new Town Hall (Fig. 10.2) and present Cathedral.

The town plan for the old centre was actually adopted only from August 1925, substantially preserving the outer roads that followed the layout of the already-demolished town walls and perimeter elements, such as the castle and Piazza Torlonia, hinge and contiguity points with the main streets of the new expansion, while "wide streets and suitable building plots" replaced the former dense urban fabric. This latter system consisted of streets, squares, long narrow blocks (comprising a double row of terrace houses, separated by a narrow central space), properties and public areas featured by the historical Avezzano with its urban morphology and "herring-bone" building typology (Fig. 10.3), common to many towns in central and southern Italy (Ciranna 2015, p. 41). In this arrangement, the church of San Bartolomeo was located at the crossing of two ancient major inter-regional routes, which can be traced back to Roman centuration and the spot where a small village germinated and grew, forming the little town of Avezzano (Saladino, 2015).

Villages of Marsica also seriously suffered from both the earthquake and radical and often tardy post-earthquake reconstruction. In villages, the first intervention was the erection of camps for emergency accommodation and almost simultaneously, building cheap permanent shelters using solid concrete blocks with rural and

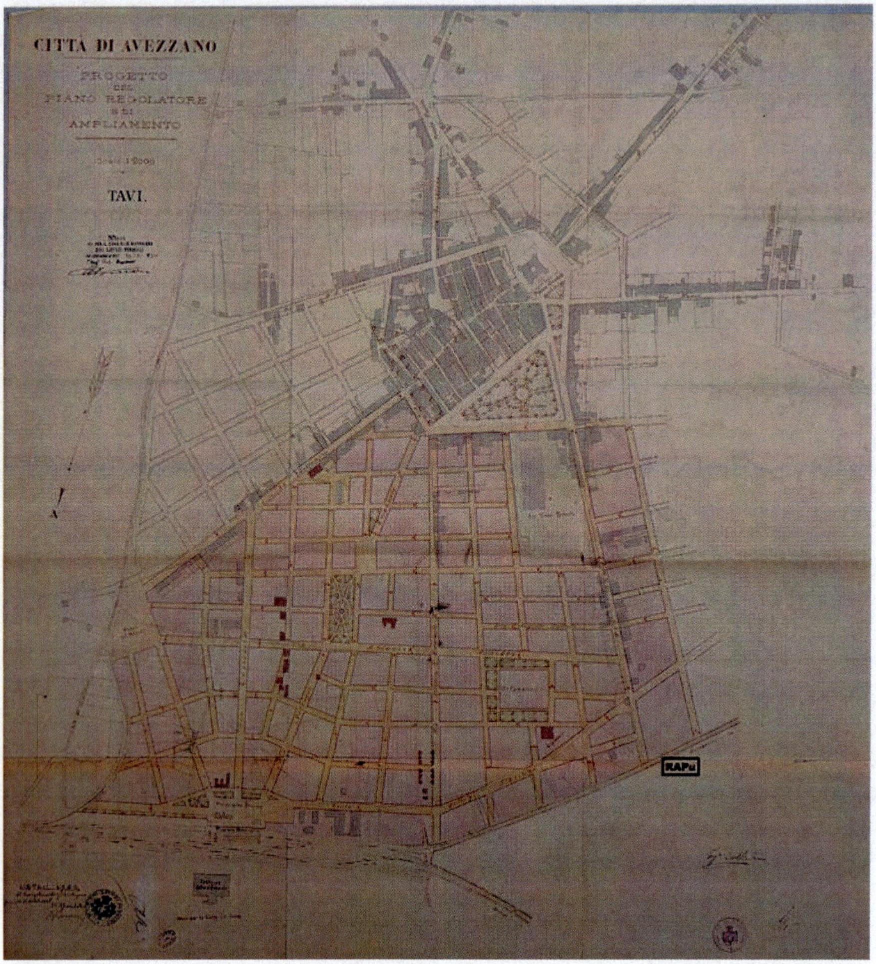

Fig. 10.1 Sebastiano Bultrini, The town plan of Avezzano of 13 October 1916 (RAPu, Rete Archivi Piani urbanistici)

inexpensive characteristics. Villages included Aielli, Celano, Paterno, Pescina, San Benedetto, Collarmele, Ortucchio, Cappelle dei Marsi, Magliano dei Marsi and Le Cese, as well as in the Cicolano and Sorano areas (Serafini 2016). Subsequent reconstruction of these villages and Avezzano, and the Abruzzo villages at the foot of the Maiella struck by another earthquake in 1933, launched by the Fascist government involved either total or partial abandonment of the old urban centre, a well-known and universal phenomenon in both ancient and recent history. This replanning largely favoured the construction of dwellings, followed by schools and administrative buildings, channelling particularly to the former a large part of the scarce funds available, penalising the production sector so fundamental to recovery.

Fig. 10.2 Sebastiano Bultrini, Main elevation of the Town Hall in the 1920 project (Historical Archive of Genio Civile Regionale of Avezzano, Collection Orlandi)

Fig. 10.3 Urban structure of Avezzano in a planimetry dated 26 September 1889 performed by the Italian Society for Water Pipelines of Rome for the project of the city's water network (Historical Archive of Municipality of Avezzano, water network, b. 14, f. 1/4)

One hundred years later, the decisions made and problems faced in the reconstruction of Marsica are still relevant and topical in reconstruction related debates in Italy, for example, the 2009 earthquake that devastated L'Aquila in Abruzzo and its reconstruction as discussed in-depth in other chapters.

In Marsica, the planning of new settlements nearby, or even adjacent to the old urban centres, and the construction of the emergency accommodations or earthquake-proof bungalows, launched growth and subsequent developments such as the so-called new towns, whereas the temporary dwelling units used around L'Aquila now affect the efficiency, economy and life of the inhabitants, thus weighing down both the present and future of the city and its territory.

In the urban model adopted by Bultrini at Avezzano, the number of plots were increased; terraced-housing—now private property—was built on the site of the early post-earthquake accommodation, while the great square, destined as the location of the new cathedral, still in 2013 has been the theme of projects and requalification, aimed at providing greater urban and civic recognisability.

Just as relevant to the subsequent reconstruction and urban development were the early initiatives taken in smaller centres and villages with severe impacts.

For example, in the village of Aielli there was a detachment between the old centre and new centre. The old centre was entrenched on its rocky spur about 1000 m high of which some areas were redesigned and rebuilt (Aielli alto), while the new centre was built on lower ground close to the station (Aielli stazione). This was already projected in the pre-earthquake village plan to expand towards the railway station. An area on ground lower than the old centre was selected for the new Aielli, close to the railway line between Rome and Pescara. Like in Avezzano, here too reconstruction accelerated the developmental transformation already projected and started at the end of the nineteenth century.

Besides guaranteeing greater safety, the different location also offered the prospect of economic development, owing to the drainage of the Fucino and the acceleration of trading with nearby towns and larger cities, determined by the recent improvements to the road network and the completion of the railway line.

Unlike Avezzano, with its nineteenth-century urban model, the little town of Aielli Stazione rose according to the rationality of the new Fascist foundations, from rural villages to the towns of the great land reclamation projects. Low asymmetric cottages were regularly spaced along short parallel roads merging onto a slightly lower area, which became Piazzale IX Maggio (Square IX May), conceived as the civic and religious centre. Indeed here were built the war memorial and the Casa Littoria (Fascism Party Headquarter) with its recreational annex, cinema, café and public baths, and the church dedicated to Sant'Adolfo (now San Giuseppe), with an ample parvis-panoramic viewpoint overlooking the plain of the Fucino. Fine materials, finishing and other architectural characteristics of the church reflects the personal commitment and contribution of the chief magistrate of Novara, Guido Letta—a native of Aielli—and the engagement of artists and qualified workforces. Designed with an earthquake-resistant structure by the engineer Giuseppe Peverelli and the architect Luigi Buffa, both of Turin, and built between December 1936 and September 1937 by S.A. Cantieri Ettore Benini of Forlì, the church also benefitted from the collaboration of institutes, artists and craftsmen such as the Regia Accademia Belle Arti of Carrara, the sculptor Arturo Dazzi and the Chini factory of Borgo S. Lorenzo (Florence). Its high façade irrupts with features of a simplified mediaeval style, reinterpreted and revised by the strong chromatic contrast of Pompeian red for the tower and porticoes, a decision probably reflecting the still-open quarrel concerning the construction of the new cathedral in Avezzano.

The long and arduous completion of Avezzano new cathedral in 1943 (later continued owing to war damage) shows significant expectations from such symbolic buildings in the reconstruction of the most important town in the Marsica region. Such expectations and approach resonate also with other reconstruction cases whether to be in Avezzano or L'Aquila after 2009 earthquake, for example, the completed works on S. Maria di Collemaggio and the still-undecided works on S. Maria di Paganica (Bartolomucci and Ciranna 2019), as also of the various small churches and town halls restored or rebuilt in smaller disaster-hit towns.

In Avezzano, the main difficulties for the construction of its church arose over finding appropriate solution for the most suitable earthquake-resistant foundations in view of the marshy soil of the only site "left" for the purpose. The debate involved the structural engineer Rodolfo Stoelcker and the designer Bultrini. The former promoted a modern and costlier foundation mat with concrete piles, and the latter clung to a more traditional system of drainage and slabs. Although Bultrini won this debate, ferocious criticism arose over the fourteenth-century stylistic architectural features particularly of the façade (Fig. 10.4).

Bultrini's solution also met with pitiless criticism from the engineer Gustavo Giovannoni who even called for a halt to the construction works of an "enormous building that, despite its exaggerated dimensions, will seem tiny, lost in the vast desolate square", while the square itself was conceived "in a 'homemade' mediaeval design, without criteria, without sentiment, matching none of the intrinsic or extrinsic conditions" (Giovannoni 1928, p. 429). This bitter rebuke was repeated the following year, this time referring to the churches of two other important urban centres in Abruzzo: Magliano dei Marsi (Marsica) and Pescasseroli (Abruzzo National Park) over which, he wrote, "looms the threat of radical transformations" (Giovannoni 1929, p. 471). These criticisms led to simplifying the figurative elements in final designs adopted in the church of Avezzano (and not only), which continued, however, to favour styles drawn from the mediaeval repertory of central Italy.

Giovannoni's criticism of the reconstruction of the towns and villages, damaged both by earthquakes (Messina, 1908 and Marsica, 1915) and by war, concerned the architectural discourse, which did not comply with the new structural systems, the local traditional styles or morphologies of new urban centres. With regard to the latter, Giovannoni drew on his expertise on theoretical debates and solutions made in France, England and Germany that were endorsed by the ideas of Camillo Sitte, Hermann Josef Stübben and Charles Buls also, his local knowledge of Abruzzo and his practical involvement in the reconstruction of the Marsica. In particular, he criticised the adoption of a grid-type morphology and the consequent uniformity and monotony of building blocks and introduced "a picturesque sense of the new towns, enhancing views of both nature and monuments, studying traffic directions

Fig. 10.4 Sebastiano Bultrini, Avezzano Project of the Cathedral of S. Bartolomeo, main elevation (Archive of the Diocesi dei Marsi drawings)

and open spaces not as geometrical figures, but as varied and living groupings" (Giovannoni 1913, p. 457). He aimed to preserve the individual character of each town or district, with recourse to gardens and monuments to break uniformity and minimise the closed-space feelings in public squares.

For villages of Abruzzo at the time, Giovannoni recommended simplicity and essentiality to match their nature. He, therefore, deems that, rather than insignificant and vulgar uniformity towns and villages should have a road network that follows the area's orography, with different groupings of houses, to enhance intimacy and spatial quality of secluded squares, like those old traditional Italian towns and villages, which utilised vegetation to mediate with the natural landscape. Such an achievement would give rise to "that lively and wholly Italian sense of art": The character of free building aesthetics cannot have precise rules, but must adapt to the natural context and environment, and every house must have its dignity, even in its simplicity (Ciranna and Montuori 2017).

Using his "romantic" conception of the Abruzzo as predominantly rural and his suggestion of urban planning to integrate contemporary values with those of history and the environment, Giovannoni prepared a plan for a small village close to the municipality of Celano and, in 1917, presenting building typologies and morphological reports (Figs. 10.5 and 10.6).

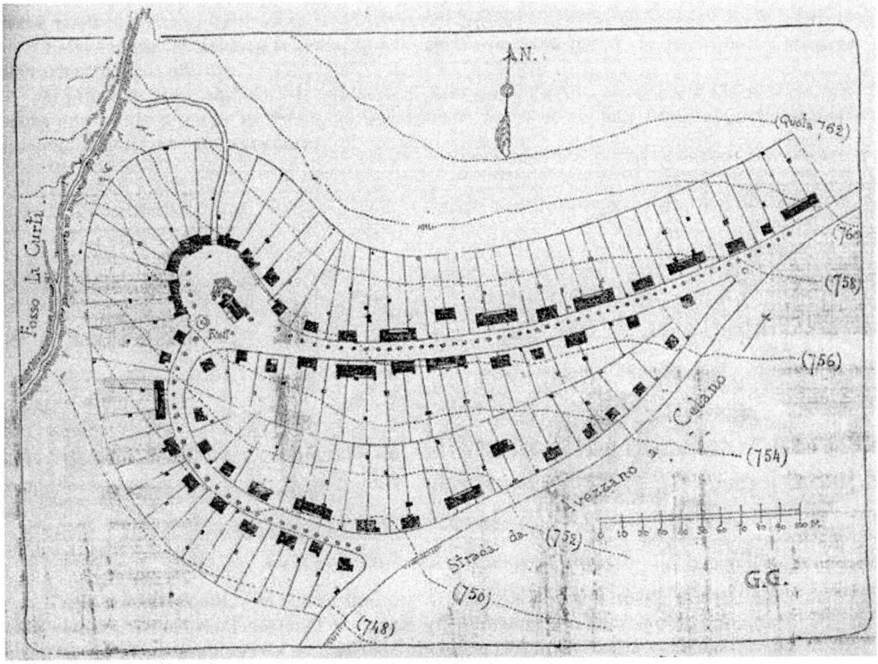

Fig. 10.5 Gustavo Giovannoni, Project for a village to be built in a place near Celano, L'Aquila. General plan (Centro di Studi per la Storia dell'Architettura, Collection Giovannoni, b. C 5, 76–b. C 2, 66)

Fig. 10.6 Gustavo Giovannoni, Project for a village to be built in a place near Celano, L'Aquila. Perspective sketches of the houses (Centro di Studi per la Storia dell'Architettura, Collection Giovannoni, b. C 5, 76–b. C 2, 66)

In Giovannoni's proposal, single-storey dwellings were raised above ground level, for simplicity and constructional safety against earthquake events, with steeply-sloping roofs because of the cold and snowfall, their finishing and architectural style guided by a criterion of simplicity. Thus, he emphasises that, unlike the excessive use of cornices, mouldings and stucco embellishments of recent reconstructions, "preference has been given to a very modest aesthetic effect, in harmony with function and the rural character, to a movement of the mass, the grouping of single features, such as projecting eaves, roofs and chimney stacks. Such essentially constructional lines may then be enlivened with appropriately chosen colours for the outside walls, varying from house to house, and with climbing vegetation and flowering plants" (Giovannoni 1917, p. 7).

10.3 Building and Restoration: Materials and Technology

Giovannoni's writings also reveal his readiness to use a reinforced concrete structure rather than traditional masonry, which often had a poor quality in Abruzzo. However, as had already occurred after the Messina earthquake, in Marsica too, the adoption of reinforced concrete was partial, with little figurative autonomy (Montuori 2019). Although the utilisation of a self-standing frame structure only became widespread after the *Regio Decreto* (Royal Decree) of 1935, it appears that economic difficulties and the lack of qualified labour caused by the World War I favoured the continuation of tradition and influenced a preference for masonry reinforced with iron rods. The use of reinforced concrete during the Marsica reconstruction was appreciated and used, albeit limited for economic reasons; it also extended to a wider scope, restoration works on the historical and monumental heritage (Montuori 2017).

A meaningful example is found in the already-mentioned activity of the U.E.N. which brought to Avezzano the experience and solutions already proven in Messina for council housing: Buildings were raised on two floors on a frame with many walls, mainly of brick masonry strengthened with reinforced concrete. They were five types variously combined with an extremely simple planimetric distribution.

The use of cement in private buildings, for which the U.E.N. played an important role in managing state funding and administration, was limited. They also used brick masonry or stonework with a structural frame or only ribs of reinforced concrete and figurative stylistic features borrowed from the historical repertory or, at the most, a tendency towards liberty style.

Although the situation in public buildings was not very different, several types of experimentation also emerged. An example is the experimental projects for standard nursery schools connected with the work of Giovannoni who designed them on behalf of a charitable institution for Avezzano, Capistrello, Luco dei Marsi, Trasacco, Aielli, Scurcola Marsicana, Massa D'Albe and Isola Liri and Fontana Liri, the latter in the Frosinone area (Ciranna and Montuori 2016) (Fig. 10.7). In these cases, although the technical commission initially recommended that reinforced concrete should not be used, due to the difficulties of getting supplies to the sites, after a careful study

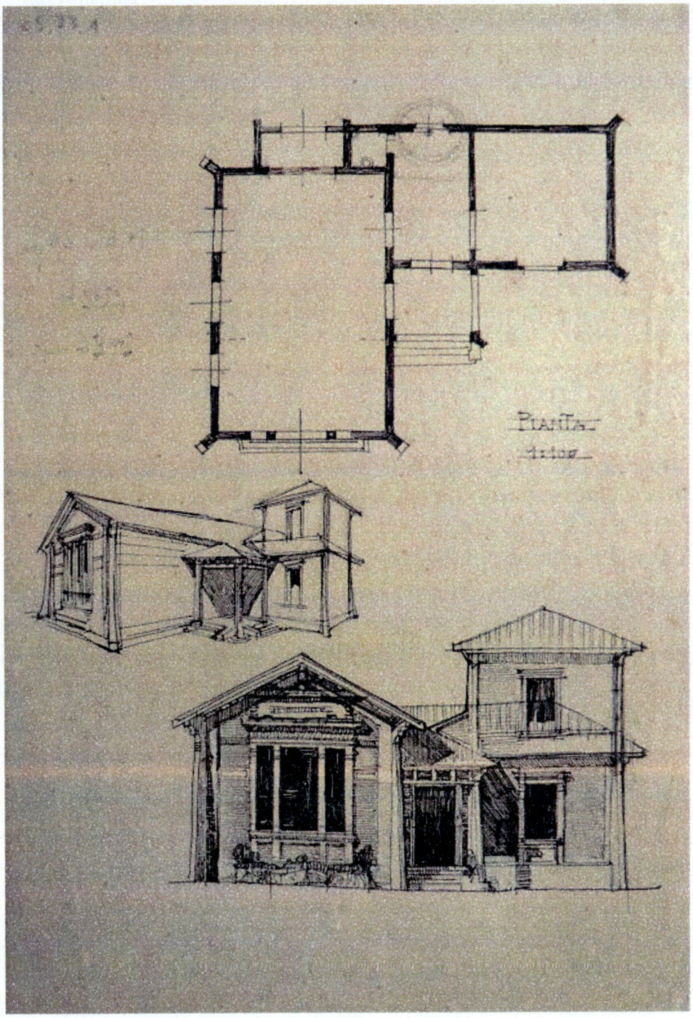

Fig. 10.7 (Left) Gustavo Giovannoni, Project for standard nursery schools in Abruzzo, s.d. (Centro di Studi per la Storia dell'Architettura, Collection Giovannoni, C5, 73)

of aseismic systems the reinforced concrete blocks were utilised. This was applied to all aforementioned nursery schools, except for those in Capistrello, Isola Liri and Fontana Liri where a reinforced concrete structure was utilised with ordinary curtain walls. Despite their structural and typological simplicity, those little nursery schools proved a significant attempt for the standardisation and prefabrication of buildings suited to post-disaster emergencies.

Another example is the use of new technologies for the foundations of the cathedral in Avezzano as of engineer Stoelcker's proposal. He suggested using a system of concrete piles, with which he was not only familiar, but also possessed the necessary tools for its application (first and foremost the pile-driver) but, as previously mentioned, against him was the project manager and the need to minimise costs. Apart from the foundations, the new cathedral's structure was designed with modern and aseismic bearing structure, achieved by "a reinforced concrete framework and iron beams". In fact, it was found cheaper to utilise a reinforced concrete structure up to a certain height, i.e. up to the impost of the arches of the nave, after which it was deemed necessary to continue with a framework of iron girders.

Besides Giovannoni, who did not exclude its use if the appearance of the fabric was unchanged, i.e. as a "hidden method" to create a "resistant framework", another fundamental promoter of reinforced concrete was Ignazio Gavini, who in 1923, as a member of the Superintendence of Monuments and Galleries of L'Aquila (Soprintendenza ai Monumenti e Gallerie of L'Aquila), expressed full faith in its employment: "reinforced concrete can facilitate the solution of innumerable static problems, especially in earthquake zones, where Italian monuments have suffered most damage. Our poor monuments, still marked by traces of multiple catastrophes, so often distorted by ancient restorers incapable of defending them against any repetition of seismic phenomena, now have the means of reacquiring their beauty and of living a long life" (Gavini 1923, p. 33).

This position was restated by the Athens Charter of 1931, which hoped for "a judicious employment of all the resources of modern technology, and particularly of reinforced concrete. Expressing the opinion that ordinarily these means of reinforcement must be dissimulated so as not to alter the aspect and character of the building to be restored".

Numerous and significant were the monuments subjected to interventions using reinforced concrete, and just as vast is the bibliography on the subject, often linked however to single cases and lacking the support of a more general documented literature of the related damages. An outstanding case is the town of L'Aquila, with its ample margin of research in this sense (Bartolomucci 2015; Fiorani 2011) with one of the most well-known restoration works on the façade of the church of Santa Maria di Collemaggio (Fig. 10.8). For the restoration of the upper left part of which was particularly damaged, Gavini made sure that the reinforced concrete supporting structure was in no way "visible from outside" (Bartolomucci 2004).

This decision, widely shared and approved, was applied in innumerable interventions by the Civil Engineers. Such interventions, however, the results of the figurative reconstruction were judged, often evidenced structural weaknesses, documented by recent earthquakes in central Italy.

10.4 Conclusion

Italy has a long history of reconstruction of historic cities and villages. This chapter provided a detailed examination of complexities, problems and solutions adopted after the 1915 Marsica earthquake; and how the reconstruction played a role in

Fig. 10.8 (Right)The façade of S. Maria di Collemaggio after dismantling in April 1915 (Bartolomucci 2004)

expert discussions about approaches on how to restore/rebuild the built heritage, and the use of new technologies and ideas at the time. It appears that over hundred years on, there are still recurrent themes and fundamental questions to be tackled in post-disaster reconstruction in such historic contexts.

Nevertheless, interventions on a post-earthquake context require the consideration of the built heritage as a synthesis of past and future (Bartolomucci and Ciranna 2019). They must involve an understanding of towns and villages, territories and landscapes, as the outcome of tangible and intangible decisions, slow stratification and rapid interruptions for restoration and reconstruction. They need to consider what happened in previous historic reconstructions, as the total destruction of prominent built heritage and the way it is tackled may ultimately cause the loss of its memory for inhabitants and urban history.

References

Bartolomucci C (2004) Santa Maria di Collemaggio. Interpretazione critica e problemi di restauro. Palombi, Roma

Bartolomucci C (2015) Gli effetti del terremoto del 1915 nella città di Aquila: i danni e gli orientamenti per il restauro. In: Ciranna S, Montuori P (eds) Avezzano, la Marsica e il circondario a cento anni dal sisma del 1915. Consiglio Regionale dell'Abruzzo, L'Aquila, pp 151–161

Bartolomucci C, Ciranna S (2019) Santa Maria Paganica a L'Aquila. Abitare la Terra, supplement n. 50, a. XVIII, p 45

Ciranna S (2015) L'inesorabile metamorfosi di un centro urbano. Avezzano 1843–1925. In: Ciranna S, Montuori P (eds) cit. pp 8–45

Ciranna S, Montuori P (2015) Tempo, spazio e architetture. Avezzano cento anni o poco più. Artemide, Roma

Ciranna S, Montuori P (2016) Legno, ferro e cemento armato. Materiali, interventi e 'presidi' antisismici tra L'Aquila, Roma e Avezzano a seguito del sisma del 1915. In: D'Agostino S (ed) History of engineering, (international conference on history of engineering, Atti del VI Convegno di Storia dell'Ingegneria, Naples 2016 April 22nd–23rd), vol I. Cuzzolin, Napoli, pp 429–438

Ciranna S, Montuori P (2017) Gustavo Giovannoni versus Sebastiano Bultrini: un contradditorio aperto tra due ingegneri della Scuola di Applicazione dell'Università Romana. Bollettino del Centro di Studi per la Storia dell'Architettura, 1 (n.s.), pp 15–32

Fiorani D (2004–2007) Rovine e 'miracoli' artistici del terremoto di Avezzano. Le architetture storiche nella piana del Fucino. Quaderni di Storia dell'Architettura, 44–50, pp 491–502

Fiorani D (2011) Il perenne ciclo del divenire nel cantiere storico aquilano. Annotazioni sul tessuto urbano, architetture e costruzione nella città dei terremoti. In: Ciranna S, Vaquero Piñeiro M (eds) L'Aquila oltre i terremoti. Costruzioni e ricostruzioni della città. Città & Storia, a. VI, n. 1, pp 239–260

Gavini I (1923) Il cemento armato nel restauro dei monumenti. Ingegneria, a. II(2):30–33

Giovannoni G (1913) Vecchie città ed edilizia nuova, Nuova antologia di lettere, scienze ed arti XLVIII(995):449–472

Giovannoni G (1917) Per le costruzioni nei paesi del terremoto marsicano. Relazione della Commissione sociale. Estratto dagli annali d'Ingegneria e d'Architettura, XXXII(4):3–13

Giovannoni G (1928) Commenti e polemiche. Le chiese nella Marsica. Architettura e Arti decorative VII 2(9)

Giovannoni G (1929) Commenti e polemiche. Per le chiese delle zone sismiche. Architettura e Arti decorative a. VIII 2(10)

Montuori P (2017) Costruzione, ri-costruzione e restauro delle chiese marsicane dopo il sisma del 1915, tra eclettismo e "uso giudizioso" del cemento armato. Il caso della chiesa di San Giovanni Decollato in San Francesco ad Avezzano. In: De Cesaris F (ed) Costruzioni dei secoli XIX-XX in Italia centrale. Palombi, Roma, pp 193–199

Montuori P (2019) Costruire e ricostruire in Italia centrale: il primo Novecento fra tradizione e modernità. *Materiali e strutture.* Problemi di conservazione a. VIII 15:87–106

Saladino L (2015) Avezzano e la sua cattedrale: dalla genesi alla perdita della memoria. In: Ciranna S, Montuori P (eds) Avezzano, la Marsica … cit, pp 15–25. Consiglio Regionale dell'Abruzzo, L'Aquila

Serafini L (2016) Abbandono e necrosi dei centri minori dopo il sismo del 1915. Argomenti per il recupero. In: Galadini F, Varagnoli C (eds) Marsica 1915—L'Aquila 2009. Un secolo di ricostruzioni. Gangemi, Roma, pp 223–234

Unione Edilizia Nazionale (1921) Unione Edilizia Nazionale, L'opera dell'istituto nel quadriennio 1917–1920. Danesi, Roma

Chapter 11
Coventry: Shell or Phoenix, City of Tomorrow or Concrete Jumble?

From Reconstruction to the Phoenix Initiative, UK

Patrizia Montuori

Abstract Coventry is an interesting case to show how the choice to modernise a historic centre devastated by war is a complex and controversial road that starts a much slower and more difficult process of reconstruction than expected, and not always leading to success. On 14 November 1940 Coventry, an important industrial town but also one of better preserved medieval English cities, was almost totally destroyed by an aerial blitz of the Luftwaffe that sarcastically borrowed the title of *Mondscheinsonate,* the sonata of Ludwig van Beethoven. Among the ruins of the ancient centre only the spier and the perimeter walls of the medieval cathedral of St. Michael were still partially standing which focused the need on the first and most immediate reconstruction. The reconstruction explored a modernist and symbolic monumentality, an old-new relationship between "shell" and "phoenix" evoking the theme of Sacrifice and Resurrection, fully expressed by the project of Scottish architect Basil Spence. Also modernist was the reconstruction of the city, starting by re-launching the pre-war plan drawn up by the English urbanist Donald E. E. Gibson; a reconstruction that contrasted a "Coventry of Tomorrow" to the idea of the historic centre "where it was and how it was", pursued in other European cities destroyed by bombing like Dresden and Warsaw. Today, Coventry is still a curious city with a "leopard skin" shape, characterised by medieval traces, concrete buildings and urban voids in the ancient centre, that has tried to regain its identity only from the nineteen-nineties, with the Phoenix Initiative, an intervention for the revival of the alienated historical centre through a path from the past to the future of the city.

Keywords Bombing · Historic centre · Post-war reconstruction · World War II · Coventry · England

---

P. Montuori (✉)
University of L'Aquila, DICEAA, L'Aquila, Italy
e-mail: patrizia.montuori@univaq.it

© The Author(s), under exclusive license to Springer Nature Switzerland AG 2021

F. F. Arefian et al. (eds.), *Historic Cities in the Face of Disasters*, The Urban Book Series,
https://doi.org/10.1007/978-3-030-77356-4_11

11.1 Introduction

Coventry, Dresden, Warsaw are cities whose names are tragically linked to the image of "modern ruins", produced not by the physiological deterioration of ancient artefacts or natural disasters, but by the devastation for which contemporary society has made itself directly and consciously responsible during the Second World War. Other ruins are added to these twentieth-century ruins during numerous recent conflicts in various parts of the world, still today, caused by the same destructive fury. They are devoid of the charm of ancient archaeological remains and, more than anything, considered a painful testimony of tragic episodes. In most cases, they have been the subject of radical interventions aimed at recovering the identity of the devastated centres, erasing the memory of the war events, as in the case of Dresden and Warsaw, totally reconstructed "where they were as they were".

These examples show how the communities affected by traumatic, extensive devastations are emotionally oriented towards this solution and often indifferent to any academic principle regarding the opportunity of a total restoration. The International Council on Monuments and Sites (ICOMOS) has introduced the legitimacy of a total reconstruction only for recent destruction due to war or natural events. In article 9 the Charter of Venice of 1964 still specified that reconstruction was to be considered an exception and every intervention on the built environment had to be based on the respect of the original substance and conducted on the basis of documentation. Conversely, in article 8 the ICOMOS Declaration of Dresden of 1982 has considerably widened the criteria for the reconstruction of the cities destroyed by wars, including in its vast casuistry also complete reconstruction, motivated by the symbolic value of the work, but always operated on exact documentation.[1]

The restoration "where it was and how it was", however, does not find application in the homeland of John Ruskin. Coventry, one of the best preserved medieval centres in England until the deadly German air blitz was to become one of the rare examples of a modernist and highly symbolic reconstruction, not only of the cathedral devastated by the bombing, but of the entire historic centre. The City Architect Donald E.E. Gibson (1908–1991) was in charge, relaunching the pre-war plan for a "Coventry of Tomorrow" (1939) and contrasting his "Coventry of Future" (1945) comprising modern buildings, elevated driveways and pedestrian spaces, to the reconstructed copies of Dresden, Warsaw and other ancient centres destroyed by bombing.[2] This choice, based on a modernist vision of the city and the object of a partial and fragmentary realisation, has produced a curious city with a "leopard skin" shape, with medieval traces, inconsistent modern buildings and large urban voids in the old part, of which Coventry has tried again to regain possession in the 1990s with the "Phoenix Initiatives" (1996).[3]

Methodologically, retracing the steps of the controversial construction process of the new identity of Coventry which began in the post-war period and is still in progress, enables us to focus on a different approach from the total restoration of ancient centres, with the aim to understand whether it is equally effective materially and psychologically for the reconstruction of cities and communities affected by war events or natural disasters.

11.2 The Shell and the Phoenix: The Reconstruction of St Michael's Cathedral

During the night of 14 November 1940 the German Luftwaffe inflicting profound devastation on Coventry in Warwickshire, an important industrial centre and one of the best preserved medieval cities in England, with a deadly air blitz that sarcastically borrowed its title *Mondscheinsonate* from the famous sonata by Ludwig van Beethoven. Satisfied with this destruction Joseph Goebbels, Minister of Propaganda of the Third Reich, described it with the neologism "koventrisieren" (to "coventrate"), used from then on with the meaning "to raze completely".

The first raid, followed by several others, reduced the area of the "Three Spires" (Fig. 11.1) to a pile of rubble. This name of the heart of the ancient centre was due to the presence of the three spires belonging to the Holy Trinity Church, the Franciscan Christ Church and the historic cathedral of St Michael, built between the late fourteenth and early fifteenth centuries. Only the external walls and the spire of St Michael remained and, at the end of the war, the need for a first and more immediate intervention concentrated around it "as a sign of faith, trust and hope for the future of the world".[4] Thus, the new cathedral became the symbol of

Fig. 11.1 The eighteenth century view of Coventry by the artist E. Mitchell preserved at Herbert Art Gallery and Museum, Coventry (https://it.wikipedia.org/wiki/Coventry#/media/File:Mitchell_ViewOfCoventry_HAGAM.jpg)

the reconstruction of the historical centre, not as a replica of the one like in other European cities damaged by bombing, for example Dresden and Warsaw. On the contrary, for Coventry the reconstruction was almost an opportunity to confront the medieval past of the city in a liberating way. Already before the war, Coventry had become, more than anything else, an industrial centre, deserted by the inhabitants in the historical part. Its destruction presented an alibi to project it towards a radical modernism, pursued by relaunching and expanding the pre-war plan drawn up by English urbanist Donald E. E. Gibson (later Sir Donald Gibson, 1908–1991) which provided for a modern, fully pedestrianised city centre with superblocks. Contrary to popular belief, Coventry's modernisation had begun long before German bombers devastated the city centre. As early as the 1930s many historic buildings, deemed non-recoverable, had already been demolished and larger streets had been built for car traffic within the medieval urban fabric.

Already in 1942, with firm opposition to the predominant logic of reconstruction "where it was and how it was" and despite public opinion, the Royal Fine Arts Commission rejected the proposal of Sir Giles Gilbert Scott (1880–1960)[5] of a new cathedral to be built on the surviving walls in a simplified gothic style. Instead, the design competition launched in January 1951 became a precious opportunity to explore a new monumentality of modernist taste, a theme at the centre of the architectural debate of the period and of numerous conferences of the International Congresses of Modern Architecture (C.I.A.M.). However, among the more than two hundred proposals, those of the young architects who totally ignored the devastated pre-existence of the Middle Ages were not favourably received. Instead, the rewards went to the three projects that had joined the old with the modern in a balanced way, the ruins of St Michael with the new cathedral, representing the binomial "shell & phoenix", the symbol of the Sacrifice and Resurrection of Coventry. The second and third prizes were assigned to two very similar solutions by the architects W. P. Hunt and A. D. Kirby respectively, who had both envisaged a new church with a longitudinal plan, parallel and totally in adherence to the remains of the ancient cathedral.[6] The winning project was the work of the Scottish architect Basil Spence (later Sir Basil Spence, 1907–1976), a pupil of Edward Lutyens. While also adopting a longitudinal layout, he had arranged the new cathedral perpendicularly to the original gothic nave. Its North–South orientation was usually considered inappropriate by the Anglican Church, but it was less invasive for the ruins of the medieval church which were transformed into a symbolic "garden of remembrances", also housing the "Charred Cross", a wooden crucifix made with the pieces of the beams collapsed under the bombs (Figs. 11.2and 11.3).

Spiritualism and Medievalism were the two aspects around which Spence conceived the new cathedral. Like a medieval building site, it was a sort of laboratory for every kind of artwork: the stained-glass windows by John Piper, the tapestry of "Christ in Glory" by Graham Sutherland, and alongside the eastern entrance the bronze sculpture of St Michael defeating the Devil by Jacob Epstein. Some critics considered the cathedral by Spence the apotheosis of medieval modernism imbued with tradition, nationalism and spirituality, typical of nineteenth-century England and also of the pre-Raphaelites and William Morris. Conversely, Rayner Banham

Fig. 11.2 The new Coventry cathedral designed by Basil Spence, image of the Shell & Phoenix combination (by DeFacto, https://it.wikipedia.org/wiki/Cattedrale_di_Coventry#/media/File:Coventry_Cathedral_2018.jpg)

Fig. 11.3 Coventry's old Cathedral ruins: at the bottom of the remains of the Gothic nave is visible the "Charred Cross", a wooden crucifix made with the pieces of the beams collapsed under the bombs (by Andrew Walker, https://it.wikipedia.org/wiki/Coventry#/media/File:Coventry_Cathedral_Ruins_with_Rainbow_edit.jpg)

underlined its character "confusing and diversionary",[7] declaring itself uncertain of becoming a model for future ecclesiastical architecture.

After its inauguration in 1962, the cathedral designed by Spence had alternating critical fortunes precisely because it was considered a "modernist" work, while it actually was a traditional building in terms of planimetry, internal and external details and its predominant material, a red sandstone similar to that of the historical walls. The reconstruction of Coventry cathedral was considered a happy work of "continuity, unity, permanence and vitality",[8] perhaps though thanks to the patina, which covered the stone walls of the new building after more than fifty years since its construction, harmonising it totally with the ancient ruins. Some even believe that against all logic, the bombings gave the medieval cathedral a new strength, thanks to the "courage to maintain the authenticity of the place" and "to altering practically nothing of the fruit of the bombing".[9]

11.3 From the "Coventry of Tomorrow" to the "Jumble of Cement"

Less fortunate was the outcome of the modernist plan by Donald Gibson, who headed the Coventry City Architect's Department since 1938 and had already in mind to transform the medieval city into a modern urban centre. The plan for the "Coventry of Tomorrow" had been presented in May 1940, about six months before the German bombing, in a homonymous exhibition at St Mary's Guildhall, organised by the local branch of the Association of Architects, Surveyors and Technical Assistants, composed largely of members of the office headed by Gibson, under the slogan "The Idea /To Avoid Chaos we must PLAN our city for all our needs".[10] The focus of the exhibition was the model of the new Civic Centre in Coventry (Fig. 11.4), which was supposed to solve the problems of overcrowding and congestion in the medieval old town, also thanks to the extensive demolition of the existing buildings. The pamphlet entitled "Coventry of Tomorrow: City of Desire" read: "How are we to build when too much is to be destroyed? Can we afford to cease work for a creative end, even though we are at war? If we do not, the open gate of defeat lies ahead and behind it the declining path of civilization and decadence. Here is the challenge! if we do not take it up we are indeed decadent, defeated".[11] However, it was the widespread war damage that enabled Gibson and his team to extend and turn the project of the

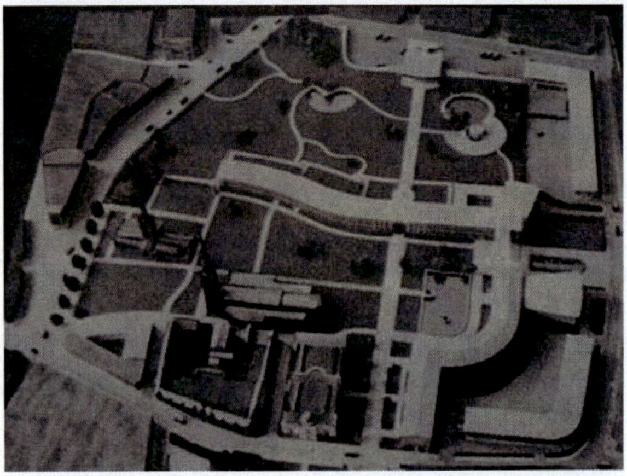

Fig. 11.4 The new civic centre in the 1939 model of Gibson (https://www.coventrytelegraph.net/news/coventry-news/coventry-blitz-world-war-ii-13721053)

modern Civic Centre into reality and, unlike in other cases, it was considered an opportunity rather than a dramatic event, at least from an urbanistic point of view. The destruction was a perfect excuse to overcome the problem of a historical and architectural heritage considered annoying and to build a new city.[12]

The influence of rationalist urbanism, already evident in the plan of 1939–40, was mediated in Gibson's subsequent proposal developed between 1941 and 1945 by the more moderate and regionalist modernism of the English urbanist Patrick Abercrombie (1879–1957), who was his professor at the School of Architecture of the University of Liverpool. In the plan for the reconstruction of the London metropolitan area (1943–1944), which he conceived with J. H. Forshaw (1895–1973), Abercrombie had foreseen various precincts, in particular in the area surrounding Westminster Abbey and the Bloomsbury district, which were primarily commercial areas, fully pedestrian and protected thanks to their total separation from vehicular traffic. While these precincts were never realised in London, they found a first and complete utilisation in the reconstruction of the historic centre of Coventry. Gibson and his team quickly extended the plan drawn up before the German blitz to a larger part of the central Coventry area. They planned commercial, administrative and recreational areas largely pedestrianised, thanks to an internal ring road that moved car traffic outside the historic centre and modern but low buildings with traditional materials, such as the red sandstone walls, creating what Gibson would later describe as the "quiet precincts where the movement of people is slow, and close and intimate".[13]

For Gibson, the cathedral destroyed in the bombing of 1940 was one of the few pre-existing buildings to be conserved, mainly as an element to rethink the historic centre and to rearticulate in modern form the close ties that once existed between religious buildings and civic life and other occupations of the city. Initially, however, he expected to insert only the surviving spire in the modern Civic Centre. He was supported in this hypothesis by Bishop Neville Gorton who would have preferred a symbolic and striking gesture of renewal to Spence's more conventional project. He would have liked to demolish the remains of the gutted cathedral and to reconstruct the building as new, reflecting in what he had stated in a public debate "You cannot have a ruin to represent the Church in your City".[14]

For Gibson, along with those of the Holy Trinity Church and Christ Church, the spire represented the essence of ancient Coventry. However, it became the visual focus of both the Upper Precinct in the area in front of it and a garden square, Broadgate (Fig. 11.5a), surrounded by new buildings on three sides and the fourth open towards it. Both of them were facing the Lower Precinct, with its two-tier commercial system (Fig. 11.5b). Next to it, Gibson also planned an "Entertainments Precinct" with a theatre, banks and offices and, in the south-east, a large "Civic Precinct" with the market area and various administrative buildings (Fig. 11.5c).

The updated version of the Coventry central area plan (Fig. 11.6) was presented by Gibson in a new exhibition entitled "Coventry of the Future", organised in 1945 at Drill Hall to coincide with the anniversary of Coventry's Charter of Incorporation. Therefore, in 1948 the first stone was laid for Broadgate House, the main building of the Upper Precinct, the headquarters of Coventry City Council and one of the first

Fig. 11.5 **a** A 1941 drawing for the proposed redevelopment of Broadgate. **b** A 1945 drawing of the "Lower precinct" looking towards the cathedral. **c** One of the 1942 "Third Model" designs for Coventry city centre that shows the "Civic Precinct" and the market area

projects completed after the war according to Gibson's forecasts. Even the Lower Precinct was built in large part as planned, except for a few accessory elements, such as the mini-play-garden in the centre and the curved garden at the head, which were omitted due to the usual lack of funds, but the new Belgrade Theatre (1955–58) was realised in the "Entertainments Precinct" as planned. Conversely, the Civic Precinct, the largest component of the plan did not have the same chance. It was never totally completed,[15] except for some buildings such as the Police Station (1954–57) and the Civic Offices (1951–57), started by Gibson and completed by Arthur Ling, City Architect of Coventry since 1955. Because a large area in the historical centre remained unbuilt or was progressively colonised by incongruous modern buildings (for example, the yellow-blue cube of Ikea built in the 2000s not far from the cathedral) the centre of Coventry had transformed into a random mix of ancient and new (Fig. 11.7).

Over the years, various commercial spaces on the upper level of the Lower Precinct were closed and abandoned, because they were more difficult to reach than those on the ground floor. Little by little, the original design of the precincts was altered with transformations and new buildings, such as the Cathedral Lanes shopping centre. It was built in the nineteen-nineties on the side of Broadgate which Gibson had left open towards the spire of the Cathedral, now hidden from view (Fig. 11.8).

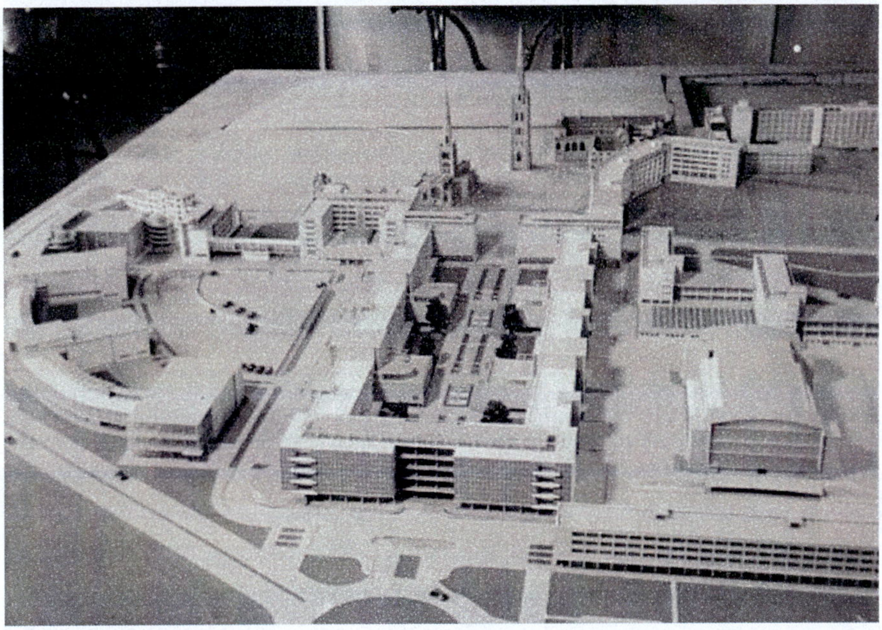

Fig. 11.6 The model of the new civic centre of Coventry planned by Donald E. E. Gibson (by the United Kingdom Government in https://commons.wikimedia.org/wiki/File:Coventry_reconstruction_model_D_15518.jpg)

Fig. 11.7 Council House & Coventry Cathedral panoramic view from Civic Centre (by Coventry City Council 20 December 2016, https://commons.wikimedia.org/wiki/File:Council_House_%26_Coventry_Cathedral_panoramic_view_from_CiCiv_Centre_4_(31804215235).jpg)

By the turn of the third millennium Coventry had become a strange city with a "leopard skin" shape, with interesting medieval parts still preserved, mixed with the interventions of the nineteen-sixties and random contemporary buildings. In tourist guides it was described mercilessly as a "jumble of cement without any attraction, where the only noteworthy building is the new modern cathedral, built next to the ruin of the original one destroyed by bombing".[16]

Fig. 11.8 The Cathedral Lanes shopping centre, built in the nineteen-nineties on one side of Broadgate (https://www.coventrytelegraph.net/incoming/gallery/cathedral-lanes-plans-12506152)

11.4 Coventry and the New Renaissance: The Phoenix Initiative

In the nineteen-nineties, in the middle of Thatcher era, the city, nicknamed "Britain's Detroit" thanks to the presence of important industrial districts, in particular where land and air vehicles were produced, had suffered a severe economic recession. This was due to the collapse of the manufacturing industry in the whole of the West Midlands, enduring peaks of unemployment reaching 20% and suffering from rampant alcoholism. By now Coventry had become a Ghost Town, especially in the central part, deserted by tourists and by the inhabitants themselves. Therefore, the city had to equip itself with a new image, able to compete with the major historical and cultural attractions of the surroundings, among them the Ironbridge, Warwick Castle and Stratford-upon-Avon. This is when the need for a new recovery of the historic centre emerged, which this time was in ruins due to the faults attributed in large part to Gibson's plan. In 1996 the City Council of Coventry and other organisations including the Herbert Museum started the "Phoenix Initiative", an urban renewal intervention that intended to combine the old with the new. Akin to the post-war period, its conception started from the remains of the previous phases of history of the city. This time round urban regeneration did not refer to the ruins of the "Moonlight Sonata Operation" in the city centre. Instead the chosen ruins dated back to the suppression of the Catholic monasteries, instigated by Henry VIII between 1536 and 1540 after the Anglican Schism, followed by the first rebirth of Coventry. The excavations were concentrated in an area west of St. Michael. Largely unbuilt it had already been the subject of archaeological investigations since 1959[17] because it was once the site of the city's first Catholic cathedral, the Priory of St. Mary[18] and of the annexed, homonymous Benedictine monastery, both destroyed during the Reformation.

These ruins were located on an ideal path connecting the ancient with the new. It linked the historic area of the Three Spires and the Cathedral of Basil Spence to the commercial districts and the Transport Museum in the northern part of the historic centre, passing through new public spaces with modern residential buildings. The goal was to revitalise the historic centre and encourage visitors to stay there longer to support the economy.

An urban design competition was organised in 1997. It was won by MJP Architects London,[19] in collaboration with the landscape design studio RRA[20] and the Public Art Commissions Agency of Birmingham. The task of the latter was to involve artists who were sensitive to the theme of reconciliation, much felt by the Coventry citizenship (Fig. 11.9).[21]

Despite MJP's critical judgment of post-war interventions, the aim of the project was not to intervene in that past but to rethink the overall use of the historic centre anew. Their way to enable the city to reclaim memories of the different phases of its history was to create a path crossing the whole of the area of intervention, starting as a traditional walkway made of wooden planks and turning into a spectacular, modern walkway with a blue-tinted glass balustrade formed by eight hundred curved slabs.

The path starts from the most ancient testimonies of Coventry's urban history, the archaeological area next to the cathedral where two gardens were created. The Priory Garden contains glass cases to protect and highlight the remains of the columns of the Priory church which are visible from an aerial walkway in burnished steel and wood. The Visitor Interpretation Centre exhibits the artefacts recovered during the archaeological excavation. Some of these finds were also used by the artist Christine Browne to create a mosaic in the centre of the garden representing the plan of ancient Coventry, "where it was and how it was" named the "Cofa's Tree" in early medieval times (Fig. 11.10).

The Priory Cloister, a smaller and more intimate garden is marked in the centre by a simple cross and surrounded by eight pillars. They evoke the columns of the ancient Benedictine cloister and transmit voices that narrate episodes from the city's past, a suggestive installation by the artist David Ward entitled "Here" or "The Whispering Voices" (Fig. 11.11).

From there the walkway leads along the "water-window", a fountain designed by Susanna Heron, that fills the small space with the roaring sound of its waterfall. The path changes from the ancient to the modern, reaching a new square, Priory Place, with new residential buildings and entertainment venues. It passes through a narrow passage and reaches the spectacular "Whittle Arch", a steel structure formed by two arches of sixty meters of light gathered in the ridge. It was erected in honour of a native of Coventry, Sir Frank Whittle (1907–1996), a high official of the RAF and inventor of the jet engine (Figs. 11.12 and 11.13). The path arrives at the Millennium Place, a fan-shaped urban space with the Transport Museum in the background, expanded, reorganised and equipped with a new curved façade. It continues with the elegant blue Glass Spiral Bridge [22] and ends in a new garden, the Priory Maze, adjacent to the old city walls and the recently restored Lady Herbert's Garden. The new garden was designed by Kate Whiteford as a portion of a circular labyrinth on the theme of international friendship. It recalls the possibility of an original maze in

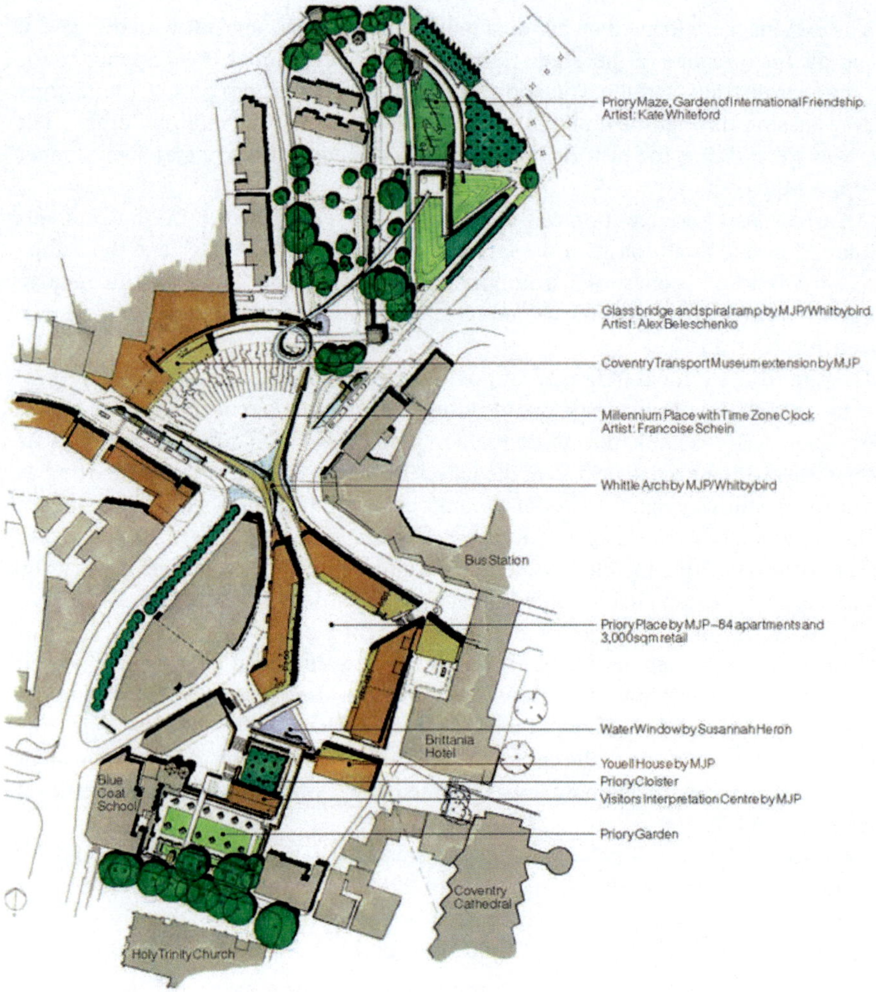

Fig. 11.9 Masterplan of the urban renewal intervention of Coventry's historic centre designed within the "Phoenix Initiative" (MJP Architects http://mjparchitects.co.uk)

the Benedictine Priory grounds depicted in early maps. The labyrinth is flanked by a curved plinth, engraved with a text from Coventry Mystery Plays,[23] by local writer David Morley (Fig. 11.14).

The renovation of the historic centre of Coventry within the Phoenix Initiative was completed in 2002 and the part closest to St. Michael in 2004. It gained the Royal Town Planning Institute (RTPI) Planning Award for City Regeneration in 2003 and the Stirling Prize conferred by the Royal Institute of British Architects (RIBA) in 2004.

Fig. 11.10 View of the Priory Garden: the visible remains of the columns of the Priory church, the aerial walkway in burnished steel and wood and, on the right, the Visitor Interpretation Centre (https://www.coventrytelegraph.net/news/coventry-news/priory-visitor-centre-coventry-closing-16315349)

Fig. 11.11 The Priory Cloister, a smaller and more intimate garden, marked in the centre by a simple cross and surrounded by eight pillars, which evoke the columns of the ancient Benedictine cloister (MJP Architects & Marc Goodwin, http://mjparchitects.co.uk)

Fig. 11.12 View from the Millennium Place toward the Priory Place with the new residential buildings, the spectacular "Whittle Arch", the Glass Spiral Bridge and in the background the spires of the cathedral of St. Michael and of Christ Church (MJP Architects & Marc Goodwin, http://mjparchitects.co.uk)

Fig. 11.13 The Glass Spiral Bridge a spectacular walkway with a blue-tinted glass balustrade formed by eight hundred curved slabs (MJP Architects & Marc Goodwin, http://mjparchitects.co.uk)

Fig. 11.14 The Priory Maze, a new garden designed on the theme of international friendship as a portion of a circular labyrinth (MJP Architects & Marc Goodwin, http://mjparchitects.co.uk)

11.5 Conclusion

Coventry is one of the few European cities that had chosen to be reborn after dramatic war destruction not as a copy of itself but by resorting to modern architecture and planning. This led to controversial results which to this day continue to arouse reactions ranging from approval to total rejection. Some thirty years ago the architecture critic Ian Nairn described the commercial area designed by Gibson as "probably the best thing of its kind in Europe".[24] Conversely, Richard McCormac of MPJ Architects, who had designed the recent urban renewal intervention in the historic centre, remarked that the Second World War had caused less disasters than the alleged modernist "remedies" implemented in the 1960s.[25] McCormac had become the spokesman for a generalised resentment against the "cementification" of British cities during reconstruction, considered the main cause of their un-liveability. However, this point of view raised dissent of various commentators. Its opponents were worried that it could lead to an indiscriminate destruction of modernist architecture of historical-architectural value, despite the protection law approved in 1987.[26] Therefore, careful mediation work was recommended between the needs of urban renewal, sustainability and conservation of ancient heritage and respect for modern heritage, as a product of another important architectural and socio-political era.

The case of Coventry shows that the choice to modernise a historic centre devastated by destructive disasters or war is a much more complex and controversial undertaking than the reconstruction "where it was and how it was", not least because it involves a much slower and more controversial process, not always destined for success. This process of modernisation is creating a real new identity, often completely different from the past. It is the result of the union between the ancient and the modern in which, however, local communities struggle to recognise themselves. It took until now for the modernist works to acquire their recognisability and integration in the image of the historical centre of Coventry and to be considered as testimonies of an important phase in the history of the city. Incorporating the most ancient traces in the "re-stitching" operation of the more recent urban renewal intervention was undoubtedly important for this integration process of the previous modernist phase of reconstruction of the city. In the near future, restorations of the modern interventions will have to be carried out. They will be aimed at the conservation of buildings, but also at their adaptation to new functional and technological needs as well as urban evolution as a whole. This "restoration of the modern" will be guided by the same theoretical critical principles of restoration and deserves the same attention applied to ancient buildings.[27]

Regarding the different Urban Renaissance interventions undertaken in England at the turn of the third millennium[28] (including the Phoenix Initiative[29]), the risk to be avoided is to confuse excellence with exception. Someone observed correctly that the search for identity should be achieved with good architecture and quality of places, instead of seeking media references, conforming to fleeting architectural trends and to real estate marketing strategies, in search of visibility and economic results.

Perhaps the wisest way to construct an identity for the destroyed historical centre—not as a copy of the lost one, but as a balanced combination of old and new—remains the project by Basil Spence for the new Cathedral of Coventry, and his own view about it: "I have not tried to create an existing building, such as can be done so readily with modern materials of construction. A cathedral should not arouse excitement but a deep emotion, and it must express the canons of the Christian faith".[30] Even today, more than fifty years after its construction and thanks to the modern but balanced design with which Spence conceived the building, the new cathedral of St Michael is considered by both inhabitants and visitors as one of the few signs of continuity between the past and the present, between the ancient and the new identity of the controversial historical centre of Coventry.

Notes

i. See ICOMOS (1964), The Venice charter 1964; ICOMOS (1982), Declaration of Dresden on the "Reconstruction of monuments destroyed by the war". For the reconstructions of Dresden and Warsaw see: Zanlungo (2018), Appelbom (2017).

ii. Gould and Gould (2016), Campbell (2007).

iii. MacCormac et al (2004a).

iv. Jones (2005, p. 82). Citation in: Ruggeri Tricoli (2008).

v. A descendant of a family of architects, he was the grandson of George Gilbert Scott (1811–1878) who among other works also designed the neogothic station of St Pancras in London, opened in 1868.

vi. About the reconstruction of St Michael's Cathedral see: Campbell (1996, 2011), Lamb (2011).

vii. Banham (1962).
viii. Seasoltz (2005, pp. 262–263).
ix. Gabriel (2019).
x. Citation in Campbell (2007).
xi. Citation in Lee (1997, pp. 139–140).
xii. Even some of the timber-framed houses that survived the bombing, in fact, will be dismantled and moved to build wider streets. See Ruggeri Tricoli (2008).
xiii. Gibson, p. 103. Citation in Campbell (2007).
xiv. See Campbell (1996, p. 43).
xv. See Gould and Gould (2016).
xvi. Else et al. (2005, p. 609).
xvii. About the archaeological excavations of St. Mary see: Rylatt and Mason (2003).
xviii. Leofric, Earl of Mercia and his wife Lady Godiva rebuilt on the remains of the nunnery to found a Benedictine monastery in 1043 dedicated to St. Mary for an abbot and 24 monks. Bishop Robert de Limesey transferred his see to Coventry c. 1095, and in 1102 papal authorisation for this move also turned the monastery of St. Mary into a priory and cathedral. The subsequent rebuilding and expansion of St. Mary's was completed about 125 years later. See: Woodhouse (2004) [1909], Ruggeri Tricoli (2008).
xix. McCormac, Jamieson and Prichard, chaired by Sir Richard McCormac, former president of the Royal Institute of British Architects—RIBA.
xx. Robert Rummey and Associates.
xxi. About the Phoenix Initiative see: MacCormac et al. (2004a).
xxii. The project for the spectacular spiral walkway was developed by MJP Architects in collaboration with Whitby Bird and Partners, an engineering firm particularly skilled in bridge construction. The glass plates of the balustrade are been engraved by the artist Alex Beleschenko, (1998–2004), who personally chose the shade of blue that would allow the bridge to blend in with the color of the sky.
xxiii. The "Coventry Mystery Plays", or "Coventry Corpus Christi Pageants", are a cycle of medieval mystery plays from Coventry and are perhaps best known as the source of the "Coventry Carol". Two plays from the original cycle are extant having been copied in the early nineteenth century from the now lost original manuscript.
xxiv. Citation in Arnot (1992).
xxv. MacCormac et al. (2004b), Citation in Ruggeri Tricoli (2008).
xxvi. While (2007, 34, pp. 645–663). Citation in Ruggeri Tricoli (2008).
xxvii. Carbonara (1997).
xxviii. Vescovi (2013).
xxix. Doubts about the effectiveness of the Phoenix Initiative are also expressed by Chrissy FitzPatrick who pointed out the too much attention to the

symbolism of the interventions, operated mainly for tourists, rather than with the goal of the creation of functional and dynamic spaces for residents. See FitzPatrick (2017).

xxx. Spence (1955, p. 151). Citation in Campbell (2011).

References

Appelbom KI (2017) Reconstruction of historic monuments in Poland after the Second World War—the case of Warsaw. In: Bold J, Larkham P, Pickard R (eds) Authentic reconstruction: authenticity, architecture and the built heritage. Bloomsbury Academic, London, pp 47–68

Arnot C (1992) Architecture: cluttering up Gibson's coventry: councillors say the fifties shopping centre is tatty, but others feel a gem is being ruined by redevelopment. Endependent 2 September 1992 [online]. Available at: https://www.independent.co.uk/arts-entertainment/art/news/architecture-cluttering-up-gibsons-coventry-councillors-say-the-fifties-shopping-centre-is-tatty-but-1548820.html. Accessed 6 Aug 2019

Banham R (1962) Coventry Cathedral-strictly 'Trad, Dad'. New statesman, LXIII, pp 768–769

Campbell L (1996) Coventry Cathedral: art and architecture in the post-war Britain. Clarendon, Oxford

Campbell L (2007) Paper dream city/modern monument: Donald Gibson and Coventry. In: Boyd Whyte I (ed) Man-made future: planning, education and design in mid-twentieth-century Britain. Routledge, London, New York, pp 121–144

Campbell L (2011) Coventry Cathedral and arts of reconstruction (chap 1). In: Lamb CA (ed) Reconciling people. Coventry Cathedral's story. Canterbury Press, Norwich

Carbonara G (1997) Avvicinamento al restauro. Teoria, storia, monumenti. Liguori, Napoli

Else D, Berry O, Dixon B et al (eds) (2005) Inghilterra. EDT, Torino

FitzPatrick C (2017) Coventry. Third time unlucky. A city destroyed by function, war then symbolism [online]. Available at: https://issuu.com/chrissyfitzpatrick/docs/coventry-_third_time_unlucky. Accessed 6 Aug 2019

Gabriel R (2019) Arte e sacro. Costruire o ricostruire? Coventry, la cattedrale attraversata dal mondo. [online]. Available: https://www.avvenire.it/agora/pagine/coventry-bombardamento-1940-cattedrale-distrutta-notre-dame-sutherland. Accessed 6 Aug 2019

Gibson D (1947) The third dimension in town planning. Report of proceedings at the town and country planning summer school

Gould J, Gould C (2016) Coventry the making of a modern city 1939–73. Historic England, London

Jones NR (2005) Architecture of England, Scotland and Wales. Greenwood Publ, Westport (CT)

Lamb CA (ed) (2011) Reconciling people. Coventry Cathedral's story. Canterbury Press, Norwich

Lee M (1997) Relocating location: cultural geography, the specific of place and the city habitus. In: McGuigan J (ed) Cultural methodology. Saga publications, London, pp 126–141

MacCormac R et al (2004) Phoenix: Architecture/art/regeneration. Black Dog Publishing, London

MacCormac R et al (2004b) Coventry phoenix initiative, achieving urban renaissance through public space and art. Paper Cityscape conference, London, 13 Aug 2004

Ruggeri Tricoli MC (2008) Anomale rovine: il caso di coventry. In: AgathÓn. Recupero e fruizione dei contesti antichi. Notiziario del dottoratodi Ricerca, Università degli Studi di Palermo Dipartimento di Progetto e Costruzione Edilizia, 2, Offset Studio, Palermo, pp 17–24

Ryatt M, Mason P (2003) The archaeology of the Medieval Cathedral and Priory of St. Mary. Coventry: Coventry, City Development Directorate, Coventry City

Seasoltz RK (2005) A sense of the sacred. Theological foundations of christian architecture and art. Bloomsbury Academic, London, pp 262–263

Spence B (1955) The Cathedral Church of St. Michael Coventry. RIBA J

Vescovi F (2013) Designing the urban renaissance: sustainable and competitive place making in England. Springer, Dordrecht

While A (2007) The state and the controversial demands of cultural built heritage: modernism, dirty concrete, and postwar listing in England. Environ Plann B Plann Des 34:645–663

Woodhouse FW (2004) [1909] The Churches of Coventry—a short history of the city & its medieval remains. Kessinger Publishing Co, Whitefish, Montana

Zanlungo C (2018) Risorti dalle rovine, La tutela dei monumenti e il destino dell'architettura sacra nella Germania socialista. FrancoAngeli, Milano

Chapter 12
Post-trauma Recovery of Monumental Buildings in Italy and the US at the Beginning of Twentieth Century

Marco Felli

Abstract This chapter compares the approaches to post-trauma restoration as pursued in the first years of the twentieth century in Italy and the US. Focusing on specific places hit by destructive disasters, such as earthquakes and floods, as well as bombardments during World War I and II, the research presented here focused on the deployment and insertion of reinforced concrete elements within existing structures. After the first experiments, this strategic approach was deployed widely in the 1930s because of the significant impact of new international theories, such as the Athens Charter for the Restoration of Monuments, whose declarations were applied as guidelines in almost all successive interventions. The introduction of different case studies for each area of interest led to the use of reinforced concrete structures all over the world during this period. Taking as examples here two restoration interventions, undertaken, respectively, in 1936 and 1937: The church of Santa Lucia, in the Marsica and the Franciscan mission of San José y San Miguel Aguayo, in San Antonio, Texas, these represent two cases in which reinforced concrete assumed an important role in the re-definition of the lost spaces and not only the replacement of structural elements. Given certain aspects that emerged during research, the conclusions aim to evaluate critically the methodology of these interventions and to provide some potential additional research strategies. The chapter contributes to the global literature on the restoration of monumental buildings.

Keywords Restoration · History of architecture · Earthquakes · Reinforced concrete · Marsica · Texas · USA · Italy

M. Felli (✉)
University of L'Aquila, L'Aquila, Italy
e-mail: marco.felli@graduate.univaq.it

© The Author(s), under exclusive license to Springer Nature Switzerland AG 2021

F. F. Arefian et al. (eds.), *Historic Cities in the Face of Disasters*, The Urban Book Series,
https://doi.org/10.1007/978-3-030-77356-4_12

12.1 Introduction

Built heritage represents the results of the historic *longue durée*, with stratification, modification and additions. In particular, architecture that has been exposed to destructive phenomena has survived to the present day because of specific interventions of recovery and restoration which modified their formal and material aspects. Technological development increased the possibility to apply modern means such as reinforced concrete which, from the end of the nineteenth century began to be used as a construction material. Monumental buildings, which tend to also represent more strongly the identity of specific places through the specific context of their architecture, have been the main edifices on which such experimentation has been conducted. In particular, an interesting phase of these restoration projects was pursued early in the twentieth century, when reinforced concrete systems started to be applied to buildings worldwide, also as a consequence of contributions to international theories of architecture and urbanism.

In these case studies, the period of analysis is the first half of the twentieth century, in which technological developments and experiments increased globally in all fields, not merely in architecture; among patents and early theories, reinforced concrete represented an innovative material because of its characteristics of functionality, short time of production and ease of application, ductile behaviour, which consequently allowed it to spread to many fields related to the field of construction, including restoration. While the material was first patented by Francois Hennebique and its frame system by the engineer Giovanni Antonio Porcheddu at the end of the nineteenth century, the first applications in preservation started at the beginning of the twentieth century, also because of the presence of the first cement companies in the hinterlands of cities (Carbonara 1995).

The use of these modern materials in some specific areas was considered advantageous due to the type of geophysical disaster when the destruction of most important buildings in a city was the case: Among them, earthquakes caused the most serious damage and, often, extensive collapse, consequently requiring rapid reconstruction over a brief period. Within this context, the insertion of reinforced concrete began to be suggested for two different reasons: Firstly, to obtain greater structural resistance for buildings, because of its mechanical properties, and its rapidity of application in construction sites. From the first years of the twentieth century onwards, experiments with this modern material were performed on buildings characterised by an urgent need of recovery, due to the serious state of damage, without any type of general guidelines able to provide application models of the new technique. Only in the 1930s, in the context of the international debates about new materials and the development of new guidelines in the fields of restoration and preservation, did reinforced concrete begin to be considered as a potential technology to be used in restoration interventions worldwide, especially because of the publication of the Athens Charter for the Restoration of Monuments of 1931, combined with the national laws and building codes, which regulated its use in preservation and reconstruction.

The Marsica, located in the hinterland of the province of L'Aquila in central Italy, and Texas, in the southern part of US, are areas, together with California, that present comparable methodological approaches to reconstruction interventions, deploying similar strategies of structurally reinforced concrete during the period of restoration, which itself followed the international release of relevant documents; also the typologies of the buildings which were treated in these restorations are comparable among the selected areas. In each context, religious architecture represents some of the most interesting case studies: In the Italian context, churches suffered serious damage and collapse through earthquakes, while the Franciscan Spanish missions of Texas and California were also damaged by floods and hurricanes.

The first part of this chapter discusses the context of the first years of the twentieth century, the development of the modern technologies and the impact of geophysical disasters on the monumental architecture of central Italy and the US, analysing common aspects among the different area and the comparable strategies in the use of reinforced concrete. The second part focuses on specific case studies. These are different architectural typologies yet restored in the same period with comparable strategies. The chapter analyses the impact of these restorations in terms of formal and structural changes.

12.2 Preservation Efforts and the Role of Reinforced Concrete

At the beginning of the twentieth century, the areas analysed here were struck by a series of events that affected their historic architecture and consequently required restoration interventions: earthquakes, floods and bombardments during World War I and II caused widespread damage to entire buildings, mainly due to weak structures. Mostly, the materials used in these kinds of buildings were characterised by elements such as brick and stone joined with poor quality mortar and therefore were fragile structures. Consequently, there was no uniform behaviour among the various structural elements during earthquakes, and cracks were revealed at the points of weakness. Another element to be considered with regard to structural behaviour is closely connected to the architectural configuration of these buildings: In particular, largely irregular forms can provoke points of cracking according to the different behaviour of material and typology due to different structural elements, such as domes, bell towers, differently sized naves, as well as additions such as modifications of architecture over time. Some of the monumental buildings, specifically the ones with defensive purposes such as castles and fortresses, were occupied by soldiers of the German army, for example, in central Italy during World War II (Felli 2018a). Occupancy by the military of specific areas, extending beyond the monumental buildings, exposed the built heritage of a town to bombardment, increasing damage and collapses. In some cases, simple abandonment of buildings was the reason for their damaged state before restoration.

Following the assessment of damage, the first efforts in the recovery process were focused on strategies to improve the structures and the re-definition of lost forms. Given the necessity for rapid recovery and reconstruction of buildings, reinforced concrete assumed a primary role, because of its functionality and ease of use. Even if the first international theories regarding preservation, with guidelines about the use of new materials, occurred in the 1930s, in previous years' interventions on historic buildings had been conducted for important works in the areas analysed here. These areas of study are characterised by the presence of authorities and engineer corps which, in large part, tried to recover the damaged buildings in such a way as to give back to the population something close to the original aspect of the architecture, using modern technologies such as reinforced concrete. The main aims of the interventions the authorities pursued were "formal preservation" and the structural improvement of buildings, without any assessment of the compatibility of the elements that were to be added within existing structures and of the traceability of these interventions. Following Porcheddu's reinforced concrete frame patented in 1894, in Italy, this modern material acquired national fame because it was applied to such symbolical architecture as the Bell Tower of San Marco in Venice, which collapsed in 1902 and was rebuilt in 1912. Similarly, in America, the application of this new technology to symbolical and important constructions, such as the Alamo of San Antonio in Texas, was highly significant.

The widespread application of reinforced concrete in restoration work entailed a vigorous international debate about the use of new materials in preservation, the final meeting point of which is represented by the release of the Athens Charter for the Restoration of Historic Monuments in 1931. The importance of this document consisted, in particular, in the collaboration between member states in agreeing the principal strategy for the preservation of art and historic monuments, and in its guidelines for the uses of new materials such as reinforced concrete, which began to be allowed in preservation efforts. These efforts focused on the preservation of elements in situ, avoiding the fakery of demolition and rebuilding. The functionality and adaptability of reinforced concrete to historic constructions ensured its use, and it became widely diffused throughout the western world also in the field of preservation.

12.3 Areas Studies: The Marsica and Texas

The territory of the Marsica, in which there are some important monumental buildings, is also recognised as an area prone to strong earthquakes: the seismic event of 13th January 1915, the epicentre of which was close to the city of Avezzano, one of the most important in the territory, destroyed entire towns, claiming a huge number of victims and causing extensive damage and the collapse of entire areas. Much monumental architecture was gravely damaged, almost to the point of collapse. Post-earthquake awareness of the vulnerability of these historic edifices conditioned the strategies for their structural recovery. During this time, efforts by the authorities responsible for cultural heritage, such as the Superintendent for Cultural Patrimony

(*Soprintendenza per I beni culturali*), became the principal actors in the definition of priorities and methods of intervention.[1] The Superintendent and its officers, such as the architect Ignazio Gavini, aimed at recovering monumental buildings through the first applications in reinforced concrete (Felli 2018b). Gavini became one of the first designers who advocated reinforced concrete as a static solution that could be used in every intervention. These intentions were clear from his text of 1923:

> Reinforced concrete structures will solve static problems in earthquake zone buildings. Seismic activity has caused enormous damages to Italian monuments, and the buildings have been damaged further by poorly trained designers and technicians; now, the buildings are able to reclaim their beauty and life, safe from irreparable damage. Deliberate and careful conservation are both social and economic work.[2]

The necessity of rapid interventions for both civil and monumental buildings involved the use of modern and "easy and rapid use" technologies, such as reinforced concrete; this went beyond its ordinary application on new residential buildings which mostly had one floor in masonry with a frame in reinforced concrete. Interesting applications of the material had been pursued in monumental constructions; among them, L'Aquila's celebrated Basilica of Collemaggio and considered one of the most important churches in the Abruzzi, which represented in this context one of the first experiments deploying modern technology for structural recovery: post-earthquake, the top level of the façade toppled over (Gavini 1915). The restoration was designed by the architects Gavini and Riccardo Biolchi of the Superintendent and aimed at recreating the uniformity of the "the façade as structural independent element", by inserting a frame in reinforced concrete, with linking beams at two different levels and completely hidden inside the existing wall, also adding two columns at the flanks, inserted inside the masonry and not visible externally (Bartolomucci 2004; Moretti 1972).

The strategy pursued at Collemaggio subsequently began to be applied to buildings in other areas hit by the earthquake. The other competent authority within the territory, the Corporation of Engineers, also played a primary role in the reconstruction works: Its officers tried to set out some guidelines for recovery, reconstruction and restoration processes for the territory, identifying and suggesting the use of reinforced concrete as an important new technology. As a consequence, the insertion of new elements became widespread in the restoration and recovery works of this area right from the earliest post-earthquake interventions. The release of the Athens Charter of 1931 also provided the basis for the creation of national codes for restoration, such as the Italian "*Carta del Restauro*" of 1932, which underlined, transposed and disseminated the guidelines of the international document into the national context (Carbonara

---

[1] *Soprintendenza ai Monumenti di Roma e degli Abruzzi.*

[2] It is translated by the author from the Italian text: "Il cemento armato potrà rendere facile la soluzione statica di innumerevoli problemi specialmente nelle zone di terremoto, ove i monumenti italiani soffersero i maggiori danni. I nostri poveri monumenti che serbano ancora le tracce di rinnovate catastrofi e spesso vediamo storpiati dai vecchi restauratori inabili a difenderli dal ripetersi di fenomeni sismici, hanno oggi il mezzo di poter riacquistare la loro bellezza e di vivere una lunga vita. La ponderata e attenta conservazione sarà così opera di civiltà e di economia nazionale" (Gavini 1923).

1995). The impact of these documents, together with the first experiments conducted successfully, encouraged the use of concrete elements in the following years and also after World War II.

Within this time frame, the gamut of realised interventions, from the recovery and partial reconstruction to the complete restoration of monumental buildings, was literally undertaken without any distinction between the different typologies of construction, even including defensive buildings. Among these, the Piccolomini castle of Celano is an interesting case study. Its restoration works were conducted by Civil Engineering Office (*Genio Civile*) under the guidelines of the Superintendent from 1940 to 1955. Religious buildings were also restored, trying to use best practice based on previous experience. The church of San Pietro in Albe represents an important case, because the stone elements were deassembled after the earthquake of 1915, followed by the creation of a continuous frame in reinforced concrete inside the existing, still standing masonry, and the replacement of the original elements. As in the previous case, the restoration works were conducted in post-World War II, from 1955 to 1957, and were designed by the architect Raffaello Delogu (Siracuse 1909–Rome 1971) of the Superintendency.

In a way comparable to the Marsica, the American case studies were characterised by the application of new elements to support existing structures, and although they were different building typologies, the interventions deployed the same approach (Gavini 1923). Depending on the area of interest, here California and Texas, the occurrences which entailed restoration work were caused by natural incidents: earthquakes, typical of the Californian area for the San Andreas' Fault, along which the twenty-one Franciscan Missions constituting the Camino Real are positioned (Fig. 12.1),

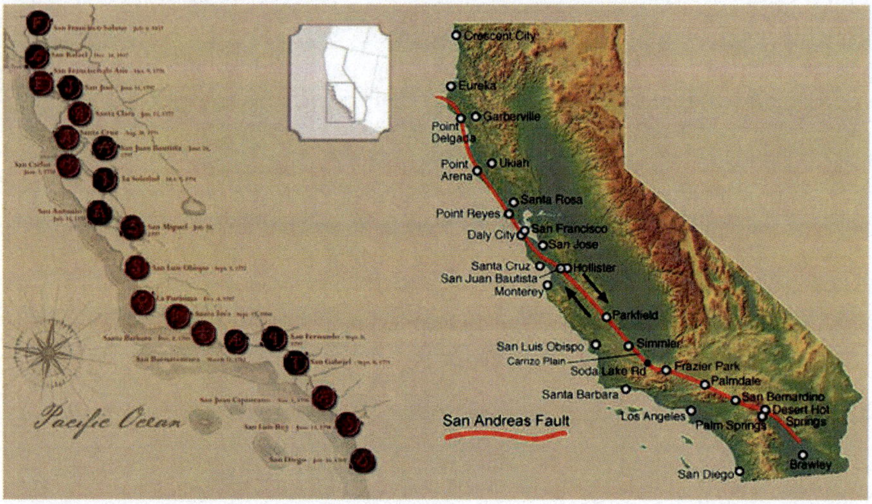

Fig. 12.1 The 21 Franciscan missions of California (left), compared to the position of San Andreas Fault (right). It is easy to understand, according to the maps, the necessity of preservation of the monumental buildings in the face of quake risks

and floods and hurricanes, found more frequently in Texan territory. The Spanish missions of the southern part of US, whose construction had been promoted by the Franciscan friars coming from Europe and Mexico, constitute groups of buildings, among which the central one is always the church, built with the aim that native people would be educated according to the standards of European society (Mission Memories 1900; Weeks-Wilson 1913). Inspired by European construction tradition, their form and characteristics reproduce many elements of Italian churches, although with irregular forms and different materials and elements and so with different mechanical characteristics (Aiken 1983; Engelhardt 1913; Sewall 2013). As with the Italian case studies, the occurrence of geophysical disasters involved comparable damage to these structures. The earthquakes of San Juan in 1906 and Santa Barbara of 1925, as the Italian case of the Marsica, brought clarity about the weak resistance of these kinds of structures, also prompting the abandonment of these buildings, a phenomenon begun with secularisation, and with the complete dereliction of settlements by the Franciscan friars in the nineteenth century, letting them in fall into ruin (Kryder-Reid 2010).

Also in the American context, the first efforts of preservation were proposed during the 1930s by associations and authorities (Kimbro et al. 2009) who had worked hard for the restoration and the recovery of monumental buildings that lay in ruins. The Civilian Conservation Corps (CCC), the federal authority for preservation and also the programme for unemployed people, constituted one of the attempts at building preservation. Other foundations, such as the Historic American Building Survey (HABS) and the Index of American Design, instead of the CCC, began the first campaign to survey monumental architecture, thanks to which a complete database of its actual state, and in some cases, also the realised preservation projects were created (Sewall 2013).

As mentioned, the first reinforced concrete applications were proposed before the release of the new international theories. The development of local building codes, including the use of new material, was permitted for existing historical structures only if damaged and "repairable". This fostered the use of concrete applications, also because of the "economical shove" of the cement factories: This might be the only difference with the Italian context, in which applications of reinforced concrete were suggested by the patent holder and theoretical development. However, the insertion of new structural elements in damaged or decayed buildings became popular with symbolical interventions, due to the importance of the buildings. Subsequently, earthquake damages from the event of 1925, causing the collapse of the upper parts of façade and the bell towers and damage in the main nave structure of these churches, resulted in the restoration of the Santa Barbara mission in California through the insertion of a continuous frame in reinforced concrete inside the area of the convent and the application of upper tie beams in the church (Felli et al. 2018). The work, designed by the architect Ross Montgomery, was completed in 1927 (Weinberg 1974). In parallel, also Texas was one of the first areas in which concrete applications were the subject of experimentation on monumental heritage before the international rules came into being. The restoration of the former mission of San Antonio de Valero, in San Antonio Texas, commonly known as the Alamo, represented an interesting restoration in the American context. The restoration project, put forward by the English architect Alfred Giles in 1921, was commissioned by the Daughters of the Republic of Texas, an association of women devoted to historical preservation, aimed at redefining the lost form of the naves, recreating their barrel vaults in reinforced concrete (Ciranna et al. 2019; Ivey 2007). In this case, in contrast to the previous case, concrete was used not only as a simple static solution, but also as a visible constructive element, recreating the supposed original ceiling of the nave.

The case studies of this research, which are the church of Santa Lucia in Magliano de' Marsi, Italy, and the mission of San José y San Miguel Aguayo in San Antonio, Texas, are different examples of restoration with the insertion of new structures and elements, and it is therefore interesting to verify how the approach to the monumental architecture is the same.

12.3.1 An Italian Case Study: The Church of Santa Lucia, Magliano De' Marsi

The church of Santa Lucia in Magliano de' Marsi, in the province of L'Aquila, was built in the fourteenth century and represent an interesting case study in Abruzzese architecture (Miarelli Mariani 1979). The result of lot of modifications over the *longue durée* the church today presents an interior—the result of the restoration following the earthquake of 1915—which makes reference to one of the first phases of the church's life (Fig. 12.2). The structure of the church is organised in a wider central nave and two narrower aisles, with three decorated entrance portals, the central one being taller and an interior with columns supporting arches. Only the nave has a roofing system of trusses, while the aisles have cross vaults with ribs. The bell tower is external from the structure of the church, positioned to the rear. From the pictures, it is possible to note the dome in the span before the presbytery.

Some early photographs record the interior and the façade of the church, facilitating the identification of various elements within and outside the building. Many exist no more because decorations such as the two windows above the lateral

Fig. 12.2 Church of Santa Lucia in Magliano De' Marsi before the earthquake of 1915. (left and centre) The main façade, with the clock tower and the two circular windows above the lateral gates, which today do not exist no more. (right) The interior of the church with the baroque decorations and the upper windows (private archive)

entrances in the façade, and on the interior, the Baroque decoration of the dome located in front of the presbytery, beyond the triumphal arch, was all removed as part of the modifications following the damage inflicted by the earthquake, which hit the city of Magliano, resulting in lots of victims and causing a lot of collapses. The church suffered much damages, with the collapse of the dome, the upper part of the main nave and the entire roofing system, and with lots of cracks to the bell tower, which was built only in the nineteenth century (Fig. 12.3). The state of damage was so great that the necessary period for taking decisions about the reconstruction and restoration of the architecture took considerable time, with the assignment of the project to the engineer Sabatini only in 1928.

The project designed by Sabatini recovered the structure, in particular the complete reconstruction of the façade, inserting the original elements including the lateral windows. According to Sabatini's project, the façade needed to be rebuilt with the same form and elements; in particular, comparing the drawing with the actual state, it is possible to see that the two windows above the lateral entrances were not reconstructed (Fig. 12.4). The intervention comprised the insertion of a new structure in a frame of reinforced concrete inside the existing walls, rebuilding the collapsed walls in bricks, the removal of all the baroque decoration of the interior and the dome and also the consequent reconstruction of the roof in a different way, with trusses in wood and tie beams in steel. Sabatini's project encountered several difficulties during its realisation, due to a lack of design and structural computations. In this way, the engineer Pietrangeli became director of the works in 1932, introducing also some modification to the structural recover of the church, trying to maintain the original purpose of the project. In this case, the use of reinforced concrete structures represents an interesting applied case in solving some structural issues.

Fig. 12.3 Church of Santa Lucia in Magliano de' Marsi after the earthquake of 1915, with the damage on the main façade, on the dome and the tower bell, and the collapse of the entire roof and the central arches (private archive)

Fig. 12.4 (left) Proposed reconstruction project of the façade of the church of Santa Lucia in Magliano de' Marsi designed by the engineer Sabatini (private archive). (right) The façade nowadays (author 2019)

The main façade was disassembled, reinforced with a concrete frame with a truss at the intersection with the new roof structure, and reassembled with the original elements, previous catalogued. The foundation level was improved with the insertion of foundation beams and plinths in correspondence to the nave columns; the arches above the columns of the nave and the cross vaults of the aisles were rebuilt with reinforced concrete structures, covered with decoration in cement simulating stone, with top beams to connect all the frames around the perimeter of the building. Another important element to be rebuilt in reinforced concrete was the roof of the main nave; in this case, the choice was strongly connected to the necessity to link the different longitudinal frame of the main nave with rigid systems (and not with sample timber trusses with steel ties): the built structure in concrete reproduced trusses, covered in wood planks (Fig. 12.5). As Sabatini had already done, Pietrangeli likewise decided to rebuild the bell tower with a mixed structure in masonry and framed in reinforced

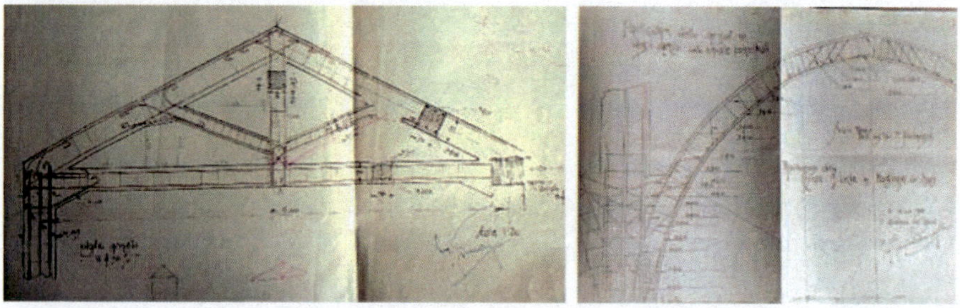

Fig. 12.5 Reconstruction project of the church of Santa Lucia in Magliano De' Marsi, designed by the engineer Pietrangeli. (left) Details of the reinforcement of the truss in reinforced concrete, with the indication also of the diameters and positions of the steel bars. (right) Details of the reinforcement of arches and columns of the main nave, with the indication of the diameters and positions of the steel bars (private archive)

Fig. 12.6 (left) Bell tower of the church of Santa Lucia in Magliano de' Marsi during reconstruction work directed by the engineer Pietrangeli (private archive). (right) The interior of the church nowadays, with the removal of the baroque elements and the dome and the insertion of trusses in the central nave and columns, arches and ribs in reinforced concrete (author)

concrete, composed of four columns at the corners of the tower, connecting beams and a staircase (Fig. 12.6). The work was completed in 1937.

12.3.2 An American Case Study: Mission San José Y San Miguel Aguayo, San Antonio

San José y San Miguel de Aguayo, the "Queen of the Missions", is the largest mission in San Antonio, established in 1720 and completed in 1782. Spanish designers, directing workers from the local Coahuiltecan tribe, built the mission using Texas limestone and brightly coloured stucco. At its height, it provided sanctuary and a social and cultural community for more than 300 Indians and was surrounded by acres of fields and herds of livestock. The mission had its own gristmill and granary, which have been preserved.

The main façade of the church is a strong reference to Spanish and European architecture, with the presence much decoration, the tower bell in correspondence with the first span, the gate, windows, the dome and the statues. The aspects of these missions, typical also in other religious buildings founded by the Franciscan friars, find their origins in the education of the friars themselves, who were the architects and designers and tried to reproduce the architecture they knew from their travels in Europe or directly out of books. This also applies to other decorative elements as well as the form of the dome, which is not hemispherical but low, and the use of coloured glass typical of Mexico and central America (Fisher 1996).

Mission San Jose, which called back to their territories of origin, was abandoned at the end of the eighteenth century with the secularisation of the Franciscan friars and started to go into decline. After many storms and hurricanes, there were two important collapses before the restoration project, which covered the period late 1860s and early 1870s. In these collapses, the church lost its roof and the main part of the north wall, as well as the dome. It is still not clear if the church suffered this damage because of the strength of the winds or because of the declining state of its structure. From the picture, it is possible to understand what caused the dome to the collapse: The northern wall at the base of the dome was severely damaged, because it was unable to withstand the toppling over forces due to the weight of the upper elements and hence collapsed. As a result, the dome, no longer having sufficient tensile strength at the base, also toppled over. Further damage appeared in other elements of the church, such as the bell tower, which was affected by the appearance of cracks in the 1920s.

The architect Harvey Smith was commissioned for the restoration project, which begun in 1932 and concluded in 1936. The restoration approach to the building was different: In particular, for the reconstruction of the collapsed part of the bell tower, the architect decided to use the same materials and techniques, avoiding the insertion of new structural elements. In this way, the bell tower would contain the same materials to achieve a uniform behaviour in the face of stresses and also the same formal restitution, with no differences from the existing and intact elements. By contrast, for the dome and the northern wall, the reconstruction approach was completely different from the bell tower. In particular, the aim of the architect was not only the recreation of the "dome" but also the insertion of an auxiliary structure, hidden inside the existing walls, able to balance the stresses out of the plane of the walls. Harvey Smith pursued this target with the insertion of different frames in reinforced concrete, with pairs of columns and linking beams at ground and upper levels, one for each bay of the church. The supporting structure of the dome, in this way, is formed by two different frames and a series of tying beams in different levels: The base plan of the dome presents an octagonal form, in which each side is composed alternately in steel and reinforced concrete; with a series of eight small columns, supported by the previous exposed base, there is a circular ring beam in reinforced concrete, which constitute the stand of the new dome, entirely realised with a slab in reinforced concrete (Figs. 12.7, 12.8). Figure 12.9 shows the current state of the building.

Fig. 12.7 Mission of San José y San Miguel de Aguayo in San Antonio during the restoration works of 1932–1936, conducted by Harvey Smith. In this picture, it is possible to see the work on the dome with the new structures in reinforced concrete (Library of Congress, Prints and Photographs Division, HABS)

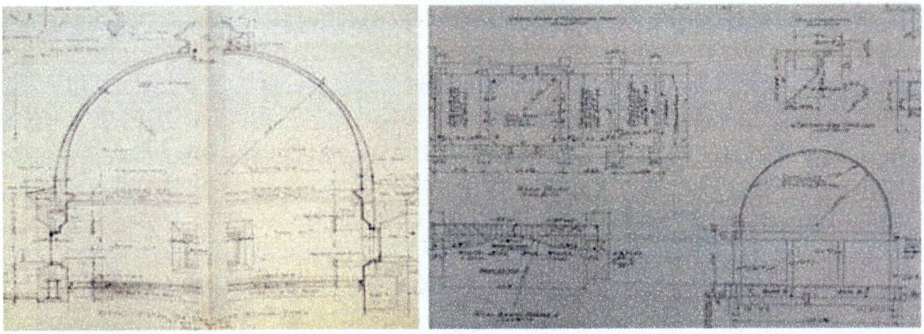

Fig. 12.8 (left) Drawing of the restoration project of the Mission of San José y San Miguel de Aguayo in San Antonio by Harvy Smith. Section of the dome, with the indications of the new structures in reinforced concrete and steel. (right) The plans and section with the scheme of placement of the new structures inside the masonry and the systems of beams at ground and top levels; section of the dome with the ring beams and columns (courtesy of the U.S. National Park Service)

Fig. 12.9 The Mission of San José y San Miguel de Aguayo in San Antonio today (author)

12.4 Conclusion

This chapter, focusing on different areas of study, analysed the methods used during a particular period in which the necessity of rapid recovery was fundamental, and technological developments had an important role in the reconstruction phases after destructive and traumatic events.

Before the critical evaluations about the interventions and the pursued strategies, it is important to remark the direct connection between the Italian and the American restoration approaches. In particular, the latter context caught the “first steps” in the preservation field directly from the European theories of the late nineteenth century. After the debate between renovation and pure conservation of the built heritage, the Italian theory of the scientific restoration declared as key points the recognisability of the interventions and the use of the modern technologies (Brandi 1957), in this way linking below the same approach the first experimentations pursued in Italy and US. The work by Giles in the Alamo, defining the new ceiling with a different coloured material from the original, highlights the differences of the element, using also the modern means made available by the construction technological developments.

The potential of concrete in particular its frame system, had been exploited in emergency situation, thanks to its good mechanical characteristics, together with its ease and quick applications in the construction sites, letting the material suitable for this kind of construction needs. In this way, it had been a natural consequence that the choice of the material followed strong and destroying earthquakes, as the ones occurred in Marsica 1915 and in Santa Barbara 1925. However, the contents of the released Athens' Charter in 1931, suggested strongly "between the lines" the use of the modern means, in particular the reinforced concrete, letting the spread use worldwide. As a result, the applications did not consider the nature of the existing architectures, applying the same methods in different contexts, inside different materials and typologies of architectures.

The restoration works of San Jose and Santa Lucia, pursued, respectively, in 1936 and 1937, certified the use of the experienced approach, thanks to the first technological developments of the century and the first application in the post-trauma recovery in the specific areas; the application of reinforced concrete frames inside the entire building, instead of local applications as in other previous cases—such as Collemaggio for example—suggested the reached awareness of the global necessity of the structural improvement. Beyond the structural evaluations, according to the figurative and formal aspect, the applications of the new elements defined also an awareness also in the material and visual recognisability of the interventions, disclosing some developments of the restoration international theories, as the ones of the 1950s.

At last, it is important also to focus on the impact of this kind of restoration on the mechanical and resistance characteristics: The application of new structural elements, apparently without any evaluations about the existing form and materials, provides the alteration of the mechanical balance of the building. The insertion of concrete elements, characterised by the higher value of stiffness, can involve torsional movements of the buildings in case of intense stresses, affecting potential damages. In this way, today, it becomes very important to evaluate the impact of these interventions, in a way to expect weak points of the structures and provide for specific improvements. This last aspect could be analysed in potential further research.

References

Anonymous (1900) Mission Memories: The Franciscan Missions of California. Kingsley-Bames & Neuner Company, Los Angeles

Aiken RS (1983) The spanish missions of alta California. Rise, decline and restoration, in Pioneer America, vol 15, no 1 (March). International Society for Landscape, Place & Material Culture

Bartolomucci C (2004) Santa Maria di Collemaggio. Interpretazione critica e problemi di conservazione. Palombi, Roma

Brandi C (1957) Uno sgomentante puzzle risolto per amore dell'arte. In: Corriere della Sera, December 27th

Carbonara G (1995) Trattato di Restauro Architettonico. Editore Liguori, Napoli

Ciranna S, Montuori P (2015) Avezzano, la Marsica e il circondario a cento anni dal sisma del 1915. Città e territori tra cancellazione e reinvenzione. Consiglio Regionale dell'Abruzzo, L'Aquila

Ciranna S, Felli M, Lombardi A (2019) The Shrine of the Alamo and its roof: history and strategies of recovering. In: World Heritage and Legacy, XVII International Forum "Le Vie Dei Mercanti", Conference Proceedings. Gangemi Editore, Roma

Engelhardt ZOFM (1913) The missions and missionaries of California. The James H. Barry Company, San Francisco

Felli M (2018a) Il restauro delle fortificazioni nel secondo dopoguerra. Tre sperimentazioni in calcestruzzo armato nell'entroterra abruzzese. In: Marotta A, Spallone R (eds) Defensive architecture of the Mediterranean, vol VIII. Proceedings of the international conference on modern age fortification of the Mediterranean Coast, FORTMED 2018. Politecnico di Torino, Torino

Felli M (2018b) The innovative role of Ignazio Gavini in restoration works in Abruzzo at the Beginning of the Twentieth Century. In: Porcinari, Vito Domenico. The XIV international congress of heritage's rehabilitation the conservation of artistic, architectural, archaeological and landscape heritage. Napoli: Luciano Editore, Session 1, pp 393–406. ISBN: 978–88–6026–245–5

Felli M, Ciranna S (2018) Franciscan Missions in California. "El Camino Real", an Identitary Reference of Architecture and Preservation. In: Capano et al (eds) The other city. History and image of urban diversity: places and landscapes of privilege and well-being, of isolation, of poverty and of multiculturalism. Federico II University Press, Napoli, pp 1499–1506 (Part IV). ISBN: 978–88–99930–03–5

Fisher LF (1996) Saving San Antonio, The Precarious Preservation of a Heritage. Texas Tech University Press, Lubbock TX

Gavini IC (1915) I terremoti d'Abruzzo ed i suoi monumenti. In: La rivista abruzzese di scienze, lettere ed arti, no 5, pp 235–240

Gavini IC (1923) Il cemento armato nel restauro dei monumenti. In: Ingegneria. Rivista Tecnica Mensile, February 1st, n. 2

Ivey JE (2007) The completion of the Church Roof of San Antonio de Valero. In: Anales del Instituto de Investigaciones Estéticas, vol XXIX, núm 91. Instituto de Investigaciones Estéticas. Distrito Federal, México, pp 125–153. ISSN 0185–1276

Kimbro EE, Costello JG, Ball T (2009) The california missions. History, art and preservation. The Getty Conservation Institute, Los Angeles, California

Kryder-Reid E (2010) Perennially New: Santa Barbara and the Origins of the California Mission Garden. J Soc Archit Hist 69(3), September 2010, University of California Press on behalf of the Society of Architectural Historians

Lee AJ (1990) Spanish Missions. J Preserv Technol APT Bull XXII(3)

Miarelli Mariani G (1979) Monumenti nel tempo. Per una storia del restauro in Abruzzo. In: Saggi per una storia dell'architettura. Carucci Editore, Roma

Moretti M (1972) Collemaggio. De Luca, Roma

Newcomb R (1937) Spanish-colonial Architecture in the United States. JJ Augustin, New York city

Sewall JM (2013) California Mission Architecture: a Survey and Sourcebook. Schiffer Publishing, Ltd

Weeks-Wilson L (1913) Monograph on the old Franciscan Missions. Pacific Coast Publishing Company, Santa Barbara, California, Santa Barbara
Weinberg NG (1974) Historic preservation and tradition in California: the restoration of the missions and the Spanish-Colonial revival, University of California

Archives Consulted

Archivio della Soprintendenza Unica Archeologia, Belle Arti e Paesaggio per l'Abruzzo, L'Aquila, Italy
Archivio di Stato, L'Aquila, Italy
Archivio Storico Ufficio del Genio Civile dell'Aquila, Sottosezione di Avezzano, Italy
Historic American Buildings Survey. US Department of the Interior, National Park Service. The Library of Congress
San Antonio Light Photograph Collection, MS 359, University of Texas at San Antonio Libraries Special Collections
The Alexander Architectural Archives at University of Texas Libraries
The Daughter of Republic of Texas' Archive, San Antonio TX
The General Land Office, Austin TX
U.S. National Park Service, San Antonio TX

Chapter 13
Historical Town Centres and Post-seismic Reconstructions: Between Functional Recovery and Heritage Value Awareness

Carla Bartolomucci

> It takes more than buildings to make a town (P. Nicolin, Tra le due città, *Quaderni di Lotus*, 2, 1983: 21).*

Abstract This chapter compares the reconstruction in progress in Abruzzo after the earthquakes of 2009 and 2016–17 with other experiences carried out in Italy concerning various seismic events over a period of fifty years, namely Belice 1968, Friuli 1976, Irpinia 1980, Umbria and Marche 1997. Such reconstructions—some of them now considered as "models"—are particularly significant in the attempt to understand different approaches to the preservation of historical architecture. During the actual emergency phases, the approaches to secure damaged areas and building heritage varied; even today, many constructions are demolished for security reasons, for example the recent demolitions at Amatrice and other villages damaged by the earthquake of 2016–2017. From a comparison of different situations, it seems that damage and loss of historical architecture are very often caused by post-seismic demolitions rather than by earthquakes. This loss, which affects subsequent reconstruction methods, derives from different circumstances. They include unawareness of the heritage and architectural values of the structures, as well as a forced choice between the will to reinstate, exemplified by the slogan "where it was and how it was" and development expectations, understood only as total renewal. Only when a prior recognition of place values has occurred does reconstruction safeguard historical and cultural identity, without jeopardising the authenticity, rebirth and development of damaged places. The seismic history and the historical–architectural heritage of Italian cities and landscapes require multidisciplinary considerations in identifying the values to be safeguarded. Sometimes, the vulnerability evaluation of historical heritage is subjected to prejudice against conservation.

*This observation underlines the failure of the "utopian reconstructions" carried out in the Belice Valley after the earthquake of 1968, because the quality of the new architecture did not correspond to the actual realisation of an idea of the city (understood as *civitas*, namely community of citizens).

C. Bartolomucci (✉)
University of L'Aquila, L'Aquila, Italy
e-mail: carla.bartolomucci@univaq.it

© The Author(s), under exclusive license to Springer Nature Switzerland AG 2021

F. F. Arefian et al. (eds.), *Historic Cities in the Face of Disasters*, The Urban Book Series,
https://doi.org/10.1007/978-3-030-77356-4_13

Keywords Post-seismic reconstructions · Sicily · Belice Valley · Friuli · Venzone · Campania · Irpinia · Umbria · Assisi · Abruzzo · L'Aquila · Italy

13.1 Introduction

Although Italy's territory is marked by significant seismicity, the historical buildings that characterise cities, towns and landscape have, over the centuries, withstood major earthquakes causing various damages to architectural heritage. Historical repairs and reinforcements are material evidence of so far little considered resilience of the related architecture, understood as its ability to withstand calamities. Similarly, the constructive diachrony of the buildings is rarely considered. Even today, artistic historiography attributes it to certain "styles" and eras, overlooking building history and the seismic events themselves. This leads to a strange contradiction. While earthquake chronicles tell of destruction and integral reconstruction, historical edifices show important evidence of their subsistence over time.

By way of example, reconstruction in Sicily (in the Noto Valley) after the earthquake of 1693 had significantly reconfigured towns and urban spaces in the new Baroque style and renovated existing buildings without substantial substitutions because the towns were not "irreparably" damaged (Trigilia 1997, p. 60). Similarly, the post-seismic reconstructions at L'Aquila over the past centuries (after the earthquakes of 1315, 1349, 1461, 1703) essentially preserved the historical urban fabric by repairing damaged buildings, introducing reinforcements and anti-seismic measures (D'Antonio 2013) and reusing existing wall structures (Fiorani 2009; Fiorani 2011; Centofanti Brusaporci 2011). Substantial correspondence between construction phases and seismic history can therefore be highlighted in the architectural heritage (Bartolomucci and De Cesaris 2009).

Conversely, several reconstructions during the twentieth century—starting with Avezzano and Marsica after the 1915 earthquake—erased the traces of previous structures, through demolitions carried out not only for safety reasons, but above all for the desire to renew the urban structure (Fiorani 2007; Ciranna and Montuori 2015; Bartolomucci 2015). This attitude often led to loss of cultural identity of towns and disorientation of their inhabitants, transferred to new towns, wholly dissimilar to the original places, without producing (except for Belice) any significant contemporary architecture.

Several cases of post-seismic reconstructions over the last fifty years in Italy, ranging from the earthquake of 1968 in Sicily to those of 2009 and 2016–17 in Abruzzo, are compared here with the aim to highlight the orientations towards reconstruction in historic town centres and the results of preserving (or not) different cultural and identity values (Fig. 13.1).

The contrast is evident between nostalgic attitudes—characterised by the abused slogan of "where it was and how it was", often misunderstood and used regardless of any question of authenticity and, on the contrary, a total indifference to, or prejudicial distrust of historical fabric and meanings. This contrast is characterised by

Fig. 13.1 An overview map of Italy with the cited locations of earthquakes

reconstruction methods ranging from the abandonment of ancient centres and the refoundation of entire new towns in other places, to in situ reconstructions, safeguarding only the most important monuments but changing the urban context by replacing buildings often of poor architectural quality and to integral remakes of buildings and historical fabrics in more or less identical forms, as if to ensure their authenticity.

13.2 Post-seismic Reconstructions Over the Last Fifty Years in Italy

Comparisons of various reconstructions over the last decades of the twentieth century have so far mainly analysed data on the costs incurred (CNI 2014) and the different policies for emergency management (Nimis 2009). Such studies are essential for an overview but say nothing about how much has been saved from destruction or how the cultural identity of the places has been safeguarded.[1]

This theme was dealt with recently at the Milan *Triennale*, by considering all the reconstructions during the twentieth century—from post-war to post-seismic ones, as well as those caused by other disasters, including that of the Vajont Dam in 1963—but without questioning the complex issues of the destruction itself (Ferlenga et al. 2018). While reconstruction practices are similar, the diversity of circumstances that caused destruction stimulates various reflections as to their meaning, thus preventing them from being regarded a priori as opportunities to "accelerate change".

Specific reflections were held on the centenary of the earthquakes of Reggio Calabria in 1908 (Valtieri 2008) and Avezzano in 1915 (Ciranna and Montuori 2015; Galadini and Varagnoli 2016), in which post-seismic reconstruction essentially coincided with post-war reconstruction.

This chapter examines the events of the main post-seismic reconstructions, with the aim to highlight results regarding the preservation of historical and architectural heritage and the maintenance of the cultural identity of the places affected. It is intended to assess whether any evolution of reconstructive practices can be identified or, on the contrary, whether every seismic event still faces new approaches regardless of previous experiences.

The reconstructions carried out during that period, as well as those in progress, are particularly significant in reflecting on how to preserve historical architecture in different contexts. In the last century, changes in construction methods led to the use of new materials (metal profiles, reinforced concrete, cement mortars), introducing significant changes to traditional construction techniques. This fact has led to a widespread distrust of historical buildings, often inadequate even from a functional point of view, which have been generally abandoned or demolished and replaced by new buildings. Conversely, renovations have often been carried out with little awareness of the values of historic buildings and the behaviour of historical structures (Coïsson 2019).

The damage observed during recent seismic events has shown that failure and vulnerability were actually caused more often by inadequate interventions (masonry breaks to insert plant or opening of new compartments, consolidations using reinforced concrete structures which have proved detrimental) than by poor building features, as is generally assumed.[2]

In particular, after seismic events, historic buildings have been demolished for reasons of "safety" or distrust, preferring new buildings, sometimes trivially reproduced in similar forms, as if this could guarantee the preservation of historical values. Never has the historical building heritage been erased by demolitions and reconstructions as much as in the last century, insensitive to the theme of conservation and unaware of the current debate on restoration.

[1] Such a reflection is started in Nimis (2009), but the comparison between past earthquakes is limited only to the "model" cases of Friuli (1976) and Umbria–Marche (1997).

[2] The claim that the collapses of the earthquake are due to "disintegrated masonry structures because they were built with very weak mortars in non-connected masonries" appears to be generic and is often contradicted by the facts (Bartolomucci 2017).

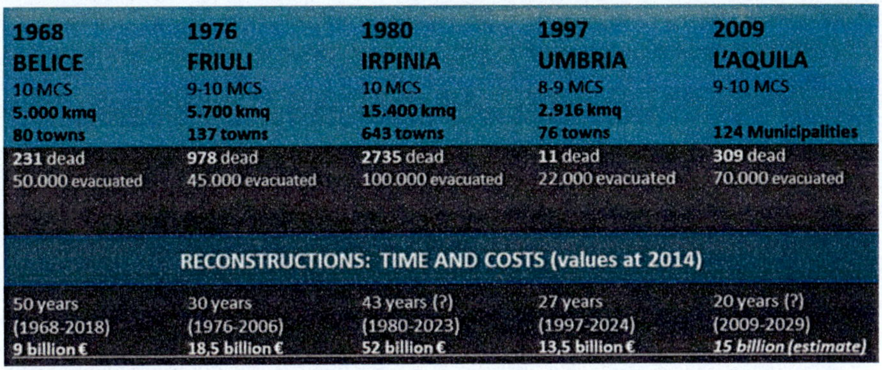

1968 BELICE	1976 FRIULI	1980 IRPINIA	1997 UMBRIA	2009 L'AQUILA
10 MCS	9-10 MCS	10 MCS	8-9 MCS	9-10 MCS
5.000 kmq	5.700 kmq	15.400 kmq	2.916 kmq	
80 towns	137 towns	643 towns	76 towns	124 Municipalities
231 dead	978 dead	2735 dead	11 dead	309 dead
50.000 evacuated	45.000 evacuated	100.000 evacuated	22.000 evacuated	70.000 evacuated
RECONSTRUCTIONS: TIME AND COSTS (values at 2014)				
50 years (1968-2018) 9 billion €	30 years (1976-2006) 18,5 billion €	43 years (?) (1980-2023) 52 billion €	27 years (1997-2024) 13,5 billion €	20 years (?) (2009-2029) *15 billion (estimate)*

Fig. 13.2 Earthquake data summarising the last fifty years in Italy (by C. Bartolomucci; data from INGV, CNI, USRC, USRA)[3]

Even today, notwithstanding a different awareness of the overall values of historic centres and urban fabrics, we see general demolitions after earthquakes, often unjustified, carried out during the emergency period or even later, as occurred in the historic centre of Amatrice (Rieti) after the earthquake of 2016 or Campotosto, L'Aquila, in 2017.

In some cases, conservation focuses on the most important monuments, neglecting the context, as occurred in several minor centres of the "seismic crater" of 2009. In other cases, safety procedures have spread to almost all historical buildings as in L'Aquila in 2009; however without any maintenance, they do not guarantee effectiveness over time.

Below is a summary of the main seismic events that have taken place in Italy over the last fifty years, with an attempt to highlight the criteria that guided urban reconstructions. However, numerical data (Fig. 13.2), although important for comparison, says nothing about the criteria and methods of reconstruction needed for our reflections here.

13.3 Belice After 1968: Negating Identity

A large bibliography on the earthquake and reconstruction in Belice (Sicily) highlights that it was "the last case of the centralistic approach" (Musacchio et al. 1981). The government started reconstruction related to the industrial development of the territory, using abstract and utopian models which, on the one hand, experimented with new design methods (Nicolin et al. 1983; Purini 2018), while on the other hand proved to be completely alien to the identity of the rural villages affected (Renna et al.

[3] INGV (National Institute of Geophysics and Volcanology); CNI (National Council of Engineers); USRC (Special Office for the Reconstruction of the Municipalities of the Abruzzo Seismic Crater); USRA (Special Office for the Reconstruction of L'Aquila).

1979; Esposito 2005; Scibilia 2016). The result was the foundation of new towns (Gibellina, Poggioreale, Salaparuta, Montevago) and the abandonment of entire areas left to destruction.

The best-known case is that of Gibellina, where the ruins of the old town were razed and buried by the *Great Cretto*, the land artwork created by Alberto Burri between 1984 and 1989 on the ruins of the old town (Ercolino 2016). The artist's aim was to pay tribute to the memory of the abandoned town, which he considered much more evocative than the new one. However, the final destruction of the ruins (after fifteen years of neglect) was perceived as a further outrage that accentuated the sense of alienation of local populations, forcibly displaced to for them abstract and unreal new towns (Fig. 13.3).

In the case of Montevago, the new town stands next to the remains of the old one, now ruins. The new Gibellina, on the other hand, was built 18 km from its original place. Even when the site of reconstruction remained the same—as at Santa Ninfa—building replacements completely confused places and landscape.

Earthquakes were thus viewed as an opportunity to undertake urban and architectural experiments that produced alienating and also unfinished results, while the historical and cultural identity of such sites was ignored and suppressed altogether. Many inhabitants emigrated because the desired economic restart and development did not happen, and living conditions in the "provisional" slums, where the population lived for over a decade, meant that the new architectural completions remained almost deserted, constituting today a new theme of conservation (Ercolino 2009; Versaci and Cardaci 2019).

Fig. 13.3 Gibellina (Trapani): An unusual view of the *Great Cretto*, created by Alberto Burri on the ruins of the old town (Ercolino 2008)

What must be highlighted here is that the negation of the identity of these sites was the concrete outcome of an absolute lack of awareness of intrinsic values; despite widespread attention to historic centres and minor architecture due to the Charter of Venice (1964), the villages of the Belice Valley were considered remnants of economic and social backwardness, rather than cultural testimonies, similar to Matera, considered in the 1950s as a "national shame" and today recognised as the 2019 European Capital of Culture.[4]

13.4 Friuli After 1976: Lights and Shadows of a "Model" Reconstruction

When Friuli in north-east Italy was hit by the dramatic seismic events of 1976, the reconstruction in Belice, eight years after the earthquake, had not yet begun. It was probably also this awareness that determined the opposite reaction of the population, immediately involved in the reconstruction process.

However, by reading what was written by those who took an active part in defending the identity of places, the experience of reconstruction in Friuli—universally narrated as a "model" of efficiency—reveals also lights and shadows, inadequacies and contradictions.

Monument conservation actions appeared to be "few, inadequate and often contradictory", due to the weakness of the protection system confined to listed buildings, while "systematic and indiscriminate demolition work" was carried out everywhere "in the inner towns and the oldest parts" (Binaghi Olivari et al. 1980, p. XIX). While the mobile cultural heritage was saved, albeit not entirely, historic buildings and all the elements connected to masonry structures (wall paintings, wooden ceilings, architectural fragments) ended up in landfills (Binaghi Olivari et al. 1980, p. 33).

The ancient settlements of Artegna, Buja and Moggio completely disappeared, as well as numerous rural buildings "with plants in many cases dating back to the fifteenth century" that characterised the agrarian landscape (Binaghi Olivari et al. 1980, p. 28). Today, Friuli's landscape has almost completely lost its historical building heritage. Entire villages have been replaced by new constructions, spread throughout the territory with considerable land consumption (Nimis 2009, p. 19). By way of example, in Osoppo in the province of Udine, the recovery of the Osoppo Fortress, a fortified site of great historical and landscape interest, shows an excessive intervention on historical architecture which appears as a "pretext" and a setting for the insertion of new architecture (Fig. 13.4).

Everywhere in the seismic area, major demolitions were carried out, justified as "removal of rubble and dangerous buildings". However, in Venzone in the same province of Udine, a group of volunteers strongly opposed what appeared to be a real "destruction industry", for speculative purposes. Although the citadel, the urban core

[4] In 1948, the political leader Palmiro Togliatti defined Matera as the "shame of Italy", owing to the poor living conditions of the people who inhabited the *Sassi*.

Fig. 13.4 The church of San Pietro near the Osoppo Fortress in Osoppo, Udine. The arrangement of the remains appears indifferent to the history and configuration of the existing building, which originally had a gable roof (author 2020)

enclosed by the wall, was listed in 1965, entire blocks were demolished "without any real reasons of public safety" (Binaghi Olivari et al. 1980, p. 84).

Ten days after the earthquake, a Cultural Heritage Recovery Committee was set up, which operated from a technical and political point of view. It was immediately clear that no list could be made of the damaged monumental buildings, because "the most serious alterations produced by the earthquake concerns the layout, the volume, the urban and civil expression of the town" (Binaghi Olivari et al. 1980, p. 81).

The decision to completely restore the citadel of Venzone was therefore an exception in the general destruction carried out after the earthquake. The committee's volunteers were instrumental in promoting awareness of the identity values of historical buildings—beyond the preservation of single monuments—and in saving what was still standing from destruction.

The results of reconstruction are well known (Cacitti 2010; Doglioni 2018) but, rather than the "philological" re-composition of the urban fabric, implemented thanks to the careful preservation of what was left, it is important to highlight the attention paid to safeguarding authenticity. The conservation of any remnants of the demolished buildings—of which the ground layouts are still visible—and, in particular, the reconstruction "by anastylosis" of the cathedral, which shows the reconstructed

Fig. 13.5 The restored cathedral in Venzone, Udine, testifies to the dramatic event by the signs of the recomposed walls, respecting authenticity and history (author 2019)

walls and leaves visible signs of collapse and gaps, reveals a real understanding of historical meaning and constructive stratifications (Fig. 13.5).

Today, although the slogan of "where it was and how it was" is often invoked to justify purely formal reconstructions (which completely ignore the material vestiges of history), this same awareness seems not evident elsewhere.

Venzone's experience is based on the awareness of the meaning of cultural identity and material values to be safeguarded. Significant in this regard was the foundation in 1978 "by popular will" of the University of Friuli, with Italy's first Degree Course of Conservation of Cultural Heritage.

13.5 Irpinia After 1980: The Endless Reconstruction

Reconstruction in the area including Irpinia, Sannio (Campania) and Lucania (Basilicata) is certainly not a virtuous example among the chronicles and accounts of reconstruction. The huge sum of money committed and squandered by criminal phenomena, as well as the infinite time taken due to the state paying reconstruction costs until 2023, is a sufficient reason to label it as a negative example to be banished to oblivion (Orsini 2018). However, this negative experience seems to have produced numerous reflections, in retrospect, on the perception of environmental identity (historical, architectural, urban, landscape) and on the change of urban image (Mazzoleni and Sepe 2005).

From the point of view of cultural premises of reconstruction and awareness of identity values to be safeguarded, the results were devastating. In a general context of "legalised destruction", it was only decided in Sant'Angelo dei Lombardi not to abandon and destroy the ancient town, but to draw up a recovery plan based on the perception of the historic centre as a "unique monument". Unfortunately, the results were not consistent with these forward-looking premises. The maintenance of existing structures was insufficient to preserve historical urban values, and above

Fig. 13.6 The archaeological park in the old abandoned village Conza della Campania in Avellino (Rossi Filangieri 2019)

all, Act 219/1981 (specially enacted) annulled conservative intentions all round. Overall, economic incentives for demolition and the possibility of increasing the residential area had devastating effects on historic centres, which were destroyed by replacements and building speculations or abandoned to decay (Verderosa 2005; Coletta 2010). Again, the criterion of relocation in favour of new settlements on more stable ground and the illusion of economic development, far from actual local productive vocations, led to territorial changes originating in the lack of awareness of the values to be safeguarded.

The ancient village of Conza della Campania in Avellino, perched on a hill overlooking the lake and characterised by centuries-old building stratifications, has today become an archaeological park[5] (Fig. 13.6), while the new village, founded further downstream, has such scattered buildings that it cannot really be defined as a "garden-city" which, moreover, is wholly alien to the identity of the site, nor does it show significant architecture.

In Teora, a small town in the province of Avellino, "hybrid" approach was experimented, in an attempt to combine the recovery of the historical fabric, implemented by reconstruction in similar forms, with a new urbanism of a rationally planned development, yet the results are mostly anonymous and poor-quality constructions.

Bisaccia Nuova in the province of Avellino is a separate village, distant from the historical centre. It has a still recognisable compact urban fabric and an interesting historical expansion developed in linear form along the ridge. Unfortunately, almost everywhere, individual building works seem to have disfigured the characteristics of the historic buildings.

In general, in the rare case when reconstruction took place in situ, buildings show significant alterations due to construction replacements. Even where the historical

[5] In the views of Pacichelli (1703), Conza was depicted as a suggestive ruin after the earthquake of 1694 (Di Mauro and Cantabene 2005). See also Carluccio (2002).

urban fabric seems to have been preserved, as, for example, in Calitri in the province of Avellino which has an valuable urban structure, the built heritage appears degraded by disqualifying building interventions.

As with previous earthquakes, in the case of Irpinia the greatest destruction was not caused by the earthquake but carried out later, misrepresenting the slogan "from reconstruction to development" that seems to characterise each similar event, often with questionable outcomes. Above all, southern Italy's complex issues cannot be resolved by solutions alien to the identity and characteristics of its sites, whose authenticity has often been safeguarded precisely by backwardness.

13.6 Umbria and Marche After 1997: The "Integrated Reconstruction"

Post-seismic reconstruction in Umbria and Marche after the 1997 earthquake is included in this study with some misgivings, considering the small extent of the disaster and damage in comparison with other seismic events (see the summary data in Fig. 13.1). Nevertheless, since the earthquake affected many small towns with a well-known historical and architectural heritage, some reflections are useful.

In this case, regulation for reconstruction outlined unified procedures to regulate contributions according to the severity of the damage (Doglioni 2000; Canti and Polichetti 2002). These guidelines were later applied to earthquakes in Molise 2002 and Abruzzo 2009. In addition to the concept of "light" reconstruction (the repair of non-serious damage) and "heavy" reconstruction (which essentially corresponds to construction replacement or anti-seismic adjustment), the idea emerged of "integrated" reconstruction, in an attempt to reconcile conservation needs with seismic improvement and urban development (Mirarchi, 2005). Integrated recovery plans were then developed in historic centres and urban or rural sites of cultural interest (Act 61/1998). However, the notoriety of affected sites and the dramatic nature of events (the vault collapse in the Upper Basilica of Assisi was filmed live on TV) put emphasis strongly on "treasures to be saved"; thus, the focus on safeguarding appeared to be addressed more to single monuments rather than to the urban fabric as a whole.

The split between restoration of monuments and recovery of minor heritage seems to emerge on this occasion. The latter appears to have been largely resolved by the application of a series of practices arising from the Recovery Manuals (Di Resta 2018). Although they have the merit of drawing attention to minor historical buildings, they tend to standardise methods and techniques of intervention, neglecting the specificity of each construction and trivialising results in "reproductions" of construction types. In essence, the results seem to safeguard the form rather than the material authenticity of the heritage they wish to preserve.

13.7 Abruzzo After 2009 and 2016–17: "Functional Return" in L'Aquila and in Minor Centres

Reconstruction in progress after the 2009 earthquake in Abruzzo registers a substantial difference between the regional capital and the smaller centres of the seismic "crater", not only in terms of timing (progress in L'Aquila appears to be more advanced than the rest of the crater), but especially with regard to awareness of the values to be safeguarded. Even in this case, the initial focus was only on monuments to be saved (Conforti 2009), but security works at L'Aquila cover all historic buildings and demolitions are localised and controlled (Altorio et al. 2013).

Cases vary in the surrounding villages, and in some of them demolitions have extended to most of the historic urban fabric (as at Sant'Eusanio Forconese, Onna, Villa Sant'Angelo, Castelnuovo).

This destructive trend was dramatically accentuated during the recent seismic events of 2016–17, for which reconstruction has not started yet. In the latter cases, post-seismic demolitions, carried out in the name of safety, were extensive and almost total in some villages (Amatrice, Campotosto, Accumoli, Arquata and Pescara del Tronto; see Fig. 13.7). Greater or lesser awareness of the value of these places seems

Fig. 13.7 Campotosto (AQ): what remains of the church after demolition (Bartolomucci 2017)

Fig. 13.8 Renovations of surface finishes in Santo Stefano di Sessanio in the province of L'Aquila. In this case, the practice of leaving some masonry parts uncovered is unrelated to the meaning of authenticity pursued at Venzone (author 2017)

to have led to very different choices, in all cases conditioned by widespread and general distrust of historical buildings, although distrust was unmotivated at times, since the highest number of victims in 2009 was in reinforced concrete buildings.

Looking at the results of reconstruction in progress, Reconstruction Plans (Act 77/2009) seem to have been rather weak tools, as well as very diverse in their solutions. Even where they have safeguarded the unitary value of historic centres, by avoiding destructive building and urban renovations—as for Santo Stefano di Sessanio, Castelvecchio Calvisio and Castel del Monte—the quality of the interventions in progress shows a lack of attention to authentic conservation (architectural surfaces being habitually renovated by re-plastering and coating; see Fig. 13.8).

In general, attributing value to individual elements rather than to architecture viewed as a layered historic complex leads to practices aimed at preserving form rather than authentic matter (Bartolomucci 2017).

The tendency to implement recovery rather than conservation practices—based on the recognition of values and knowledge of historical building stratifications—seems fairly prevalent. Above all, recovery is understood in a functional sense as restitution of usability and not as renewed attention to building characteristics, according to the proposals of the Recovery Manuals in previous decades, adopted for reconstruction in Umbria. Restoration generally appears misrepresented, because it is incorrectly perceived as the reinstatement of previous situations or limited only to exceptional episodes or individual formal elements (such as decorative equipment).

The results of such criteria are a disavowal of history and its meanings understood as the constructive history of buildings, but also of the events that marked town and territory. This led to a clear prevalence of systematic renovations of surface finishes (plasters, colours, coats), as well as of structural elements. Roofs have been mostly completely replaced and in minor building vaults and floor slabs have often been demolished and remade. The tendency is thus to neglect material authenticity, to erase all traces of the dramatic event and to restore functionality instead to any building without recognising its historical significance (Bartolomucci 2018b).

This purpose—which attributes a predominant "property" value to architecture—clashes paradoxically with any effective return of inhabitants to historic centres. Above all, the "reversal of values" of historical building in favour of their current use and their restitution to owners overlooks any reflection on the meaning of cultural heritage transmission (Dalai Emiliani 2010; Bartolomucci, 2019; Bravaglieri et al. 2019).

13.8 Conclusion: "Identity" Does not Mean "Identical"

The misleading contrast between two opposing positions—the one tending towards restoration of the previous situation by erasing any trace of the dramatic event and the other in favour of an imagined future development regardless of context—is an antithesis that does not correspond to the reality of facts. However, this opposition has a significant impact on reconstructions; especially, immediate demolitions are crucial in orienting the results of reconstruction, generally carried out several years after the dramatic event.

For these reasons, the guidelines[6] drawn up after the seismic events of 2016 recommend greater prudence "in order not to lose an essential constituent part of the area's landscape, urban and architectural heritage", based on awareness of the "overall values of historic centres and scattered settlements". The document also calls for "the preservation of what remains, the integration of what is damaged, the reconstruction of what has been lost" and, albeit in favour of an in situ reconstruction, considers it necessary to preserve the remains even when it is impossible to rebuild. The guidelines recommend a series of coordinated actions, such as selective collection of rubble, cataloguing and storing fragments, provisional works including protection scaffoldings and controlled demolitions,[7] in order to devise reconstruction according to "critical-conservative restoration" criteria.

Arriving too late to influence the demolitions after the seismic events of August–October 2016 and January 2017 which were already completed in June 2017, these recommendations, although based on legitimate principles, appear mostly generic. While affirming the validity of anti-seismic improvement of historical buildings (as

[6] *Methodological and Technical Guidelines for the Reconstruction of the Cultural Heritage damaged by the Earthquake of 24 August 2016 and following* (DDG del 30.11.2016—Rep. 651), Roma, 6 June 2017.

[7] See the *Directive for the Procedures for the removal and recovery of debris of protected property and historic buildings* of 15 September 2016, issued by the Directorate General for Archaeology, Arts and Landscape of MIBAC.

opposed to the widespread prejudice in favour of their demolition because any adjustment is deemed impossible), they nevertheless seem to admit demolitions and reconstructions in similar forms. They call for the use of "old materials" and traditional techniques and recommend the preservation of "standing walls" to avoid the complete demolition of such walls "which can possibly only be removed during reconstruction, having finished their function as documentation". Yet, the recommendations seem to limit the meaning of such masonry to documentation only, thereby justifying banal reinstatement in identical forms. Thus, the ambiguity between conservation and recovery seems to persist, although the generic "complete reconstruction in accordance with the values of existing building" may appear inappropriate, because it is not always necessary. Any voids, perceived as *lacunae*, can be nevertheless significant compared to any reintegration "made with different methods, from analogue to the critical interpretive ones". Gaps themselves form equally part of continuity with historical context.[8]

In conclusion, an earthquake is an opportunity to deepen knowledge of buildings, materials, building techniques, construction history and transformations generally. In that connection, it is necessary to document preliminary studies, restoration projects and the implementation of works. Documentation activities have not only a didactic purpose, but also technical–scientific and cultural significance in a broad sense. All such documentation must be transmitted to the future, together with cultural heritage, through publications and open-access information systems.

Post-seismic reconstructions cannot be accomplished only through material repairs and restoring functionality to buildings. It is essential to consider the meaning of places and to preserve their cultural and identity values, also as perceived by the inhabitants affected by earthquakes. Particular attention must be paid to wider historical urban fabrics and not just to the most important architectural monuments.

Cultural demands are just as important as material ones; safeguarding the authenticity of places is a right of all citizens, including future generations.

References

Altorio E, Basti S, Marchetti L (a cura di) (2013) MISAQ: Messe in sicurezza all'Aquila. Dopo il terremoto del 6 aprile 2009. Officina multimedia Abruzzo, Avezzano

Bartolomucci C (2015) Gli effetti del terremoto del 1915 nella città di Aquila: i danni e gli orientamenti per il restauro. In: Ciranna S, Montuori P (eds) Avezzano, la Marsica e il circondario a cento anni dal sisma del 1915. Città e territori tra cancellazione e reinvenzione. Consiglio Regionale dell'Abruzzo, L'Aquila, pp 151–162

Bartolomucci C (2017) La ricerca nel restauro come risposta al disastro. Il terremoto in Abruzzo: priorità, prospettive, sfide e occasioni (sinora) mancate. In: Fiorani D (coord) Ricerca/Restauro, vol. 3A. Quasar, Roma, pp 705–715

[8] See the signs of historical "*lacunae*" still recognisable today in the urban fabric, such as the residual walls of buildings (presumably collapsed in the past centuries) that support raised gardens. Today, these "voids"—evidence of past collapses that have not been reinstated—have an important function as testimony and "warning" for the future.

Bartolomucci C (2018a) Terremoti e resilienza nell'architettura aquilana. Persistenze, trasformazioni e restauro del palazzo Carli Benedetti. Quasar, Roma

Bartolomucci C (2018b) Il restauro nell'epoca dell'apparenza: quando intervenire diventa rinnovare. In: Biscontin G, Driussi G (eds) Intervenire sulle superfici dell'architettura tra bilanci e prospettive. Edizioni Arcadia Ricerche, Venezia-Marghera, pp 65–74

Bartolomucci C (2019) Committenza privata e interesse pubblico: la ricerca di un difficile equilibrio. In: Restauro. Conoscenza, Progetto, Cantiere, Gestione, Proceedings of II Convegno SIRA (Bologna, 21–22 settembre 2018). Quasar, in corso di stampa, Roma

Bartolomucci C, De Cesaris F (2009) L'Aquila: Palazzo Carli-Benedetti. Fasi costruttive e storia sismica, «Arkos» 20:71–77

Binaghi Olivari MT et al (1980) Le pietre dello scandalo: la politica dei beni culturali nel Friuli del terremoto. Einaudi, Torino

Bravaglieri S, Furioni S, Zenoni E (2019) L'insostenibile leggerezza del "com'era, dov'era." Ananke 87:73–83

Cacitti R (2010) "Reaedificabo illud sicut in diebus antiquis" Friuli 1976: le scelte per il patrimonio culturale e la ricomposizione del centro storico di Venzone. In: Nicita P (ed) L'Aquila: questioni aperte. Il ruolo della cultura nell'Italia dei terremoti. Atti del convegno (Roma, 10 dicembre 2009), pp. 67–76

Canti M, Polichetti ML (2002) Il patrimonio culturale dall'emergenza sismica del 1997 al piano di ripristino recupero e restauro: il caso delle Marche. Silvana editoriale, Cinisello Balsamo

Carluccio M (2002) Conza della Campania: il parco archeologico Compsa. De Angelis, Avellino

Centofanti M, Brusaporci S (2011) Il disegno della città e le sue trasformazioni. In: Ciranna S, Piñeiro V (eds) L'Aquila oltre i terremoti. Costruzioni e ricostruzioni della città, "Città & Storia", vol VI, no 1, pp 151–187

Ciranna S, Montuori P (eds) (2015) Avezzano, la Marsica e il circondario a cento anni dal sisma del 1915. Città e territori tra cancellazione e reinvenzione. Consiglio Regionale dell'Abruzzo, L'Aquila

CNI (eds) (2014) I costi dei terremoti in Italia. Centro Studi Consiglio Nazionale Ingegneri, Roma

Coïsson E (2019) Riduzione del rischio sismico negli edifici storici in muratura. Maggioli Editore, Sant'Arcangelo di Romagna

Coletta T (2010) I centri storici minori abbandonati della Campania: conservazione, recupero e valorizzazione. Edizioni scientifiche italiane, Napoli-Roma

Conforti C (2009) L'Aquila: una città d'arte da salvare. Carsa, Pescara

D'Antonio M (2013) Ita terraemotus damna impedire. Note sulle tecniche antisismiche storiche in Abruzzo. Carsa, Pescara

Dalai Emiliani M (2010) Conclusioni. In: Nicita P (ed) L'Aquila: questioni aperte. Il ruolo della cultura nell'Italia dei terremoti. Atti del convegno (Roma, 10 dicembre 2009), pp 195–203

Di Mauro L, Cantabene G (2005) Il valore storico. In: Mazzoleni D, Sepe M (eds) Rischio sismico, paesaggio, architettura: l'Irpinia, contributi per un progetto. Università degli Studi di Napoli Federico II, Napoli, pp 24–30

Di Resta S (2018) Umbria e Marche 1979, 1997. I "Manuali del Recupero". In: Ferlenga A, Bassoli N, Galli J (eds) Ricostruzioni: architettura, città, paesaggio nell'epoca delle distruzioni. Silvana editoriale, Cinisello Balsamo (MI), pp 111–116

Doglioni F (2000) Codice di pratica (Linee Guida) per la progettazione degli interventi di riparazione, miglioramento sismico e restauro dei beni architettonici danneggiati dal terremoto umbro-marchigiano del 1997. In: Bollettino Ufficiale della Regione Marche, a. XXXI, n. 15, 29 settembre 2000, pp 15–252

Doglioni F (2018) Friuli 1976. Venzone com'era e dov'era. In: Ferlenga A, Bassoli N, Galli J (eds) Ricostruzioni: architettura, città, paesaggio nell'epoca delle distruzioni. Silvana editoriale, Cinisello Balsamo (MI), pp 83–91

Ercolino MG (2009) The Grande Cretto in Gibellina. Issues related to the conservation of art, memory and landscape. In: Art d'aujourd'hui, Patrimoine de demain: Conservation et restauration des oeuvres contemporaines, 13° Journées d'étude de la SFIIC, Paris, 24–26 juin 2009. Institute National du Patrimoine, Paris, pp 268–275
Ercolino MG (2016) Tra conservazione e 'restauro partecipato'. Riflessioni sul grande Cretto di Burri a Gibellina. In: Driussi G, Biscontin G (eds) Eresia e ortodossia nel restauro. Progetti e realizzazioni. Edizioni Arcadia Ricerche, Venezia-Marghera, pp 323–333
Esposito R (2005) Ricostruzione come cancellazione: il Belice. In: Sepe M (eds) Rischio sismico, paesaggio, architettura: l'Irpinia, contributi per un progetto. Università degli Studi di Napoli Federico II, Napoli, pp 187–193
Ferlenga A, Bassoli N, Galli J (eds) (2018) Ricostruzioni: architettura, città, paesaggio nell'epoca delle distruzioni, Catalogo della mostra (Milano, La Triennale, 30 novembre–10 febbraio 2019). Silvana editoriale, Cinisello Balsamo (MI)
Fiorani D (2007) Rovine e 'miracoli artistici' del terremoto di Avezzano: le architetture storiche nella piana del Fucino. In: Sette MP, Caperna M, Docci M, Turco MG (eds) Saggi in onore di Gaetano Miarelli Mariani, "Quaderni dell'Istituto di Storia dell'Architettura", 44–50 (2004–2007). Bonsignori, Roma, pp 491–502
Fiorani D (2009) Edifici storici, stratificazioni e danni nell'aquilano: una panoramica. Arkos 20 (n.s.):8–17
Fiorani D (2011) Il perenne ciclo del divenire nel cantiere storico aquilano. Annotazioni su tessuto urbano, architetture e costruzione nella città dei terremoti. In: Ciranna S, Piñeiro V (eds) L'Aquila oltre i terremoti. Costruzioni e ricostruzioni della città, "Città & Storia", vol VI, no 1, pp 239–260
Galadini F, Varagnoli C (eds) (2016) Marsica 1915-L'Aquila 2009: un secolo di ricostruzioni. Gangemi, Roma
Mazzoleni D, Sepe M (eds) (2005) Rischio sismico, paesaggio, architettura: l'Irpinia, contributi per un progetto. Università degli Studi di Napoli Federico II, CRdC-AMRA, Napoli
Mirarchi F (2005) Ricostruzione come recupero di un patrimonio: l'Umbria. In: Mazzoleni D, Sepe M (eds) Rischio sismico, paesaggio, architettura: l'Irpinia, contributi per un progetto. Università degli Studi di Napoli Federico II, Napoli, pp 211–220
Musacchio A et al (1981) Stato e società nel Belice. La gestione del terremoto: 1968–1976. Franco Angeli, Milano
Nicolin P, Minardi B (eds) (1983) Dopo il terremoto. Belice 1980: laboratorio di progettazione. Electa, Milano
Nimis GP (2009) Terre mobili. Dal Belice al Friuli, dall'Umbria all'Abruzzo. Donzelli, Roma
Orsini F (2018) Irpinia 1980. Un terremoto dimenticato. In: Ferlenga A, Bassoli N, Galli J (eds) Ricostruzioni: architettura, città, paesaggio nell'epoca delle distruzioni. Silvana editoriale, Cinisello Balsamo (MI), pp 92–97
Purini F (2018) Belice 1968. Un'avventura siciliana. In: Ferlenga A, Bassoli N, Galli J (eds) Ricostruzioni: architettura, città, paesaggio nell'epoca delle distruzioni, Catalogo della mostra. Silvana editoriale, Cinisello Balsamo (MI), pp 72–76
Renna A, De Bonis A, Gangemi G (1979) Costruzione e progetto: la valle del Belice. CLUP, Milano
Scibilia F (2016) The reconstruction of Gibellina after the Belice earthquake of 1968. Int Plan Hist Soc Proc [S.l.] 17(2). https://journals.open.tudelft.nl/index.php/iphs/article/view/1245. Accessed 22 July 2019
Trigilia L (1997) Le città ricostruite dopo il terremoto siciliano del 1693. Alcune riflessioni sullo stato degli studi e sul ruolo delle "varianti" locali. In: Casamento A, Guidoni E (eds) Le città ricostruite dopo il terremoto siciliano del 1693. Tecniche e significati delle progettazioni urbane. Kappa, Roma, pp 56–64
Valtieri S (ed) (2008) 28 dicembre 1908: la grande ricostruzione dopo il terremoto del 1908 nell'area dello Stretto. Roma, Clear, p 2008
Verderosa A (2005) La legge di ricostruzione n. 219/81, la distruzione legalizzata dei centri storici. In: Mazzoleni D, Sepe M (eds) Rischio sismico, paesaggio, architettura: l'Irpinia, contributi per un progetto. Università degli Studi di Napoli Federico II, Napoli, pp 316–330

Versaci A, Cardaci A (2019) Restauro e problemi di conservazione: la valle del Belice in Sicilia. In: Driussi G, Biscontin G (eds) Il patrimonio culturale in mutamento. Le sfide dell'uso. Edizioni Arcadia Ricerche, Venezia-Marghera, pp 1185–1194

Chapter 14
Integrating Green Solutions into Post-earthquake Recovery of Bam, Iran

Ameneh Karimian, Alireza Fallahi, and Seyed Hassan Taghvaei

Abstract The 2004 Indian Ocean tsunami drew attention to the need to incorporate environmental and ecosystem safeguards in disaster risk reduction strategies. Shortly after the disaster, several green approaches emerged within post-disaster reconstruction policies. These strategies were taken into account one year earlier in Bam as a response to the 2003 earthquake. Conservation of environment and cultural landscapes was acknowledged as one of the top priorities due to the exceptional urban fabric of the city. In practice, however, it seems that integrating too many complex principles into this multidimensional reconstruction caused a setback in achieving initial prospects such as safeguarding the unique characteristics of the city. This chapter presents the experiences and the adoption of green solutions in the Bam post-disaster reconstruction with a special focus on the revitalisation of gardens and *qanats* (the traditional water system), wastewater treatment and reuse of solid waste in the geographical scope of garden-houses. This chapter shows that in the newly built houses in Bam both green solutions and sustainable traditional techniques were overlooked, and only structural measures were adopted, whereas in the vernacular houses of the region, environmental practices can be traced, and it seems that there is an opportunity to integrate further green measures into already existing ecological approaches. This will lead to more resilience and sustainable housing and concurrently ensure the safeguarding of the traditional garden-houses. There are potentials to combine environmentally responsible practices with the principles of vernacular architecture to significantly improve the results in post-disaster housing reconstruction. Accordingly, some strategies for green reconstruction of garden-houses are proposed and elaborated within the chapter.

A. Karimian (✉) · A. Fallahi · S. H. Taghvaei
Shahid Beheshti University, Tehran, Iran
e-mail: ameneh.karimian@gmail.com

© The Author(s), under exclusive license to Springer Nature Switzerland AG 2021

F. F. Arefian et al. (eds.), *Historic Cities in the Face of Disasters*, The Urban Book Series,
https://doi.org/10.1007/978-3-030-77356-4_14

Keywords Green recovery · Traditional knowledge · Bam earthquake · Garden-house · Qanat · Bam · Iran

14.1 Introduction

The greater inclusion of environmental sustainability objectives in recovery and reconstruction documents and guidelines has been steadily rising since 1994 through, for example, the Code of Conduct for Disaster Relief (1994). These objectives have been further reinforced with incorporation into a number of global policy agreements including Sendai Framework for Disaster Risk Reduction, the United Nations Sustainable Development Agenda and the Sustainable Development Goals and the Paris Agreement on Climate Change. This has also resulted in developing several approaches such as Ecosystem-based Adaptation, Ecosystem-based Disaster Risk Reduction, Green Infrastructures, Green Recovery and Reconstruction, Going Green and Build it Back Green (Renaud et al. 2016; WWF 2005; DOE 2009; Sabto 2011). Although they have a different interpretation of the term "green", they all target minimising environmental degradation.

Taking a step back, vernacular architecture is claimed to comply with the environment due to the traditional knowledge that has been accumulated and passed on. Over the centuries, indigenous communities have developed various coping mechanisms and adaptation strategies to collectively respond to change (Lungharwo and Trenk-Hinterberger 2015). Thus, the traditional knowledge systems embodied in local construction and management systems can not only prevent or mitigate the impact of disasters but also provide sufficient coping mechanisms to deal with post-disaster situations (UNESCO WHC 2010). In Bam, the urban fabric has its roots in ancient water irrigation system of *qanats* that were genuinely developed in the region as a response to climate change being in complete harmony with nature. The house, the garden and the *qanat* create a dynamic sustainable system that has endured for hundreds of years and enabled the local populations to adapt to the severe climate of the region. In a community with rich indigenous knowledge, employing the traditional techniques that are more economically and socially sustainable than modern methods will provide a solid foundation for incorporating further new green solutions. This chapter draws on a master's thesis by the first author. The chapter explores combining traditional techniques with contemporary sustainable practices in garden-houses to address the issues of rebuilding resilient housing during post-disaster reconstruction.

14.2 Background: Green Solutions

The term "green" has been widely being used with no clear common definition. The World Bank uses "green" to describe a world in which natural resources are sustainably managed and conserved to improve livelihoods and ensure food security (World

Bank 2012). In disaster risk management literature, this term occasionally refers to energy efficiency (DOE 2009) and mitigating greenhouse gas emissions (Urban et al. 2010). The Green Recovery and Reconstruction Toolkit (GRRT) presents a more comprehensive definition for "green reconstruction": "It aims to improve the quality of life for communities and affected individuals whilst minimising the negative impacts of reconstruction on the environment and maintaining the long-term biological diversity and productivity of natural systems (WWF 2005, p. 3)". These items should be considered in all different phases of the project management cycle and various aspects of reconstruction, including site selection, project design, construction, waste management, livelihoods, education, monitoring and evaluation, water and sanitation, etc.

It seems that in recent years green practices have been frequently transformed from ecological approaches to high-tech solutions, ignoring the need to be cost-effective and applicable in deprived areas. It should be noted that every situation is unique and measures must be adopted in accordance with the setting. The city of Greensburg in the USA, for instance, was built from scratch after the 2007 devastating tornado and became a pioneer for the incorporation of modern renewable energy technologies into the town's rebuilding efforts. As a result, new houses use in average 40% less energy, many LEED-certified buildings were constructed, a 12.5-megawatt wind farm was built and Greensburg Sustainable Comprehensive Master Plan was developed (DOE 2012). Sri Lanka, on the other hand, enjoyed more affordable sustainable practices after the 2004 Indian Ocean tsunami that mostly relied on traditional knowledge and small-scale ecologically friendly technologies. As one of the pilot locations in which the GRRT was articulated, the programme helped tsunami-affected families adopt environmentally friendly waste management through community-based home gardening, composting and recycling, among many other activities (Environmental Foundation 2010). The first approach seems more infrastructure-based and substantial, but it is likely to be more of a desirable ideal rather than an achievable reality in developing countries. In these settings, drawing on indigenous knowledge and cost-effective practices are more pragmatic. This would also highlight the value of "partnering up with communities rather than just inviting their participation" (Sabto 2011). In Table 14.1, a number of adopted measures as part of the green reconstruction and recovery approaches are presented. All the mentioned green practices are feasible to be employed in small communities. The key factor is public awareness and mobilisation to include these techniques in their house or community.

A brief review of previous experiences demonstrates that it is easier to integrate green practices during design and construction, instead of trying to retrofit them. Hence, the reconstruction phase provides good opportunities to promote best practices (WWF Nepal 2016). The green solutions should be selected very carefully though, to be fully in harmony with the setting, socially, culturally and ecologically. As Garnier et al. (2013) suggest, new technical proposals should be based on the local building cultures to have a chance to be reproduced by local populations and to promote an approach based on common sense and science while revisiting lessons from the past. This would also facilitate broader acceptance and engagement by the community. According to Koohafkan and Altieri (2011), there are real opportunities

Table 14.1 Small affordable green solutions (WWF and American Red Cross 2010; UNEP and SKAT 2007; Arslan 2007; Browning 2006; Lungharwo and Trenk-Hinterberger 2015)

Case study	Measures	Strategies
Bangladesh, after the 2007 Cyclone Sidr	Using concrete structure and walls with local thatch matting allowing for easy repair	Proper consideration of community needs, consideration of the life cycle of the building, and designs that reduce future disaster risk
Pondicherry, India, after the 2004 Indian Ocean tsunami	Using ërat trapí[1] method of masonry structure that conserved 25% of production materials, recycling and introducing fly ash as a resource	The efficient use of the material and reducing their production
Turkey, after the 1999 earthquake	Disassembling many unused temporary housing units and reusing and recycling their materials to construct longer-term ëredesignedí houses	Speeding up the recovery process, a quicker transition to normalcy for affected populations and reducing waste
New Orleans, after the 2005 Hurricane Katrina	A six-day salvage operation, nail and debris removal, and full-site cleaning leading to salvage roughly 40% of the damaged home or 150 cubic yards of material	Mobilisation of the local community, reducing the amount of debris, mitigating the need for costly new materials
Aceh, Indonesia, after the 2004 Indian Ocean tsunami	Constructing household-level artificial wetland for wastewater treatment	Reducing wastewater discharge, mitigating irrigation water in need
Baan Pru Teau Township, Thailand, after the 2004 Indian Ocean tsunami	Using a septic tank and wetland treatment system to treat the wastewater from a newly built township with 80 households	Applying low-cost, robust, ease of operation and appropriate technologies, promoting the recovery and reuse of wastewater
Vangma, Lao PDR, responding to drought	Establishing forest conservation and planting mixed varieties of tree crops	Developing a 'green community' and providing an additional organic food source that improves their food security in the aftermath of a hazard
Bhotang, Nepal, responding to water scarcity	Building ponds based on local knowledge and technologies to store water and replenish groundwater	Water resource management and promoting biodiversity

[1] In this method all the bricks are laid on their sides instead of flat. The headers are laid across the wall to form a similar pattern on the other sides. As a result, the stretchers effectively form two parallel walls with a gap between them (WWF and American Red Cross 2010).

for investing in local communities and their resources, indigenous knowledge and social innovation, rather than relying on excessive external inputs and often inappropriate and unsustainable technologies from outside. This chapter tries to identify some of these opportunities in the setting of Bam's garden-houses to integrate green solutions.

14.3 Bam Reconstruction

Bam is an oasis on the southern fringe of the Lut Desert in the south-east of Iran. Reported as the deadliest earthquake of the year, the 6.6 magnitude earthquake (USGS Website) hit the Bam area on 26 December 2003. In Bam city, 91.2% of total surveyed places and 95.5% of residential houses were severely damaged (SCI Website). Bam and its cultural landscape were inscribed on the UNESCO's World Heritage (WH) List and the List of WH in Danger simultaneously within seven months of the disaster. The massive excavation in the aftermath of the earthquake revealed that the region of Bam was inhabited since the 4th millennium B.C. (Adle 2007) and the development of its territory represents a concrete testimony to human ingenuity in establishing and sustaining a habitable settlement within a harsh desert environment (Taniguchi and Fardanesh 2006). This full harmonisation between the local culture and the natural environment is a process of centuries of efforts to increase the resilience, reduce vulnerability and make full use of the locally available resources. Accordingly, the water behind the fatal strike-slip fault was managed to be extracted by means of *qanats* as a climate change adaptation strategy in the first millennium B.C. (Adle 2007) which led to flourishing locally adapted agricultural systems and the vernacular settlements. Before the earthquake, the gardens together with the *qanat* network and adobe constructions were illustrative of the traditional visage of Bam organic garden city.

Shortly after the disaster, several codes and guidelines were developed for recovery and reconstruction in which environmental concerns were addressed together with a number of other issues. In the *BamSustainable Development Manifesto* (2004), the adoption of green principles of building recommended making Bam a role model for the country. Preserving the identity of the city and household participation in the process of rebuilding were two of the main principles acknowledged by this manifesto (Fallahi 2007). The second volume of the *Special Structural and Strategic Plan for Bam* (2004), prepared as part of a suite of guidelines for the reconstruction, is dedicated to natural resources and environmental studies which, for instance, included stabilising the groundwater and prevention of digging new water wells. Another document, the *Urban Design Guidance Framework* (2006), provides both a theoretical framework and urban design codes for the city and proposes Bam as a green system of palm trees in both public and private realms. The *Comprehensive Management Plan (BamCMP)* (2008–2017) aims to highlight tangible and intangible cultural heritage, the natural heritage and environmental settings of Bam as vectors for sustainable development (ICHHTO and UNESCO 2008).

It should be borne in mind that the Bam reconstruction was the second experience in the country for the urban fabric and the first in an iconic cultural heritage setting. The environmental concerns were addressed for the first time officially, though not entirely practically. Sixteen years after the disaster, however, Bam is not a pioneer in boosting green buildings. In the case of preserving the identity and promoting the condition of the garden city, Arefian (2016) claims that the initial approach towards the objective of safeguarding the architectural fabric was broad and idealistic. This downplayed the post-disaster situation and the existing realities of housing development at the time. Despite reaffirming the safeguarding of the environment, measures were adopted in large and small that highly degraded the fragile environment of the city and had severe drawbacks on its identity. These were:

- Some palm groves and citrus gardens perished in the post-earthquake reconstruction phase, when residents took the opportunity to increase their building area and considerably reduce the size of gardens and yards (Parsizadeh et al. 2015). According to interviews conducted by Meskinazarian (2011), forty per cent of the interviewees believed that the reconstruction process caused extreme damage to the palms owing to their replacement with houses (Fig. 14.1).

- Overlooking the necessity of Environmental Impact Assessment led to poor site selection and endangering the environment and particularly the groundwater table. For instance, the exhibition of construction materials and techniques was established in a 7 ha land over the city's aquifer, same as some illegal construction

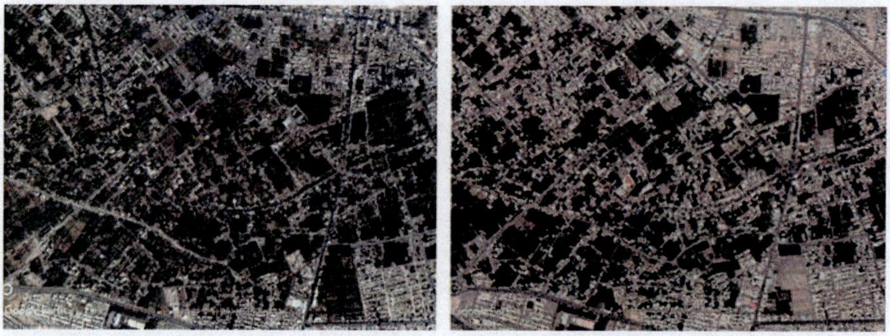

Fig. 14.1 Transformation of gardens to buildings in Bam from 2004 (left) to 2019 (right) (Google Earth 2019)

in the south of the city which led to polluting groundwater. Moreover, several large camps were constructed in the city fringe by the Ministry of the Interior in the *qanats'* buffer zone. The reconstruction also led to disposing a high amount of construction waste in the seasonal river bed and consequently exposing the ecosystem to higher risks (Bahmanpour 2006).

- Based on the law of fair water distribution, it is legal to replace a drying *qanat* by a pumped well. Following the earthquake, the *qanat* owners resorted to drilling a well. It led to mushrooming of the pumped wells and abandoning the *qanats* which could still be refurbished by extending their galleries (Semsar Yazdi and Labbaf Khaneiki 2005). According to JICA report (2005), prior to the earthquake, 20–30% of the agriculture sector employed deep wells. In the aftermath of the earthquake, as an emergency situation, 40–45% were using this resource to irrigate the farms.
- The annual water consumption rate rose about two and half times in comparison with the pre-disaster conditions which put extra pressures on already weakened water resources. It was because the affected populations were not charged for their water consumptions for several years after the earthquake and it tempted them to exploit potable water for even irrigating the plants in their yards (Bagheri et al. 2010).

Following the disaster, the local population lost their dynamic valuable connection with their surroundings. This might be a result of focusing on the physical reconstruction, overlooking the recovery of social bonds and employing a holistic approach to revive the relationship between human and nature. Without undertaking an integral approach in terms of physical, social and cultural aspects, there will be little chance to find a suitable and sustainable solution (Fallahi 2008). The reconstruction process should not be limited to restoring built environments but should consider recovering the social and cultural fabric of communities and strengthening links between populations and their environment and revive human dignity, beyond the repair of built structures (Garnier et al. 2013).

14.4 Traditional Sustainable Solutions in Bam's Garden-Houses

A traditional garden-house in the region consists of the house, the garden and the yard. Water circulates the garden-house in form of the *qanat*, interlinking the elements together and creating a living system. Golpayegani and Einifar (2007) present two main types of traditional and semi-modern garden-houses, in their comprehensive study on the housing types in the Bam area: one in the form of the courtyard and the other in villa (free-standing) style (Figs. 14.2). What distinguishes a garden-house of Bam from other typical garden-houses is the fact that the garden has economic values for the households.

Using sun-dried or baked bricks, thick walls, dome roof, wind-catcher, semi-open spaces, vaulted passageway and high ceilings is a common technique in the region to adapt to the extremely arid climate (Fig. 14.3). It is not about using brick and other materials and techniques as a decorative tool to resemble the vernacular architecture, but it is about grasping the notion of traditional constructions and combining them with suitable modern methods to improve the overall sustainability and resiliency of the communities and the natural and built environment. Considering the relationship

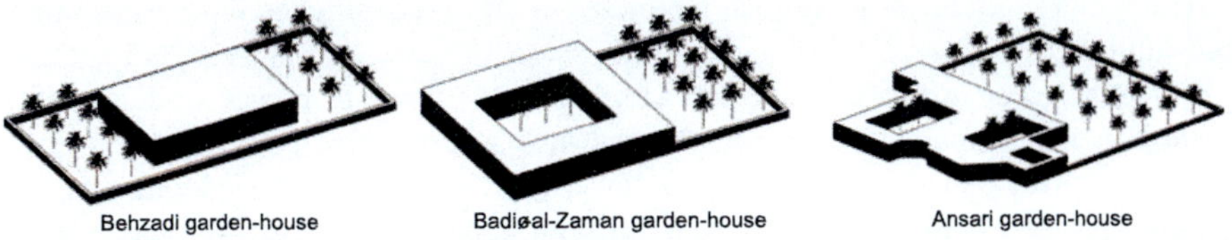

Fig. 14.2 Relationship between garden and house in three traditional garden-houses (Golpayegani and Einifar 2007)

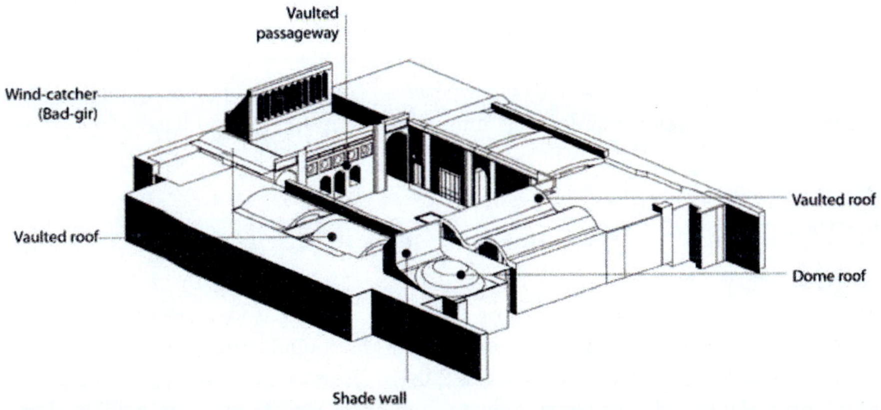

Fig. 14.3 Vakil-Bashi garden-house and its traditional sustainable elements (modelled by Karimian based on original drawings from ICHHTO)

between the building and the open space, Golpayegani and Einifar (2007) list items, which should also be taken into account:

- Possibility to hold social activities and resting place with sunshade,
- Plant diversity,
- Economic benefits from garden and possibility for food production and poultry raising,
- Water circulation and efficient water management system,
- Multi-functionality and resiliency of spaces, and
- Materials in harmony with the traditional urban fabric.

In ancestral agricultural systems, the citrus trees are planted in the shadow of date palms and the herbs and vegetables cover the ground. In addition to reaching the highest levels of diversity, productivity and efficiency, this stepped orchard together with the surrounding adobe walls results in a microclimate that provides a more pleasant atmosphere. The yards and gardens are gathering places for various family occasions and ceremonies in Bam and play a key role in social activities. Another practice is to build a temporary shelter (locally called *kavār*) by reusing dead fronds of the date palm. That is because in the south of the country all parts of palm trees are usually used carefully as they are highly honoured and valued. In the summer, spraying water on the fronds leads to air circulation and provides cool weather. Presently, this shelter is more used as a parking lot or for keeping the water tank safe from sunlight.

Known as green infrastructure (Taghvaei 2006), the *qanat* brings groundwater to the surface using gravity which indicates that it is a sustainable system by its very nature. The cultivated area is usually on a lower level and can be flooded with water. In traditional courtyards, the water runs through pool house and/or fills the water reservoir. This circulation of water has a significant effect on the air condition. Around 70% of the people living in Bam and surrounding villages are directly or indirectly engaged in agricultural activities (Siadat 2007) and therefore highly dependent on the irrigation system of *qanats*. In addition to economic and environmental benefits, *qanats* also contribute to the social sustainability of the communities. In fact, their existence rests on community cooperation for regular maintenance and passing oral indigenous knowledge. This sense of cooperation strengthens the social consensus that paves the way for sustainable development in the region (Labbaf Khaneiki 2019).

14.5 Green Reconstruction Challenges in Garden-Houses

Based on the GRRT definition, adopting environmental measures and safeguarding the natural systems are two key objectives in the green reconstruction which would benefit the livelihood of the community and their lifestyle. In Bam area, the natural ecosystems and cultural heritage are interwoven. Thus, the identity and livelihood of residents depend on the human-made ecosystem which itself relies on the dynamic relationship of human and nature. To facilitate introducing green solutions, the following challenges are highlighted as the most primary ones in the landscape of a traditional garden-house: (a) revitalisation of gardens and *qanats*, (b) wastewater treatment and (c) reuse of solid waste. In the following sections, these challenges are discussed, and as a result, strategies and associated green solutions are presented.

14.5.1 Challenge 1: Revitalisation of Qanats *and Gardens*

The economy of the Bam region has been traditionally based on agriculture, which in turn is built upon the strict sociocultural system of water management using *qanats* in this arid environment (ICHHTO and UNESCO 2008). As mentioned earlier, the role of water resources in the region goes beyond the economic value of agricultural production and contributes a great deal to the social structure and institutions of the local communities; e.g. each *qanat* is shared between several farmers and landowners (UNDP 2008). Besides cultural and socio-economic values and climate adaptation, the gardens are used to bring households self-sufficiency by letting them growing crops, keeping poultries and producing various handicrafts with date palm waste. A SWOT analysis is prepared as follows to identify the key internal and external factors and generate strategies based on them (Table 14.2).

Strategy for challenge 1: In addition to the rehabilitation of gardens and *qanats* in the city, their revitalisation in the scope of garden-houses should be addressed. More concentration should be placed on the relationship between the population and their environment rather than merely focusing on physical restoration. Considering the strong cooperation among communities and vast wealth of indigenous knowledge on water and soil management in *qanat*-based oases, the revitalisation of gardens and *qanats* should be evoked based on the existing social structure to revive the beneficial interaction between households and their environment. Koohafkan and Altieri (2011) suggest the empowerment of smallholder family farming communities to continue to conserve their traditional agricultural systems and create economic value for the conservation of biodiversity so that nature and people can prosper together.

Table 14.2 SWOT analysis for qanats and gardens

Qanats and gardens	**Strength**	**Weakness**
	1. Rehabilitation of damaged *qanats* in the immediate aftermath of the earthquake 2. Continuation of many *qanats* on irrigating the gardens 3. The survival of a few semi-modern garden-house with their traditional performance 4. Broad traditional knowledge and experience on soil and water management 5. Stepped orchards that enjoy diversity, productivity and efficiency	1. Degrading the traditional management system of *qanats* 2. Focusing on physical structure instead of empowering social structure 3. Overlooking the importance of gardens revival in garden-houses 4. Not employing local masters for the rehabilitation of *qanats*
	Opportunity	**Threat**
	1. No extreme change in the traditional social structure behind the *qanat* 2. Inscription of Bam Cultural Heritage Landscape in 2004 and the Persian *Qanat* in 2016 in the UNESCOís WH List 3. Sense of attachment and belonging to gardens and *qanats* by some local residents	1. Burning palm trees to free up area for construction and diminishing the size of gardens and yards 2. Drying up the *qanats* to dig water wells 3. No comprehensive documentation of the indigenous knowledge regarding *qanats* and traditional agricultural systems as intangible cultural heritage which is at peril

In this regard, drawing on principles from permaculture, agro-ecology, community gardening and similar sustainable agricultural approaches is needed. Households should be empowered through training and community-based interventions to learn about new efficient methods like combining drip irrigation with *qanat*.

14.5.2 Challenge 2: Wastewater Treatment

The recovery and reconstruction phase following a disaster represents an opportunity to improve water and sanitation system and help affected communities move beyond pre-disaster conditions towards long-term resilience and a better natural environment (WWF and American Red Cross 2010). In the aftermath of the Bam earthquake, tremendous effort was placed on restoration of the water supply system. Many international NGOs and UN agencies took part along with governmental organisations and implemented projects to address this issue, *restoration of water supply infrastructure systems for earthquake-affected smallholder date palm plantations in bam* by UNDP and *formulating long-term reconstruction plan for water supply system* by JICA, to name a few. Less attention, however, was paid towards the possibility of wastewater treatment and infrastructural development. A SWOT analysis of the condition of wastewater is presented in Table 14.3.

Strategy for challenge 2: Several eco-friendly technologies and techniques are available for wastewater treatment. Among them, using a septic tank for primary filtration and constructed wetlands for secondary treatment is more efficient, affordable and feasible. Drawing on the previous experiences around the globe, it appears that these systems are applicable in the aftermath of a disaster in developing countries in the reconstruction phase. The recycled water can be utilised for the irrigation of private gardens or public green spaces, and the surplus can recharge the groundwater table. Johannessen et al. (2012) acknowledge that for adopting sustainable sanitation options, awareness should be raised to facilitate the implementation of more eco-friendly techniques.

Table 14.3 SWOT analysis for wastewater

	Strength	Weakness
Wastewater	1. A high tendency towards water recycling, despite no current comprehensive wastewater management plan 2. Ease of collecting wastewater as houses have limited stories (one, in case of garden-houses)	1. Low quality of tap water in the city 2. Groundwater depletion due to water over-extraction 3. Groundwater pollution due to poor site selection
	Opportunity	**Threat**
	1. Possibility of using existing channels to use recycled water for irrigation 2. Enough open space in yards to implement a wastewater treatment system	1. Wrong policies leading to using potable water for irrigation 2. Serious water shortage in the region 3. Lack of a comprehensive wastewater management plan

14.5.3 Challenge 3: Reuse of Solid Waste

In a post-disaster setting, the capacity of solid waste management systems can be quickly exceeded and ultimately impact the environment and the people who depend on it. In many developing countries, solid waste management systems do not function well, and the disposal of building materials places an extra burden on these systems (WWF and American Red Cross 2010). According to UNICEF (2004) after the Bam earthquake, the very limited solid waste and garbage collection facility was a serious problem in all of the affected areas. Although steps were taken to encourage survivors to separate usable materials from debris and get paid by delivering them, it was limited and not part of a comprehensive solid waste management programme.

The possible recycling of materials is usually called into question because reprocessing requires a great deal of energy (Vegas et al. 2014). Thus, the most environmentally sustainable option for resourcing construction projects is the reuse of waste building materials in their existing state without downgrading and reprocessing into new products (WWF and American Red Cross 2010). This approach mitigates the need to buy new materials and bearing additional load on the environment and simultaneously prevent the consumption of energy in moving debris to landfill areas. Table 14.4 shows the SWOT analysis of the solid waste in Bam.

Strategy for challenge 3: Based on GRRT, at the design stage, the project planners should consider how their material choices can optimise environmental sustainability, for example, through the use of materials that include recycled content or have a higher life cycle, the reuse of disaster debris and the construction materials that fit the local conditions better. In the aftermath of a disaster, materials like bricks, wood, concrete, stone and metal sheets are best to be reused or recycled. Concrete, used bricks and stones can be employed as fill material to construct roads (UNEP and SKAT 2007). Furthermore, as recommended in Bam CMP, bricks can replace the traditional mud layer walls around the gardens which were immensely damaged after the earthquake.

Table 14.4 SWOT analysis for solid waste

	Strength	Weakness
Solid waste	1. Implementing cash incentive programmes to engage local communities in debris removal and separation process 2. Allowing households to keep the temporary accommodations as subsidiary supporting spaces[2] 3. Presence of owners at the time of debris removal	1. Absence of recyclable waste centres in all the region 2. Lack of a comprehensive solid waste management plan
	Opportunity	**Threat**
	1. Reusing bricks as a material in harmony with Bam urban fabric 2. The suggestion of reusing salvaged bricks as a replacement for damaged surrounding mud layer walls in Bam CMP 3. Reusing and recycling steel structure	1. Disposing debris on the river bed leading to polluting groundwater and surface water and posing a major threat on the environment 2. Disposing of debris in the archaeological sites

[2] The average size of Bami houses was over $200m^2$ whereas the reconstruction finance was calculated for $80m^2$. To address this issue, design guidelines for reconstructing permanent houses requested to consider the supposedly in situ temporary accommodations. They were seen as the subsidiary supporting space for families even after the completion of permanent houses (Arefian, 2018).

14.6 Integrating Green Solutions and Traditional Vernacular Systems

To explore the missing potentials of integrating green solutions and traditional vernacular systems in Bam unfortunately, no historical "living" garden-house could be identified as many houses have become a simple cube in a small land lot, with no resemblance of traditional sustainable practices (model numbers 1 and 2 in Fig. 14.4). Moreover, the newly built units in the yards highly impacted the interaction between human and landscape and the integrity of the house (no 3 in Fig. 14.4) (Karimian 2016). Finally, house of Davoud Yousefzadeh was examined. The owner started a cultural centre in his house after disaster. The property is an active free-standing semi-modern-style garden-house in the south of Bam, with a 400 m^2 building area in a 3000 m^2 land lot and provides a suitable example of the integration of green solutions with traditional vernacular systems. Constructed in 1972, the 2003 earthquake left minor damage on the house, and after further assessment by experts, it was stabilised and restored. Built on a platform, the house enjoys a large date palm grove in the south and a smaller garden in the north (Fig. 14.5). Davoud is a shareholder of Qanbar Abad Qanat, one of the oldest *qanats* with the highest amount of discharge; once a week, a sub-branch of the *qanat* circulates his garden and the plants enjoy the water for 30 minutes (Karimian 2016).

The north façade of the building is designed with multiple fractions, to avoid direct sunlight. To provide enough natural light for the central hall, the roof rises in the centre and small windows are placed in four directions. A *kavār* (sunshade) is assembled in the yard to shelter the water tank from harsh weather. The two temporary settlements that were positioned following the disaster are still in the garden, merged and transformed into a working place, as a true demonstration of a successful sustainable reuse practice (Fig. 14.6) (Karimian 2016).

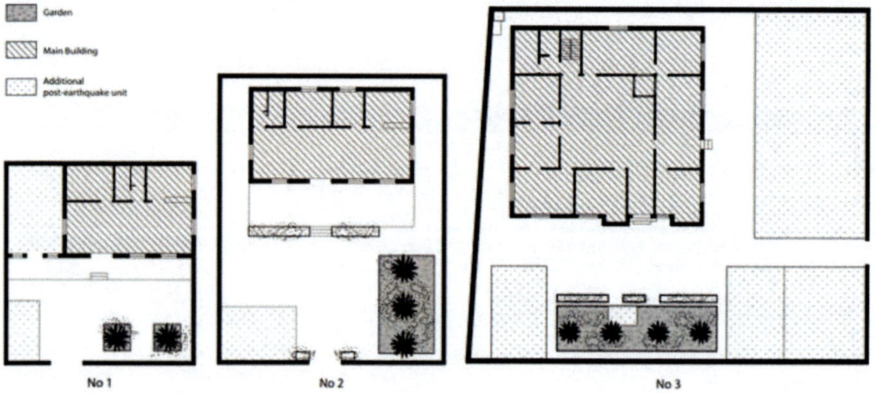

Fig. 14.4 Housing model numbers 1 and 2 are built from scratch after the earthquake, but house number 3 was only restored (Karimian 2016)

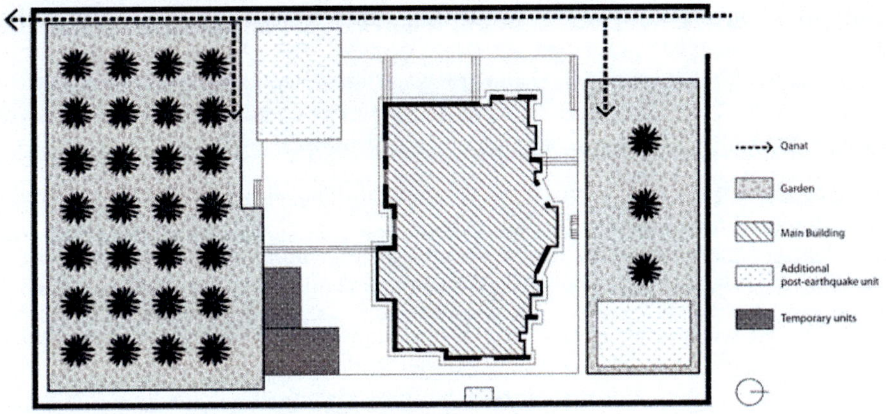

Fig. 14.5 Davoud's garden-house site plan (Karimian 2016)

Fig. 14.6 Two temporary housing units in Davoud's house (left: exterior, right: interior) (Ameneh Karimian 2016)

As mentioned earlier, the garden of the proposed house enjoys the water of a well-preserved *qanat*. Following the earthquake, the owner turned part of his house into a community centre, providing a vibrant atmosphere for social bonding. This also provides a proper setting for strengthening the social structure for *qanat* and garden maintenance and rehabilitation. According to the owner, in the Bam region, households used to provide the daily foodstuff in their garden. This resembles the emerging concept on community gardening and urban farming in modern societies. To revive this sense of self-sufficiency and address the food scarcity in the aftermath of a disaster, the small north garden can be dedicated to food production. This small farming system can improve the synergy between residents and their environment and improve the economic value of the garden-house to some extent. To encourage the households to develop small farming and be involved in participatory gardening based on the social structure that revolves around *qanat*, vacant lands along the *qanat* course can be devoted to community gardening as well. This can also reinforce the social bonds and strengthen the interaction between human and nature as one of the identity elements of the city. Principles of stepped orchard should be taken into account to use the land optimally.

For wastewater (both blackwater and greywater) treatment, a vertical flow planted filter (constructed wetland) is efficient, while for a more comprehensive approach a two-stage treatment system can be implemented as follows:

(a) The primary treatment for physical pollutant removal, by sedimentation and flotation through settling (septic) tank,
(b) The secondary treatment for the removal of organic matter and reduction of hazardous loads through horizontal flow planted filters (constructed wetlands).

The treated water can directly irrigate the garden or be stored in a tank and used as necessary. Although the annual rainfall of the city is quite low (60.4 mm), considering the large size of the flat roof and proper slope and drainage, installing a simple rainwater harvesting tank under the roof gutter will bring an additional amount of water to the house.

Regarding the solid waste management, the undamaged bricks can be exploited for surrounding wall restoration. Further and more importantly, the temporary accommodations are already well insulated and in use (Fig. 14.7). To reduce the high temperature inside these prefabricated units, local people cover the roof with dead fronds of date palms. This material could also be employed to construct a simple resting chamber, *kavār*, as mentioned in the previous sections.

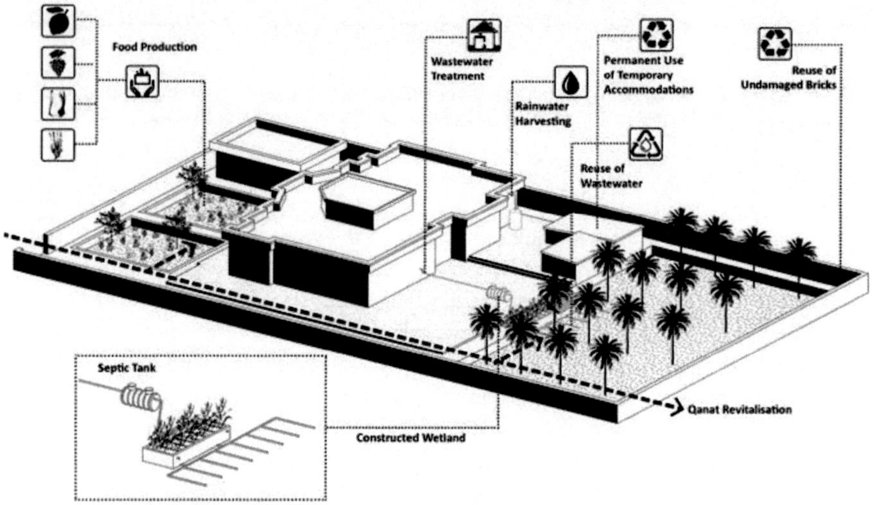

Fig. 14.7 Diagram of integrated green solutions into the pilot garden-house (modelled by Karimian)

14.7 Discussion

In the cultural landscape of Bam, the beneficial interaction between human and nature in a harsh desert environment is shaped upon the traditional knowledge and forms one of its overall universal values. At present, it is widely acknowledged that indigenous knowledge can contribute to the conservation and management of heritage and connect cultural heritage to the specific natural environment in which it is located (Wijesuriya 2020, p. 14). Since traditional knowledge and cultural heritage are not static and have changed and evolved by generations over time, their incorporation into the reconstruction policies and measures should address new challenges and reconcile with modern knowledge. This chapter tried to shed some light on the possibility of integrating green solutions into the traditional vernacular systems in a post-disaster situation in a cultural landscape. The green options have value not just because they provide a cost-effective way of securing climate adaptation and disaster risk reduction gains, but also due to the considerable development co-benefits that they generate in terms of value-added and costs avoided to other economic sectors and processes (Emerton et al. 2016, p. 38).

The study reveals that the relationship between local people and their environment became fragile after the 2003 earthquake, especially in the scope of the garden-houses where households took the opportunity to rebuild their houses larger and overlooked the landscape. It highlights the vitality of reviving this relationship by addressing its major components, i.e. the *qanat* and the garden through empowering already existing social structure revolving around *qanat*. By limiting the study to three primary challenges in garden-houses, it provided more in-depth information and proposed pragmatic solutions. These green solutions were inspired by the experiences of the post-earthquake recovery in developing countries that were proven to be feasible and environmentally friendly and then were localised to draw on the participation of the local community. This participatory approach based on age-old knowledge will lead to a more holistic and cross-sectoral reconstruction. The study demonstrated the potentials of the garden-houses in Bam, including their active water management (*qanat*) and ancestral agricultural system (stepped orchard), and proposed integrating new techniques like drip irrigation and permaculture. Implementing an ecological wastewater treatment system and rainwater harvesting container will mitigate the water stress in the agricultural field while safeguarding the groundwater and *qanat* networks. Dealing with a massive amount of debris in the aftermath of a disaster is a complex problem which can be alleviated by developing reuse and recycle policies. In the pilot project, the temporary accommodations are insulated and transformed into a permanent working space.

These small green solutions in the scale of a garden-house can start a positive sustainable movement and take it to the next level by developing more comprehensive strategies, e.g. enforcing regulations to prohibit households from diminishing the size of their courtyards and gardens and ban drilling wells in the buffer zone of *qanats*. Suitable wastewater treatment and solid waste management should also be included in disaster risk reduction plans. Furthermore, the realm of traditional knowledge in the region is more diverse, and some of them are still not identified and acknowledged by the heritage practitioners. Therefore, it is crucial to document oral knowledge and indigenous practices before they perish. This should lead to a paradigm shift in national policies to acknowledge the necessity of integrating traditional ecological knowledge in reconstruction and recovery measures.

14.8 Conclusion

Even though an abundance of research has been dedicated to Bam and its reconstruction, there are still more lessons to learn from this multidimensional complex case. The focus of this chapter was mainly on the relationship between human and nature in the scope of a garden-house and associated traditional knowledge. Since garden-house is one of the components of the dynamic ecosystem of the garden-city, their recovery will directly revive the disturbed identity of the city. The study acknowledges the fundamental role of vernacular architecture which is genuinely in complete harmony with its surrounding and can provide a solid ground to anchor other techniques and solutions. It is essential to choose modern environmental approaches in accordance with the local context and with the acceptance of the community members as knowledge-holders. Raising awareness and community-based interventions are needed to mobilise the community towards adopting new sustainable measures. Furthermore, based on the overall objective of reviving the landscape of garden-house and the dynamic interaction of households and their environment, new green solutions are proposed in a living garden-house in the city with three identified main thematic issues. This pilot project can provide valuable lessons for other cities around the world on employing indigenous knowledge together with modern eco-friendly techniques and community participation to bounce forward and achieve sustainable and resilient development in the setting of living heritage.

References

Adle C (2007) Qanats of Bam: an archaeological perspective, irrigation system in Bam, its birth and evolution from the prehistoric up to modern times. In: Honari M, Salamat A, Salih A, Sutton J, Taniguchi J (eds) Qanats of Bam: a multidisciplinary approach. UNESCO Tehran Cluster Office, Tehran

Arefian FF (2016) Getting ready for urban reconstruction: organising housing reconstruction in Bam. In: Arefian FF, Iran Moeini SH (eds) Urban change in Iran. The urban book series. Springer, Cham, Switzerland

Arefian FF (2018) Organising post-disaster reconstruction processes. The urban book series. Springer, Cham, Switzerland

Arslan H (2007) Re-design, re-use, and recycle of temporary houses. Build Environ 42:1

Bagheri A, Darijani M, Asgary A, Morid S (2010) Crisis in urban water systems during the reconstruction period: a system dynamics analysis of alternative policies after the 2003 earthquake in Bam-Iran. Water Resour Manag 24:2567–2596

Bahmanpour S (2006) Part of the stakeholders' activities in Bam reconstruction, Hamyaran-e Ghada, unpublished (in Persian)

Browning P (2006) The story of 2118 Dumaine Street. https://reliefweb.int/report/united-states-america/usa-story-2118-dumaine-street. Accessed 22 Oct 2020

Emerton L, Huxham M, Bournazel J, Kumara MP (2016) Valuing ecosystems as an economic part of climate-compatible development infrastructure in coastal zones of Kenya & Sri Lanka. In: Renaud FG, Nehren U, Sudmeier-Rieux K, Estrella M (eds) Ecosystem-based disaster risk reduction and adaptation in practice. Springer, Cham, Switzerland, pp 23–45

Fallahi A (2007) Lessons learned from the housing reconstruction following the Bam earthquake in Iran. Aust J Emerg Manag 22(1)

Fallahi A (2008) An interdisciplinary analytical study on the risk preparedness of Bam and its cultural landscape, a world heritage property in danger in Iran. Aust J Emerg Manag 23(2)

Foundation E (2010) Green recovery program: reducing vulnerability to disasters by building sustainable communities based on healthy ecosystems. Sri Lanka (unpublished document), Environmental Foundation

Garnier P, Moles O, Caimi A, Gandreau D, Hofmann M (2013) Natural hazards, disasters and local development, CRAterre Editions

Golpayegani A, Einifar A (2007) Typology and design guide for housing in Bam. Ministry of roads and urban development, Office of architecture and urban planning, Tehran (in Persian)

ICHHTO and UNESCO (2008) Bam and its cultural heritage landscape world heritage property comprehensive management plan (2008–2017), Iranian Cultural Heritage, Handicraft and Tourism Organisation & UNESCO Tehran Cluster Office

JICA and TDMMC (2005) The comprehensive master plan study on urban seismic disaster prevention and management for the greater Tehran area in the Islamic Republic of Iran, Japan International Cooperation Agency (JICA) and Tehran Disaster Mitigation and Management Centre (TDMMC)

Johannessen A, Patinet J, Carter W, Lamb J (2012) Sustainable sanitation for emergencies and reconstruction situations. Factsheet of Working Group 8. Sustainable Sanitation Alliance (SuSanA). https://www.susana.org/en/knowledge-hub/resources-and-publications/library/details/797. Accessed 22 Oct 2020

Karimian A (2016) Green reconstruction in the garden-houses of Baravat-e Olya after the 2003 Bam earthquake, unpublished master thesis. Shahid Beheshti University, Tehran (in Persian)

Koohafkan P, Altieri MA (2011) Globally important agricultural heritage systems a legacy for the future. Food and Agriculture Organisation of the United Nations, Rome

Labbaf Khaneiki M (2019) Territorial water cooperation in the central plateau of Iran. Springer, Cham, Switzerland

Lungharwo T, Trenk-Hinterberger C (2015) Local actions, solution to global challenges initiatives of indigenous peoples in climate change adaptation and disaster risk reduction based on traditional knowledge, Asia Indigenous Peoples Pact (AIPP). Chiang Mai, Thailand

Meskinazarian A (2011) Social resilience of post-earthquake Bam. PhD thesis, Department of Geography, King's College London, London, UK

Parsizadeh F, Ibrion M, Mokhtari M, Lein H, Nadim F (2015) Bam 2003 earthquake disaster: on the earthquake risk perception, resilience and earthquake culture—cultural beliefs and cultural landscape of Qanats, gardens of Khorma trees and Argh-e Bam. Int J Disaster Risk Reduct 14(4):457–469

Renaud FG, Nehren U, Sudmeier-Rieux K, Estrella M (2016) Developments and opportunities for ecosystem-based disaster risk reduction and climate change adaptation. Ecosystem-based disaster risk reduction and adaptation in practice. Springer, Cham, Switzerland, pp 1–20

Sabto M (2011) Build it back green, ECOS, Issue 159, CSIRO Publishing. www.ecosmagazine.com/?paper=EC10067. Accessed 22 Oct 2020

Semsar Yazdi AA, Labbaf Khaneiki M (2005) A survey on the qanats of Bam: from technical and engineering point of view. UNESCO-ICQHS, Tehran

Siadat A (2007) Qanats of Bam: a socio-economic perspective. In: Honari M, Salamat A, Salih A, Sutton J, Taniguchi J (eds) Qanats of Bam: a multidisciplinary approach. UNESCO Tehran Cluster Office, Tehran

Statistical Centre of Iran (SCI) Website:www.amar.sci.org.ir/Data.aspx?No=201629&Ln=F&S=GW. Accessed 13 June 2018

Taghvaei SH (2006) Oasis, ecology and sustainability aspects in environment and landscape processing. Soffeh 16(45):6–23 (in Persian)

Taniguchi J, Fardanesh F (2006) Advocating a cultural response to natural disasters lessons learnt in Bam (December 2003) & Borujerd (March 2006) Earthquakes, International Conference on Integrated Disaster Management, Tehran, 13–14 February 2006

UNDP (2008) Restoration of water supply systems infrastructure in earthquake affected smallholder date palm plantations in Bam, Project Final Report, UNDP, Iran

UNEP and SKAT (2007) After the Tsunami: sustainable building guidelines for South-East Asia, United Nations Environment Programme and SKAT, Sadag, France

UNESCO WHC (2010) Managing disaster risks for world heritage: world heritage resource manual, Paris, UNESCO

UNICEF (2004) Crisis appeal earthquake in Bam, Iran

Urban F, Mitchell T, Silva Villanueva P (2010) Greening disaster risk management: issues at the interface of disaster risk management and low carbon development. Strengthening Climate Resilience Discussion Paper 3, Brighton: IDS

U.S. Department of Energy (DOE) (2009) Rebuilding after disaster: going green from the ground up, National Renewable Energy Laboratory (NREL)

U.S. Department of Energy (DOE) (2012) Integrated deployment: disaster recovery, National Renewable Energy Laboratory (NREL)

USGS website. Today in Earthquake History: On December 26th. https://earthquake.usgs.gov/learn/today/index.php?month=12&day=26#:~:text=6%20%2D%20Southeastern%20Iran%2C%202003,destroyed%20in%20the%20Bam%20area. Accessed 22 Oct 2020

Vegas F, Mileto C, Guimaraens G, Navalón V (2014) Defining sustainable architecture. In: Correia M, Dipasquale L, Mecca S (eds) Versus: heritage for tomorrow, vernacular knowledge. Firenze University Press, Firenze

Wijesuriya G, Court S (2020) Introduction. In: Wijesuriya G, Court S (eds) Traditional knowledge systems and the conservation and management of Asia's heritage. ICCROM & the Cultural Heritage Administration of the Republic of Korea

World Bank (2012) Toward a green, clean, and resilient world for all: a World Bank Group environment strategy 2012–2022, Washington, DC

WWF (2005) Green reconstruction policy guidelines for Aceh, WWF-Indonesia, Jakarta, Indonesia

WWF and American Red Cross (2010) Green recovery and reconstruction toolkit for humanitarian aid. WWF US and American Red Cross, Washington DC, USA

WWF Nepal (2016) Building back safer and greener: a guide to sound environmental practices for disaster recovery in Nepal. WWF Nepal, Kathmandu, Nepal

Chapter 15
Reconstruction of Heritage and Spirit: Mending the Scars of Aleppo

Mounir Sabeh Affaki

Abstract Reconstruction processes tend to generate prolonged controversies especially in the aftermath of intercommunal conflicts where complications accumulate due to the sensitivity that the word reconstruction can carry. In this case, the process of post-war reconstruction has the potential either to advance social recovery or to hinder it. This concerns every expert and decision-maker, who hope for the sustainability of a fragile peace and of what is to be reconstructed, with fears of failure and relapse. These two possibilities refer to the public reception of the process and the new reality it implies. Therefore, it must be addressed in order for the reconstruction process to transcend from being merely physical to also becoming social. This chapter intends to go through this discussion and investigate the contribution of built heritage and the restoration of historical centres to social recovery and the resilience of cities. It employs Aleppo as a case study, a divided city that is in a state of precarious peace. Its historical centre, the Old City of Aleppo, is a World Heritage Site that was severely destroyed in the Battle of Aleppo (2012–2016). Its reconstruction has been facing numerous challenges, including the lack of coordination and funding, issues of ownership and legislations, as well as bureaucratic procrastination. These problems became more evident through the quality of the implemented reconstruction works. In response, questions of what to rebuild and how, of priorities and methods, and the value of public participation are discussed in this chapter. In addition, a context-specific reconstruction strategy is proposed. It aims to contribute to social recovery by addressing the sociocultural and economic issues of the city. The discussion of this chapter, which tackles a contemporary case study, brings new considerations for upcoming reconstruction strategies, in Syria and elsewhere, that integrate physical and social reconstruction processes.

Keywords Post-war reconstruction · Urban policy · Social recovery · Built heritage · Aleppo · Syria

M. S. Affaki (✉)
University of Coimbra, Coimbra, Portugal
e-mail: mouniraffaki@hotmail.com

© The Author(s), under exclusive license to Springer Nature Switzerland AG 2021

F. F. Arefian et al. (eds.), *Historic Cities in the Face of Disasters*, The Urban Book Series,
https://doi.org/10.1007/978-3-030-77356-4_15

Acronyms

AKCS	Aga Khan Cultural Services
AKTC	Aga Khan Trust for Culture
DGAM	Directorate-General of Antiquities and Museums
ERP	Emergency Response Plan
GCES	General Company for Engineering Studies
OUV	Outstanding Universal Value
UNESCO	United Nations Educational, Scientific and Cultural Organization

15.1 Introduction

Controversy over reconstruction processes is inevitable regardless of the cause of destruction, was it induced by natural or human-induced incidents. However, in post-war situations, controversy is significantly higher due to the challenge of securing public approval in divided societies (Martz 2010). In this case, the process faces pressure from the conflicting sides. It becomes a reference point in history, either positive or negative, as important as the destructive event itself. It shapes a new social reality as it accompanies the reestablishment of normalcy and the return to the point where peace was dominant. Apart from being physical, the reconstruction of built heritage, in particular, serves to restore the disrupted cultural continuity, as well as the shared identity to which people can relate despite their divisions. From this perspective, it has the potential to promote a shared future that is based on coexistence and tolerance. On the other hand, following natural disasters, people tend to demonstrate unity and solidarity. Hence, they are able to move on relatively faster, while in the aftermath of a conflict, the society can in fact suffer more division. Social recovery has also proven to be relatively slower due to the human and cultural loss (Barakat 2005; Junne 2010). Therefore, reconstruction needs to go hand in hand with peacebuilding processes, for it can contribute to the physical aspects of healing and reparation processes.

This chapter introduces the Old City of Aleppo as its case study, where it investigates the potential contribution of the restoration of its built heritage to the recovery of the Aleppine society. It begins with an overview on the urban history of the city, its social context and the *status quo* of the Old City in particular. Then, it explores current reconstruction works, discussing the adopted approaches and the numerous challenges and obstacles of realising these projects in such a devastated city. To explore further, two case studies are briefly discussed: the restoration of the Umayyad Mosque and *al-Saqatiya* market, focusing on issues of funding and executing, the main challenges, visions, and motives, the adopted approaches and, finally, the public reception of these works as retrieved from social media platforms, as well as local journals and news networks. Highlighting the lack of a reconstruction vision and uncertainty about its strategies and methods, this work attempts to provide a set of guidelines for a strategy with a vision of a sustainable reconstruction from the socio-cultural and economic aspects. It emphasises on the value of public engagement in social recovery and cohesion while considering the economic feasibility of reconstructing a historical city centre that occupies an area of 364 hectares in a country that is trying to recover from an ongoing crisis since spring 2011.

15.2 The Lost Heritage of Aleppo

Aleppo is considered one of the oldest continuously inhabited cities in the world. According to Sauvaget (1941), its urbanisation dates back to the Hellenistic era (333 B.C–395 B.C). The Hellenistic city had a gridiron urban pattern that was transformed after the arrival of the Arabs in 637 into an organic layout, with a tree-shaped, compact urban form. Around the thirteenth century, urbanism covered the *intra muros* area which then required expanding the walls and establishing the *extra muros*. During the Ottoman rule and the French mandate, new districts were built with modern planning principles inspired by contemporary European town planning and modern building techniques.

There were several interventions to modernise the Old City during the twentieth century in order to facilitate car access to the *khans* and *Suqs*[1] for commercial purposes (Bianca 1980). In 1954, French architect André Gutton proposed a master plan with two main roads that penetrate the ancient fabric of the Old City. His vision was to expose and emphasise the main monuments of the Old City by cleaning their surroundings *"dégager pour mettre en valeur"*, resulting in the demolition of tens of hectares (Ibid.). In 1969, another plan by the Japanese planner Gyōji Banshōya was proposed with slight modifications to reduce the demolition (Ibid.). Following these plans, wide streets were opened and tall buildings were built on their sides that murdered the privacy of the neighbouring traditional two-storey houses and led to their abandonment (David and Hreitani 2005).

By 1979, conservationists were able to stop the implementation of these plans and push the Department of Antiquities to register the *intra muros* part of the Old City as a national monument in 1978 (Bianca 1980). Finally, in 1986, the Old City was listed as a World Heritage Site by the United Nations Educational, Scientific and Cultural Organization (UNESCO) for its Outstanding Universal Value (OUV) due to the preservation of its cohesive and unique urban fabric (David 2018). Following the listing, legislations were issued in order to preserve it. The problem is that those legislations aimed literal replication of the pre-existing built environment, which has repeatedly resulted in the lack of authenticity of infill designs and restoration works. Several restorations took place, the latest being the rehabilitation of the perimeter of the Citadel by the Aga Khan Trust for Culture (AKTC), between 1999 and 2008.

The identity of the Old City was drastically transformed with the demographic change that happened following the independence of Syria in 1946. The rich intellectual families that originally lived at the centre moved to either the newly built

[1] Arabic for hostels and markets, respectively. Suq is also spelled souk.

suburbs of the city or abroad seeking a modern life. Following this migration, new rural settlers filled this vacuum benefitting from the low real-estate prices due to low demand. This demographic change has altered the culture, traditions and level of education of the population. Consequently, the physical status of the old city had increasingly deteriorated with many violations and urban encroachments due to the lack of awareness of the value of built heritage among the newcomers. In addition, the modernisation attempts that initially were to stop and reverse the migration of the original residents have backfired and led to further migration. The new designs were neither physically nor culturally appropriate. These include the mentioned tall buildings as well as newly imposed functions that deformed the character of the historic centre and overwhelmed it with traffic, such as the administrative buildings located at the perimeter of the Citadel (David and Hreitani 2005). The demographic change entrenched class-based social division in the city. This led to an urban segregation between the centre and the modern city that later extended to include the informal settlements established by the newcomers east to the Old City (David 2002). The centre that already lacked public spaces became associated with lower-class inhabitants and considered "too popular" (Boissière 2014). It became uncomfortable and even dangerous for others to pass through some areas except for the tourism zone.

The Syrian War reached Aleppo in summer 2012 and lasted until the end of 2016, in what is known as the Battle of Aleppo. This did cause not only further social division and the loss of cultural elements of music, art, handcrafts and traditional industries due to the forced migration, mortality and poverty, but also an unimaginable damage to the built historical core of the city. This led the UNESCO to list the Old City as a World Heritage Site in Danger in 2013, only after one year of conflict (UNESCO 2013). During the war, the compact urban form of the Old City became a nurturing environment for the armed groups and an optimum place for fortification, camouflage and sniping. The ability of building-to-building penetration through shared walls, manoeuvring through the narrow and acute streets and underground detonation through ancient tunnels were all factors that complicated and sustained the street war for around 4 years. At one point, the response of the government was to bombard some areas. These actions left the Old City in an overwhelming state of destruction. The damage assessment by the General Company for Engineering Studies (GCES),[2] in 2019, shows that about 20% of the Old City is heavily to totally damaged, 60% is slightly to moderately damaged, and 20% is mostly intact (see Fig. 15.1).

The battle of Aleppo did affect not only the built environment of the city but also its social fabric and the psychological well-being of its citizens. It has further entrenched the social divisions of class, sect and political affiliation, which became physical through the line that divided the city between Eastern and Western. The line also ran through the Old City, separating the newly built, commercial and wealthy areas from the old, compact and conservative ones. People are now traumatised not only for losing their loved ones or their properties, nor for being victims of violence,

[2] In Arabic الشركة العامة للدراسات الهندسية. A public construction company that follows the Syrian Ministry of Public Works and Housing.

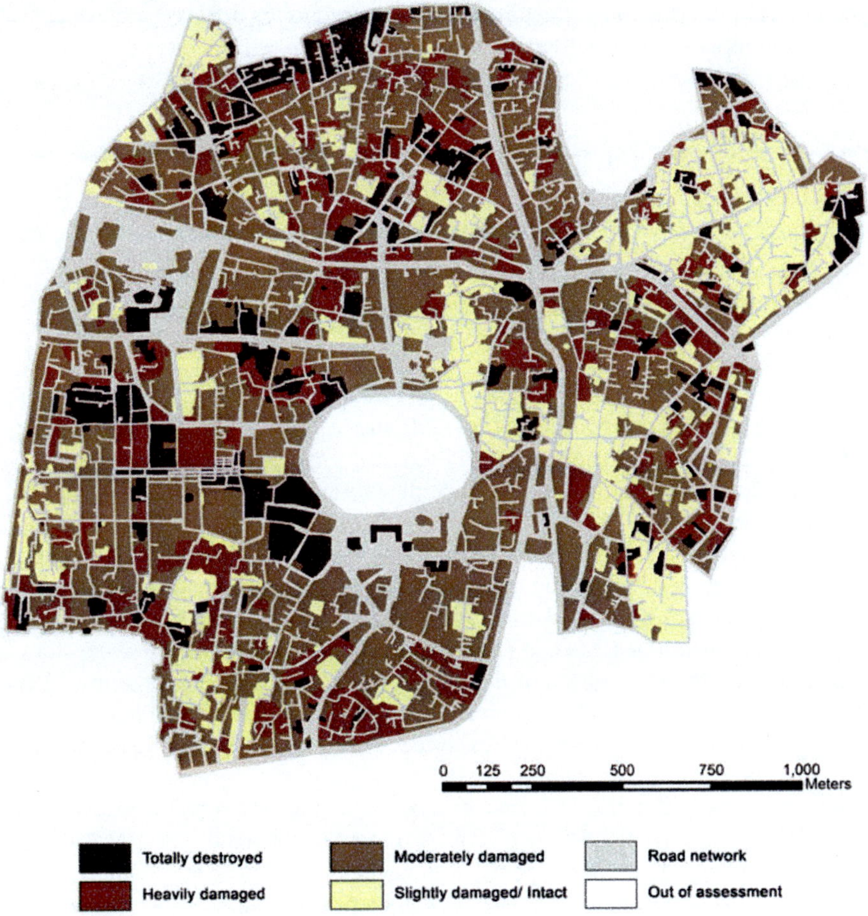

Fig. 15.1 Damage assessment of Old Aleppo (author based on GCES 2019)

but also for the destruction of their city as they knew it. The destruction of the Old City and the loss of an important part of their identity and source of pride had a further influence on the people for the reasons behind it. Besides collateral damages, Haj Ismail and Morezzi (2014) categorised the motives behind the destruction of symbolic buildings which are mostly induced by extreme religious beliefs. Destruction was used as a tool to frame the other side, punish them, sap their morale, or erase historical proofs of their existence. These motives and their consequences, represented in mass urbicide, have caused an obsessive feeling among Aleppines that their existence and identity were targeted and, at stake, that the destruction was systematically serving these purposes with the direct participation of some countries by funding, arming and recruitment, or with the total silence and neglect of others (Haj Ismail and Morezzi 2014). Accordingly, because the destruction of the Old City has caused all these psychological effects among Aleppines, its reconstruction is believed to have some effects, either positive or negative.

15.3 Current Reconstruction Works in the Old City

The reconstruction of the Old City began soon after Aleppo was recaptured by the Syrian army. In January 2017, an Emergency Response Plan (ERP) was adopted, which proposed modifications on the general organisational scheme of the city, in addition to several interventions and reconstruction projects (Ferrier 2020). A "National Steering Committee for the Restoration of the Ancient City of Aleppo" was established, led by the Ministry of Culture and represented by the Directorate-General of Antiquities and Museums (DGAM) and its branch, the Directorate of the Old City. The reconstruction of the Old City is still limited to the removal of debris and the restoration, rehabilitation and documentation of 210 buildings considered the most important.

The 2018 report by UNESCO categorised reconstruction projects into four groups: first is the restoration of shops and houses, carried on by citizens either at their own expense or by receiving financial aids. Still, it has to be according to the directions of *The Urban Legislation of the Old City of Aleppo,* after the acquisition of a restoration permit from the Directorate of the Old City. Second is the restoration of key monuments carried out under the supervision of the Directorate of the Old City, with the help of international funding and expertise. Third is the restoration of specific building typologies, such as schools, mosques and parts of the Ancient *Suq*, carried by the Directorate of Religious Endowments, the Chamber of Commerce and that of Industry. Fourth are the small-scaled restoration works carried by local NGOs and civil groups, either at their own expense or financed by donations and funding, as in the restoration of *Bab al-Nasr* and the Clock Tower of *Bab al-Faraj*.

According to UNESCO (2018), the implemented reconstruction works have demonstrated a lack of coordination between the several entities, groups and individuals, involved in the process. There is also a lack of a shared vision or a comprehensive strategy, or even general guidelines, which raises serious concerns regarding the preservation of buildings and the restoration of the historic fabric (Ibid.). Furthermore, the damages are increasing because some buildings are suffering further decay due to structural damages and bad weather conditions affecting the now exposed materials of mud, hay and wood (UNESCO 2019a). The urban legislations have not been updated to follow the latest restoration methods, and they certainly were not designed for the conditions of a post-war context. With the absence of official control, most private initiatives have proven disastrous. A report by local newspaper *Tishreen* states that, at that time, 67% of the 55 restoration licences granted to owners to rehabilitate their properties infringed the legislations (Nabulsi 2018a). They demonstrated a lack of structural analyses and incorporated the use of inappropriate materials, such as black cement and concrete blocks. This refers to the abuse of small-scale restoration permits, licensed according to *The Urban Legislation of the Old City of Aleppo*, as a cover for large-scale interventions. With the lack of monitoring cadres, owners felt free to violate. Shortage of traditional materials and skilled craftsmanship, together with ignorance and lack of public concern, have also contributed to this situation (Nabulsi 2018a).

Together with the lack of coordination among stakeholders, bureaucratic inertia and lack of awareness among the society, the reconstruction in Aleppo is facing a perplexing amount of challenges. They are mainly about finding adequate financial resources for removing debris, rehabilitating the infrastructure, providing traditional materials for restoration, employing experts and developing a comprehensive management plan. The lack of funds is probably the most important challenge facing the whole country. According to Syrian officials, the allies of the government, including Iran, Russia and China, will have the priority in the process of reconstruction and investment (Sputnik 2019). Meanwhile, large-scale reconstruction processes are on hold until the war is completely over and the currency rate becomes stable in order to prevent fluctuation losses (Ibid.). This can have serious implications on the quality of reconstruction and its harmony. The Umayyad Mosque and *al-Saqatiya* market are examples of how the reconstruction process is put in practice, addressing various issues and challenges around them, such as visions, motivations, approaches and funding and the public reception of these works.

15.3.1 The Restoration of the Umayyad Mosque

The Umayyad Mosque is the biggest and one of the oldest mosques in the Old City, located on the *Decumanus* axis that connects the Antioch Gate with the Citadel. It is an architectural emblem by all accounts. Its extreme value derives from the fact that it has been rebuilt, transformed and renovated many times through history. Its location used to be the *Agora* during the Hellenistic period, the *Forum* during Roman times, a pagan temple during the Byzantine rule and a Cathedral for Saint Helena during the Christian era, all before becoming a mosque in 715 (Sauvaget 1941; Archnet 2019). Prior to its severe damage in 2013 and the destruction of its 45-m minaret that dated back to the eleventh century, the mosque was an important gathering point, a pilgrimage destination and a tourist landmark. Together with the *Suq*, they formed the vital core of Old Aleppo.

Aleppines are looking forward to the inauguration of their beloved mosque so that they can resume Friday prayers and gatherings after several years. The reconstruction study of the Umayyad Mosque was carried out by the University of Aleppo, audited by the GCES and is being executed by the Military Construction and Implementation Corporation with the financial help of the Chechen Ministry of *Awqaf* (endowments) (Evin 2017). In an interview by Sputnik News Arabic (2018), one of the main architects operating on the site stated that the adopted approach is to preserve the bullet marks on some of the walls, as a witness for the destruction that happened, while covering the rest with wooden panels for the next generations to decide what to do in future. As for the minaret, the stones were collected, documented and numbered with

Fig. 15.2 Restoration of the Umayyad Mosque (anonymous 2018)

the help of the AKTC experts in order to rebuild it like it was. Before its reconstruction, which began in May 2020, the mechanical resistance of the collected stones was measured to determine the possibility of reuse and eliminate those with low resistance (Evin 2017; Moussa 2018). The missing stones are now being replaced with new ones of the same material (Fig. 15.2).

15.3.2 *The Restoration of* Suq al-Saqatiya

Located on a one-kilometre-long axis between the Citadel on the east and the Antioch gate on the west, the Ancient *Suq* is one of the biggest historical covered bazaars in the world with an area of sixteen hectares. It is the longest in the world with a total length of seven kilometres. Through its maze of narrow alleys, it features a journey of around 14 kms around 1625 stores (Zain Al-Abidin 2006). Before its destruction, the *Suq* was a vital public space where people of all backgrounds and religions used to gather for the common purpose of trade and commerce. It is an irreplaceable part of the identity of Aleppo since its early history, centuries before its emergence as the commercial capital of Syria and the establishment of *Sheikh Najjar*, the biggest industrial city in the Middle East. The fire that devoured the Ancient *Suq* was argued to be a form of punishment directed at the merchants of Aleppo for their political affiliation (Haj Ismail and Morezzi 2014). As a result of bombings and underground detonation, 8 out of 39 *Suqs* are totally destroyed while the rest is heavily damaged (Nabulsi 2018b). Nevertheless, former antiquities Director-General has declared that 70% of the *Suq* is salvageable (Evin 2017).

The reconstruction of the *Suq* has a symbolic meaning of resilience and going back to the good old times when life was prosperous. Unfortunately, this operation is facing many challenges of funding and ownership. Consequently, recent reports have revealed that the owners are expected to restore the interiors of their stores (Alsayed 2019), which poses a two-sided problem: tracking down all the owners and forcing them to restore. Furthermore, there are some difficulties in proving the ownership of some properties due to the loss of official documents and the migration of owners and inheritors. This will eventually lead to prolonging the restoration process and the loss of the general sense of place. However, the Old City Directorate and the Chamber of Commerce have already rehabilitated several parts of the *Suq*, including *Suq al-Juma*, *Khan Kher Beik* and *Khan al-Olabieh.* Recently, Aga Khan Cultural Services (AKCS) in Syria took on the restoration of *Suq al-Saqatiya,* finalised by July 2019, and by the time of writing, they are restoring of *Khan al-Harir* (Fig. 15.3).

The 98-m-long section of the Ancient *Suq*, known *Suq al-Saqatiya*, was selected by AKCS as a pilot project for its centrality, its historical and cultural value and economic feasibility, as well as its relatively slight level of damage and small size, making its reconstruction a quick process (Ibrahim 2020). According to the project's chief architect, the adopted approach aimed at both correcting the previous violations and "*erasing the marks of war*" (Beshara 2019). While the small *Suq* is reopening

Fig. 15.3 Restoration of *Suq al-Saqatiya* (commissioned by author 2020)

and coming back to life and despite the precision of the restoration work and its aesthetic outcome, the denial of war memory is an approach that has been widely criticised in previous international cases. For instance, this recalls the reconstruction of the Beirut Central District carried by Solidère. The Solidère approach on erasing the war memory provoked endless controversy for promoting an amnesia of the brutality of the Lebanese Civil War (1975–1990), yet the approach was replicated in Libya as discussed in another chapter of this book.

15.4 Developing a Reconstruction Strategy

With the many problems in the Old City of Aleppo, both pre-existing or emerging, there is a pressing need to contemplate a comprehensive vision that takes into consideration all the obstacles and potentials reconstruction has. The process of post-war reconstruction should be employed as a tool for social transformation and recovery, not only for physical restoration. In order to achieve that, it should be part and parcel of a comprehensive social reparation programme. As Vandeginste (2003) states, these programmes should be based on inclusiveness, appropriatenessand effectiveness. Therefore, reconstruction can play an active role in the process of social recovery when following these three principles.

(a) Inclusiveness
The inclusiveness of reconstruction implies the engagement of all parts of the society in it. Their participation can be either through being consulted for their opinion or expertise, through their donations, investments and financial support, or through their volunteer or payed labour. This has a great potential to stimulate cooperation among the society, overcome its divisions and eventually promote tolerance and advance social cohesion. In a post-war context as of the case of Aleppo, such inclusivity and public approval in decision-making can contribute to the sustainability of peace and, therefore, the sustainability of what is to be reconstructed. Public consultation at early stages of decision-making can contribute to the legitimacy of the planning process and add an immense value to the final product. Furthermore, periodic assessments and public surveys should be made during the process in order to ensure civic approval and evaluate the efficiency of the work done. The case of Solidère Beirut demonstrates the consequences of both excluding local professionals and the establishment of an exclusive centre. The project lost an important opportunity for social recovery when it overlooked the potential of reconstructing the *Suqs* that once formed a common ground and fostered social interaction between people of all backgrounds, thus promoting social cohesion. The transformation of Beirut's *Suqs* into an exclusive luxurious commercial centre has further highlighted the divide between socio-economic classes instead and widened the gap.

(b) Appropriateness

Reconstruction is a very sensitive term and process where public approval is fundamental. This implies that its approach should be very cautious. Otherwise, it might deepen the wounds and even provoke further conflict. The process should be feasible, custom-made and tailored to fit the conditions of the case. It should take into consideration all the legal procedures and bureaucracy, cultural backgrounds, and values, as well as the economic possibilities and resources, analyse them and find their implications, limitations and potentials, in order to achieve the best possible outcomes. Among the limitations is the economic difficulty of securing funds. Hence, the reconstruction of the Old City is expected to be a long, gradual process.

Responding to this reality, it is advised to set priorities and divide the process into phases. Reconstruction priorities should be deeply thought. For instance, the case of Mostar shows the consequences of rushing to reconstruct the historical centre while disregarding the devastated infrastructure and the need for emergency housing. A similar tendency is noticed in Aleppo, with funds being scarce and directed to the reconstruction of monuments, while the comprehensive process is still on hold until the war is over and the economy is more stable. Moreover, architectural designs should be culturally and historically appropriate and subtle. This also includes acknowledging war as part of history. Therefore, it is recommended to find a balance between memorialisation and the restoration of cultural and historical continuity that got disrupted by the physical destruction that took place. It is also recommended to use reversible designs for they allow future alteration needed in case the reconstruction method proved to be inappropriate. The full documentation of the damage and the restoration project, as well as the use of slightly different materials and colours and the application of minimal structural changes that allow dismantling the new structures, are considered appropriate actions that demonstrate a high level of consideration towards the cultural and historical value of the site.

(c) Effectiveness

The restoration of buildings with high cultural and social value promotes the revival of traditions and cultural practices and the restoration of cultural continuity and the lost sense of identity associated with the place. Responding to the economic limitations and the potential of reconstruction towards healing and social recovery, the reconstruction of the Old City should be punctual. If applied properly, it can act in a similar manner to "urban acupuncture" which, according to Frampton (2000), consists of "catalytic, small-scale interventions, [...] capable of achieving a maximum impact, with regard to the immediate surroundings" (pp. 29–30). This strategy also implies treating problems step by step according to priority. This allows assessing the effectiveness of each intervention, which builds more experience and capacity and creates funds for the next one. For instance, the restoration of key monuments and mosques would potentially reactivate cultural tourism and pilgrimage and form a source of income. Similarly, the rehabilitation of the ancient *Suqs* would revive trade and commerce, form another and so on. These new incomes can be directed to the reconstruction of less-prioritised buildings and zones in the Old City as well as the entire city.

15.4.1 Setting Priorities

"We are inside the Old City, we need to know the historic archaeological value, the traditional, even the national, of each stone, each building and corner" (David and Hreitani 2005, p. 87). Figure 15.4 demonstrates the Old City and the categorisation of properties according to their historical value. However, in addition to the historical and cultural value of buildings, post-war reconstruction should respond to the broader

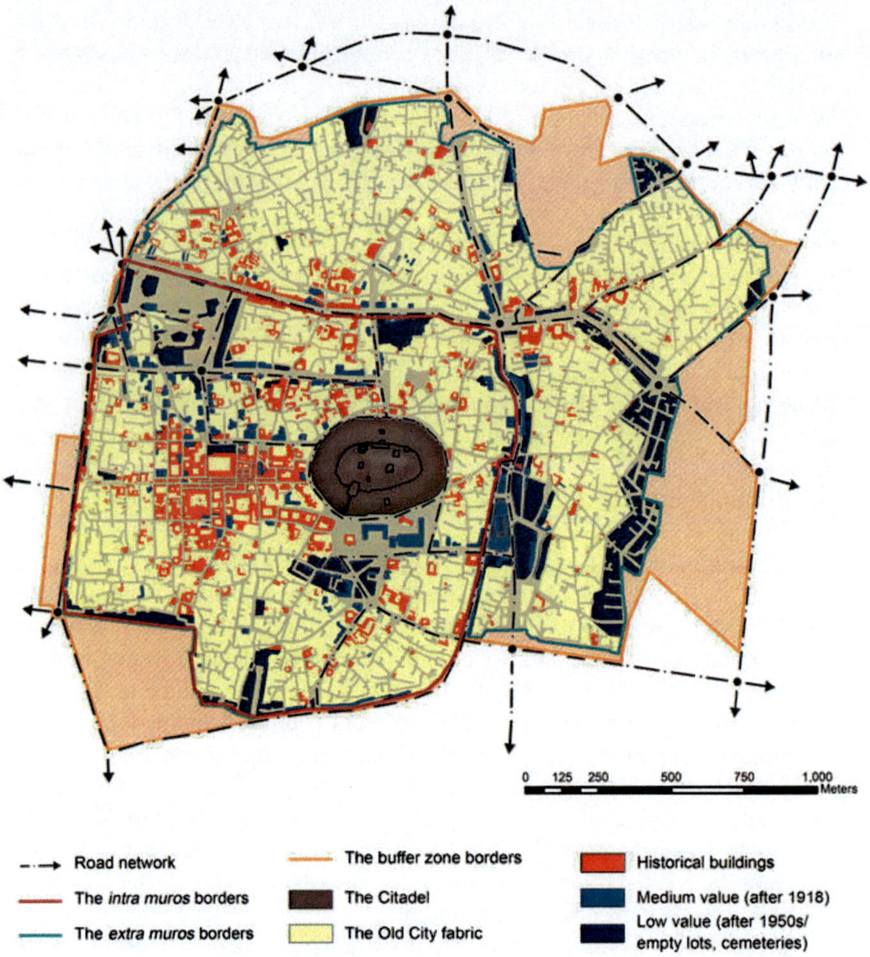

Fig. 15.4 Historical value of the buildings of the Old City (author 2020)

social, economic and political value. It should take into consideration the level of damage, as well as the economic feasibility and profitability of the reconstruction of selected buildings and areas, whether through cultural and war tourism revenues or by encouraging commercial and cultural investments. This assessment should also question the potential of reconstruction in activating social relations, including the role certain buildings and areas played during the conflict, either positive or negative, as well as their level of accessibility. For instance, monuments and prominent public spaces are usually highly accessible, while traditional houses are more difficult to access because their entrances are located at the end of narrow alleys. Finally, the selection of buildings and areas should demonstrate diversity in their function, style, and value to different ethnic groups. These criteria would enable us to extract a reconstruction priority list that contains the names and numbers of properties ordered according to their collective cultural, social, political and economic value, the level of damage, expected funding possibilities and a timeline of reconstruction. The result would be a more detailed list for each of the following priority categories for:

(a) First-grade priority
This category includes monuments and symbolic buildings and complexes that carry an important social or distinct ethnic value. Their reconstruction is essential as a solid manifestation of memory, pride, belonging and social interaction. They once played a key role in the daily life of Aleppines and created a deep emotional attachment being the public realm of the city and the containers of social interaction. It is pressing to give a special attention to them because the quality of their reconstruction is a key to social recovery and cohesion. A combination of national and international funding and expertise is recommended in order to reconstruct this category.
(b) Second-grade priority
The identity of historic city centres like the Old City cannot be reduced to its monuments. Therefore, their reconstruction is not enough for social recovery and the restoration of social relations. The true catalyst of social relations in the Old City is its exquisite urban fabric. Moreover, the restoration of the sense of place is achieved by restoring the original form of the urban fabric and the continuity of streets. This implies reconstructing the buildings that are located on the main axes of the urban structure, which form this category. Reconstruction in this category can also benefit from grass-roots funding campaigns and volunteering to promote public participation.
(c) Third- and fourth-grade priority
The third category includes the valuable buildings that are not located on the main axes and have low accessibility in addition to the ones located in the *extra muros* area, for is relatively newer as its oldest neighbourhood *al-Jdeideh* dates back to the fifteenth century. The fourth category includes buildings of a medium value, which means the ones that are less than one hundred years old. The reconstruction of these two categories should be put on hold until funds are available. For the moment, they can host commemoration events, ceremonies,

galleries and exhibitions which can also contribute to social recovery through acknowledgement and enabling storytelling.

In addition to the above, there are private properties, which reconstruction naturally depends on their owners. The immigration of owners over the years is a factor that affects the speed of planning and reconstruction processes. It is a very sensitive issue, because a reconstruction that aims for social recovery and cohesion must respect the ownership of properties. Otherwise, it may provoke further problems if not immediately or in the short run, then in the near future. There are also the buildings that are not considered worthy of reconstruction, for being either severely damaged or completely destroyed, for having low to no architectural or aesthetic value, or for negatively impacting the fabric and the sense of place, such as the new buildings and encroachments.

The step to follow the priority setting is phasing. The first phase is the removal of unsalvageable rubbles and debris and the reinforcement of buildings that are prone to collapse in order to prevent further damages and human injuries. Debris management is already taking place with the volunteers' assistance, including women and youth. This is a good indicator of the importance of cultural heritage to Aleppines as indicated by UNESCO (2019a) and a good incentive for people's participation and community building. Another decision is to determine which quarters of the Old City to be reconstructed first. Here, it is recommended to reflect the urban evolution of the city and prioritise the historical value. Previous studies on the city evolution and heritage are helpful for this. Going back to the study provided by Sauvaget (1941), the Citadel and its perimeter will be the starting points, followed by the Hellenistic zone on its west, after that would be the Byzantine, the Arab, the Ottoman and then the French.

15.4.2 Sustaining the Reconstructed

The UNESCO (2019b) has emphasised the importance of respecting and preserving the coherence of the urban fabric of the Old City in future developments. Accordingly, reconstruction should be considered an opportunity that allows correcting previous mistakes. This includes demolishing violations and encroachments and restoring of the continuity of the streets as much as possible, as well as removing unsuitable functions and securing the needed ones. The reconstruction process should also avoid repeating these mistakes by having a mechanism to provide materials at competitive prices to encourage the compliance with recommended standards in the restoration of private properties. The urban retrofit of the Old City also implies relocating those functions that once created problems because of their locations, the adaptive reuse of some properties and the creation of new public spaces. There is also a need to limit excessive car access and traffic, which requires removing the administrative functions from the centre. It is also recommended to develop the buffer zone to filter vehicle access and limit it to the main axes and dead-end parking lots.

Besides contemplating effective on-the-ground solutions to enhance and preserve the image of the historic city and reverse its deterioration, there is a need to educate the public and raise awareness on heritage value. Working on this objective, Al-Adeyat Archaeological Society of Aleppo has been organising informative tours around the Old City soon after the government recaptured Aleppo in December 2016. It is a promising step for Aleppines to gather and understand the true value of their common heritage. The historical centre also needs to become more welcoming and inclusive to people from all backgrounds. Diversity can be achieved by encouraging and attracting investments to the centre and activating its commercial axes. This approach can create a new social configuration of a mixed population that can generate new dynamics and stimulate more interaction, leading to cultural and social amelioration.

15.5 Conclusion

The Old City of Aleppo had been destroyed multiple times during the history. Each time, it was rebuilt and restored to accommodate the needs of the lifestyle of that time. However, the urban interventions that took place during the last century, in addition to the recent war, raise serious concerns about losing its distinct urban fabric. Post-war reconstruction is a sensitive operation, especially when it follows an intrastate conflict. In this case, it carries a special significance as a key turning point for the society and its government. It becomes a political tool to enhance the image of the government and verify its legitimacy and a subject for criticism on the performance of a post-war government that has just lost the excuse of war, which is commonly used to justify the inability of meeting the needs of the population (Charlesworth 2006; Wennmann 2019).

The current reconstruction projects demonstrate scattered efforts and a lack of a comprehensive vision on what reconstruction should look like. The adopted approaches are ranging between keeping some marks of destruction as seen in the Umayyad Mosque and completely erasing them as seen in *Suq al-Saqatiya.* There seems to be no public consultation nor a concern for the population that is expected to live in the historical centre and its characteristics. This situation calls for an immediate action to study the social factor and develop a reconstruction strategy that can contribute to social recovery and cohesion, based on inclusiveness, appropriatenessand effectiveness.

References

Alsayed A (2019) واقع أعمال ترميم سوق السقطية بحلب The reality of the restoration work of al-Saqatiya souk in Aleppo [website]. Available from https://www.facebook.com/alaa.alsayed.927/posts/10157257764476908. Accessed 28 May 2019

Archnet (2019) Jami' al-Umawi (Aleppo)—Description [website]. Available from https://archnet.org/sites/1804. Accessed 29 May 2019

Barakat S (2005) After the conflict: reconstruction and development in the aftermath of war. I.B.Tauris, London

Beshara L (2019) Syria's ancient Aleppo souk poised to regain its bustle. https://gulfnews.com/world/mena/syrias-ancient-aleppo-souk-poised-to-regain-its-bustle-1.62586381#carousel-article-b5b6b64b-7321-4dce-8781-049491479897. Accessed 20 May 2019

Bianca S (1980) The conservation of the old city of Aleppo. UNESCO, Paris

Boissière T (2014) From the souk economy to globalization [De l'économie de souk à la mondialisation]. In: David J, Boissière T (eds) Alep et ses territoires: Fabrique et politique d'une ville (1868–2011). Presses de l'Ifpo, Beirut-Damascus, pp 297–317

Charlesworth E (2006) Architects without frontiers: war, reconstruction and design responsibility. Architectural Press, Oxford

David J-C (2002) Public spaces in Aleppo since the end of the 19th century. Urban planning and user practices [Les espaces publics à Alep depuis la fin du XIXe s. Urbanisme et pratiques des usagers]. Géocarrefour, vol 77, no 3. L'espace public au Moyen-Orient et dans le monde arabe, pp 235–244

David JC e a (2018) Five years of conflict: the state of cultural heritage in the ancient city of Aleppo. http://unosat-sdn.web.cern.ch/unosat-sdn/unesco/syria/aleppo_publication/UNESCO_UNITAR-UNOSAT_2017_Publication_Full.pdf. Accessed 19 May 2019

David J-C, Hreitani M (2005) المدينة القديمة فى حلب: التدهور ومحاولات الإحياء [The Old City of Aleppo: Deterioration and Revival Attempts]. Ray Publishing & Science, Aleppo

Evin F (2017) Les nouvelles technologies au chevet d'Alep, « la scintillante » mutilée. https://www.lemonde.fr/architecture/visuel/2017/11/10/les-nouvelles-technologies-au-chevet-des-tresors-d-alep-la-scintillante-mutilee_5213190_1809550.html. Accessed 15 May 2019

Ferrier M (2020) Rebuilding the city of Aleppo: do the Syrian authorities have a plan? European University Institute Research Repository, Fiesole, Cadmus

Frampton K (2000) Seven points for the millennium: an untimely manifesto. J Archit 5(1):21–33

Haj Ismail S, Morezzi E (2014) Post conflict conservation or reconstruction: analysis, criteria, values of the recent Syrian cultural heritage. Florence, La Cultura del restaure e della valorazzione, Alinea Editrice SRL

Ibrahim S (2020) Decision-making methodology between revitalisation and rehabilitation of world heritage city centers. Case study: the ancient city of Aleppo (Syria). The international archives of the photogrammetry, remote sensing and spatial information sciences, vol XLIV-M-1–2020, pp 255–262

Junne G (2010) Designing peace: bricks and mortar of reconciliation. Broker 20:30–33

Martz E (2010) Introduction to Trauma rehabilitation after war and conflict. In: Martz E (ed) Trauma rehabilitation after war and conflict: community and individual perspectives. Springer-Verlag, New York, pp 1–26

Moussa F (2018) إعادة بناء جامع حلب الكبير بالصور والحقائق [The reconstruction of the Great Mosque of Aleppo with pictures and facts]. https://arabic.sputniknews.com/arab_world/201811061036575421. Accessed 28 May 2019.

Nabulsi I a-D (2018a) تقرير يكشف الأوضاع الكارثيـة التى ستؤدى إلى خسارة الآثار والمدينة القديمة فى حلب [A report reveals the catastrophic conditions that will lead to the loss of antiquities and the old city of Aleppo]. http://tishreen.news.sy/?p=140330. Accessed 30 May 2019

Nabulsi I a-D (2018b) الحرب دمرت (٠١) خانات و(٨) أسواق، قرارات رفع الإيجارات تؤثر سلباً على عودة الأهالي وأصحاب المحلات إلى حلب القديمة [The war had destroyed 10 Khans and 8 markets, decisions to raise rents negatively affect the return of residents and shop owners to Old Aleppo]. http://tishreen.news.sy/?p=228682. Accessed 30 May 2019.

Sauvaget J (1941) Alep: Alep. Essai sur le développement d'une grande ville syrienne des origines au milieu du XIXe siècle. Librairie orientaliste Paul Geuthner, Paris

Sputnik (2019) دمشق تذكر ثلاث دول مسموح لهم المشاركة فى إعادة إعمار سوريا [Damascus mentions three countries allowed to participate in the reconstruction of Syria]. https://sptnkne.ws/mt4g. Accessed 8 Aug 2019

Sputnik News Arabic (2018) مراحل ترميم الجامع الأموى بعد دمار الحرب فى سوريا [Stages of the restoration of the Umayyad Mosque after the destruction of the war in Syria]. https://sptnkne.ws/gKHJ. Accessed 28 May 2019

UNESCO (2013) Syria's six world heritage sites placed on list of world heritage in danger. https://en.unesco.org/news/syria%E2%80%99s-six-world-heritage-sites-placed-list-world-heritage-danger. Accessed 19 May 2019

UNESCO (2018) State of conservation—ancient city of Aleppo. https://whc.unesco.org/en/soc/3795. Accessed 19 May 2019

UNESCO (2019a) State of conservation: ancient city of Aleppo. https://whc.unesco.org/en/soc/3866. Accessed 19 Apr 2020

UNESCO (2019b) Ancient city of Aleppo. https://whc.unesco.org/en/list/21/. Accessed 19 May 2019

Vandeginste S (2003) Reparation. In: Reconciliation after violent conflict: a handbook. International Institute for Democracy and Electoral Assistance (IDEA), Stockholm, pp 145–162

Wennmann A (2019) A peacebuilding perspective on post-conflict reconstruction in Syria. https://blogs.lse.ac.uk/crp/2019/02/14/a-peacebuilding-perspective-on-post-conflict-reconstruction-in-syria/. Accessed 14 Jun 2019

Zain Al-Abidin M (2006) حلب: عمارة المدينة القديمة - نماذج وتجارب [Aleppo: The Architecture of the Old City—models and experiences]. Dar Al-Anas, Damascus

Chapter 16
Beyond the Damage, the Reconstruction of L'Aquila

Simona Bravaglieri, Elia Zenoni, and Silvia Furioni

Abstract Twelve years after the earthquake hit L'Aquila, it is time to evaluate its reconstruction from a qualitative point of view. The case of L'Aquila demonstrates how the persistence of the physical sense of a place, the numerous episodes of historical–artistic interest and the quality of the built heritage characterise the historic centre as an irreproducible identity nucleus, on which to base the material and immaterial reconstruction. Particularly, nowadays this identity is compromised by interventions supported by anachronistic hopes of "recovery of the ancient splendour", which reveal uncertain cultural assumptions and a misunderstood continuity with the past. With these premises, the research aims to investigate how the actors involved in the reconstruction of L'Aquila have chosen to shape the historic centre as if the earthquake had never happened. From the analysis of 13 case studies, a selective memory approach has emerged that aims to remove the painful recollection of the catastrophe as an ordinary attitude in a post-disastrous phase. This approach is of reluctance to the architectural project as a tool of reinterpretation and renewal of the historic architecture, while it deals with the seismic vulnerability only through an engineering-oriented approach. A selected past is displayed while liquidating the value of the architectural addition, thus generating a mere copy of a presumed original city. The plethora of palimpsests is therefore lost—including the earthquake's trace—as a result of the human's secular inability to accept sudden changes. In L'Aquila, this emerges through the reconstructed monuments managed by the public institutions that have embraced this effortless procedure as a tool to prove their efficiency. On the other hand, one could wonder whether an "architectural absence" would be more meaningful or at least more reversible.

S. Bravaglieri (✉) · E. Zenoni · S. Furioni
Politecnico Di Milano, Milan, Italy
e-mail: simona.bravaglieri@polimi.it

E. Zenoni
e-mail: elia.zenoni@polimi.it

S. Furioni
e-mail: silvia.furioni@polimi.it

© The Author(s), under exclusive license to Springer Nature Switzerland AG 2021
F. F. Arefian et al. (eds.), *Historic Cities in the Face of Disasters*, The Urban Book Series,
https://doi.org/10.1007/978-3-030-77356-4_16

Keywords Reconstruction · Conservation · Historic centre · Architectural addition · L'Aquila · Italy

16.1 Introduction

Twelve years after the incredibly strong earthquake that hit L'Aquila on 6 April 2009, the city has become a real laboratory for architectural, social and cultural reconstruction. The research presented here aims to narrate the reconstruction of the historic centre of the city from the point of view of the architectural heritage that has been lost. What choices have guided the restoration of the monuments that were severely damaged by the earthquake? What is the image that the historic centre is assuming as the buildings are slowly being returned to their citizens? Here, twelve buildings were analysed, chosen from the list published by the Ministry for Cultural Heritage and Activities on 25 May 2009, of 45 monuments to be restored together with their respective census and damage assessment files. This study was based, in the first instance, on the identification of monuments located in the historic centre of L'Aquila and subsequently on the selection of those buildings that witnessed substantial collapse with consequent loss and transformation of the pre-earthquake reality. This chapter is based on the results of the authors' master's thesis combined with further investigations carried out in the field, including interviews with institutions and local operators directly involved in the reconstruction process, as well as bibliographic analyses and recent surveys of the media and local information (Bravaglieri et al. 2016, 2019).

What was to be investigated and reflected upon is in fact precisely the way in which one could react and make a response when confronting the traumatic reality of an irreparably damaged and partly destroyed heritage among the indeterminacy of what remained. If, on the one hand, it was observed how the extensive building site of L'Aquila was an opportunity to test new technologies and construction practices, aimed at improving the resistance level of historic structures, it is argued that, unfortunately, the possibility of conducting such experimentation also in the field of architectural design and the evolution of the the image of the city has been largely ignored. In fact, this study reveals how the general approach currently undertaken has resulted in the vain attempt to bring the image of the city back to the pre-earthquake one, perceived almost as a relic rather than a living and evolving identity. Why is this approach, focused on the mere reconstruction of an image, often put into practice after the loss of cultural heritage? And why in L'Aquila was the dialogue with the past resolved in the single attempt to copy it rather than turn it into an opportunity for enrichment through the adoption of new languages and architectural narratives?

16.2 The Eternal Recurrence of the *Authentic*

To understand the origin of these choices, it is necessary to question the concept of authenticity and the meaning it can assume in a post-traumatic context such as that of L'Aquila. The numerous and enthusiastic declarations issued by the experts at the termination of construction sites[1] reveal how much an idea of *formal authenticity* is still deep-seated, which leads to an identification of "the authentic not with original but with origins: authentic would be everything that pertains to the initial moment, and authenticity in a work would essentially be determined by the absence of modifications or alterations compared to the initial forms" (Dezzi Bardeschi Authenticity 2017a, p. 16). This concept of *authenticity*, which would be more difficult to find approval for in the context of an "ordinary" architectural conservation intervention, instead becomes an arbitrary aim in orienting an "extraordinary" intervention such as that of the restoration of a monument severely damaged by the earthquake. The recomposition of the "authentic" image, pertaining to an authenticity that in the case of L'Aquila sees its original moment in the pre-earthquake state (no matter how false that claim is), is therefore nothing more than reasserting the validity of an approach that refuses to read, understand or acknowledge built objects as stratified, denying recognition of the *overall authenticity* of the work, of which the traumatic and transforming event of the earthquake has, like it or not, become part.[2]

The results are reconstructions which claim continuity by producing an evident discontinuity in the history of the city through the removal of the painful traces of the earthquake. As Mager states, such architectural reconstructions disintegrate the historical relationships inherent in the original building (Mager 2015). Indeed, historicist-style reconstructions can be read as attempts to stop the inevitable processes of time and decay "reordering time and processes, countering a physical assault upon a natural evolution with a philosophical assault on notions of time and memory, risking the distortion of both" (Bold et al. 2018, p 16). This is evident when

[1] To mention a few: "a splendid city which rises upon the fifth post-seismic reconstruction of its history" (Antonini O. Mons. 2019); "L'Aquila 'city of art' […] is the activation and management of a network system of places and cultural assets to be implemented on an experimental basis, thus starting the implementation of a common strategy of cultural enhancement aimed at transforming the city into a centre of excellence of culture and art." (Avellani 2019); "the ancient city emerged from the excavations of the yards […] precious frescos and decorations are visible again […] while the construction works continue to offer exciting archaeological discoveries" (*L'Aquila 2009–2019*).

[2] Interestingly, the Italian Cultural Heritage and Landscape Code (Codice dei Beni Culturali e del Paesaggio 2004) omits the ambiguous and debated word "authenticity". According to art. 29, restoration is a "set of operations aimed at the material integrity" of the artefact and not at the formal one. Moreover, there is no mention of "material", as it prefigures an idea of preservation bounded to the physical feature of the artefact, rather than the aestheticising vision which has always been favourite (Capurro 2017).

reconstruction coincides with a sort of normalisation process desired by local authorities to rebalance the emergency conditions in which people live in the aftermath of a disaster. The reopening of Palazzo Pica in L'Aquila, for example, has been described as "another piece that returns to its place, a contribution to the conquest of everyday life that we have lost [...], walking through these spaces now that they are back as they were, I feel pervaded by beautiful sensations" (Lombardo 2018). Reactions to the destruction appear to be as strong as the event itself, especially when a large audience is watching. The reconstruction thus becomes a gesture supported by the political affirmation of the city's identity and the understanding of how places should look (Bold et al. 2018). All this is simultaneously combined with the motto "we will give people what they want".

However, rebuilding how it was and where it was is not always the solution everyone wants and is often not what is needed, especially when a public debate on possible alternatives is substantially non-existent.[3] Such awareness should originate from the observation of architectural remains through an archaeological study, an essential phase of the process of gathering knowledge that ought to precede any reconstruction of the "life of things" and of the people who used them (Di Biase 2015). Paradoxically, the conditions in which the buildings of L'Aquila were in before the 2009 earthquake have been studied more than their architectural remains, followed by the deployment of "the eternal return of the authentic", an easy political choice that is certainly not risky. However, culturally speaking, it seems much more a non-choice, a seriously lost opportunity for the historic centre of L'Aquila and its citizens.

16.3 L'Aquila and Further Reconstruction Approaches

Immediately after the earthquake, the case of L'Aquila was exploited for political purposes and managed mainly as a media event. The emergency was entrusted exclusively to a Commissioner for Reconstruction, supported over time by 14 new special administrative structures that have proven not to be able to cooperate together. In response to the immediate housing emergency, 19 new permanent settlements (C.A.S.E. project) were built quickly and recklessly, while assessments of the conditions of structural stability of existing buildings were still underway, thus bypassing the question of reclaiming some existing housing for reuse and ignoring the actual need for housing. The new settlements led to the abandonment of the historic centre, reinforced by the establishment of the "red zone"—a sizeable portion of the city where access by civilians was quite limited—and within, the buildings were propped

[3] The reconstruction *à l'identique* suggests a reductive simplification "summed up in the motto kalòs kai agathòs" that identifies a passive role of communities which due to information asymmetry—and here the expert's task comes into play—express the need to rediscover the system of values linked to the urban landscape in 'as it was, where it was'" (Moioli 2019, p. 77).

up with mammoth metal pylons and then abandoned for years. The places and buildings that were symbolic for its citizens have been inaccessible for years, first because of the earthquake and then because of the emergency management policies, which led to the creation of an "event city" (Andreassi 2012) rather than to the reuse of the existing city and the restoration and recovery of urban spaces. The closure of the historic centre, although negative in many respects, nevertheless provided an opportunity to evaluate the approach to intervention in terms of a "long" restoration time.

A different approach was adopted after the earthquake that hit Central Italy in 2016, in which the rush to act led to generalised distrust in the structural resistance of the surviving buildings—excluding very few specific cases—and to the clean slate that therefore came into being. Here, the last (false) hopes of conservation relied on the rubble that was collected, catalogued and left awaiting possible use. Other examples have also been known for the haste shown in restoring tangible evidence of memory and collective identity. Only a decade after its destruction, the Mostar bridge was rebuilt and the adjacent Old Town restored (Kalman 2017), representing, according to UNESCO, a real rebirth. Similarly, in L'Aquila the reconstruction removed all physical evidence of destruction, excluding any possibility of the seismic event being remembered by posterity. Rebuilding is often considered by the international community as a default approach in the context of post-war mitigation, as it embraces both social and material reparations (Kalman 2017). This was the approach chosen and adopted institutionally also in L'Aquila, passed off as the only method to reconstruct and confront the past.

In some cases, the entire urban fabric became part of the process of reconstructing memory and the social appropriation of its history. In Dresden, following the destruction caused by World War II, the long-standing reputation of the "German city of art" together with the incomparable tragedy of the aftermath of its bombing was claimed as part of remembering the city as a symbol and "above all the product of political exploitation of experience, memory, and commemoration" (Rehberg and Neutzner 2015, p. 102). The idealisation of Dresden began immediately after the bombing by the allies, when the city was described by the Nazi regime as "Florence on the Elbe". During the Soviet occupation, the image of art and beauty remained unchanged and Dresden became, albeit in the form of an idealised memory, an "imagined city" built on the basis of an assumed utopia that looked to the past. Even in L'Aquila, the historical images preceding the traumatic event of 2009 became products of a "history of oneself" that compose a sort of institutional self-biography (Rehberg and Neutzner 2015). Here too, another "story of sacrifice" after the earthquake claimed the image of an "art city", diverting attention away from the disastrous event and intending to preserve and recreate the signs and symbols of the city.

Reconstruction practices justified by such distorted concepts of authenticity are profoundly debatable, for example, when, in addition to the tangible characteristics of the artefact, the construction techniques also are "resurrected". A clear case is the Katarina Church in Stockholm, which burned on 17 May 1990 and was rebuilt with tools and techniques reproduced in turn to imitate the original ones, despite their use having been lost in the meantime as often happens in Western culture.

The controversial use of "original techniques" to create "authentic replicas" (Weiler 2017) was adopted, for instance, for the reconstruction of the historic centre of Warsaw, where a special government agency had the task of disseminating traditional technical knowledge, which had been lost, with the aim of putting it back into practice. However, this is even more questionable when, conversely, new techniques are used to achieve the *formal authenticity* of ancient buildings. This was intended to be precisely the case in L'Aquila, where innovative techniques have been used in many conservation and restoration projects to reproduce the same image of the building prior to the earthquake. A similar approach, adopted following the 2010 Chilean earthquake, was criticised above all because the reconstruction models in historical areas were based on a superficial understanding of the concept of heritage, leading to the recreation of historical formal elements through the use of contemporary materials, even when there was no evidence of these in previous buildings or when the buildings did not even foresee the presence of such elements (Devilat Loustalot 2013). Despite these spurious attempts, there is no authentic reconstruction, a definition recently used by scholars as an apparently oxymoronic notion (Bold et al. 2018).

In addition to the use of local historical buildings as a catalogue for an uncritical imitation of traditional architectural elements, the experience at L'Aquila of the term restoration *à l'identique* was also used in terms of technological choices. The professionals involved in the reconstruction of the city used this expression to indicate the compatibility of the chemical–physical characteristics of the materials, the use of traditional construction techniques in continuity with the existing ones, the preservation of the original structural behaviour and any technique that does not falsify the characteristics of the historical building and does not experiment with little-known technologies. This causes a terminological confusion that causes misunderstandings in the design approach based on "slippery and elusive notion[s]" (Bold et al. 2018, p. 317) such as an identical reconstruction or *formal authenticity*.

This research reveals that, in the cases of heavily damaged buildings, the predominant choice was that of massive reconstruction which negatively affected the authenticity of the artefacts. In particular, it precludes the evolution of the building itself since without permanence there can be no transmission of culture, and without transformation over time—in terms of both the reasons and the choices made, whether wanted or sufferingly endured—there would be no history (Dezzi Bardeschi 2009). At L'Aquila, the "heavy reconstruction with improved energy efficiency, wire networks and seismic strengthening" cannot be considered "mainly performed in an authentic manner", as stated by Lemme (Bold et al. 2018, p. 229). In terms of resources (which we remember being limited, perishable and irreproducible), what is lost is lost forever and there is certainly no professional able to recover it (Dezzi Bardeschi 2009).

16.4 Twelve Restoration Projects in L'Aquila

The 12 case studies examined below include a variety of restoration work in various states of progress in the historic centre of L'Aquila. Among the buildings restituted to the community after the completion of work are the churches of San Pietro a Coppito, Santa Maria del Suffragio—known as Holy Souls (Anime Sante)—and the basilicas of Collemaggio, San Bernardino and San Giuseppe Artigiano. In contrast, the churches of Santa Maria della Misericordia, Santa Giusta, Sant'Agostino and the cathedral of Santi Massimo and Giorgio—better known as the Cathedral of L'Aquila—fall into the category of works still in progress, with varying levels of completion—the Spanish Fort and the Government Palace. Finally, the restoration project of the church of Santa Maria Paganica, still paralysed among the scaffolding erected as a safety measure, has not yet been defined. The following analysis divides the case studies according to the results of the respective restoration interventions and the themes they highlight (Fig. 16.1).

In the first group, consisting of San Pietro a Coppito, San Bernardino, the Holy Souls, Sant'Agostino and San Giuseppe Artigiano, the most striking results of the

Fig. 16.1 Map showing the twelve analysed cases in the historic centre of L'Aquila (authors)

restoration of the pre-earthquake image of the monuments of L'Aquila emerge. In this case, these buildings owned by the Curia—a factor that inevitably influenced the design choices adopted—were in some cases considered symbols of the very rebirth of the city and its community, religious or otherwise. The second category includes the churches of Santa Giusta and Santa Maria della Misericordia, and although these fell within the same design logic of the above-mentioned interventions, they also revealed some inconsistencies in the exhumation process that highlighted a profoundly selective and arbitrary design nature. The third group, on the other hand, contains the projects of three buildings of fundamental importance for the city that are still in progress: the cathedral, a symbol of great delay and the absence of architectural debate; the Government Palace, whose ruins undoubtedly have become one of the iconic images of the earthquake; and the Spanish Fort, the first place around which life on the edges of the historic centre of L'Aquila was revived.[4] Finally, the fourth group consists of a pair of monuments: the basilica of Collemaggio, whose status as municipal property has permitted the establishment of a design path whose outcome is the only one—among the cases analysed—that differs slightly from the previous ones, and the church of Santa Maria Paganica, apparently abandoned to an uncertain future and whose restoration in the recent debate has been timidly identified as a possible alternative to what seems to be the destiny of the large majority of monuments in the city.

16.5 Aesthetical Reconstructions

The imminent end to the restoration work at San Pietro a Coppito, the splendid principal church of its parish in L'Aquila, was announced with great fervour by the local media in 2018. The building had been severely damaged in the 2009 earthquake, resulting in the collapse of the top portion of the facade as well as the belfry. The church is owned by the Curia, but restoration work was financed by donations from the Municipality of Trieste, the Department of the Presidency of the Council of Ministers (Dipartimento della Presidenza Consiglio dei Ministri) and funds from the Superintendency and today appears "even more beautiful, if possible, than pre-earthquake" (Il Capoluogo 2018). In fact, comparing images of the building's exteriors before the earthquake with today's one could easily assume, at first glance, that they are exactly the same stones, the same facade and the same belfry, which would seem to have simply been subjected to a ferocious surface cleaning (Fig. 16.2).

The same fate was reserved for the bell tower of the church of San Bernardino, the dome of the Holy Souls, the lantern of the church of Sant'Agostino and the facade of the church of San Giuseppe Artigiano, whose material and formal integrity had largely been lost during the earthquake. Naturally, apart from the recovery of

[4] Mainly around the temporary-born auditorium, by Renzo Piano Building Workshop, built under the architect's desire to contribute to the reconstruction of the city and avoid the removal of citizens from its historic centre.

Fig. 16.2 Church of San Pietro a Coppito, comparison between the facades before the earthquake (left) and after the restoration (right) (authors)

some ashlars through anastylosis operations carried out more out of pity than for real conservation reasons, these are not "the same stones". This is because above all they are technologically improved reconstructions, as it should be for a restoration carried out in an area with a high seismic risk. From this point of view, there is no doubt that the aim of the restoration, in addition to "preserving" material integrity, is also that of "improving" (and not adapting) the structure of the entire building, as prescribed by the Regulations of the Superintendency for cultural heritage (Codice dei Beni Culturali).

But what are the reasons for which the design effort is entirely dedicated to the solution of a problem that is treated as exclusively technical, aimed solely at combining safety and material compatibility, without there being any broader discussion of the other integral issues, substantially architectural, of the problem? Why is image design reduced to a mere academic exercise of copying the "original"? In other fields, such as in fashion or design, we would speak of "retro" or "vintage" or even "pastiche" and "fake".

Such an approach was also affirmed by the announcement of the competition for the consolidation of the bell tower of the basilica of San Bernardino. Here, it was planned to reconstruct the collapsed part as it was and where it was, removing (because obviously unwanted) any visible trace on the monument of the tragic event. The project that is the basis of the competition even foresees dismantling the only surviving "sail" (vela) of the belfry, albeit damaged, but fortunately still standing "disturbingly" out of plumb (therefore harnessed in previdential and temporary garrison works). The idea was to reconstitute two sides by recuperating part of the existing original material (the parts of the sails that remained in situ and had been recovered from the ashlars on the ground: now assembled on the ground in the cloister of the monastery awaiting cataloguing and selection). For the other two sides, the competition expected that they would be rebuilt in the traditional way, using new but local lithic materials; thus, a soothing and reassuring (but fatally approximate) reconstruction was requested which, by removing all traces of the tragic event, would erase its disturbing historical memory (Dezzi Bardeschi 2014).

Numerous proposals for San Bernardino's bell tower suggested revising the usual interpretation of reconstruction to ban all the temptations of restoring or falsification. One of those is Marco Dezzi Bardeschi's proposal, in which the silhouette of the bell cell is recalled by wire in a Corten structure.

Another example is Gnosis Architettura's project, which re-proposes the cell with a metallic mesh skin that at a distance allows it to be perceived in the same texture and colour as the remaining wall without having to resort to dangerous reconstructions with stone materials that, years later, will risk being read as coeval with the ancient ones (Gnosis Architettura 2012). Despite these proposals, the outcome of the call favoured (again) the proposal accordant to the initial intentions most evocative and mimetic. Now, a "modern" copy of the San Bernardino' bell cell soars in memory of what never happened (Fig. 16.3).

One of the most important restoration interventions, in terms of both media coverage and because of the complexity of the intervention, was that carried out on the dome of the Santa Maria del Suffragio, better known as Anime Sante. The reconstruction of the portions of the collapsed drum and dome was carried out with blocks of limestone and lime mortar laid according to the construction techniques visible in the surviving elements, while the lantern was rebuilt following the original profile, but with techniques and materials that permitted the load to be significantly

Fig. 16.3 Church of San Bernardino after the reconstruction of the bell tower (authors)

lightened. The result is a hoax: the recreation of forms and decorative elements has brought the pre-earthquake image back to life, so much so that in order to distinguish a current photograph of the church from one prior to 2009 one is forced to observe the adjacent buildings and identify the decisive presence of cranes and scaffolding in the background (Fig. 16.4). And what will one be able to recognise when even the last crane has disappeared? On the other hand, it is practically impossible to distinguish the interventions of restoration inside the church, despite the designers' attempts to create a "gap"—or rather an "identifiable difference" of a few centimetres—that delimit the preserved portion (but reintegrated and cleaned of any sign of age) from the reconstructed one (absolutely indistinguishable in the treatment of the surfaces and shapes of the decoration). The solution was defined by the director of works himself, the architect of the Superintendency, Franco De Vitis, as "a very modest difference in height of a few centimetres, barely noticeable" (De Vitis 2018, p. 24). It will then be the task of the tour guides to point out to future visitors, probably with the support of special binoculars, the invisible scar that perhaps one day will become part of the city's curiosities (Fig. 16.5).

Fig. 16.4 Church of Santa Maria del Suffragio (Anime Sante). Comparison between the facades before the earthquake (left) and after the restoration (right) (authors)

Fig. 16.5 Church of Santa Maria del Suffragio (Anime Sante). Comparison between the domes before the earthquake (left) and after the restoration (right) (authors)

The same fate, that of formally re-proposing the pre-earthquake state, was reserved for the collapsed lantern of the church of Sant'Agostino and the top portion of the basilica of San Giuseppe Artigiano, previously known as San Biagio d'Amiternum. The latter project—rapidly concluded in 2012 because of the direct assignment of restoration work[5] by the Curia and funded by the Rome Foundation (Fondazione Roma), a private non-profit organisation that supports culture, art and education sectors—managed to draw attention to itself especially for the discovery, during reconstruction, of some frescoes dating back to the fifteenth-century phase of the building, brought to light through openings made for research in the walls of the apse that were subsequently not closed up. The restoration of the top portion of the facade as and where it was can be considered less controversial, although within the ecclesiastical world the aesthetic annoyance of seeing a portion of the "unfinished" facade being rebuilt, called for a more decisive intervention according to this group (Il Capoluogo 2012).

16.6 Selective Reconstructions

The examples analysed so far would lead one to think that the process of restoring the pre-earthquake state of a building is something acritical and totally objective, above all because the restoration of collapsed portions involved elements of undisputed value, above all aesthetic (witness the idea of rediscovered beauty that resonates in public discourse). There is therefore, apparently, no doubt that it was worth rebuilding these monuments just as they were.

The case of the church of Santa Maria della Misericordia is a good reminder of how the restoration of an image of a building can represent an opportunity to make a selective and perhaps better choice about the image itself. Here, the project carried out provided for the demolition of a severely damaged (but not entirely collapsed) volume which, leaning against the church wall, served as a distributive element of the rooms annexed to it. Although previously historicised and considered an integral part of the architectural and functional evolution of the building, this part was considered a later addition[6] that significantly altered the rigorous symmetry of the eighteenth-century façade.

---

[5] The assignment preceded the Decree n.125/2015, which required religious buildings also to be included in the process of assigning works of public buildings and therefore to determine the assignment of the works by public procurement under the jurisdiction of the Ministry of Cultural Heritage and Activities (MiBACT). The category excludes those buildings' part of a private aggregate (as it would have been for the basilica of San Giuseppe Artigiano).

[6] "Superfetazione", Italian term meaning a part added to a building after its completion and such as to spoil the aesthetic appearance of the building itself or even the surrounding environment. The authors refer to Dezzi Bardeschi, according to whom the term is to be avoided as evokes the implicit invitation to the removal/abolition of that undesirable intrusion (Dezzi Bardeschi *Superfetazione* 2017c).

Fig. 16.6 Church of Santa Giusta. Comparison between before the earthquake (left) and after the restoration (right) (authors)

Another example in which the recovery of the image was not completed is that of the church of Santa Giusta, for which only a partial evaluation can be provided because just one part of the intervention begun in 2015 and involving only the apse and transept has been completed. Precisely at the south end of the transept, the restoration decided not to re-propose the very vulnerable collapsed bell tower, being limited instead to restoring the roof and the portions of collapsed masonry (Fig. 16.6).

16.7 Facadist Reconstructions

Three major projects await completion: the Spanish Fort, of which one has been able to see the portion of the reconstructed facade for some time; that of the Government Palace, still in its initial phase but about which some reflections regarding the restoration project of the sadly known façade can be made; and finally that of the Duomo, dedicated to Saints Maximus and George, the recently begun reconstruction which was preceded by years of inertia, which aroused much controversy due to, among other things, the lack of provisional coverage of the collapsed portion of the transept.

The restoration of the Spanish Fort was conceived in such a way that the only feasible option was the "mimetic reconstruction" of the main facade. In fact, the attention of public opinion and subsequent administrations has been directed from the outset almost exclusively to the destiny of the interior spaces that housed the National Museum of the Abruzzi, whose works are still partially housed in the new headquarters opened in 2015 in the restored rooms of the former municipal slaughterhouse of L'Aquila (Fig. 16.7).

Different attention was paid instead to the facade of the Palazzo della Prefettura, better known as the Government Palace, whose entrance colonnade in ruins has become a photographic icon of the earthquake and a perfect backdrop for photographs and selfies of Italian and international politicians visiting the damaged city. In fact, the executive project of the works envisages an intervention similar in intention to

Fig. 16.7 Spanish Fort. Comparison between before the earthquake (left) and after the restoration (right) (authors)

that of the interior of the dome of the Holy Souls. As mentioned, as the report, to avoid on the one hand,

> any form of falsifying restoration, on the other any easy yielding to ruinism, [...] we opted for a differentiation of finish between the rebuilt part and the regenerated part. The final image will retain the indelible trace of the injury suffered allowing, on the one hand, the recognition of the built portion and, on the other, not giving up a clear legibility of the potential unity of the work.

Here too, pending dismantling the scaffolding, the proposed solution seems to be more "pictorial" than architectural. The choice of "reintegration of the gaps in assonance" reserved for the external facades, whose decorative stone elements have been lost and will be replaced with "similar elements with a mixture of acrylic resin and stone dust", contrasts with the multiple reintegrations of "dissonant gaps" in the internal spaces of the building. In this case therefore, the desire to conceal the great transformations of the building becomes explicit. This will change its function also as it will host the future headquarters of the province behind this rather staid facade which basically follows the previous one: the celebrated colonnade will be reassembled through an accurate anastylosis which will not, however, aid in avoiding the alteration of the pre-earthquake facades of Piazza San Marco (Fig. 16.8).

Fig. 16.8 Government Palace. Comparison between the main facades before the earthquake (left) and in 2019 (right) (authors)

An enormous gap, not visible from Piazza Duomo, is that of the cathedral, the apse laid bare because of the collapse of one of the four pillars of the crossing, with the consequent collapse of the presbytery and transverse arches and also of the end of the southern transept. Here, the project, the process of which has been enormously slowed down due to the tug-of-war between the Curia and public bodies trying to establish responsibility for the intervention,[7] provides a solution in terms of the "detachment" between consolidated and reconstructed portions. Such a detachment should, as in the case of the Holy Souls, act as a symbolic trace that separates two architectural sectors whose stratigraphy would otherwise not be known. The chapel lost in the collapse of the southern end of the transept will, instead, be rebuilt in the image and likeness of its opposite number that survived.

16.8 Alternative Reconstructions

The restoration project of Collemaggio basilica is the only case that differs from the results already analysed within the historic centre of L'Aquila. It is no coincidence that this is because the building is owned by the municipality and a fruitful collaboration was established between the administration, the Superintendency, numerous universities and the business world (in this case the energy giant Eni). Restoration began in December 2015 and ended two years later with the reopening of the basilica to the public (Fig. 16.9).

The 2009 earthquake caused the collapse of the vaults and the roof of the transept, as well as the three arches of the lateral and high altar chapels. However, the restoration project was essentially limited to the reconstruction of "load-bearing" structural elements, such as the two pylons, the chapel arches and the roof, avoiding any attempt at recomposing the image of the lost decorative apparatus of the vaults. The intervention on some pillars of the nave, which were completely disassembled and rebuilt to correct their seriously compromised stability, was rather invasive. Nonetheless, the visible outcome of the operation is significant: the pillars, the triumphal arch and the roof—although reconstructed without any clear intention to distinguish them from the collapsed originals—promote a methodology of minimum intervention that creates an image of the "unfinished" which is highly significant in the context of L'Aquila, because it offers a convincing alternative to the slavish recomposition "of the potential unity of the work" or, worse still, of the "return to the ancient splendour" implemented in almost all other cases.

Finally, it is precisely in the context of this idea for an alternative solution that the last of the cases is analysed here. The saddest and most potentially significant of all examples is the church of Santa Maria Paganica, a highly representative building

---

[7] The Curia attempted to manage the assignment of the works under private intervention procedures, receiving a proposal estimated €35 million. However, while pending the granting of funds, the proposal was questioned by the Decree n.125/2015 (refer to note 5). Afterwards, a state attorney allowed a derogation for those cases where a project had been drawn up before 2015.

Fig. 16.9 Collemaggio Basilica. Comparison between before the earthquake (left) and after the restoration (right) (authors)

which was among the most severely damaged and which has lain in a state of neglect for a decade. The church was completely uncovered and left open to the elements because of the collapse of the dome and roofs, as well as left mutilated by the collapse of the bell tower and the vaults of the chapels. For years, it has been left to be seen as a ruin at the mercy of time above which the imposing structure of the provisional roof rises. Here, it is not relevant to enter into the merit of the events behind this paralysis; in reality, they are very similar to those already mentioned in the case of the Duomo. Rather, it is interesting here to situate the case of Santa Maria Paganica as a manifest and final great opportunity for the city of L'Aquila to open up a debate on the alternatives that can be put forward in the face of the hitherto undisputed tendency to reject an explicit dialogue between the conservation of the existing and contemporary design as an opportunity to create surplus value that is not considered a threat to an identity that is more presumed than real.[8]

Unfortunately, the recent history of the city reveals a highly questionable, almost non-existent relationship, between contemporary and historical architecture. This emerged clearly immediately after the earthquake when the auditorium designed by Renzo Piano, located at the foot of the Spanish Fort, became a new point of reference and meeting place for the citizens of the city, but also much criticised by them. More recently, the new building designed by 2Studio in Piazza Santa Maria Paganica, near

[8] However, the authors are aware that a project for the reconstructive restoration of the church already exists, drawn up by professionals appointed by the Curia—owner of the building—and in standby substantially due to lack of funds which will have to be huge, as is very often the case in projects involving a reconstruction "as it was where it was".

Palazzo Ardinghelli and in front of the ruins of the church, has been contested by public opinion and yet defended as "the first serious attempt by the contemporary to compete with built heritage of which the historical centre constitutes material testimony"[9] (News Town 2017).

Probably, the reason for this absence of debate lies in the limited knowledge and understanding of the public whose opinion is highly conservative, together with the substantial extraneousness on the part of the University of L'Aquila's school of architecture with respect to the dialogue between ancient and modern architecture as the teaching of architectural design and restoration at the University of L'Aquila is done by the Department of Civil and Environmental Engineering. There are few buildings dating after the great demolition season of the 1930s that fit without sadly blending into the context of the historic centre. This is, of course, due to the protection of the historic core, but also predominantly because of the intrinsic rejection of openly contemporary architecture being added to the historicised context.

16.9 Conclusion

It is evident that a reconstruction that reassured and mitigated against the catastrophe prevailed in L'Aquila as a response to the persistent need (or will?) of the inhabitants to preserve and recover. While the attempt to undo loss is understandable, what does the architectural heritage signify when it is just a replica (Mager 2015)? Is it possible to recover cultural identity and re-know history through reconstruction? After the earthquake, a specific moment in the past was deliberately chosen and the built environment was restored to that condition. These reconstructions are a forced gesture that results in the city being preserved through a "continuous act of will [with] explicit attention to the activity that is no longer an integral element of the house, [...] transmission of the artefacts, in the fullness of their [in] authenticity" (Kealy 2015, p. 80).

Instead of moving into the future, the city clock was turned back, just as it was in the past in other contexts. This tendency is not new in post-war Germany till the recent years where, despite the promising premises for urban improvements after the bombings, only a few cities seized the opportunity to renew themselves. The case of the Market Square of Mainz—quite extreme compared to L'Aquila—represents the peculiar trend that has recently emerged, with post-war facades being pulled down in the last decades to be replaced by copies of their predecessors, destroyed in World War II or fallen victim to post-war redevelopment (Schmidt 2013). These types of new buildings are thus incorporated into an ancient reconstructed image, becoming timeless material presences, time capsules, objects produced to transcend

[9] It should be emphasised that it was a matter of replacing the existing building, which was also modern and endowed with a certain compositional dignity, with its own stylised and renewed contemporary version, with the same volumes and the same rhythm of the former facade. The materials are the only aspect in which the designers have been able to vent their creative vein.

the present and become, in a future moment, a legible trace of the past, objects that hide the void (Gordillo 2014). Through replication, architecture is transformed from a historical source into a result of our perception of history, failing to provide a reliable account of the past (Mager 2015). Thus, L'Aquila will go from being a real temporary open-air museum of the earthquake—the vast construction site it currently is—to a frozen in time museum of its (false) pre-earthquake image (Lemme 2018). Even the stakeholders involved in the reconstruction grasp the fact that the image of the city no longer exists, becoming unreliable due to its obliteration (Rehberg and Neutzner 2015).

In the city, the authorities were quick to provide the population with what they wanted and not what they necessarily needed. The materiality of the objects of cultural heritage, on the contrary, is not a necessity, especially given that the authentic object has been lost. Occasionally, indeed, the disappearance of heritage can/could further contribute to the identity and memory of communities. That happened in Liège, where the void left by the prolonged destruction of Saint Lambert Cathedral relentlessly seemed a "place of absence" until the new century, when a temporary installation evoking the silhouette of the lost main apse was cherished by the local community, as if a lost past suddenly had been brought back to life (Coomans 2012).

Given the current and dominant conservation paradigm according to which every element of cultural heritage is intrinsically valuable but the value of any newly created "heritage" cannot replace that of the existing heritage (Holtorf 2015), how can it be imagined that the value of lost heritage can be recovered through the mechanism of reconstruction? If it is agreed that the values of an architectural artefact are not lost even if this is no longer physically existing (Holtorf 2015), why not take the opportunity to create a new expression of intangible characteristics (and add new ones) of the lost building?

Despite the long period of time available, the opportunity was never taken to transform the historic centre of L'Aquila, which lay in tatters, into an important project laboratory through which to work with different proposals. Instead, naive self-deception avoided mourning through immediate action and hence also avoided the commitment to responsibly plan the future (Dezzi Bardeschi *Com'era/Dov'era* 2017b). The image resulting from this choice is homogeneous, preventing those forms of diversity and repetition that normally always exist within the historic centres of old cities, the home of conservation efforts (Di Biase 2015).

What emerges from the panorama at L'Aquila reveals quite a reluctance towards the idea of the project as a tool for reinterpreting and renewing the meaning conveyed by historical architecture. This is evident above all in the projects that insist on the key monuments of the city, such as Holy Souls or the basilica of San Bernardino. The possibility has been lost of transforming a loss, in the context of mourning—such as the collapse of the lantern and the dome in the former and the partial collapse of the bell tower in the latter—through architectural projects that aim at overcoming such loss not with removal and reconstruction, but through reworking: a hybrid approach between contemporary and historical architecture that is able to accept the challenge of the project, which is to convey an idea of the future in continuity with the past.

This is not the same as a mere image to be contemplated, but a heritage that is both alive and lived.

References

Il Capoluogo. Palmerini G. (2012) San Biagio: recupero insoddisfacente. https://www.ilcapoluogo.it/2012/07/23/san-biagio-recupero-insoddisfacente/. Accessed 23 May 2019

Il Capoluogo. (2018) San Pietro a Coppito, torna a splendere la chiesa capoquarto. https://www.ilcapoluogo.it/2018/10/27/san-pietro-a-coppito-torna-a-splendere-la-chiesa-capoquarto/. Accessed 22 Aug 2019

Abruzzoweb. Lombardo L (2018) L'Aquila. Riapre Palazzo Pica Alfieri, festa e mostra per un gioiello che risplende dopo il sisma del 2009. https://www.abruzzoweb.it/contenuti/l-aquila-riecco-palazzo-pica-alfieri-festa-e-mostra-per-un-gioiello-che-fa-commuovere/668558-365/. Accessed 2 Aug 2019)

Andreassi F (2012) La città evento. L'Aquila ed il terremoto: riflessioni urbanistiche. Aracne editrice, Roma

ANSA. Antonini O. Mons. (2019) L'Aquila sarà splendida. Nunzio apostolico,pensare ricostruzione per un turismo culturale. http://www.ansa.it/abruzzo/notizie/2019/04/06/mons.-antonini-laquila-sara-splendida_5a2b67f9-81ec-4dea-9c63-3ee843effc4b.html. Accessed 17 Apr 2019

Bold J, Larkham P, Pickard R (2018) Authentic reconstruction. Authenticity, architecture and the built heritage. Bloomsbury.

Bravaglieri S, Furioni S, Zenoni E (2016) L'Aquila rinasce (con fatica). Ananke, vol 79. Altralinea Edizioni, Firenze, pp 96–117

Bravaglieri S, Furioni S, Zenoni E (2019) L'Aquila dieci anni dopo: ancora senza progetto. Ananke, vol 87. Altralinea Edizioni, Firenze, pp 78–84

Capurro R (2017) Codice dei Beni Culturali e del Paesaggio. In: Dezzi Bardeschi C (ed) Abbeceddario minimo 'Ananke. Cento voci per il restauro. Altralinea Edizioni, Firenze

Coomans T (2012) Reuse of sacred places: perspectives for a long tradition. In: Coomans T, De Dijn H, De Maeyer J, Heynickx R, Verschaffelt B (eds) Loci Sacri. Understanding sacred places. KADOC studies on religion, culture and society, vol 9. Leuven University Press, pp 221–241

De Vitis F (2018) Conservare e ricostruire. In: Simone G (ed) Santa Maria del Suffragio. La Rinascita. Miniprint Srl, L'Aquila, pp 24–25

Decreto Legislativo 22 gennaio 2004, n. 42, Codice dei beni culturali e del paesaggio, ai sensi dell'articolo 10 Legge 6 luglio 2002, n. 137

Devilat Loustalot B (2013) Beyond the appearance of heritage: reconstruction of historic areas affected by earthquakes in Chile. ArchNet-IJAR: Int J Archit Res 7(3):24–39

Dezzi Bardeschi M (2009) Restauro: punto e a capo. Frammenti per una (impossibile) teoria. Franco Angeli, Milano

Dezzi Bardeschi M (2014) L'Aquila, due concorsi per San Bernardino. Ananke, vol 71. Altralinea Edizioni, Firenze, pp 126–155

Dezzi Bardeschi C (2017a) Autenticità. In: Dezzi Bardeschi C (ed) Abbeceddario minimo 'Ananke. Cento voci per il restauro. Altralinea Edizioni, Firenze

Dezzi Bardeschi M (2017b) Com'era/Dov'era. In: Dezzi Bardeschi C (ed) Abbeceddario minimo 'Ananke. Cento voci per il restauro. Altralinea Edizioni, Firenze

Dezzi Bardeschi M (2017c) Superfetazione. In: Dezzi Bardeschi C (ed) Abbeceddario minimo 'Ananke. Cento voci per il restauro. Altralinea Edizioni, Firenze

Di Biase C (2015) Small towns in inland areas: shared landscapes. In: Crisan R, Fiorani D, Kealy L, Musso SF (eds) Conservation, reconstruction. Small historic centres. Conservation in the midst of change. European Association for Architectural Education, Hasselt

Gnosis Architettura (2012) Relazione del progetto esecutivo sulla base del progetto definitivo a base di gara (proposte migliorative) per i lavori in messa in sicurezza, consolidamento e restauro
Gordillo G (2014) Rubble: the afterlife of destruction. Duke University Press, Durham
Holtorf C (2015) Averting loss aversion in cultural heritage. Int J Herit Stud IJHS 21(4):405–421
Kalman H (2017) Destruction, mitigation, and reconciliation of cultural heritage. Int J Herit Stud IJHS 23(6):538–555
Kealy L (2015) Dialectics. In: Crisan R, Fiorani D, Kealy L, Musso SF (eds) Conservation, reconstruction. Small historic centres. Conservation in the midst of change. European Association for Architectural Education, Hasselt
Lemme A (2018) The path of reconstruction of the city of L'Aquila after the earthquake of 2009. In: Bold J, Larkham P, Pickard R (eds) Authentic reconstruction. Authenticity, architecture and the built heritage. Bloomsbury
Mager T (2015) Architecture reperformed: the politics of reconstruction. Routledge, London
Moioli R (2019) L'insostenibile leggerezza del "com'era, dov'era". Ananke, vol 87. Altralinea Edizioni, Firenze, pp 73–77
News Town. Avellani N (2019) L'Aquila Città della Memoria e della Conoscenza: ecco il manifesto di iniziative pensate per il decennale. https://news-town.it/cultura-e-societa/24637-l-aquila,-citt%C3%A0-della-memoria-e-della-conoscenza-ecco-il-manifesto-di-iniziative-pensate-per-il-decennale.html. Access 13 May 2019
News Town. (2017) Santa Maria Paganica, l'intervento che fa discutere: un altro punto di vista. https://news-town.it/cultura-e-societa/17730-santa-maria-paganica,-l-intervento-che-fa-discutere-un-altro-punto-di-vista.html. Accessed 14 Sept 2019
Rehberg KS, Neutzner M (2015) The Dresden Frauenkirche as a contested symbol: the architecture of remembrance after war. In: Stig Sørensen ML, Viejo-Rose D (eds) War and cultural heritage: biographies of place. Cambridge University Press, New York
Schmidt L (2013) Back to the future—forward to the past: replacing modernist building with reconstructions. Hist Environ 25(1):32–45 (Australia ICOMOS)
Weiler K (2017) Aspects of architectural authenticity in Chinese Heritage Theme Parks. In: Gutschow N, Weiler K (eds) Authenticity in architectural heritage conservation: discourses, opinions, experiences in Europe, South and East Asia. Springer

Chapter 17
The "Solidere" Effect and the Localisation of Heritage Reconstruction in Post-war Transitions, Libya

Nada Elfeituri

Abstract In the wake of mass urban conflicts across the Middle East and North Africa (MENA), the most pressing question has been how reconstruction can be achieved on a large scale. Cities such as Benghazi, Aleppo and Mosul have witnessed widespread destruction that will take billions of dollars and decades to repair. With such a momentous task, many governments in the region appear to be looking towards the "Solidere" model of reconstruction that was applied in downtown Beirut after the Lebanese civil war, a model built around "disaster capitalism" in which new laws facilitate the role of private companies to lead the process. While there have been countless criticisms of the effect that this process had in Beirut, the region offers few other examples of successful reconstruction projects. Indeed, with the current climate of authoritarian rule, it is the central governments rather than residents themselves who have decision-making power to shape the reconstructed city. These decisions are driven not only by economic opportunities but by socio-political strategy, namely what should be forgotten and what will remain in the post-war city. Within this process, the efforts of citizens and local actors—often the first to initiate reconstruction of their neighbourhoods—are often overlooked or ignored. There have been increasing calls for more locally led reconstruction processes that are driven by people rather than profit, within the wider shift towards more participatory processes in urban development. These processes can be seen as more inclusive and sustainable than the "Solidere" model of reconstruction, but there is limited literature regarding how these local mechanisms operate in reconstruction contexts in the MENA region, and how they can fit into wider political processes. The aim of this chapter is to investigate local reconstruction efforts and how they play out in heritage centres. It focuses specifically on the case of downtown Benghazi's reconstruction in Libya after the 2014 civil war. It will conclude by attempting to answer the question of what place local reconstruction should have in national visions of urban redevelopment in cities affected by conflict.

N. Elfeituri (✉)
University College London (UCL), London, UK
e-mail: ucfunel@ucl.ac.uk

© The Author(s), under exclusive license to Springer Nature Switzerland AG 2021

F. F. Arefian et al. (eds.), *Historic Cities in the Face of Disasters*, The Urban Book Series,
https://doi.org/10.1007/978-3-030-77356-4_17

Keywords Post-conflict reconstruction · Heritage · Civil society · Benghazi · Libya · Beirut · Lebanon · Syria · Iraq

17.1 Introduction

Cities in the Middle East and North Africa (MENA) have seen a sharp rise in urban warfare in the past ten years, following the uprisings in 2011 and subsequent overthrow of many national governments. The political, economic and social instability that followed led to violent armed conflict which has primarily taken place in urban areas. Cities such as Aleppo, Mosul and Benghazi have seen widespread destruction as a result of heavy artillery used in built-up areas, leading to mass displacement, limited service delivery and economic collapse. This escalation of events has placed the topic of reconstruction on the forefront of many planning and policy agendas for national and international actors (Human Rights Watch 2019; Isakhan and Meskell 2019; UNDP 2018) .

However, reconstruction is a complex and contentious process. While often framed as a "post-conflict" activity, rebuilding often occurs long before the end of the fighting (Shwayri 2008). It is a process that offers new opportunities in shaping the built environment and can be lucrative for both local and international stakeholders. But it also creates tensions between competing visions of what the rebuilt city should look like. This tension is often heightened around the issue of reconstruction in and of historic districts, because it raises questions of what should be preserved and what should be removed and, more crucially, which actors have the right to make those decisions (Ghandour and Fawaz 2010). These historic districts form the core of most cities in the region and are often symbolic of the cultural identity and memory of its inhabitants (Carrión 2005). Their cultural significance, combined with their usual strategic locations, acts as potential magnets for foreign investment and is often the first to encounter redevelopment plans that threaten their historic urban fabric.

One of the most prominent examples of this contentious and complex process is the case of Beirut's city centre. Destroyed during the 1975 Lebanese civil war, the city centre was rebuilt by the "The Lebanese Company for the Development and Reconstruction of Beirut Central District", commonly known as Solidere. A private company set up by the government, Solidere, was tasked with the complete reconstruction of central Beirut, and the result is a process that is often touted as a failed example of post-war reconstruction (Hourani 2015). It was fraught with corruption and built on a vision of globalised urbanisation that overlooked the needs of the local population. It also contributed to further destruction of much of the city centre's heritage sites (Leclair-Paquet 2013).

The resulting lesson from reconstruction efforts like that of Beirut has been an emphasis on more locally led, grassroots initiatives for rebuilding damaged neighbourhoods, with the aim of reconstructing more inclusive spaces while also protecting their cultural heritage (Hilhurst et al. 2010). This has largely taken the form of supporting the work of civil society.[1] Agencies such as UNESCO have been leading

[1] The term "civil society" encapsulates a range of movements, organisations and people's collectives formed around ideological, faith-based or cultural goals, among others. They can have both formal

training and advocacy courses on empowering civil society organisations to take on this role (Isakhan and Meskell 2019). Yet, many governments in the region today are looking towards the Solidere model to reconstruct their own war-torn cities, foregoing long-term grassroots efforts in favour of rapid results that is expected to bring in investment opportunities and exert authoritarian hold (Heydemann 2018).

The aim of this chapter is to investigate these two models of reconstruction with the aim of contributing more nuance to the wider debates around reconstruction in the region. It seeks to problematise the notion of "local" reconstruction and argue against the binary of private sector vs grassroots reconstruction processes. To do this, it will employ Jeremy Till's notion of "the myth of community architecture" (Till 1998) to show that local initiatives are themselves not necessarily representative of the desires and aspirations of the larger community, and that community initiatives must plug into the wider political processes around urban development. This analytic approach is also underpinned by Haim Yacobi's conceptualisation of "the NGOisation of space", which critically examines the role of civil society around planning in relation to the responsibilities of the state (Yacobi 2007). Yacobi provides a critical assessment of the role that NGOs play in urban development, arguing that while NGOs positively contribute to spatial planning through highlighting social inequalities, their work is often disjointed and does not feed into the wider socio-political context. This framework will be applied to the case of reconstructing Benghazi's city centre, which is currently being led by local organisations.

Benghazi has been chosen as a case study for several reasons. Firstly, Libya remains an understudied country in the MENA region and so this research aims to contribute to its limited scholarship. Secondly, Benghazi offers a unique insight into the reconstruction process, because, while the conflict has largely ended in the city, the political legitimacy crisis on the national scale persists. This has prevented international companies and organisations from fully accessing the country, which has amplified the efforts of local actors.

The research data consist primarily of news articles, interviews with local stakeholders and the author's own personal experience of working on reconstruction projects in the city. The analysis will highlight the tensions that exist even on the local level and argue that there is a need to rethink reconstruction of heritage sites in the MENA region away from the binary rigidity of these models.

It is important to mention that this study is limited in its view of both top-down and bottom-up approaches to specific parts of the MENA region. The reason for this is that the current political and social climate in the region is affected by the uprisings of 2011 and subsequent efforts towards political transitions. This has created a distinct model of both reconstruction and community mobilisation which is specific to certain cities in the region, particularly those which have experienced prolonged political instability and armed conflict such as in Libya, Lebanon, Iraq, Syria and Yemen. There are other reconstruction models which have taken on different dimensions, but do not necessarily apply to this specific context.

structural and loosely formed informal groups. For the purpose of this chapter, civil society is defined as the non-governmental organisations made up of local groups who work in a given area.

17.2 The Solidere Model of Reconstruction

Before analysing the case study, it is necessary to further define what is meant by the "Solidere model". The reconstruction of Beirut's city centre by Solidere is described as a form of disaster capitalism (Hourani 2015), in which neoliberal policy reforms are enacted in the wake of a disaster with the aim of rapidly responding to the aftermath, but which are often unpopular and would not have been accepted in normal circumstances (Klein 2007). In the political context of the MENA region, these reforms are often imposed by authoritarian regimes built on patron-client relationships, focused on the economic benefit for small groups of people over the general public good (Heydemann 2018). The "Solidere model" that will be explored here can be described as "*authoritarian disaster capitalism*", a prominent example of which is Beirut's reconstruction.

The armed fighting in Lebanon that began in 1975 is often described as a civil war, although it also involved many regional actors. By the time a peace agreement was reached in 1990 to end the conflict, many parts of Beirut—Lebanon's capital—were destroyed. The city was divided into the Christian East and the Muslim West by the Green Line, which acted as the front line in the fighting between the various armed groups (Leclair-Paquet 2013). This line ran through the city's central business district (CBD), and as a result this area witnessed some of the most severe destruction.

During the period of recovery, the presence of the state was barely felt due to its fragile institutions (Shwayri 2008). With no real military and no natural resources, political leaders pursued neoliberal policies to open the country's economy to the global market in order to facilitate the reconstruction process, seen as the only way to rebuild the country (Hourani 2015). This meant that the reconstruction of Beirut's CBD—a key hub in the country's connection to the global economy—was one of the top priorities for reconstruction planning.

From this mission, Solidere was born. It was created by the Lebanese government in 1994 as a private company tasked with the complete reconstruction of the CBD, with many of its shares owned by government officials. In order to facilitate its work, the company was exempt from taxes for 10 years (Shwayri 2008), and Law 117 was passed which allowed the company to expropriate property from owners in exchange for Solidere stock shares (Leclair-Paquet 2013). These steps were taken in order to ease and simplify what was set to be a very complex reconstruction process, and the justification of these reforms—and indeed of Solidere's lead in the process—was that the government was unable to handle issues such as compensation, investigation of property ownership or attracting foreign investors and funding (Hourani 2015).

Solidere's role was to implement the government's reconstruction plan, which was modelled on a vision of re-instating Beirut's place on the world stage as key economic and touristic node in the Middle East, envisioning it as a global city. The company was responsible for financing, developing, (re)constructing and managing the CBD, including negotiating with local citizens and property owners. The policies enacted by the government gave the company considerable implementation power. The reconstruction was set to span a 25-year period in three phases, although these plans were disrupted by the 2006 war with Israel and the 2011 revolutions that resulted in the mass displacement of Syrians into Lebanon, as well as legal and political delays.

The resulting reconstruction process is one that has been heavily criticised by local and international actors because of its negative impact on the economic, social and cultural fabric of Beirut and the perception that it favoured profit over people (Hourani 2015; Shwayri 2008; Ghandour and Fawaz 2010). Among the most salient issues that have been brought forth are the unfair compensation process of property owners, the demolition of cultural heritage sites that survived the conflict and further fragmentation of Beirut's urban fabric.

The compensation process was carried out in a manner than would allow for Solidere to acquire land and property ownership of the CBD in order to implement the reconstruction. While there was already outrage over the passing of Law 117 to compensate property owners with stock shares rather than cash, this was further exacerbated when these owners were compensated for far less than the actual value of the property. Those who refused to "sell" their property to Solidere often found themselves forcefully coerced into doing so (Leclair-Paquet 2013).

The second issue was that of cultural heritage destruction. The conflict had destroyed much of the historic buildings and landmarks in the city centre. However, there were still some remaining heritage sites that were only minimally damaged and could have been renovated. It has been argued that, rather than renovate these sites, Solidere had demolished them in order to make way for newer, more "modern" buildings that fit with the new "global city" vision (Leclair-Paquet 2013).

Lastly, the reconstruction process by Solidere was accused of fragmenting the urban fabric of the city, creating an "enclave" that targets affluent residents and tourists at the expense of the rest of the population (Ghandour and Fawaz 2010). Despite Solidere's aim to remove the "East–West" division of the conflict, it created a division between the centre and the rest of the city instead. Once a vibrant economic district that reflected Lebanon's diversity, the reconstruction of the CBD has left it largely empty due to the unaffordable costs of renting residential and commercial spaces.

Despite the backlash against Solidere, there have also been voices who have defended the company and its reconstruction of Beirut's city centre (Bollens 2012). These voices argue that—without Solidere—the city centre would never have been reconstructed and would have remained in ruins for years. Others argue that the complexity of the reconstruction process would have made it difficult for any entity to rebuild in a manner that was suitable for all stakeholders. Others still have argued that regional political and military instability are the causes behind the limited success of the project.

- ***Revival of the Solidere Model After 2011***

The top-down, market-driven process of reconstruction has been criticised for—among what has been mentioned previously—being unsustainable and responsible for creating new tensions rather than addressing grievances (Sharp 2018). Yet, despite these critiques, the current reconstruction plans being drawn up by governments in parts of the MENA region appear to follow the example of the Solidere process.

In Syria, the government has passed a number of urban planning policies to facilitate the reconstruction of destroyed cities with the aim of implementing new development projects (Human Rights Watch 2019; Heydemann 2018). Among the most controversial of these projects is "Marota City", a development in Damascus which will see the construction of luxury neighbourhoods targeted at tourists, as a means to attracting foreign investment to the country (Al-Lababidi 2019). The Damascus Cham Holding Company was created by the Syrian government to manage this project by establishing partnerships with the private sector, in a move that echoes Solidere. From its inception the project has been criticised for corruption, confiscating private land and favouring profits over the real needs of Syrians after the war (Al-Lababidi 2019).

In Iraq, a proposal was put forth by the regional governor and private investors to redevelop the old city of Mosul after the conflict against ISIS destroyed much of the city. The proposal outlines a vision of a "new city" which would see much of the heritage destroyed to make way for modern high rises, as a means of bringing in the investment needed to start repairing the city.[2] Like in Syria and Lebanon, it would also require the acquisition of private land owned by families in the city.

In Libya, Solidere itself visited Benghazi and was invited by government officials to present a reconstruction plan to a number of state institutions in Libya (Akhbar Libya 2018). While it is not clear whether the same financial and administrative model that was used for Beirut would be duplicated for Benghazi, the proposal would adopt the same model of developing a culturally significant and geographically profitable part of the city to attract foreign investments.

For all these governments, the Solidere model is attractive because responsibility is shifted away from the public sector—which has been weakened due to the conflict—to a private sector actor. It also offers opportunities for officials to benefit financially through backroom deals with private companies and allows for the "winner" of the conflict to consolidate the effects of war and rebuild the city in the image they desire (Ghandour and Fawaz 2010). In the case of Syria, this means that those who fought against the government would have their property confiscated, while in Benghazi it would be a chance for unelected authorities to gain legitimacy by rapidly "erasing" the effect of the war.

[2] Ibrahim, A., and Al-Rubaie, A., (2019), "*Modern revamp threatens to raze riverfront in Mosul's Old City*". Aljazeera. [Available at: https://www.aljazeera.com/indepth/features/modern-revamp-threatens-raze-riverfront-mosul-city-190713100636619.html].

17.3 Locally Led Reconstruction as an Alternative Approach

As conflicts continue to take place in urban areas, a new approach has been formulated in the past few decades which advocates bottom-up reconstruction efforts (Hilhurst et al. 2010). This approach is primarily advocated for by international humanitarian and development actors and extends to issues such as the safeguarding of cultural heritage after a conflict and repairing physical infrastructure. There has been a renewed call by these actors for this type of bottom-up reconstruction approaches in the MENA region in recent years in Libya, Syria, Iraq and Yemen[3].[4]

This trend can be seen as fitting into the wider calls for more participatory processes in urban design and planning (Sanoff 2008). Indeed, community involvement in design has been shown to empower local residents to address entrenched inequalities in the build environment. If done in partnership with local and national authorities, it can also build trust between people and their governments and strengthen democratic processes. For this reason, many development organisations have been applying a participatory approach to post-conflict reconstruction.

"Reconstruction from below"—as it is sometimes referred to—is a participatory form of reconstruction in which local people lead the process. It can involve the physical repair of buildings and clean-up of open spaces, as well as interventions that target the social, economic and psychological impact of the war. The mechanism of participation can be both through "formal" institutions such as civil society organisations, local authorities, architects and academics, as well as "informal" groups such as neighbourhood groups and tribes (Al-Harithy 2015). These groups are able to mobilise funding from a wide array of sources, including crowdfunding, private sector donations and grants. It is important to note that the distinction between formal and informal groups is often blurred on the ground, and there is frequent collaboration between all stakeholders as a "community" rather than as distinct entities.

Locally led reconstruction is also sometimes characterised by the transnational partnerships that are formed with international organisations. The project "Revive the Spirit of Mosul" was led by UNESCO in partnership with local groups to repair heritage sites in Mosul (Isakhan and Meskell 2019). In Syria, UNDP has been working on "social infrastructure" projects to train artisans, artists and other local groups in order to empower their role in reconstruction and contribute to overall stability in conflict-affected areas.[5]

[3] World Bank, (2020). "*Rebuilding for Peace. Reconstruction for Security, Equity and Sustainable Peace in MENA.*" World Bank Group. [Available at: http://documents1.worldbank.org/curated/en/747201593601797730/pdf/Building-for-Peace-Reconstruction-for-Security-Equity-and-Sustainable-Peace-in-MENA.pdf].

[4] Apap, J. and Cesluk-Grajewski, M. (2020). "*A joint EPRS-EUI policy roundtable.*" European Parliamentary Research Services Blog. [Available at: https://epthinktank.eu/2020/02/13/the-middle-east-and-north-africa-region-mena-what-future-for-stabilisation-and-reconstruction/].

[5] UNDP Syria (2020). "*Supporting the Resilience of Local Communities*". [Available at: https://www.sy.undp.org/content/syria/en/home.html].

However, the notion that local reconstruction efforts are the alternative to government-led neoliberal processes has been challenged as a false binary (Lizzarralde et al. 2010). While local communities are often the first to initiate reconstruction in their neighbourhoods by repairing their homes and properties, this rarely crystallises into a cohesive and large-scale effort with clear leadership. International organisations that promote bottom up reconstruction often emphasise the importance of empowering the "locals" to take the lead on reconstruction, but rarely do they describe who these locals are, what mechanisms are in place to facilitate this process, or where the funding will come from.

In his critical analysis of participatory approaches to issues of the built environment, Jeremy Till argues that those advocating for community-led development often disregard the fact that the community is not homogenous, or that the process often ignores the political and economic forces around local initiatives (Till 1998). He also argues that those who act on behalf of "the community" are not always representative of the diverse views and backgrounds. Furthermore, Yacobi highlights that, while these initiatives can amplify the voices of groups who were previously unheard, they still work within the existing status quo of unbalanced power relations and struggle to challenge them (Yacobi 2007). This analytic lens will be applied to community-led reconstruction in the next section, by looking specifically at the case of Benghazi, Libya.

17.4 The Case of Benghazi and "Bottom-Up" Reconstruction

In 2014, the tensions between various armed factions in Benghazi culminated in a military confrontation that led to a full-scale civil war in the city. The overthrow of the Gadhafi regime in 2011 had led to a high level of precarity in the city, characterised by lawlessness and a rise in religious extremism, and this contributed to the escalation leading up to the war. The conflict had paralysed the city and resulted in the displacement of hundreds of thousands of people from their homes, the destruction of infrastructure and buildings and the worrying rise of humanitarian issues such as unexploded ordinances and a strain on the health sector.

While the fighting was primarily limited to the city's peripheral neighbourhoods, it also took place in the city centre due to its strategic location next to the city's port and the vantage points offered to snipers and guards. As the fighting abated in the rest of the city by 2017, the city centre—known as downtown Benghazi—was the last district to witness an end to the fighting. The narrow streets and high buildings had hindered the armed confrontation, and while one of the warring factions was effectively surrounded and "trapped" in the area, the port meant that they could continue to receive supplies of aid and weapons. Figure 17.1 highlights the location of the district.

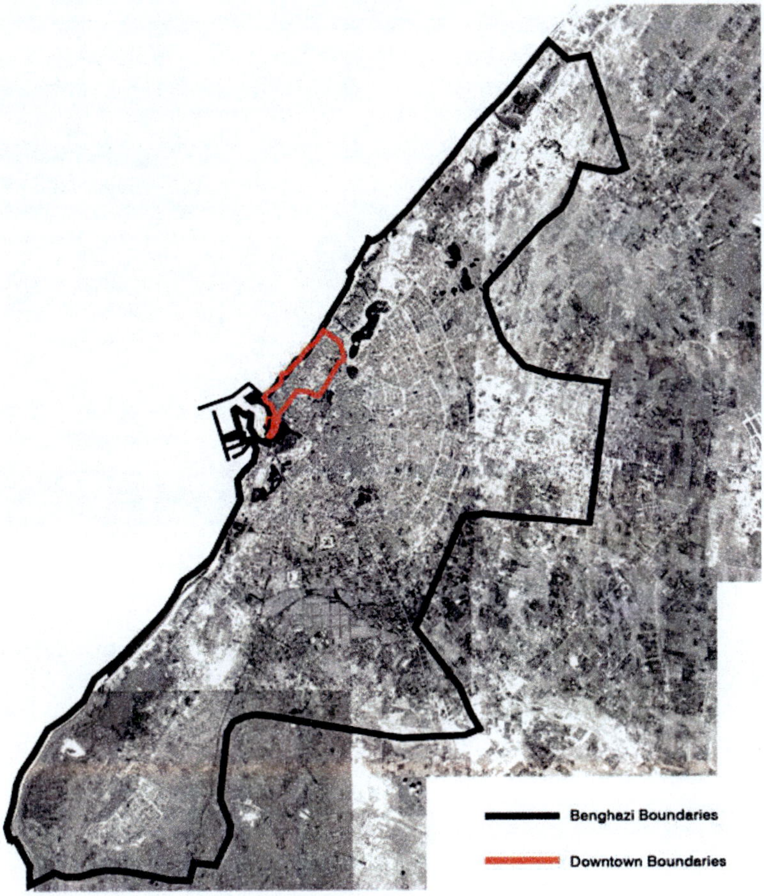

Fig. 17.1 Location and size of Benghazi's downtown district (illustration by author)

Downtown Benghazi was officially declared "liberated" by the military in July 2017 (BBC News 2017) but was not opened to civilians until many months later. The extent of the damage in the area and the risks to civilians was severe. Many buildings had been razed to the ground or were near collapse. The damage to infrastructure meant that many streets were flooded with sewage water from burst pipes. Mines were planted in many parts of the district and unexploded ordinances dotted throughout the area. Because downtown Benghazi is also the site of the historic old city, many heritage buildings were lost or heavily damaged. Unlike other parts of Benghazi, the destruction of the city centre was particularly bitter for residents, because the area was seen as embodying the "soul" of the city. Figure 17.2 highlights some of the destruction.

Because of this perceived historic and cultural importance, reconstruction efforts by citizens and local organisations began almost immediately after the area was

Fig. 17.2 Destruction in downtown Benghazi (author)

re-opened. These initiatives began with clean-up campaigns to remove the rubble and clear the streets to allow for further access into the district. Those who had the financial means to renovate their properties began to rebuild, and civil society organisations worked on clearing up and repairing public spaces. These initiatives were often done in coordination with the neighbourhood council, the municipality and the military de-mining unit.

One of the earliest tangible rebuilding efforts was the revival of the "Plaza of the Tree", a historic public space in Benghazi. Before the conflict, the plaza was renovated by a civil society organisation (CSO) called "*Benghazi Al-Amal*" (بنغازى الأمل) which translates to "Benghazi the Hope"; a charity organisation founded during the war by activists from the city. The plaza and surrounding buildings were destroyed by the heavy artillery used during the conflict, and Benghazi Al-Amal mobilised their resources and volunteers to repair the damage. The facades of the buildings facing the plaza were renovated, and the plaza itself was re-paved. The fountain located in the centre was also repaired and new plantings were placed around it.

When I interviewed the organisation about the project, they said that the repairs were a symbolic act to "bring life back to the heart of Benghazi" and to show the public and authorities that reconstruction efforts with minimal resources was possible. The project was funded almost entirely by private donors, and the organisation makes a point to emphasise that they receive no funds from government agencies. One member of Benghazi Al-Amal said that they have also been carrying out clean-up campaigns and other performative acts of reconstruction to make a point about the government's absence.

While members of the organisation are not residents of the area itself, the decision to work in the downtown area came from the perception that it reflects the identity of the city, due to the historic nature of the area. The CSO works closely with the local community and neighbourhood council and carries out a number of projects in the area, including helping people to renovate their homes and cleaning up rubble, but it is the renovation and revival of public spaces that the organisation is focused on. During the months of Ramadan in 2019 they coordinated with other actors to hold a "Ramadan festival" to revive the marketplace of the downtown area (Alghad 2019).

Speaking to members of the neighbourhood council, who coordinate with these groups as they implement their projects, there were mixed reactions to the work of CSOs like Benghazi Al-Amal. While they admired and appreciated civil society efforts, they also felt that there were bigger priorities than renovating public places. The council was facing pressure by residents who were still displaced and unable to return because the infrastructure was still not repaired. They felt that renovating the electrical power stations, sewage lines and other services that would allow people to return to their homes was more crucial. However, they also recognised that these types of projects required more resources than what civil society organisations had the capacity to achieve.

One member of the neighbourhood council expressed admiration for reconstruction projects such as the one undertaken by Solidere. He felt that the scale of destruction in Benghazi required the efforts of an international company because, in his opinion, local and national actors did not have the capacity to repair the city centre. When asked about the heritage aspect of the reconstruction, he said that he would like to see some of the historic buildings protected, but also wanted "modern" buildings and urban design in his district, and felt that focusing only on preserving what had existed before the war would "keep Benghazi back from developing like the rest of the world".

Another key actor in the city centre's reconstruction is the Authority of the Protection of the Old City (APOC), an official government body that regulates all aspects of the built environment in downtown Benghazi. It is primarily made up of architects, planners and cultural heritage experts. Construction and demolition projects are required to first seek approval from APOC, although after the war and lack of funding from the central government, the authority has found it difficult to enforce its mandate.

When interviewing a member of APOC, she also expressed her admiration of civil society and their ability to mobilise people but was frustrated that the reconstruction of heritage buildings was not done according to the guidelines for renovating these types of buildings. She blamed this on the lack of conservation experience in Libya and said that it was necessary for large international companies to get involved to provide this expertise. She felt that CSOs were crucial in "reconstructing the people" but that the physical repair of the built environment required large scale efforts that needed the state, the private sector and international assistance.

At the time of writing this chapter, the only international actors that have been able to implement reconstruction projects in Benghazi are International Non-Government Organisations (INGOs) and UN agencies. The largest effort has been undertaken by the United National Development Programme (UNDP), who have been implementing a multi-million-dollar project entitled the "Stabilisation Facility for Libya"

(SFL) which conducts reconstruction in various Libyan cities (UNDP 2018). In Benghazi, the SFL has renovated a school and a clinic and is planning to repair infrastructure services as well as the corniche along the beach front. These projects are done through approval by the neighbourhood council and local residents in "town hall" style meetings.

However, like with CSOs, local stakeholders felt that these efforts were not enough to address the city-wide issues of infrastructure and the return of displaced families. A member of the SFL team who was interviewed acknowledged that UNDP would not be able to carry out a "real" reconstruction of the city centre, because the level of destruction required specialised companies, particularly around repairing water and power lines. While the SFL project has over $5 million marked for reconstruction in Benghazi—raised through the contributions of UN member states—the extent of the damage is estimated to be in the billions of dollars.

As the national political crisis continues in Libya, it seems unlikely that multinational companies are willing to work in Benghazi due to the high level of risk it poses to reconstruction projects. This crisis has impacted the ability of local government institutions to achieve more than small-scale initiatives such as the light reconstruction of public buildings and the replacement of streetlights. The municipality, recognising this weakness, has held a number of reconstruction conferences that have aimed to attract international interest (Zaptia 2018). Solidere is the only international company that has so far expressed an interest in working in Benghazi and presented plans to state actors (Akhbar Libya 2018).

17.5 Challenges

The previous section examined local efforts in reconstruction, by looking at the dynamics of various state and non-state actors. As was shown, locals groups and institutions each had different visions for what the reconstruction of downtown Benghazi should look like. From these interviews, it was clear that tensions also existed between these different groups. In particular there was some hostility towards APOC from the community actors, which was seen as hindering the process of renovating historic buildings, due to their strict guidelines and insistence on professional conservation processes.

The most frequent issue that came up from all the interviewed stakeholders was that grassroots efforts were not enough. Local actors did not possess the technical skills, the resources or the leadership to reconstruct downtown Benghazi due to the extent of destruction in the area and the need for a new development vision. Because of the fragility of the state and limited local private sector companies, all actors felt that the city should reach out to international construction and consultation companies in order to support the process. However, there was no clear idea of what the mechanism of cooperation should look like, or what the role of the local groups should be within these larger international plans for the city.

While all actors agreed that the downtown was an important heritage site that should be preserved, each group had a different idea of how the reconstruction should take place. Some expressed the desire to see a Solidere-style reconstruction process because they felt that Benghazi should develop into a global city, similar to those in the Gulf region. This could be attributed to the fact that Libya—also a major oil producer—had the potential and resources to implement globalised development. None of those interviewed expressed a desire to see the city centre rebuilt exactly as it was before the war.

It is clear that the "local community", while able to work together, are not a homogenous group with shared priorities and ideals, in keeping with Till's notion of the "myth of community architecture". There are disagreements around how heritage preservation should be carried out, what kind of new development should be built and the ultimate architectural vision for the new city centre. In the locally led programming approach of most international organisations, there are few provisions or tactics around how to address these disagreements and differences of opinion, and how to bring together different visions for a more nuanced and complex process within the context of post-conflict cities in the MENA region.

The community itself also recognises its limits in what it can achieve and the importance of wider economic and political processes in the reconstruction of downtown Benghazi. Despite the insistence of international development actors to "empower the locals", the community interviewed here did not seem to require "empowerment". They were very conscious of their own capacity and their limitations.

Perhaps even more significantly is that—while "community-led development" is meant to help local people rise above the incompetence or unresponsiveness of the state—the community members that were interviewed still held the Libyan government to account for fulfilling their social contract with citizens. In writing on the increased "NGOisation of space" in the MENA region, Yacobi highlights the concern that placing the responsibilities of spatial planning onto local groups risks absolving the state of its role towards its citizens, often a result of neoliberal policies that favour privatisation. This scenario is, to some extent, what has happened in Beirut. After the Lebanese civil war, civil society grew substantially in the country even while reconstruction was being privatised (AbiYaghi et al. 2019).

In Benghazi there is a notable resistance to the complete NGOisation of space. Indeed, many of the actions of the CSOs working in the city centre highlighted the government's failure to implement reconstruction efforts rather than celebrate their own successes. This political awareness is considered by Till to be a crucial element in successful community development that helps move the process away from the binary of state vs. local development towards more collaborative processes between the different sectors. (Till 1998, p. 41). Indeed, the wider literature around participatory spatial development often emphasises the need for dialogue between the community and state actors (Sanoff 2008).

The analysis of Benghazi's reconstruction by local actors is in keeping with the emerging literature that problematises "community-led reconstruction" (Hilhurst et al. 2010; Lizaralde et al. 2010). While the participation of the community is crucial to ensuring that the reconstructed city is accessible and inclusive to all its residents, it is important to consider the local power dynamics at play and to understand the "community" away from simplistic and neo-colonial perspectives that frames it as purely a target for development programming. It is also important to recognise that desires for globalised urban development are not limited to government planning but also to residents of these cities who want to see their cities grow through modernisation projects. What remains a crucial question also for this research that underpins the current reconstruction projects in Benghazi and elsewhere in the MENA region, is what role the state must adopt in relation to the local community, and what mechanisms can facilitate the coordination between all relevant stakeholders.

17.6 Conclusion

This study has attempted to analyse two common approaches to reconstruction in parts of the MENA region; top-down plans which are implemented by a national government and rely on private sector funding and bottom-up participatory processes led by communities and often with support from international organisations. It has focused on reconstruction in the central heritage zones of urban areas because these "historic towns centres" are often the first to witness development planning due to their potential to bring in foreign investments, and because of their importance as a symbol of culture, identity and memory to residents of the city.

Each approach has been shown to achieve some success. The Solidere model is able to mobilise financial resources and technical knowledge on a global level, allowing it to operate on a large scale and tackle difficult issues such as the repair of sewage systems, the construction of new buildings and compensation of damages. On the other hand, locally led reconstruction processes are able to respond more rapidly and address basic needs with little to no resources, as well as facilitate partnerships between and with various local authorities. In Benghazi, local groups were most successful in clearing out rubble, helping residents to access their homes and renovating public spaces.

However, both top-down and bottom-up approaches have glaring shortcomings which limits the extent to which reconstruction heals and promotes recovery. The Solidere model is often built on corruption between government officials and private sector companies, and the "big-picture" vision of the reconstructed city is geared towards foreign tourists and companies, which can result in unused real estate that divides the city. As seen in Beirut previously and now in Syria, government-led reconstruction is also used as a tool to consolidate power, often through the passing of laws and policies that dispossess residents of their properties. It is a process that can lead to further grievances and exacerbate tensions in an already fragile post-conflict environment.

On the other hand, bottom-up reconstruction relies on the efforts and resources of local community groups, many of whom are vying to put in place their own vision and further their own group's interests. While collaboration does occur, there are also points of conflict, and the label of "community-led" does not necessarily mean that it is inclusive of all points of view. These groups also lack the technical knowledge or access to finance to implement more than just small-scale interventions. The involvement of international actors as sources of funding or capacity-building also raises the question of whether the process is truly locally led or if community participation is a tokenistic measure to appease donors.

Both approaches should not be seen as an "alternative" to one another but rather as missing pieces of a multi-scaled interdisciplinary approach to reconstruction. Recovery from the physical impact of war is a complex process that requires long-term commitment and collaborative governance on the scale of the street, neighbourhood, city and state. Local communities rely on the resources and outreach that national governments can provide, but they can also act as a watchdog and hold officials accountable. Projects that advocate community involvement must explicitly state how this approach relates to existing national processes, while also ensuring that different voices within the community are heard. These voices must not just be heard only as a performative exercise but should inform planning and reconstruction policies.

In the MENA region where conflict and instability have crippled national governments, a reconstruction strategy at the level of the city has potential. As seen in the case of Benghazi, there already exists a relationship between both local government and civil society actors. While positively strengthening this relationship is crucial, it also requires a broader network of actors that include intergovernmental organisations, NGOs and technical expertise of architects, planners and engineers, among others. It is not an easy task, but by taking these points into consideration, reconstruction can be a tool for recovery rather than division.

References

AbiYaghi M, Yammine L, Jagarnathsingh A (2019) Civil Society in Lebanon: The Implementation Trap. Civil Society Knowledge Center, Lebanon Support. Available at: https://civilsociety-centre.org/paper/civil-society-lebanon-implementation-trap. Accessed on 13 Oct 2020

Akhbar Libya (2018) 'تقدم عرضاً لإعادة إعمار بنغازى "سوليدير" ...على خطى بيروت' Following the Beirut Plans ... Solidere Presents an Offer to Reconstruct Benghazi. Available at: https://libyaakhbar.com/libya-news/798003.html. Accessed on 2 Sept 2019

Alghad (2019) 'رمضان يعيد الحياة إلى الأسواق الشعبية فى بنغازى' Ramadan Brings Like Back to the Local Markets of Benghazi. Available at: https://www.alghad.tv/رمضان-يعيد-الحياة-إلى-الأسواق-الشعبية/. Accessed on 3 Sept 2019

Al-Harithy H (2015) The participative discourse: community activism in post-war reconstruction. In: Saliba R (ed) Urban design in the Arab world: reconceptualizing boundaries, pp 39–50

Al-Lababidi M (2019) Damascus businessmen: the pantoms of Marota city. European University Institute. Available at https://cadmus.eui.eu/bitstream/handle/1814/62227/MED_2019_07_EN.pdf?Accessed 3 Sept 2019

BBC News (2017) Libya eastern commander Haftar declares Benghazi 'liberated'—BBC News. Available at: https://www.bbc.co.uk/news/world-africa-40515325. Accessed on 17 March 2019

Bollens S (2012) City and soul in divided societies. Routledge, Oxfordshire

Carrión F (2005) The historical centre as an object of desire. City Time 1(3):1

Ghandour M, Fawaz M (2010) Spatial erasure: reconstruction projects in Beirut. Architecture Publications, Iowa State University, p 48

Heydemann S (2018) Beyond fragility: Syria and the challenges of reconstruction in fierce states. Faculty Publications, Middle East Studies

Hilhurst D, Christoplos I, Van Der Haar G (2010) Reconstruction 'From Below': a new magic bullet or shooting from the hip? Third World Quarterly 31:7

Hourani N (2015) People or profit? Two post-conflict reconstructions in Beirut. Hum Organ 74(2):174–184

Human Rights Watch (2019) Rigging the system: government policies co-opt aid and reconstruction funding in Syria. Available at: https://www.hrw.org/report/2019/06/28/rigging-system/government-policies-co-opt-aid-and-reconstruction-funding-syria. Accessed on 29 June 2019

Isakhan B, Meskell L (2019) UNESCO's project to 'Revive the Spirit of Mosul': Iraqi and Syrian opinions on heritage reconstruction after the Islamic State. Int J Heritage Studies

Klein N (2007) The shock doctrine: the rise of disaster capitalism'. Random House of Canada.

Leclair-Paquet B (2013) Beirut divided: the potential of urban design in reuniting a culturally divided city. DPU Working Paper No. 153, UCL

Lizarralde G, Johnson C, Davidson C (2010) Rebuilding after disasters. From emergency to sustainability. Spon Press, London

Sanoff H (2008) Multiple views of participatory design. Int J Archit Res 2(1):57–69

Sharp D (2018) Lebanon and the fog of reconstruction. In: The politics of post-conflict reconstruction, pp 44–50. Carnegie Middle East Center

Shwayri S (2008) From regional node to backwater and back to uncertainty: Beirut, 1943–2006. In: Elsheshtawy Y (ed) The evolving Arab City. Routledge, London, pp 69–98

Till J (1998) Architecture of the impure community. In: Hill J (ed) Occupying architecture: between the architect and the user. Routledge, New York, pp 34–42

UNDP (2018) Stabilization facility for Libya annual report. Available at: https://reliefweb.int/sites/reliefweb.int/files/resources/Annual%2520Report%25202018%2520FINAL%252020-5-2019.pdf. Accessed on 15 July 2019

Yacobi H (2007) The NGOization of space: dilemmas of social change, planning policy, and the Israeli public sphere. Environ Plan 25:745–758

Zaptia S (2018) Benghazi Rebuild conference opens with participants from 18 states. Libya Herald. Available at: https://www.libyaherald.com/2018/05/08/benghazi-rebuild-conference-opens-with-participats-from-18-states/. Accessed on 3 Sept 2019

Chapter 18
Bell Towers Under (Seismic) Attack: Saving a Symbol, Once It Became a Menace

Eva Coïsson, Lia Ferrari, and Federica Ottoni

Abstract The numerous high and slender bell towers that characterise the skyline of historical city centres turned out to be particularly vulnerable on the occasion of the 2012 seismic shocks in Northern Italy. The danger that these constructions would collapse, besides representing a technical and structural problem, also had strong social implications, generating fear and uncertainty in the population with the imposition of extensive and clearly demarcated off-limits areas. As a consequence, after the Emilia earthquake, many mayors requested the authorisation for the demolition of these peculiar buildings, because this looked like the most rapid and cheapest option in order to reinstate normal activities in their city centres. However, this solution was more complex than expected, not only operationally, but also from an ethical point of view. The Regional Offices of the Ministry for Cultural Heritage (*Direzione Regionale per i Beni Culturali e Paesaggistici dell'Emilia Romagna—Soprintendenza*) rejected most of the applications for demolition, redirecting these institutions towards first-aid interventions. Nevertheless, the lack of adequate knowledge and experience about this type of intervention soon emerged, and the *Soprintendenza* decided to contribute by appointing a commission of experts in restoration and building techniques from the universities of Parma, Bologna and Padova. The commission analysed many bell towers, defining for each case some guidelines for rapid intervention aimed at combining conservation and safety. This chapter analyses some of the most meaningful examples of emergency structural interventions on bell towers after the 2012 Emilia earthquake, comparing the indications supplied by the regional *Soprintendenza* with the actual work executed and discussing the implications that these interventions, often carried out during the emergency without time for an adequate reflection, may and did have on the definitive intervention, both in terms of economy and function.

E. Coïsson · L. Ferrari (✉) · F. Ottoni
University of Parma, Parma, Italy
e-mail: lia.ferrari@unipr.it

© The Author(s), under exclusive license to Springer Nature Switzerland AG 2021

F. F. Arefian et al. (eds.), *Historic Cities in the Face of Disasters*, The Urban Book Series,
https://doi.org/10.1007/978-3-030-77356-4_18

Keywords Masonry bell towers · Seismic reinforcement · First-aid interventions · 2012 Emilia earthquake · Italy

18.1 Introduction

High and slender bell towers have characterised the skylines of historical city centres for centuries, becoming a symbol of cities and even more of their communities (Fig. 18.1). These monuments represent some of the most impressive results of the never-ending quest for the "art of construction" and master-builders' construction techniques aimed at defying gravity and even resisting the risks of earthquakes.

These constructions have gradually improved thanks to ever bolder constructive stratagems and progressive refining of rules and proportions, thus challenging the laws of statics and becoming ever higher and slenderer. While these proportions are perfect against gravity, they have a significant drawback which becomes evident in the case of earthquakes. Over the centuries, most of these structures have been ruined because of seismic activity, revealing the errors hidden in their construction: some collapses have indeed been caused by incorrect proportioning processes, whereas some others were the result of incorrect building techniques.

The failure of these structures (Fig. 18.2) in both cases implies serious consequences connected not only with the loss of their structural and monumental value, but also with the loss of memory and cultural identity, a social problem that is even more important and difficult to solve. Indeed, bell towers not only are historical documents as all monuments: more than others, they represent a reference point and a symbol for the community, soaring above the cities. Moreover, the danger of these constructions collapsing during a seismic event—or immediately after it—generates fear and uncertainty in the population, thus becoming not only a technical and structural problem, but even an issue of public safety and logistics that concerns the entire community. The potential collapse of a high tower also concerns a large number of surrounding buildings, imposing the demarcation of extensive off-limits areas.

The issue of securing these structures was not an easy task, and there was not a general solution (Tandon 2018). During the emergency phase, characterised by a lack of time and money, the demolition of these particular buildings, when strongly

Fig. 18.1 Skylines of historical city centres characterised by the presence of bell towers (https://www.eventiesagre.it/Da+Visitare_Visite+Guidate/21175494_La+Camminata+Le+Antiche+Torri+Di+Parma.html; https://rolandodondarini.wordpress.com/2011/04/28/latredicesima-porta/; https://www.donnedellavite.com/single-post/2018/11/02/san-gimignano-trastoria-arte-e-cultura-enoica-sabato-15-dicembre)

Fig. 18.2 Historical drawing of a damaged bell tower (https://www.sicilians.it/ilterremoto-del-1783-22885/)

damaged by the seismic event, often seemed to be the most rapid and cheap option aimed at being able to reinstate the normal activities in these cities. Any other conservative solution—even if more correct from an architectural and a cultural point of view—would require a proper study of these historical structures (e.g., material properties, geometry, construction techniques, boundary conditions), thus resulting in significantly more expensive, time-consuming and operationally difficult conservation strategies.

The characteristic reaction to seismic events of this structural typology continues to represent a critical issue, and this research aims to provide some answers by investigating the original methods used for the construction of these great towers: the empiricism, and thus the observation, of the actual behaviour of these historical structures during and after the 2012 earthquake. In particular, an examination of the damage suffered by a series of bell towers in the recent Emilia earthquake (Candigliota et al. 2012) constituted an occasion to set up a sort of "catalogue of damage" for these structures on which a reliable strategy for emergency interventions could be based. Indeed, the operational choice between the options of demolition or provisional and emergency restoration of the damaged structures involves not only structural considerations, but also cultural, historical and logistical ones.

The final aim here is to present and analyse the methodology adopted in these instances, which can constitute possible operational guidelines when a significant number of buildings of this type are involved in an effective and rapid plan for intervention.

18.2 Bell Towers: Seismic Behaviour and Damage

The seismic risk assessment of masonry towers is a growing concern, as dramatically demonstrated by the partial or total collapses that occurred during past earthquakes. These structures were built to withstand mainly vertical loads and, consequently, due to their typological and morphological features, horizontal seismic actions may have severe effects on their stability.

Actually, the dynamic response of these structures is more complex. Unexpectedly, slenderness seems to be a positive factor if the structure is well-made and free to oscillate. Thanks to this feature, indeed, bell towers are characterised by low own frequencies, far from the seismic ones, thus not undergoing high seismic oscillation. For the same reason, the deterioration of soil characteristics improves the seismic behaviour of the towers, further lowering their natural frequencies by virtue of the deformability of the constraint at the base (Di Tommaso et al. 2010). Indeed, their global seismic behaviour can be determined by observing their frequencies and by taking into account the interaction between ground motion and masonry characteristics.

However, these data are not easy to determine in the case of historical structures: the lack of significant knowledge of their mechanical parameters still makes it difficult to establish a reliable numerical modelling for historical towers (Bartoli and Betti 2012). Nevertheless, a simplified structural analysis can be rather useful to evaluate the pros and cons implicit in the choice of a first-aid intervention or, on the contrary, the demolition of these buildings (once they are irredeemably damaged by an earthquake) . This kind of simplified analysis can be carried out by identifying the most probable mechanisms of collapse which depend on the conformation and the architectural peculiarities in each individual case. Indeed, even if masonry towers are characterised by a wide variability of construction techniques and material properties, they maintain similar geometric configurations, and this allows us to investigate more general considerations starting with the observation of architectural elements and boundary conditions (Lionello 2011).

18.2.1 Italian Guidelines: Possible Collapse Mechanisms

Despite the generally satisfactory global behaviour of isolated bell towers under seismic loads, local damage can constitute a serious problem for the stability and the future conservation of these structures. Such aspects have been taken into account in the official Italian guidelines for the preservation of cultural built heritage from seismic risk (Linee Guida per la valutazione e riduzione del rischio sismico del patrimonio culturale, DPCM 2011): there is a specific chapter devoted to the structural typology of churches, in which there is a return to the empirical approach of previous campaign studies (Doglioni et al. 1994). In particular, the attachment with the graphic list for church damage (*Abaco dei meccanismi di collasso delle chiese*) describes the main mechanisms of collapse, providing a graphic scheme of the different, potential local damage to the functioning of the architectural configuration with regard to the position and dimension of openings, masonry texture and quality, presence of pinnacles and adherence to other buildings.

In the specific case of freely oscillating bell towers, compared to thick masonry towers, their better global seismic behaviour has been demonstrated, by virtue of their greater slenderness and consequent lower natural frequencies. On the other hand, these structures generally present local damage concentrated near the openings in façades: due to the oscillations of the structure in phase or counter phase, vertical cracks can appear on perimeter walls, which, respectively, cause shear and tensile stresses in the masonry (Blasi 2013). The lower resistance of the intermediate zone of the walls (because of the presence of openings, stairs and weaker masonry) causes the cracks to concentrate in the middle of façades. In this way, the original structure separates into two (or four) vertical macro-elements, each with a sensibly lower inertia with respect to the previous undamaged situation. This mechanism can be hindered by the presence of tie rods or flooring, especially if these are well connected to the perimeter walls.

In some cases, horizontal seismic actions may cause the ejection of the corners (Fig. 18.3), on which the load of the roof and the upper belfry sit, often unevenly. The presence of thrusting vaults, frequent in the lower part of the towers, can worsen this mechanism. For this reason, it is important to survey the texture of the walls, with particular attention to the corners, in order to detect possible differences in stiffness which could cause fractures and partial collapses.

Bell towers with spires have proved to be particularly vulnerable: once affected by seismic events, the detachment and collapse of the spire are the most common mechanism (Fig. 18.4). The fracture line on the spire can be inclined at 45° for shear stress (DPCM 2011), or more frequently is horizontal, when the top portion

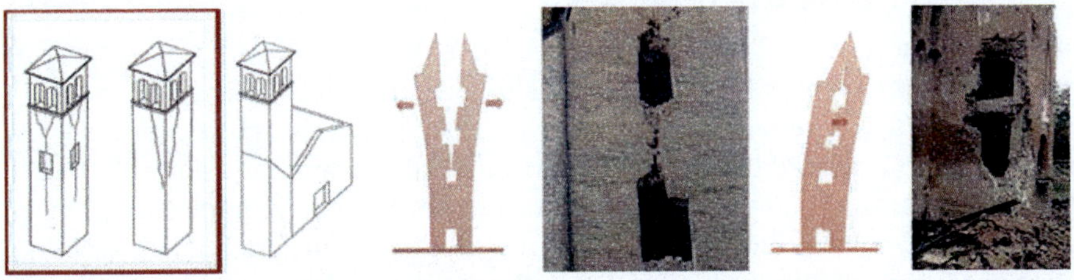

Fig. 18.3 Seismic behaviour of an isolated bell tower (courtesy of Agenzia Regionale per la Ricostruzione)

Fig. 18.4 Seismic behaviour of belfries and spires (courtesy of Agenzia Regionale per la Ricostruzione)

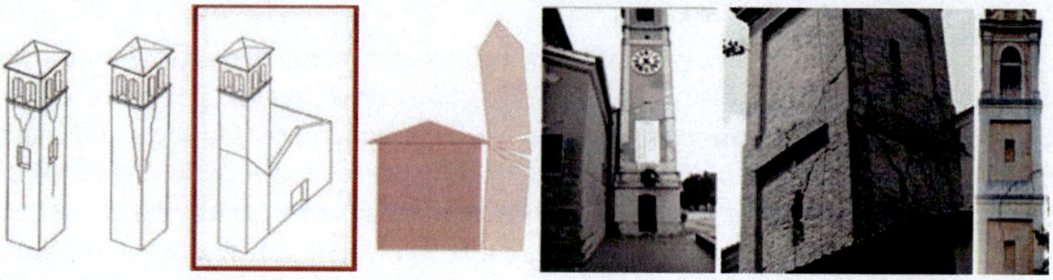

Fig. 18.5 Seismic behaviour of a bell tower in adherence with other structures (courtesy of Agenzia Regionale per la Ricostruzione)

slides off or topples over. The slenderness of the spire masonry itself is a further vulnerability, which makes this structural element, together with the thin pillars of the bell chambers, particularly dangerous in the event of an earthquake.

Most of the bell towers are very close or attached to other structures (usually churches) and are therefore not free to oscillate. In these cases, serious tensile phenomena can occur on the bell tower portions that rise above the connection level, causing their collapse (Fig. 18.5). Undoubtedly, this last case represents not only the most common problem, but is also the most difficult to solve because of the safety issues connected not only to the intervention in itself, but also to the stability of adjacent buildings and, therefore, to their use.

18.3 The 2012 Emilia Earthquake: "Empiricism" or Learning from Damage

The 2012 seismic event in Northern Italy—although dramatic—also represented an opportunity to test acquired knowledge about these structures, the collapse mechanisms and the available techniques for facing up the emergency, not only from a technical-structural point of view, but also (and even more importantly) from a managerial and cultural one. After the seismic shocks and the preliminary surveys of damage, many mayors requested the authorisation to demolish numerous bell towers that were heavily damaged and in danger of collapse, also due to the lack of adequate experience about the required first-aid interventions. The *Soprintendenza* rejected most of the applications for demolition and appointed a commission of experts in restoration and building techniques from the universities of Parma (prof. Carlo Blasi) , Bologna (prof. Angelo Di Tommaso) and Padova (prof. Claudio Modena), to set out guidelines for rapid interventions aimed at combining conservation and safety issues. Only in one case, (Poggio Renatico) was it necessary to demolish the twentieth-century bell tower with explosives, whereas in two others situations (San Martino in Buona Compra and Santi Senesio e Teopompo Martiri in Medolla), the bell towers were partially or completely dismantled and the pieces numbered for future reconstruction. Regarding the other structures, the commission indicated what they considered the most suitable interventions, selecting from various possibilities. The most common solution consisted in encircling the trunk of the tower with different types of materials (metal tie rods, composite materials, polyester bands); in some cases, a structural scaffolding or a steel lattice structure was inserted to avoid toppling over collapses; more infrequently, the masonry was strengthened with sprayed concrete (*spritz-beton*). Clearly, each of these technologies has pros and cons; therefore, some of the most significant case studies of emergency structural interventions on bell towers after the 2012 earthquake are analysed and discussed here, also comparing the guidelines with the real work that was then approved and realised. This discussion aims to deeply analyse the proposed first-aid interventions, in order to underline the consequences, both in terms of cultural, architectural, conservation and cost that were the result of the actual interventions, carried out after the emergency phase.

18.4 The Bell Tower of the Beata Vergine Del Rosario in Finale Emilia (MO)

The church of the Beata Vergine del Rosario is located towards the north-west border of the ancient city centre of Finale Emilia (MO). The main building dates back to 1580, whereas the bell tower was built only in 1856, in the north-east corner of the complex. The tower has a square base, incorporated within the church structure; it is 24 m high, and it terminates with a pyramidal spire. The main cracks, caused by the May 2012 earthquake, are concentrated at the bottom of the tower and have a horizontal and inclined development (Fig. 18.6). The damage mechanism was triggered by the interaction of the bell tower with surrounding buildings. Immediately after the earthquake, even before the files for the damage survey were filled in, the municipality of Finale Emilia, in cooperation with the Civil Protection (*Protezione Civile*), designed and undertook first-aid work, aimed at ensuring the safety of the structures involved and of the surrounding roads. The upper part of the trunk of the bell tower was encircled with tensioned polyester bands as a preliminary measure, interposing wooden elements on the masonry, in order to ensure adequate safety conditions for subsequent operations. Usually the lengthy time required for the design, authorisation and realisation phases of definitive interventions on these historical buildings

Fig. 18.6 Location of the bell tower with respect to the church, crack pattern and damages (Ferrari 2020)

suggested avoiding the use of polyester strips, as they have high strength, but they are also subjected to viscous phenomena, thus requiring frequent interventions to reinstate the correct state of tension. But in this case, they were applied only for a short period, during which a steel tower was designed to sustain the bell tower. This provisional structure was composed of four lattice trusses connected together and anchored to a reinforced concrete foundation with 16 micro-piles. At a height of 4.5 m, the connections between the trusses act as a protective plane for the underlying passage of vehicles and people. At a higher level, the same connections serve to encircle the bell tower and belfry: thanks to the connections with the masonry, loads and thrusts are transferred to the lattice structures. Moreover, where the bell tower and the church separate, Dywidag steel bars were inserted within the masonry, with crosses on a 45° angle, like a wind-bracing, in order to improve the in-plane behaviour. These first-aid interventions cost €87.554,38.

The metal structure built in the emergency phase (Fig. 18.7) allowed the bell tower of the Rosario Church to be strengthened despite its complex architectural configuration, which made it difficult to reach for the operators. The modularity of the temporary structure ensured the safety not only of the bell tower, but also of the surround area and passageways. The choice of steel instead of classical timber struts enabled the reduction of its encumbrance volume (which nevertheless remains considerable) and helped to maintain its efficiency over a longer period of time; on the downside, the costs were higher.

The extreme urgency that characterised the first-aid work did not permit the intervention, which turned out to be quite invasive, in particular considering the irreversible insertion of bars inside the masonry or the heaviness of the metal sustaining structure, to be optimised. Moreover, the structure could not be reused in the final strengthening phase, notwithstanding this was foreseen in the initial plan, thus considerably increasing the total costs: removing the foundations of the provisional metal

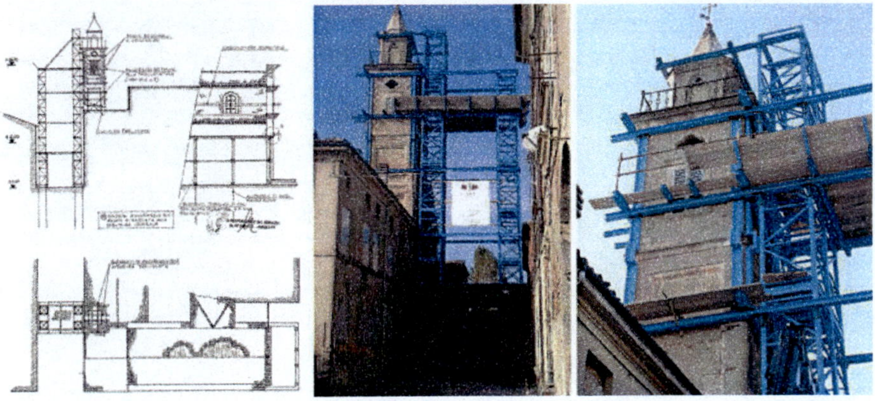

Fig. 18.7 Project and realisation of the provisional device (courtesy of Agenzia Regionale per la Ricostruzione)

structure required about one-fourth of the amount destined for the definitive strengthening of the bell tower. In total, the construction and subsequent demolition of such provisional structures cost €107.444,03, compared to €61.181,63 used for the final strengthening of the bell tower: thus, the external provisional structure cost 70% more than the inner definitive structure. This is not necessarily surprising, because the emergency structures have prevented heritage loss and possibly reduced further damage, so that restoration could be subsequently undertaken, but it seems likely that some more economies could have been obtained planning a better integration between first-aid and definitive works.

18.5 The Bell Tower of San Lorenzo Martire in Casumaro Di Cento (FE)

At the beginning of the sixteenth century, in Casumaro, a new church was built close to a pre-existing, fifteenth-century oratory. The church was constructed together with a guard tower, which was transformed in 1835 into the present, higher, bell tower. Originally isolated, the bell tower nowadays adjoins the vicarage on two sides. The masonry structure has a tapered base, a trunk with a clock and a belfry topped by a conical spire set over a low drum (Fig. 18.8).

The three wide cracks that developed diagonally from the contact point between the bell tower trunk and the vicarage are clearly a consequence of the tension between the two structures, characterised by different levels of stiffness and different seismic behaviours. Moreover, the torsional effect caused by uneven constraints induced a slippage of the masonry. The structural disorders (cracks and being out of plumb) caused of the bell tower to become high unstable, which, under the effects of additional seismic actions could collapse on the church or into the nearby road: these

Fig. 18.8 Location of the bell tower with respect to the church, crack pattern and damages (Ferrari 2020)

worrying conditions resulted in the decision that demolition would be the best option, at least from a functional point of view. The damage survey outlined the dangers of the safety operations, and the possibility of demolition is reported as being preferred by the fire brigade. In order to avoid the complete loss of the monument, the above-mentioned report of a technical expert resulted in a proposal, even before the damage file was completed, of a solution to avoid the the bell tower's collapse.

The first provisional intervention was the insertion of timber struts, executed with extreme urgency by a specialised firm, in order to ensure the safety of subsequent operations. The lattice structure, covering the entire cracked part of the trunk, was loaded at its base with counterweights and secured at the top to hinder possible movement and collapse stemming from any new seismic action. Neoprene buffers were inserted between the lattice structure and the masonry to reduce the possibility of tension and avoid an excessive increase in stiffness of the masonry. This first-aid phase permitted the reopening of some of the shops and a pedestrian passage close to the bell tower. Nevertheless, in this phase, collapse was still a threat, as it was not possible to install struts on the two adjacent sides.

The second phase was aimed at reinstating the building with its previous structural properties, through the combination of different strengthening techniques applied on the external walls where reachable from the exterior: the cracks were sealed with quick-setting liquid mortar containing anti-shrinkage additives (injected by an operator on a cherry picker); after the removal of the plaster, two vertical strips of reinforced carbon fibre were applied to the masonry surfaces, two levels of systems of horizontal hooping, made of steel cables were installed. This urgent provisional intervention was authorised for a foreseen cost of €54.450,00; but one year later (in 2013), it was necessary to fund a new intervention, aimed at the reopening of all of the area and nearby roads, for a cost of €90.250,62.

On February 2016, an on site visit revealed the precarious condition of the supporting structures, due to their long exposure to weathering, which caused the timber elements to decay and the metal elements to rust (about 30% of the elements), so that many elements were no longer in contact with the masonry. Three years after their application, a further €28.050,00 was spent for the maintenance of the provisional interventions, substituting the deteriorated parts and improving their formal compatibility with their surroundings structures.

The severe damage required particular precautions for the safety of people and workers, conditioning the choices for intervening. Another significant constraint was the presence of the vicarage, which hindered interventions on two sides. Despite the extreme urgency, the provisional intervention was conceived in relationship to the planned future definitive intervention and foresaw the completion of the encircling cables on the sides that could not be accessed, the realisation of a seismic joint between the two adjoining structures, the reduction of the aesthetic impact of the composite strips with new plaster and the substitution of the visually distracting external tie rods with internal ones, that are less visually invasive.

In the final intervention, only the scaffolding structure (Fig. 18.9), partially integrated, was reused to carry out the work, while both the encircling cables and

Fig. 18.9 Realisation of scaffolding as a provisional device (courtesy of Agenzia Regionale per la Ricostruzione)

composite strips were removed and substituted by a much more drastic intervention, which covered up all the masonry surfaces. From an economic point of view, the provisional intervention (including maintenance) cost is 20% less than the definitive intervention. It also should be noted that the inner structure could not have been realised safely without the external sustaining structure. A special mention should be made about the extremely limited cost for the creation of the seismic joint, whose previous absence generated the majority of the damage described. It is, therefore, clear that cheap preventive interventions like this one can avoid the development of severe problems, which are very costly to be solved.

18.6 The Bell Tower of Sant'Anna in Reno Centese (FE)

The church of St. Anna is located in the centre of the town of Reno Centese and was built at the end of the eighteenth century, with a bell tower embedded within its structure. In 1874, the bell tower was severely damaged by lightning: eight years later, a new bell tower about 29 m high was built separate from the church, in order to avoid similar problems in the future. Nowadays, the bell tower is at the centre of the main square (Fig. 18.10). The base has the shape of a truncated pyramid, while the trunk is characterised by protruding cornices, creating rectangular frames with round apertures (*oculi*) inside. The belfry, with four large openings, terminates in a conical spire. During the 2012 earthquake, the trunk suffered severe shear cracks on all four sides. In this case, the cracks did not follow the classical X shape because, after the first inclined crack developed, the upper part of the trunk slid, avoiding the development of the second inclined crack in the opposite direction (Di Tommaso and Casacci 2013) . Therefore, a single macro-crack developed on each side, about 8 m high, causing the twisting of the upper part of the bell tower.

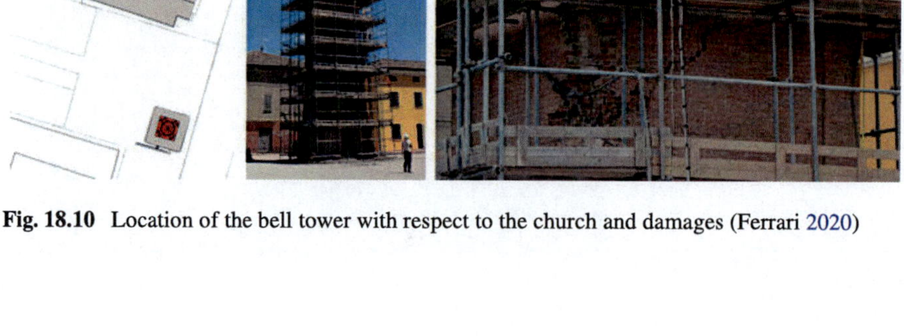

Fig. 18.10 Location of the bell tower with respect to the church and damages (Ferrari 2020)

Despite isolated bell towers normally displaying good seismic behaviour, in the case of Reno Centese, the level of damage was significant, probably due to the elevated vertical component of the seismic action that occurred close to the epicentre (Finale Emilia) . Moreover, the scaffolding that was in place at the time of the seismic event could have interacted with the masonry structure, with possible tensile effects which could have worsened the crack pattern. Local damage was also caused by the presence of internal ties, with anchoring set below the external layer of masonry (Blasi 2014): although the tie rods showed good behaviour, the anchoring on the south-east corner damaged the brick masonry, causing the expulsion of material. Moreover, the thin round crack on the upper part of the spire indicates the activation of a damage mechanism to this particularly vulnerable part.

Immediately after the earthquake, given the high risk of collapse for the structure, a wide "red" off-limits area was defined, within which all activity was forbidden, and the houses were considered unfit for occupation, even if they had limited or no damage. The inhabitants of Reno Centese, thus, became hostages of their own bell tower and exerted pressure for a quick reinstatement of safe conditions in their historical centre, also in view of possible aftershocks. In particular, the mayor requested authorisation for the demolition of the bell tower but, by the end of June, the *Soprintendenza* for the Cultural and Landscape Heritage of Emilia Romagna approved the preliminary interventions proposed by the technical commission, that were urgently carried out by the fire brigade.

Given the impossibility of safely getting close to the bell tower for the first phase of intervention, it was necessary to choose a technique that preserved the safety of the operators: taking the cue from technology deployed for stabilising rocks in tunnel excavations, the wide cracks were saturated with a fibre-reinforced cementitious mortar (FRCM), sprayed from a 52 m long nozzle (Fig. 18.11). The chosen mortar, characterised by low shrinkage and high mechanical properties, both in tension and compression, and by a high fracture energy, was reinforced with polymeric fibres. This allowed the mortar not only to close the cracks, but also to create a sort of buffering jacket with a confinement effect on the substrate (Di Tommaso and Casacci 2013).

Fig. 18.11 Project and realisation of the provisional device (courtesy of Agenzia Regionale per la Ricostruzione)

In the second phase of intervention, working from the scaffolding that were already on site for the restoration works begun before the earthquake, the trunk of the bell tower was wrapped, both in horizontal and vertical directions, with pre-impregnated composite strips: on the four corner lesene, two layers of glass and carbon fibres were applied, while three strips of fibres with the same characteristics were placed over them horizontally, approximately every 2.5 m. The technical commission suggested this solution as it was the most rapid and effective technique that also could be applied to a rough surface such as the one obtained through the first phase of the intervention. Indeed, despite the fact that composite materials are able to adapt to non-planar surfaces, in this case before applying the vertical strips, it was necessary to apply a layer of mortar to regularise the uneven surface created by the *spritz-beton*. On the contrary, the horizontal strips did not require this preliminary regularising intervention because the corner lesene was protruding and left the horizontal layers detached from the masonry surface, thus making it easier also to remove them in the successive phases.

Thanks to these cheap preliminary interventions, the safety conditions were improved enough to let the operators access the inside of the bell tower, where they were able to survey properly the damage and define more precisely the next interventions required to strengthen the structure. The sprayed concrete technique adopted permitted a quick intervention on a low budget, while also guaranteeing the safety of the operators. The subsequent application of composites reinstated strength to the structure and permitted the reopening of the surrounding area, hitherto off-limits. The chosen technique was easy to apply, even in the case of parts that were irregular and difficult to reach, caused by the surrounding scaffolding. On the down-side, however, this type of intervention cannot be considered definitive or adaptable to a future reuse: an entirely new intervention would be needed to substitute this one, and it would need to aim for more integrated conservation. In particular, from this point of view, the removal of the *spritz-beton* caused some concern, and some tests were carried out in order to define the best solution for the cleaning operations. Most of it could be removed manually, with fairly good conservation of the brick masonry surface. Nevertheless, some traces of the cementitious mortar remain, together with some chippings on the bricks, requiring a subsequent restoration intervention for a complete cleaning, filling gaps and treatment to render the surface uniform. In the limited areas where the removal of the *spritz-beton* left more severely damaged masonry, the local substitution of the brick masonry was necessary. From an economical point of view, the removal of the *spritz-beton* required more or less the same amount of money (€10.652,60) as required for its application (€9.863,62); but overall, the provisional intervention was very cheap: less than 4% of the final restoration costs.

18.7 The Bell Tower of Sant'Egidio in Cavezzo (MO)

The church of Cavezzo originates from a twelfth-century oratory, but its bell tower was rebuilt in 1781 and restored in 1921. The belfry is entirely in brick masonry, with four single arches on the four sides and a balustrade. In 1830, the original dome was substituted by the present-day spire on an octagonal base. The main cracks are concentrated between the base and the trunk, in correspondence with the contact point with the reinforced concrete cornice of the adjacent building, clearly as a consequence of the tensile effect between the two structures (Fig. 18.12). More cracks appeared at the extremities of the belfry corner pillars, due to the oscillation of the upper part. The arches' tie rods were effective in balancing the horizontal forces during seismic activity, as demonstrated by the deformations of their anchorages, yet the top part of the spire collapsed, and the whole church and its surrounding area were declared off-limits.

In June 2012, the technical commission proposed a solution for reinstating the safety conditions of the bell tower, composed of two phases. The first one was aimed at ensuring the safety of the operators, removing the precarious elements on the spire and in the belfry using a cherry picker and then applying four polyester strips to keep together the parts of the trunk separated by the cracks. The second phase directly focussed on the structural retrofit of the tower, recovering the vertical continuity of the masonry by means of injections of pre-mix, rapid setting and low shrinkage mortar. Moreover, L-shaped steel bars were placed at the corners and connected with both horizontal and inclined steel rods, thus creating a flexible framework with moderate stiffness (Fig. 18.13). A similar intervention also was carried out on the belfry pillars, while a non-cementitious *spritz-beton* was applied over the crumbled masonry of the spire, and this was covered by a waterproof membrane. Finally, demolishing the concrete cornice of the adjoining building, to leave a gap of at least 10 cm, was suggested.

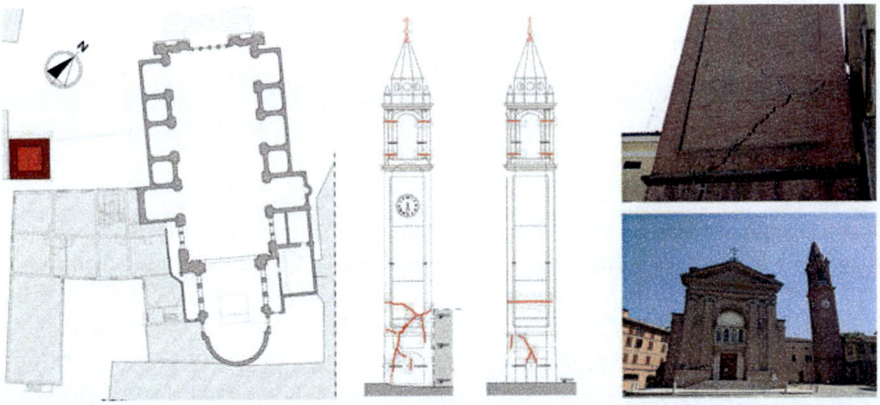

Fig. 18.12 Location of the bell tower with respect to the church, crack patterns and damages (Ferrari 2020)

Fig. 18.13 Project and realisation of the provisional device (courtesy of Agenzia Regionale per la Ricostruzione)

This solution could not avoid entirely a potential shear or toppling over collapse of the upper part of the bell tower over the lower part, as it was not possible, for safety reasons, to enter the adjacent buildings. The technical report concluded that first-aid interventions could be included at least partially in the final design's strengthening techniques, such as covering the injected mortar with new plaster and extending the corner steel bars in the lower part. The first-aid works that allowed the reduction of the off-limits areas, cost €156.799,60: thus, quite expensive, considering that they had to be removed and completely substituted by an internal lattice structure in the definitive intervention, causing an increase in the overall costs. In any case, given that the definitive interventions have not been realised yet, the provisional ones have proven to be effectively designed in order to remain functioning in the medium to long term period, limiting ongoing interference with the activities in the surrounding areas and the need for further maintenance.

18.8 Conclusion

The aforementioned case studies show the complexity of managing the post-seismic emergency phase, of the difficulties in keeping a balance between safety and conservation (Bartolomucci 2013). Following the Emilia earthquake and caught off guard, solutions were adopted that were not optimised for the specific case, or that sometimes proved even to have negative effects (from a technical or economical point of view) in the long run. However, even for a seemingly simple case like a single bell tower, the available techniques to be deployed are numerous and variable.

The first issue that has to be faced is the safety of workers: during the first-aid interventions on damaged bell towers, often carried out when aftershocks are still ongoing, the main operational difficulties derive from precarious safety conditions. This makes it difficult to get close to the buildings and to work on their structures (Blasi 2014), strongly influencing the choice of the strengthening technique. For example, after the Emilia earthquake, the encircling cables intervention of the tower trunks was basically excluded in most cases where belfries were close to or part of church structures: the high level of damage and the complex articulation of the building hindered access to surrounding structures in order to undertake the strengthening on all sides of the towers. Given these preliminary considerations, it is clear that the first intervention required is always the removal of rubble and unstable elements, usually using cherry pickers, as the first aim is always to guarantee adequate safety conditions for workers and the citizens daily activities.

Once appropriate safety conditions have been restored, it is possible to proceed with the installation of provisional devices to prevent damage progressing. The case studies above demonstrate that techniques need to be chosen according to the type of activated damage mechanism, the damage level and the conditions of accessibility to the structure. Generally speaking, in order to hinder the toppling over mechanisms, the most widespread technique used is the provisional encircling of the trunk with steel cables or bars, whose ease of application usually allows the operations to be carried out from a cherry picker. By contrast, when the vertical continuity of the masonry is interrupted by horizontal cracks (for tensile effects with adjoining buildings or global mechanisms), conservation work becomes more complex and often requires more invasive and costly interventions, like propping up the trunk with heavy-load structural scaffoldings, or surrounding the bell tower with a steel cage that, if well designed, could be reused during the definitive work of conservation.

As always in the field of restoration, each case is unique, given the specificity of each individual historical building in terms of evolution, construction features and damage. Identifying these conditions through an on-site survey and historical analysis is necessary in order to identify the most suitable first-aid intervention, which, in such occasions, must be found quickly from among the different available technical solutions. It is thus extremely important to analyse these relations in times of "peace", in order to be immediately operational and ready to secure damaged structures as soon as possible after there is an earthquake, possibly designing the interventions with the awareness of the situation, taking into account also the future definitive restoration.

Acknowledgements The authors would like to express their gratitude to Carlo Blasi and Angelo Di Tommaso for sharing their experience and material about the work done within the Technical Commission. Moreover, Dott. Dir. Enrico Cocchi, Arch. Antonino Libro, Ing. Davide Parisi and all the working group of the Agenzia Regionale per la Ricostruzione are gratefully acknowledged for their contributions in terms of discussions and materials.

References

Bartoli G, Betti M (2012) Il progetto Ri-SEM: rischio sismico negli edifici monumentali. In: De Luca A, Spinelli P (eds) Proceedings of the workshop on design for rehabilitation of masonry structures, pp.221–230. Edizioni Polistampa, Firenze

Bartolomucci C (2013) Structure and architecture: the illogical results of considering them two separates entities, after the 2009 earthquake in L'Aquila. In: Cruz P (ed) Structures and architecture: concepts, applications and challenges (2nd international conference on structure and architecture), pp1621–1628. Taylor & Francis Group, London

Blasi C (2013) Architettura storica e terremoti—protocolli operativi per la conoscenza e la tutela. Wolters Kluwer, Milano

Blasi C (2014) Il consolidamento dei campanili danneggiati dal sisma: riflessioni su conservazione e sicurezza. In: Di Francesco C (ed) A sei mesi dal sisma—Rapporto sui Beni Culturali in Emilia-Romagna. Minerva Soluzioni Editoriali, Bologna, pp 222–237

Candigliota E, Carpani B, Immordino F, Poggianti A (2012) Damage to religious buildings due to the Pianura Padana Emiliana earthquake. Energia, Ambiente e Innovazione 4–5:58–68

Di Tommaso A, Casacci S (2013) Sopravvivenza di Torri e Campanili. In: Seminario Internazionale sull'Evoluzione nella sperimentazione per le costruzioni. CIAS, Crete

Di Tommaso A, Lancellotta R, Focacci F, Romaro F (2010) Uno studio sulla stabilità della Torre Ghirlandina. In: Cadignani R (ed) La Torre Ghirlandina—Storia e restauro, pp 204–217. Luca Sossella Editore, Roma

Doglioni F, Petrini V, Moretti A (eds) (1994) Le chiese e il terremoto. LINT, Trieste

Ferrari L (2020) Messa in sicurezza di chiese e campanili: analisi tecnico-economica degli interventi post-sisma 2012 per la definizione di linee di indirizzo. PhD Thesis, Università degli Studi di Parma

Lionello A (ed) (2011) Tecniche costruttive, dissesti e consolidamento dei campanili di Venezia. Corbo e Fiore Editori, Venezia

Ministero dei Beni Culturali (2011) Direttiva del Presidente del Consiglio per la Valutazione e Riduzione del Rischio sismico per il patrimonio culturale. DPCM 26.02.2011, G.U.n.54

Tandon A (2018) First aid to cultural heritage in times of crisis. ICCROM, Roma

Chapter 19
Ancient City of the Future: Notes on the Reconstruction of Beirut

Federico De Matteis

Abstract The reconstruction of Beirut's city centre after the Lebanese Civil War (1975–1990) represents a case of the application of neoliberal planning practice on historic urban contexts. The results of this operation have been widely criticised with the once thriving urban core today largely rejected by the city's inhabitants. This chapter investigates the affective dimension of Beirut's urban space to understand how design decisions related to the restoration have affected the inhabitants' emotional response. This study describes some specific characteristics of both the wartime destruction and the post-war reconstruction, observing how they have impacted the lived space of the Lebanese capital's historic centre.

Keywords Civil war · Reconstruction · Conservation · Urban atmospheres · Beirut · Lebanon

19.1 Introduction

More than most other cities that have undergone the trauma of destruction and the fulfilment of reconstruction, Beirut has a dense and intertwined history, layered with alternate and inconsistent narratives, multiple and contradictory spaces. The roots of this complexity can be traced to well before the Civil War that ravaged Lebanon between 1975 and 1990, extending deep into the pre-colonial era, with the stratification of religions and cultures that for centuries has been the region's hallmark. The very idea of "identity", with its burden of reductions and misunderstanding, can here only be understood as a plural. Nevertheless, the reconstruction of Beirut's city centre, an ongoing process that is about to enter its fourth decade, seems to have been constantly pointing in the direction of crystallising a single image, an artificial construct spatially expressed by urban spaces and buildings (Fricke 2005, p. 169; Nagle 2017, p. 2).

F. De Matteis (✉)
University of L'Aquila, L'Aquila, Italy
e-mail: federico.dematteis@univaq.it

© The Author(s), under exclusive license to Springer Nature Switzerland AG 2021

F. F. Arefian et al. (eds.), *Historic Cities in the Face of Disasters*, The Urban Book Series,
https://doi.org/10.1007/978-3-030-77356-4_19

In the almost thirty years following the armistice, the gargantuan effort of rebuilding Beirut has been intensively monitored by scholars and activists, observing in real time the physical results of the process. However, even as the construction workers have left, the gates have been reopened and the reconstruction can be considered complete, we must consider that the process by which human life may again unfold within the space of a wounded city is more complex and covers a much longer timespan. This can call into question dynamics that lie beyond the reach of architects and planners. If this misalignment between reconstruction and re-habitation may be considered a general rule when geophysical- or human-induced catastrophes strike a city, it is even more true for Beirut. Surrounded by a tormented geopolitical landscape, where a reliable political stability has never returned after the Civil War, the turmoil of global finance has made the country's economy weak and unpredictable. Reconstruction has taken place within a shifting and labile context, and some of its shortcomings can be ascribed to this instability.

Walking through the streets of Beirut today, we are faced with an urban space that is a direct evidence of many underlying dynamics, even before assessing the wider political, social and economic conditions of the city to help clarify critical points and why to a large extent the reconstruction has failed in its intentions. The city's "historic" centre with its many names—*Balad, Bourj, Centreville, Central District, Solidere*—is all but rejected by Beirut's inhabitants, who have severed any affective liaison with its streets and public spaces. If some of the causes for this response are obvious—the heavily guarded enclosures, the glamorous and overpriced gated communities—we may argue that other reasons lie directly embedded in the qualities of spaces that have been restored after the wartime destruction. More specifically, even the way the restoration works have been carried out, with the intent of reviving the damaged historic core, has in fact contributed to this rejection on the side of the city's residents.

This chapter intends to produce a *phenomenography* of Beirut's central district: a description of its lived spaces as it emerges from the encounter of its inhabitants with the physical scaffolding of the city (De Matteis et al. 2019). This chapter is not seeking to provide an objective aetiology through its "medical history", analysing the treatment the city received to assist recovery from wartime wounds: the chapter rather considers the events unfolding in space as symptoms of a *current* condition that may be a consequence of decisions taken by authorities and planners during the course of recent history, but is also telling of the way Beirutis respond to space in its contingent manifestation, apprehending it through their affective corporeality. The method this chapter espouses is based on first-person direct experience, by means of an ethnographic account of the spatial situations within the city centre, with a particular focus on the architectural configuration that has emerged during the reconstruction of the historic fabric.

19.2 Investigating the City's Lived Space

Among the various practices of describing urban space, some contemporary trends based on phenomenological methods tend to place the experiencing subject at the centre of concerns, while other approaches privilege the material configuration of the built environment or the social and economic processes that subtend its operation. This study explores the *affective* dimension of urban space in terms of the subject's emotional response to the contingent situations they encounter.

The theoretical scaffolding for this method of investigation can be traced to several different strains, all revolving around the fundamental notions of centrality of the subject's perception and the relevance and transversality of their affective response. From the late 1960s onwards, German philosophers Hermann Schmitz (Schmitz 1969, 2011, 2016; Schmitz et al. 2011) and Böhme (2013, 2017a, b, 2018) have elaborated the notion of *atmosphere* as the spatial dimension of feelings that become accessible to all present subjects through their lived-body sensations. Space is thus intended as a contingent manifestation that emerges *between* the environment with its shifting situational character and the subject that encounters it through their corporeal disposition. This phenomenological model attributes great relevance to the immaterial components of space, those that are not connected to the object-related dimension of the environment. Böhme's work, in particular, with many contributions directly dedicated to the built environment, has become seminal for contemporary architectural theory.

Geographer Jürgen Hasse, drawing on Schmitz's theory of atmospheres, has developed a method of description specifically dedicated to urban and architectural spaces with a strong focus on movement, perception and the subject's direct experience (Hasse 2012, 2015a, b). The term "phenomenography" that he borrows from previous literature points to the relevance of the modes of appearance of things over their formal and material constitution, thus highlighting the situational character of lived space.

A more recent, parallel approach concentrates on the shared anthropological dimension of emotions: authors such as geographers Anderson (2009) and Gandy (2017) focus less on the embodied layers of experience, while investigating how spatial dynamics are influenced by the affective qualities of atmospheres, the political connotations of urban life and how these influence the subjects' agency.

A third, more empirical strain is that of French urbanist and sociologist Thibaud (2011, 2013, 2015), whose research on the notion of *ambiances* aims at providing a description of urban spaces assigning great weight to the extended sensorial aspects of experience. In particular, he proposes a method of description relying on the tradition of walking as an explorative practice, with the live comments of the subjects that are recorded to provide a direct assessment of an urban environment's qualities.

These various approaches share several methodological intersections: they privilege the subject's pre-reflective experience; in the centrality of the subjects' affective response intended both in the individual corporeal stirrings and in the collective dimension of shared emotions; in the non-linguistic nature of the expressive qualities that are encountered in the built environment; and in the contingent character of the description, which considers the dynamic unfolding of the urban space's situational articulation. Leveraging on these theoretical foundations, it is possible to return a "thick description" of lived urban space that does not solely rely on distancing mechanisms but rather includes the subjective sphere as a primary tool for the understanding of reality.

19.3 Anamnesis of a Destruction

The Lebanese Civil War, a tight sequence of internal and external conflicts that seamlessly spanned fifteen years, left most of the country's territory in a profoundly damaged state, with infrastructure all but destroyed and the national capital a vast expanse of ruins. While before the war Beirut's social structure was largely mixed, with citizens of various confessions sharing urban areas, the sectarian clashes brought to a strong polarisation (Fregonese and Minca 2004, p. 121), particularly with an East–West division running along the rue de Damas, eventually re-baptised as the "Green line" (Ragab 2011, p. 109). Throughout the conflict, the street and its surroundings were extensively shelled and entirely abandoned: although no part of Beirut was spared from harm, this and other areas were most direly affected by the destruction, since the physical damage summed with the "natural" decay caused by the lack of inhabitants. The line was defined "green", in fact, not because of some military jargon, but due to the fact that thick vegetation grew within this area that remained inaccessible for several years (Fig. 19.1).

The destruction inflicted on the city's various neighbourhoods was uneven, mirroring the political divisions caused by the evolving warfare. In some cases, the urban configuration or even individual buildings would influence the nature and extent of damage: exemplary cases are the two infamous towers that were chosen by snipers to control the surrounding areas from high vantage points: the Murr tower, in the Al Kantari district, and the Holiday Inn Hotel, just north of the former, overlooking St. Georges bay (Fig. 19.2).

Both buildings were the stage of gruesome battles and—perhaps unsurprisingly—still today lie as ruins amid the reconstructed city (Naeff 2016, p. 231; Nagle 2017, p. 8). During the war, their presence as menacing panopticons determined a site-related topography of destruction and decay, given that anything within their sight range was constantly under the threat of fire.

Other sectors of the city, where the traditional urban fabric had not been modified through the insertion of contemporary architecture, underwent a different type of damage. The central market area, or *Souk*, was a labyrinthine mesh of streets and alleys, home to a lively urban scene revolving around the vibrant commercial activity. Wartime destruction affected this area setting the stage for its eventual demolition: the intricate web of social and economic relations that had sustained the Souk's life for centuries was disrupted by the sectarian polarisation of the city, as well as by the

Fig. 19.1 "Green" line in 1982 (James Case, https://commons.wikimedia.org/wiki/File:Green_Line,_Beirut_1982.jpg)

fact that during the war many shops had relocated to safer areas, if their owners—like many other Lebanese—had not altogether left the country. The market's "hardware"—buildings, stores, covered streets, etc.—was largely left devoid of its human material, and the thriving business life that once unfolded there would have been unable to return perhaps even if the area had not been later slated for demolition.

Throughout Beirut, some specific streets, buildings or areas accounted for an invisible affective topography, an overlay that only partially matched the intensity of destruction. In terms of the experience of lived space, these immaterial contents remained inscribed in the inhabitants' relationship with the urban scenario even long after the end of the clashes. Furthermore, the still-looming presence of a conspicuous number of war-scarred buildings dotting the city's central area afford even the casual visitor a haunted re-presentation of the war's trauma, even more so to those who were physically present at the time.

The space of cities, laden with affective content, acts upon its inhabitants and visitors not only through its contingent, immediate qualities, but also by means of an invisible "layer" of time. A community of local residents, such as the shop owners and customers of the old Souk, bears an embodied memory of what the place was like before and experience a different affective response from the occasional visitor, who is entirely unaware of this further layer (De Matteis 2015). Both the presence and the absence, in this particular circumstance, can shape the subject's response to the experience of space.

Fig. 19.2 Former Holiday Inn Hotel on St. Georges Bay (Wusel007, https://commons.wikimedia.org/wiki/File:Hotel_Holiday_Inn_Beirut_front.JPG

19.4 Rebuilding Beirut

With the cease-fire of 1990, the reconstruction of Beirut became a national priority for Lebanon. At stake was not only the need of reinstating the country's sole political, cultural and economic hub, but also the staging of a spatial narrative of national reconciliation (Fricke 2005, p. 169). In a conflict with no winners or losers, the capital city's role of representing this pacification through its urban configuration, recovering if not overcoming the splendour of that short-lived period during which it was known as the "Paris of the East", extended beyond a simple planning and architectural effort. Rafic Hariri, the Lebanese entrepreneur who had amassed a large fortune as real estate developer in Saudi Arabia, would mastermind the reconstruction as Prime Minister, in a position that went well beyond a simple understanding of "conflict of interest".

Solidere[1]—the company he created to manage the reconstruction, was granted by law tremendous power, single-handedly centralising the process and eschewing any public confrontation (Hourani 2015).

In many cases, destruction and reconstruction in post-catastrophe cities are not two entirely separated moments. Destruction may extend into the reconstruction phase, if the political and technical paradigms embedded in planning culture subjugate public discussion to implicit economic rationales (Rizzi 2011, pp. 18–19). The Beirut Hariri envisioned was not a mere re-production of the pre-war capital, rather a glamorous capitalist metropolis modelled after booming Gulf hubs such as Dubai and Abu Dhabi. Reverting to the charming yet old-fashioned city of yesterday, animated by familiar residential areas and mom-and-pop shop,s would not have attracted the foreign—especially Arab—investors whose money was crucial to sustain the Premier's dream. Gleaming skyscrapers, fashionable shopping malls and exclusive clubs catering to the wealthy Middle Eastern elite had to replace the old urban fabric, supposedly integrated with the remnants of restored historic buildings: thus came Solidere's branding of Beirut as an "Ancient City of the Future" (Leclair-Paquet 2013, p. 20).

History, as we know, is not an artificial construct. It cannot be reinstated once it is gone, only recovered from its actual remains, for it is solidly bound to the notion of authenticity. Nevertheless, there is no shortage of cases where administrations have heralded the rebirth of history in cities that had once been historical, if not in places where nothing truly ancient has ever been present (Fu and Hillier 2018). These constructs, however, are simulacra and differ from what is original in their spatial configuration, material constitution and in the way they irradiate their presence in lived space. Their aim is not that of creating an actual urban space, complete with its social, cultural and economic life, but rather an *image*, a conceptual standard of post-modern urban design practices (Harvey 1997).

Solidere's plan for the reconstruction of Beirut clearly pointed in that direction (Ilyés 2015, p. 14), with specific guidelines for the restoration setting out the extreme cleansing and homogenisation of cladding materials, while providing great freedom in the transformation of interior spaces (Ragab 2011, p. 110). In many of the restored buildings, in fact, all that remained after Solidere's "rejuvenating" interventions were the facades, while the interior structures were entirely substituted and redesigned to accommodate contemporary activities (Fig. 19.3).[2]

[1] Société Libanaise pour le Développement et la Reconstruction du Centre-ville de Beyrouth.

[2] From Solidere's website: "The restoration of Beirut city centre's historic fabric was a real success confirming the sustainability of traditional districts and heritage buildings and their great potential for creating value, once adapted to the needs of contemporary life and business. [The buildings] were faithfully restored in accordance with a set of rules and restoration guidelines established by the Company. International experience in urban and architectural conservation was adapted to local materials and know-how [...]. Buildings were rejuvenated through the use of skylight atriums, roof gardens or glazed roofs. The interiors were fitted to comply with modern requirements for functionality, flexibility, comfort and efficiency." https://solidere.com/city-center/solidere-developments/real-estate/solidere-restored-buildings, accessed 09/07/2019.

Fig. 19.3 A street in Centreville, showing the historical facades cloaking strongly modernised buildings (Y. Rizk)

Yet even before prescribing restoration practices, the Solidere reconstruction plan fulfilled the destruction of the ancient city that had been begun by the Civil War. Although never declared, the damaged centre provided an occasion to modernise the historic city, with the intent of bringing it up to par with other supra-national elitist financial hubs, as Monte-Carlo or Hong Kong (Fregonese and Minca 2004, p. 141). To this end, a "critical" revision of the centre's remaining urban fabric was carried out, slating for demolition a vast number of buildings, many of which could in fact have been recovered, such as the commercial area of the Souks (Fricke 2005, p. 171).

The new shopping complex of the South Souks, built after an international design competition between several world-renown architects (Ilyés 2015, p. 18), represents a drastic change in scale and type from the previously existing fabric (Fig. 19.4). Although its architectural configuration presents some ingenious solutions, it only

Fig. 19.4 The new Beirut Souks Shopping Centre by Rafael Moneo facing the historical facades of Centreville (author)

preserves a thin veneer of local character through the choice of cladding materials and some isolated formal elements, tending towards a standard, globalised commercial model catering to international brands rather than local shops. The northern part of the complex, currently under construction, relinquishes even the bland typological approach of the earlier building, opting for an oversized, blob-like department store.[3]

Each of these buildings, in their more or less outlandish and dystopic architectural character, relates to the restored part of the urban fabric that has in the meantime been "polished" to remove most traces of damage and weathering to the facades (Fig. 19.5). The "mismatch" between the material form of the city, that in some of Centreville's streets is rooted in historic buildings, and the actual manifestation of a certain urban ambiance is patent. It is not only bound to the forceful gentrification of the area, with the substitution of the original residents and urban activities with an insipid mix of globalised venues and urbanites: it precedes such sense of topological alienation, deriving from the very architectural qualities of the rebuilt city.

Beirut's rebuilt historic centre has been "completed" nearly a decade ago. Several key areas, however, remain to be defined, most notably Martyr's square, once the

[3] In the dramatic events of 2020, the building site of the Zaha Hadid-designed department store was ravaged by fire before completion.

Fig. 19.5 Deep-cleansed restored facades in Centreville (Y. Rizk)

city's central symbolic place. More than any in other space, here one can observe the clash between the various political, religious, social and economic forces that are shaping the city's contemporary configuration. Side-by-side sit the modern-style buildings showcasing Beirut's global ambitions; the massive, historicising al-Amin mosque, willed by the late Rafic Hariri, and completed after his death as a symbol of the Sunni community (Vloeberghs 2012); the restored facades of Beirut's colonial heritage; the archaeological remains of its Roman past, and the faux-picturesque buildings of the new upscale, semi-gated residential areas. As the city's largest public space, still today characterised by a conflictual, unclear identity, it is perhaps unsurprising that ever since the 2005 anti-Syrian uprising following Hariri's assassination, Martyr's square has become the central location of public protest (Ilyés 2015, p. 19). The conflict and tension among these elements, as expression of the diversified claims towards the square's space, keep it in a "limbo" restraining its reconfiguration: still today, its central areas lie largely empty, used as a parking lot or for occasional public events.

In purely quantitative terms, the Solidere-led reconstruction of Beirut's centre has been very effective. The lunar landscape of the war-torn city has largely disappeared beneath new buildings and cleansed facades. However, a handful of structures has fortuitously survived demolition, mostly due to the intervention of local activists,

Fig. 19.6 Barakat building, top (author)

who have rallied to prevent the removal of some remnants that, despite their precarious conditions, are particularly relevant or symbolic. Among these is the Barakat building, a 1920s residential structure overlooking the Green line that was occupied by snipers throughout the conflict (Fig. 19.6). Profoundly damaged, it was due to be demolished in 1997, but a fierce civic battle saved it from its fate (Nagle 2017, p. 13). After a long and complex restoration process, the building has now become a city museum: but what is most striking is that, unlike the Solidere interventions, here the main facade has preserved all the holes, perforations and marks left by the wartime gunfire and shelling, thus serving as a "live" memorial of the conflict.

Another celebrated case is that of the City Centre Building, better known as "The Egg", a modernist complex designed in 1965 by Lebanese architect Joseph Karam (Fig. 19.7). Its futuristic appearance, with an egg-shaped cinema hovering above a commercial mall, made this a favourite of Beiruti architects and preservation activists, and although partially demolished, it was eventually saved from being entirely removed by Solidere. Today, the complex lies empty and fenced, and despite the multitude of proposals for its reuse, so far no clear plan has been laid out for its future (De Matteis 2013).

These two celebrated buildings, along with a few others, have become uncanny landmarks in Beirut's historic centre. Their presence acts as a sort of counter-memorial, opposing the scarred surfaces to the finely dressed facades of the Solidere buildings. Through both their stark difference and uncanny appearance, they exert a powerful influence on those who come upon them, modulating their affective response in a disturbing way.

Fig. 19.7 City centre building known as "The Egg" (author)

19.5 The Urban Ambiance of Beirut's Rebuilt Centre

In its first decade after reconstruction, Centreville has experienced various phases of life, mostly as repercussions of the geopolitical events in the Middle East. Once more freely accessible, the area's pedestrian precinct around Place de l'Etoile, where the country's main public institutions are assembled, has become heavily guarded, with military blockades controlling the entrance of visitors. This form of "threatening" enclosure alters the subjects' perception of space not only when entering, but also while roaming through the area's streets. The expensive shopping milieu of Centreville renders it exclusive rather than inclusive, establishing a first "filter" (Nagle 2017, p. 11). The effect is further increased by the controlled access, and the resulting streetscape is most of the time scarcely populated, with few human presences to animate the area (Fig. 19.8).

The ambiance of a segregated, economically exclusive urban space may verge towards low intensity is rather obvious. However, in the case of Centreville, the approach to the restoration of historic buildings was carried out in accordance with Solidere guidelines further increasing a certain alienating effect. This nuance may emerge only to the eyes of a trained observer, whose expectation of a "historic" neighbourhood includes a certain range of material qualities, varied rather than homogeneous, weathered and rich in patina rather than cleansed, or with a deep articulation

Fig. 19.8 Desolate streetscape in Centreville (author)

of conditions resulting from the spontaneous life of the city: but it is nevertheless there and influences the spatial perception of any visitor. Here, all facades, materials and colours look surprisingly the same, as if they had been cast from a single mould at the very same moment (Fig. 19.9). This condition, its visual impact, as well as the expressive qualities of these surfaces that atmospherically tinge space, make the observer feel not so much the way we would expect in a "true" historic neighbourhood, but closer to what we experience in the nearby post-modern pastiche residential developments, where historicity is only simulated, and everything always looks new and pristine (Fig. 19.10).

Fig. 19.9 Material and chromatic homogeneity of renovated buildings, top (Y. Rizk)

Fig. 19.10 Saifi Village, a historical *pastiche* bordering Centreville (author)

The Western modernist "obsession" for newness, that has pervaded many contemporary conservation practices (Salvo 2016), leads to what is old to look very much like what is new, substantially shortening the gap between the appearance of historic and contemporary objects. If from Centreville we cross Martyr's square into nearby neighbourhoods such as Gemmayzeh or Mar Mikhael, we encounter a very different situation. Despite having been heavily gentrified, the buildings here have not been subjected to the Solidere "method" and still largely display the spontaneous variety of living historic neighbourhoods. These areas are not segregated as Centreville, and together with their material expressiveness, this allows the unfolding of an intense vitality.

The urban experience in Centreville is intense, in a negative way. In authentic urban centres, life unfolds throughout the day, with activities of different character taking place in the various hours. Beirut, as a deeply Mediterranean city, is extremely vital, and most of its diverse neighbourhoods are filled with a lively population that thanks to the climate exploits street space in an intense way. Centreville is perhaps the sole exception, since its monotone commercial activities are almost all that happens there: when the shops close in the evening and the few customers leave, the sensation is that entire city has been shuttered down, as happens in a shopping mall.

Yet even more than this discontinuous character, it is the very material quality of restored buildings, the total lack of patina, that produces the strongest alienating sensation. Weathered building materials characterising an historic urban district do not only symbolise historicity: they *irradiate* it towards the experiencing subject, being apprehended before the interpretive sphere. Conservation practices that seek a "purified" version of historical buildings, removing all traces of weathering and reinstating a presumed originality, usually end up emanating a petrified coldness. In Centreville, Solidere's restoration guidelines have extended this principle to the urban scale, creating an atmosphere of unreal suspension.

19.6 Urban Affects After the Reconstruction

The reconstruction of Beirut, in all its complexity, contradictions and shortcomings, represents an exemplary case of the crisis emerging from the application of contemporary neoliberal planning practices to historic urban contexts. As Solidere's vision quite plainly states, history is commodified as a value-creating asset, reduced to a veneer coating an otherwise contemporary urban system. More specifically, the underlying economic mechanism espouses the notion of a pacified district, into which amnesia of the war's trauma has been embedded (Nagle 2017, p. 2). Instead of ideally returning the city's space to all the various communities and factions that confronted each other during the conflict, all ideological content has been deracinated, producing an affective vacuum, where all one feels is an alienated emptiness. The neatly polished facades of the "new" historic buildings are not only symbols of this process: through their aseptic cleanliness, they directly express the deliberate erasure of a traumatic past, which, if present, would disrupt the anaesthetised comfort of the global elite's leisure time.

Describing the centre of Beirut after the reconstruction highlights how the urban subject's experience is deeply affected by the spatial qualities of the city. The difference becomes particularly striking as we compare the neighbourhoods where the "cleansing" has been most intense with those that were largely spared from the process. The ambiance one experiences—the overarching affective tone that can be observed through our corporeal responses—somehow changes "temperature": from the cold, distancing appearance of the sanitised portions of the Place de l'Etoile, to the warmly Mediterranean "climate" one still encounters elsewhere. One could claim that each type of urban atmosphere caters to a different segment of the city's population: perhaps the wealthy, globalised cosmopolitans may feel more comfortable in the lacquered streets of the upscale renewed Souks, while the natives find the still run-down areas more familiar and authentic. Yet this attempt at "designing" the atmosphere, determining the subjects' experience beyond the simple urban container where it unfolds is a form of manipulation of affects that can risk damaging cities in an irreversible way.

A city's trauma, particularly in a place like Beirut, is not easily eradicated. The few remaining ghost-like buildings, shelled facades and bullet-pocked walls that still haunt the city's space serve as a reminder of the fire that once engulfed the streets. The deadlock over these structure's fate has so far preserved them from demolition, and—if they will somehow survive—it could be their very presence to provide a more earnest monument to Beirut's dramatic history.

19.7 Conclusion: A New Reconstruction?

Between this chapter's first version, the editorial process and the publication, the Beirut's history seems to have accelerated. Massive anti-government protests, fuelled by a deep economic crisis, blocked the city in autumn 2019. The COVID-19 pandemic, which had initially spared the country, later brought to high numbers of infections. And, most dramatically, on 4 August 2020, a gigantic explosion in the port devastated a huge portion of the centre, all but razing entire neighbourhoods. Historic sections of the city, such as Gemmayzeh or Mar Mikhael, were profoundly affected: the urban situation described in this study now belongs to the past, and a new reconstruction will have to be started. The country's dire economic conditions will probably make the process extremely complex, as market forces have immediately set in to take advantage of the situation,[4] further jeopardising the urban heritage that had so far survived. But as the population recovers from this last catastrophic event, one can see how the controversial Solidere reconstruction has created a wide

[4] See: https://www.nationalgeographic.com/travel/2020/09/how-beirut-is-rebuilding-after-explosion/; https://www.pri.org/stories/2020-09-03/beirut-not-sale-besieged-residents-fend-developers-investors, accessed 12 Oct 2020.

public awareness of the damage that can be done in such a process. This is a lesson learnt which, perhaps, will orient the new efforts for upcoming reconstruction in a wiser way.

References

Anderson B (2009) Affective atmospheres. Emot Space Soc 2(2):77–81
Böhme G (2013) Atmosphäre: essays zur neuen Ästhetik. Suhrkamp, Berlin
Böhme G (2017a) The aesthetics of atmospheres. Routledge, London and New York
Böhme G (2017b) Atmospheric architectures: the aesthetics of felt spaces. Bloomsbury, London
Böhme G (2018) Leib: Die Natur, die wir selbst sind. Suhrkamp, Berlin
De Matteis F (2013) Bonjour Beirut. Realtà e sovrascrittura libanese. In: Marini S, Giancotti A (eds) Alter-azioni: Note oltre la realtà, pp 96–105. Nuova Cultura, Rome
De Matteis F (2015) Le sponde del tempo. In: Reale L, Fava F, López Cano J (eds) Spazi d'artificio, pp 47–54. Quodlibet, Macerata
De Matteis F, Bille M, Griffero T, Jelić A (2019) Phenomenographies: describing the plurality of atmospheric worlds. Ambiances 5:1–22. https://doi.org/10.4000/ambiances.2526
Fregonese S, Minca C (2004) Réve(il) de la ville: geografie della ricostruzione a Beirut. In: C. Minca (ed): Orizzonte mediterraneo, pp 102–145. CEDAM, Padova
Fricke A (2005) Forever nearing the finish line: heritage policy and the problem of memory in postwar Beirut. Int J Cult Prop 12(2):163–181
Fu S, Hillier J (2018) Disneyfication or self-referentiality: recent conservation efforts and modern planning history in Datong. In: Ding Y, Marinelli M, Zhang X (eds) China: a historical geography of the urban, pp 165–191. Springer, Cham
Gandy M (2017) Urban atmospheres. Cult Geogr 24(3):353–374
Harvey D (1997) The New Urbanism and the communitarian trap: on social problems and the false hope of design. Harv Des Mag 1:1–3
Hasse J (2012) Atmosphären der Stadt: aufgespürte Räume. Jovis, Berlin
Hasse J (2015a) Der Leib der Stadt: phänomenographische Annäherungen. Karl Alber, Freiburg
Hasse J (2015b) Was Räume mit uns machen—und wir mit ihnen: Kritische Phänomenologie des Raumes. Karl Alber, Freiburg
Hourani NB (2015) Post-conflict reconstruction and citizenship agendas: lessons from Beirut. Citizsh Stud 19(2):184–199
Ilyés II (2015) Rebuilding downtown beirut. Central European University, Budapest
Leclair-Paquet BJ (2013) Beirut divided: the potential of urban design in reuniting a culturally divided city. DPU Working Papers, London
Naeff J (2016) Disposable architecture—reinterpreting ruins in the age of globalization: the case of Beirut. In: Lindner C, Meissner M (eds) Global garbage: urban imaginaries of waste, excess, and abandonment, pp 221–236. Routledge, London and New York
Nagle J (2017) Ghosts, memory, and the right to the divided city: resisting amnesia in Beirut city centre. Antipode 49(1):149–168
Ragab TS (2011) The crisis of cultural identity in rehabilitating historic Beirut-downtown. Cities 28(1):107–114
Rizzi R (2011) L'Aquila: S(c)isma dell'immagine. Mimesis, Milan
Salvo S (2016) Restaurare il Novecento: Storia, esperienze e prospettive in architettura. Quodlibet, Macerata
Schmitz H (1969) System der Philosophie III.2: Der Gefühlsraum. Bouvier, Bonn
Schmitz H (2011) Der Leib. De Gruyter, Berlin and Boston
Schmitz H (2016) Atmospheric Spaces. Ambiances 2–10

Schmitz H, Müllan RO, Slaby J (2011) Emotions outside the box—the new phenomenology of feeling and corporeality. Phenomenol Cogn Sci 10(2):241–259

Thibaud J-P (2011) The sensory fabric of urban ambiances. Sens Soc 6(2):203–215

Thibaud JP (2013) Commented city walks. Wi: Journal of Mobile Culture 7(1)

Thibaud J-P (2015) The backstage of urban ambiances: when atmospheres pervade everyday experience. Emot Space Soc 15:39–46

Vloeberghs W (2012) The politics of sacred space in downtown Beirut (1853–2008). In: Ababsa, M, Dupret, B, Denis E (eds) Popular housing and urban land tenure in the Middle East, pp 137–168. The American University in Cairo Press, Cairo and New York

Part III
Society, Governance and Collective Resilience

Chapter 20
Bonding Between Urban Fabric and Capacity of Collective Resilience: The Case of Talca Historic Centre, Chile

Giulia De Cunto

Abstract This chapter addresses the relationship between community's ability to develop mechanisms for collective responses to a catastrophe and the pre-existing settlement model, using as a reference the case study of the historic centre of Talca, destroyed by the earthquake that hit south-central Chile in 2010. Before the earthquake, the historic centre of Talca was the core of the city and was characterised by a certain heterogeneity of artefacts, inhabitants and activities. All around the centre, a series of satellite neighbourhoods began to appear from the 1980s, mainly erected by private enterprise without a clear public design, characterised by strong monofunctionality, clear distinctions by income levels, large commercial buildings and large road infrastructures. Reconstruction after the earthquake has encouraged the development of these new districts, while the centre is being rebuilt in an uneven way, partly to replace destroyed areas and partly by densification interventions. Although the historical nucleus is no longer clearly recognisable today, the liveliness of the centre's public life has succeeded in resisting, for a while, through the mobilisation of the inhabitants in defence of some symbolic buildings. Of particular interest is the process of citizen mobilisation around the value of a state and comprehensive schools (Escuelas Concentradas) that led the school buildings to keep the same function and form it had before the earthquake. The results of the study highlight the connection between the settlement pattern and the inherent relationship among the inhabitants towards it; how a process of collective awareness post-catastrophe is closely linked to the pre-disaster scenario, as amply emphasised in the literature; and how the complexity of daily practices and relationships that develop in the historic city are, at the same time, a tool of mitigating social risk.

Keywords Reconstruction · Collective resilience · Urban policies · Talca · 27F · Chile

---

G. De Cunto (✉)
Università Degli Studi Di Milano-Bicocca, Milan, Italy
e-mail: giulia.deq@gmail.com

© The Author(s), under exclusive license to Springer Nature Switzerland AG 2021
F. F. Arefian et al. (eds.), *Historic Cities in the Face of Disasters*, The Urban Book Series,
https://doi.org/10.1007/978-3-030-77356-4_20

20.1 Introduction

On the 27 February 2010, an earthquake with 8.8 magnitude, followed by a tsunami, known as 27F, strongly affected central-southern Chile. The epicentre of the earthquake was recorded in Cobqecura, in the region of Bío-Bío which, together with the Maule was the most affected area. There were 521 fatalities and 56 missing persons. The material damage to the houses amounted to 370,000 destroyed or damaged houses and 2 millions of affected people (Gobierno De Chile 2010).

The day after the catastrophe, President Bachelet declared a state of emergency for the two most affected regions and launched a campaign to raise funds for immediate remedial action, which managed to finance 30,000 emergency dwellings. From 11 March 2010, the presidency of the Chilean Republic passed to Sebastián Piñera who, as soon as he was elected, had to deal with the emergency of a catastrophe of this magnitude, and he tried to manage the reconstruction with efficiency and speed. The working method for drafting the General Reconstruction Plan[1] was to entrust the decisions to the existing ministries and to the institutions already present in the region, while a Committee for Reconstruction was in charge of directing and coordinating the plans drawn up by each ministry. The physical reconstruction of a large part of the destroyed building heritage was the responsibility of—Ministry of Housing and Planning (MINVU).[2] Its specific plan aimed to move from emergency to reconstruction as quickly as possible, in line with the policy carried out by the Government from the outset. At the local level, the reconstruction plans for each affected area were not necessarily drawn up by the municipalities. Very often they were delegated to third parties, including private ones, and were of an indicative and non-binding nature. This basically produced differentiated effects on the territory from one municipality to another.

Talca, the case study of this research, is the capital of the Maule region. After the earthquake the city has undergone substantial transformations beyond the control of the public actor, which have emptied the historic centre and produced sprawl all around the city. This chapter investigates what the inhabitants of the historic centre have lost and felt in this process, what meanings some of the collective actions undertaken by the inhabitants could take in defence of some symbolic places and what effects these transformations have produced in the collective dynamics of the future city. The study was conducted through direct observation of the affected areas, a series of field interviews of activists, academics and local politicians and a bibliographical research concerning the sociological aspects of the disaster and the literature produced on the specific case of Talca.

[1] Plan General de Reconstrucción.

[2] Ministerio de Vivienda y Urbanismo.

20.2 The Talca Case Study

20.2.1 Talca Before the Earthquake

Talca is the most important urban centre in the Maule region in central Chile, of which it is also the capital. The city was founded in the middle of the eighteenth century in the central valley of the country, between the Cordillera and the coast, in the communication corridor corresponding to "Routa 5" at present, following the usual geometric grid composed of *manzanas* (blocks) of 120 × 120 m. At the beginning of the nineteenth century, the city began to take on relevance in the national scene thanks to the political fervour generated by the struggles for Chilean independence. In the course of that century it passed from being an urban centre, linked to the church, to a small town with many services and commercial activities due to agricultural exportation and a lively cultural life. At the beginning of the twentieth century, three important interconnected factors marked a rapid growth of the city: industrialisation, migrations from rural areas to the city and reconstruction after the 1928 earthquake which had destroyed more than 75% of the existing buildings.

The new industrial establishments located themselves around the fringes of the historical centre. They consisted mainly of manufacturing production encouraged by state resources, which involved the use of technologies adequately developed for the time. The industries attracted workers from rural areas which increased the demand for housing and services. In the midst of this process the 1928 earthquake took place. The reconstruction was a long process, which attracted new workers, but at the same time, it was in charge of answering the request for accommodation of the new arrivals. In fact, numerous public housing complexes were built inside the historic centre modifying its structure by dividing the size of a *manzana* into smaller housing units. As a result of this structural change, due to the layout required by the new seismic regulations (which favoured the tiling of the buildings, creating two-storey curtain walls) and given the freedom of architectural design granted by the reconstruction legislation, the historic centre changed its face (Bustamante Silva 2011), becoming more populated and more heterogeneous than before.

Nevertheless, the structure of the historical centre did not undergo substantial radical changes until the new earthquake of 2010, but outside the centre the city began to expand horizontally starting from the nineteen-seventies, when the urban policy of the Military Regime established that urban land was not "a scarce commodity" and its price had to be fixed by the real estate market (Letelier 2015). Thus, new large-scale urban complexes arose, such as shopping malls, highways and mega-parking lots. Neighbourhoods were built on the outskirts for low-income families while neighbourhoods for the upper middle class were built in separate sectors around the city centre. These different and non-connected interventions were all designed according to a new logic, according to which it was no longer public planning that drove development but private interest. This new city, with its new conception of mono-functional spaces, ended up becoming a space of segregation, a series of "non-places", which contrasted with the fifteen neighbourhoods of the historic core, where

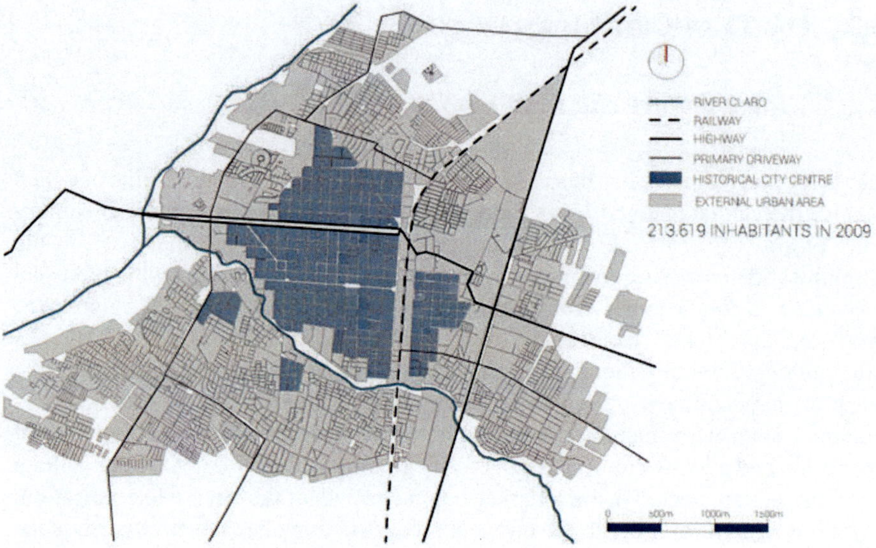

Fig. 20.1 Map of Talca, 2009 (author's elaboration)

urban quality continued to be high, despite the change of political vision. The historic centre, which in 2010 represented only 20% of the urban area, was still the centre of public investments, services and cultural activities; public space, in particular green areas, kept high standards and above all the social fabric continued to be strongly heterogeneous (Letelier and Rasse 2013). The 2002 census indicated that half of the inhabitants of the historic centre belonged to income groups D and E,[3] families in economic difficulty or in poverty. For these families, living in the nerve centre of the city was a precious resource. It meant having direct access to the opportunities that are physically positioned in the centre, as well as a possibility of integration given by the diversity of social backgrounds that coexisted in the same place, combining diversity of models, of information, of experiences (Letelier and Rasse 2013). As in the rest of the Chilean cities, there was a problem of housing deficit in Talca that led lower-income families to live in sublet premises, or in conditions of "*allegados*", that is people living in houses belonging to friends or relatives, without formal recognition of this condition (Fig. 20.1).

[3] Socio-economic groups are classified in Chile as: AB high class (average monthly income per family 4.386.000 Chilean Pesos corresponding to 5911.77 €), C1a upper middle class (2790.10 € monthly per family), C1b emerging middle class (€1851.98 per family per month), C2 typical middle class (€1091.78 per month per family), C3 lower middle class (€677.98 per month per family), D vulnerable (€413.80 per month per family) and poverty (€212.96 per month for family). Source: http://www.emol.com/noticias/Economia2016/04/02/796036/Como-se-clasifican-los-grupos-socioeconomicos-en-Chile.html.

20.2.2 Earthquake and Reconstruction

On the night of 27 February 2010, an earthquake capable of altering the inclination of the earth's axis by eight centimetres, upset central Chile, but the subsequent tsunami that struck the coasts from the fifth to the ninth region created even more damage. This earthquake was among the five most intense ever recorded, 8.8 degrees on the Richter scale.

The city of Talca was among the settlements most affected by the earthquake. 24.2% of the population saw their homes destroyed and fortresses were damaged (Ministerio de Planificación 2010). The historic centre recorded the highest level of damage, also in terms of building degradation. In the official land registry of the damage recorded by the Ministry of Housing and Planning (MINVU),[4] 3500 houses were damaged and 8457 people affected (El Bosque and Urbana 2013). According to studies carried out by the Universidad de Talca and by the NGO Surmaule, a correction should be made of the amount of damage that would bring the damaged houses to at least 6000 and the families affected which were not owners of the houses to about 2000 (Letelier 2015).

After the first emergency actions, the debate on reconstruction in Talca was started by the citizens. A discourse with increasingly political content was emerging, supported by some NGOs and other associations in the area, which proposed a different way of thinking about reconstruction, opposed to the government's focus on speed. This counter-discourse materialised in the proposal by the Talca College of Architects and some technical associations to form a local consortium that would take over the reconstruction plan of the city, a proposal that was well accepted by the administration in the first instance. However, after a week the denial came, with the news that Talca's plan would be developed, as in other cities, by a real estate group,[5] in this case El Bosque. The initiatives that sought a change of perspective with respect to the reconstruction in progress intensified and became larger, although they never amounted to a mass movement. Nevertheless, their actions succeeded in overcoming the regional boundaries and became known as the national movement for just reconstruction,[6] bringing petitions to the parliament that claimed a right to the city which was not protected. In the meantime, the masterplan for a sustainable strategy of reconstruction (PRES)[7] of Talca was drafted by the private company as intended. Besides not being binding, it only listed some specific projects, i.e. the recovery of the central market building, the establishment of a park on the river Claro, road repaving and, as the only two urbanistic features, a recovery programme for the historical buildings with a continuous façade, as well as densification of some sectors of the centre. The historical building programme for the recovery of the typical buildings of the centre did not have a very favourable outcome. In fact, very

[4] Ministerio de Vivienda y Urbanismo.

[5] From an interview with Stefano Micheletti, ONG Surmaule, Universidad Católica del Maule, (25/10/2017).

[6] Movimiento Nacional por la Reconstrucción Justa.

[7] Plan Maestro de Reconstrucción Estratégica Sostenible.

Fig. 20.2 Configuration of Talca, 2019 (author's elaboration)

few inhabitants chose to reconstruct according to the suggestions given, while the densification programme was seen by the municipality as an opportunity to redevelop and by developers as an interesting opportunity to transform some parts of the city significantly.

Little of the historic centre has been reconstructed by now (2020) and those who rebuilt it did so as they preferred. Instead, the city as a whole grew after the earthquake outside its historical nucleus and on the model of the neoliberal city that was already developing before the earthquake. Many of the inhabitants of the centre have moved to the new neighbourhoods outside, because the economically accessible real estate offer for the lower classes had moved to the newly built neighbourhoods. If it is true that today these inhabitants have, in some cases, a more technically suitable houses than the previous one in the city centre, it is also true that they had to give up the integration opportunities provided by the previous location and access to services and culture that living in the historic centre offered them. Although after the earthquake Talca lost much of the public space which characterised the centre in the past, public life has continued for some time for citizens who persisted in their battles for a more careful reconstruction and in defence of some symbolic places of the city, such as the one represented by the buildings of the Comprehensive Schools: "Escuelas Concentradas" (Fig. 20.2).[8]

[8] Comprehensive schools.

20.2.3 Citizens Action in Defence of the Escuelas Concentradas

The state schools for girls and boys (Escuela Superior de Niñas[9] Presidente José Manuel Balmaceda y Fernández and Escuela Superior de Hombres[10] Carlos Salinas), known in the city as Escuelas Concentradas were located on one of the central lots of the historic centre of Talca, because historically, since 1928, different parts of society had been concentrated there.[11] While the educational system in Chile tended to be increasingly privatised, the Escuelas Concentradas offered a good level of state education and above all an opportunity for social integration that reflected the qualities of the place where they were located.

The 2010 earthquake severely damaged these two buildings, and when parents and teachers realised that the reconstruction of these buildings would not start soon, they began to organise themselves, meet local politicians, seek the support of NGOs and collect signatures to prevent the buildings from being demolished. In the meantime, the mayor Juan Castro had expressed the desire to allocate the school site to a civic centre, which combined with the reactivation of the adjacent market would have constituted "an important centre of development for the regional capital" (Letelier and Boyco 2013). PRES, the masterplan proposed to move the school, suggesting the design of a new infrastructure for that site. The citizens sought the support of the Ministry of Education, succeeding in obtaining a repair fund that enabled them to start work in September 2012. In October, the works stopped but did not raise concern in the first instance as it was considered a normal construction site delay. The surprise came when, through various media, the citizens learned that the mayor had announced that he would call for a vote in the Council to revoke the agreement with the Ministry of Education and that he intended to demolish and rebuild the schools as more modern ones. In February, the Municipal Council ordered the actual demolition of the buildings but the movement of parents and teachers, which had by now managed to involve even those not directly concerned, succeeded in blocking the works by submitting the case to the Talca Court of Appeal, then turning to the Supreme Court and starting to make contact with the National Monuments Council. The real victory came thanks to their strategy of having the Escuelas Concentradas recognised as a national monument in April 2013, thus obtaining that the reconstruction should take place in a manner consistent with the existing one and above all that the school could neither be delocalised nor its intended use changed. Today, a reconstruction project is ready and the funds are allocated to restart the regeneration of the schools, but the works have not yet started.

The community that formed around the defence of the school buildings was able to keep the mobilisation alive for about three years but above all to animate the public space in front of the schools. This space has indeed become a reference in

[9] Secondary school for girls.

[10] Secondary school for boys.

[11] From an interview with the person, in charge of the reconstruction in Talca SERVIU Maule from 2014 and activist of NGO Fundación Superación de la Pobrez (18/10/2017).

Fig. 20.3 Public realm in front of Escuelas Concentradas awaiting regeneration (author)

the current life of the semi-deserted historic centre. Strolling through the Cienfuegos square where a market is located, it is possible to meet groups of youngsters talking, smoking and playing. An unexpected sense of normality characterises this space, despite the imposing building at the bottom of the square showing evident signs of the earthquake's destruction (Fig. 20.3).

20.3 Collective Action as a Tool for Reconstruction

The mobilisations of the citizens of Talca's historic centre in defence of Escuelas Concentradas have not only had a political value in the post-earthquake scenario. Indeed, this collective action reveals some aspects of no small importance. First of all the citizens of the historic centre were able to clearly recognise the social value that the schools represented for the city and to mobilise effectively to defend it, demonstrating not only capacity of self-organisation but also a level of consciousness with respect to collective problems that cannot be taken for granted. Secondly, the mobilisation led to the recognition of the Escuelas Concentradas as a national monument, not only for the architectural value of the buildings but for the symbolic value that these spaces had assumed for the community concerned, opening a reflection on the meaning of conservation and transformation in a post-earthquake context. Starting from these considerations, we are investigating its roots and implications.

20.3.1 Settlement Model and Collective Opportunities

As previously mentioned, in the process of rebuilding the city of Talca, many of the inhabitants moved from the historic centre to the new residential districts. The families who did not own their own homes were particularly affected by this phenomenon. Those who had lived in conditions of tenants and *allegados* before the earthquake had only access to reconstruction subsidies for constructions on new sites during reconstruction. The real estate offers accessible to them were outside the consolidated city, where real estate companies had already acquired land before the earthquake. Even for families who owned their properties, it was not easy to rebuild their homes where they were and with the characteristics they had enjoyed before the earthquake. Most of the destroyed houses were large, and the state subsidy was not sufficient to rebuild their homes while maintaining their size. Thus, some of these families decided to slowly rebuild one part at a time, others decided to rebuild a smaller house, others to move and dwell in totally different houses from those they had before, outside the historic centre (Letelier and Rasse 2016). Of the families which received a subsidy for the reconstruction in Talca, one in four was relocated (Ministerio del Interior y Seguridad Pùblica[12] 2014) A substantial part of the inhabitants went from living a complex, layered life characterised by a strong functionally and socially mixed settlement model, to a simpler and more schematic one, made up of mass produced houses, wide road arteries, large commercial complexes and little integration. Would it be possible for the inhabitants to implement a mobilisation like the one in defence of the Escuelas Concentradas in the city's new configuration?

The answer to this question is not absolute, but there are important reflections that could give us strong doubts that this is not the case. The complexity of urban forms also highlights the complexity of the processes that generated them, in the activities that keep them alive and therefore in the relationships, opportunities and practices that they underpin. Implementing an urban model that aims at homogeneity, offering standard housing solutions and reconstruction based on the housing needs alone, also means giving up the values of complexity and to favour exclusion. As Richard Sennett (2018) writes:

> Exclusion … also involves simplifying the look and construction of the place so that the place fits one kind of person, but not others. Mixed forms and uses invite mixed users. While in a stripped-down environment, the more form becomes simple, clear and distinct, the more it defines who belongs there and who does not.

In the case of the Escuelas Concentradas the historic centre community was one that defended the value of integration. It was a community accustomed to live in an

[12] Ministry of Interior and Public Security.

area of integration that defended the opportunities that can be found in growing up and being educated in a context of diversity. The communities that are settling down, the young people whose formation takes place in the spaces of segregation, may not find themselves spontaneously defending a value that does not belong to the place in which they live.

This aspect appears particularly relevant in the post-earthquake context because the tendency to simplify is a risk that links easily to the need to provide quick answers to the settlement problems generated by the earthquake. It is in fact recognised in such a way that the earthquake generates processes of acceleration of the obsolescence of central areas (Letelier and Rasse 2016).

The risk that the inhabitants of Talca run in the loss of complexity during the transition between pre-and post-earthquake, is the same for the inhabitants who are living in L'Aquila, a city in central Italy, strongly devastated by the earthquake in 2009. Already before the earthquakes in the last 50–60 years, L'Aquila was experiencing a process of urbanisation of the peri-urban area outside the historic city, which in the meantime was also slowly being emptied of administrative functions and services. After the earthquake, a joint set of circumstances meant that this phenomenon accelerated (Olori and Ciccozzi 2016). Decisive was the prolonged safety measure, the "red zone", with various modifications that lasted about ten years which has prevented the inhabitants from the fruition of their historical centre. To this was added the fragmentation of emergency solutions and their uneven location on the territory. In particular, the experimentation of the Antiseismic Sustainable Eco-friendly Complexes (CASE)[13] residential modules contributed to the proliferation of the sprawling city. This model of emergency housing was not entirely temporary but not planned as new complex settlements either. It was located in areas not yet urbanised and therefore clearly outside the primary urban nucleus and its opportunities. These buildings were in fact built with permanent foundations, but they are thought of as a temporary solution for the earthquake victims, during the period in which their houses were being rebuilt. In reality, this temporary housing the destination of which is not clear after the emergency may well become permanent on the territory and no longer removable (Forino 2012). Even if only temporarily this type of housing carries still the risk of sedimentation, thus some of the inhabitants of the historic centre of L'Aquila find themselves in the same condition as those of the old town of Talca who have not only lost their homes in the reconstruction process, but also the profound essence of their collective life.

20.3.2 *The Value of Continuity*

Starting from the mobilisations in defence of the Escuelas Concentradas, it can be observed how the possibility of triggering similar processes of urban mobilisation is linked to the pre-existing social context of the historic centre of Talca, characterised

[13] CASE: Complessi Antisismici Sostenibili Ecocompatibili.

by a substratum of relations built up over 150 years of history (Gac and Micheletti 2013). Literature on disasters in the sociological field generally agrees in affirming that every disaster is the result of a previous development of specific dynamics and social choices, which have implications in the moments resulting from the impact of the post-emergency and reconstruction (Lucini 2017). There is therefore a direct relationship between before and after the disaster, even if the impact of this on the structure of the place causes a rupture, a discontinuity, on the level of the social customs of the community and in physical terms in the forms of the built environment.

In the processes that are triggered between the before and the after the catastrophe, the construction of a link between the past and the future is played out, both on the immaterial level and on the material level. On the immaterial level, communities may be able to reorganise and respond to the difficulty. Fois and Forino (2014) define the "reaction from below, spontaneous, to an external shock" as community resilience, connecting the response to the catastrophe with the autonomy of the community, with local resources and with the capacity to organise. It is not automatic that all communities develop a capacity for resilience. The possibility of implementing this process of "*self-repair*" (Spagnuolo 2017) in this space of discontinuity depends in fact on the possibility for the communities to have access to the resources that they need, ranging from knowing how to exploit ties to maintain networks and, more generally, to the awareness of being able to make decisions for themselves.

On a physical level, Cavalli (2005) shows how communities deal with discontinuity mainly by operating a removal mechanism, in some cases of the past before the catastrophe, reconstructing with new forms and with the will to start again from a promising future, in other cases by removing the catastrophe, with the will to rebuild everything as before on the ideal of a glorious past. Then, there is a third way, the "*selective reconstruction*", which implies the faithful reconstruction only of certain symbolic elements around which to reconstruct the collective identity. In this case, discontinuity becomes an opportunity to redefine identity while taking into account both the past and the future of the place. The choice to keep some elements as memory takes place intentionally and is part of the process of recovery.

In the case of the Escuelas Concentradas, around this building and the collective values that it represented, the community of the historic centre has reconstructed its link between before and after. They wanted to bring back a key element of their past into the future configuration of the city, thus elaborating the catastrophe, choosing to keep that precise architectural element as it was. This brings us back to a reflection on the value of material heritage in the context of reconstruction. The clash between transformation and conservation of buildings is the discourse that captures the scene of most debates on reconstruction but, usually it concerns buildings or even just built elements of architectural value. When addressing the reconstruction question ourselves on whether and what to maintain of the material heritage we consider it rightful to ask ourselves what is the collective and symbolic value that the heritage assumes, not only from the point of view of architectural value but also in its functional and symbolic characteristics. The inhabitants of Talca wanted the schools to remain exactly what they were from an important point of view, and in fact, they did not ask only for their physical recovery but above all for the recovery of their original functions.

20.4 Conclusion: Complexity in Settlement Models as a Measure of Reduction of Social Risk

Post-earthquake reconstruction often works as accelerator of socio-territorial dynamics already in place in the territory (Olori and Ciccozzi 2016). The central areas of the cities of Talca and L'Aquila, which had already been affected before the catastrophe by phenomena of emptying the meaning of the historical centres, in favour of the city spreading around, underwent an extreme of this dynamic with the earthquake. The impoverishment of the role of the centre, which coincided with the lack or too slow physical reconstruction of it, has strengthened and enlarged the city sprawl. The settlement pattern made up of complex forms of the historic centre, whose value in relation to the quality of life of the inhabitants was clarified in chapter, was abandoned in favour of simpler settlement patterns, which do not encourage integration and collective action. The processes of spontaneous activation of the community linked to physical space, as in the case of Talca's Escuelas Concentradas, constitute a valuable resource in post-catastrophe scenarios in countering the effects of this territorial dynamics. At the same time, these spontaneous processes make it possible to establish a connection between the life of the place before and after the catastrophe, and they should therefore be encouraged.

However, the possibility of implementing this kind of collective response is closely linked to the previous situation in the territory, in terms of quality of relationships and social organisations. In the contexts in which reconstruction is conducted with simpler settlement models than those characterising the city before the earthquake, the future possibility of developing mechanisms of collective resilience in response to a catastrophe is reduced. Therefore, maintaining complexity in the reconstruction constitutes a measure of reduction of social risk in anticipation of new possible disasters.

References

Bustamante Silva E (2011) Evolución y desarrollo urbano del centro histórico de Talca, Universidad del Bio – Bio Facultad de Educación y Humanidades

Cavalli A (2005) Tra spiegazione e comprensione: lo studio delle discontinuità socio-temporali. In: Borlandi M, Sciolla L (eds) La spiegazione sociologica. Metodi, tendenze, problemi. Il Mulino

El Bosque SA, Polis Arquitectura Urbana (2013) Pres Talca, Comuna de Talca, Región del Maule

Fois F, Forino G (2014) The self-built ecovillage in L'Aquila, Italy: community resilience as a grassroots response to environmental shock. John Wiley and Sons, Disasters

Forino G (2012) Riflessioni geografiche sul disaster management all'Aquila. Semestrale di Studi e Ricerche di Geografia. Roma—XXIV, Fascicolo 1, Jenuary–June 2012

Gac D, Micheletti S (2013) El riesgo y el derecho de (re)contrucción social de la ciudad. Los sin tierra en Talca. UCMaule—Revista Académica N°44

Gobierno De Chile (2010) Plan de Reconstrucción Terremoto y Maremoto del 27 de febrero de 2010. https://www.preventionweb.net/files/28726_plandereconstruccinagosto2010.pdf

Letelier F (2015) Talca. Relocalización de familias damnificadas en Talca, in Columbia-Chile Fund De Columbia Global Center Santiago Y Conicyt, Learning from 27 F. A comparative assessment of urban reconstruction process after the 2010 Earthquake In Chile. https://globalcenters.columbia.edu/sites/default/files/content/santiago/PUBLICATIONS/Learning%20from%2027F%20gm.pdf

Letelier F, Boyco P (2013) Talca a tres años del terremoto: aprendizajes colectivos para la acción en la ciudad. Temas Sociales, SUR Corporación de Estudios Sociales y Educación

Letelier F, Rasse A (2013) El proceso de reconstrucción de viviendas en el centro de Talca: fotografía a dos años de la catástrofe. Revista Invi 77 Interior.indb 139

Letelier F, Rasse A (2016) Política de Reconstrucción y Desplazamiento: el caso de las familias de bajos ingresos del centro de Talca. Revista de Urbanismo N°35, Departamento de Urbanismo – FAU - Universidad de Chile

Lucini B (2017) Il terremoto a L'Aquila, la resilienza sociale e territoriale nel post sisma. In: Mela A, Mugnano S, Olori D (eds) Territori vulnerabili. Verso una nuova sociologia dei disastri italiana. Franco Angeli s.r.l, Milano

Ministerio del Interior y Seguridad Pùblica (2014) Diagnóstico estado de la recontrucción terremoto y tzunami 27 de febrero de 2010. Delegación Presidencial para la Reconstrucción, Santiago. http://www.observatorioreconstruccion.cl/wp-content/uploads/2014/06/Informe-27F.pdf

Ministerio de Planificación (2010) Encuesta Post Terremoto: Principales resultados. Efectos en la calidad de vida de la población afectada por el terremoto/tsunami. http://www.desarrollosocialyfamilia.gob.cl/pdf/upload/3_ENCUESTA_POSTTERREMOTO_RMS.pdf

Olori D, Ciccozzi E (2016) L'Aquila città in frantumi: la ricostruzione come acceleratore delle dinamiche socio-spaziali. In: Castriganò M, Landi A (eds) La città e le sfide ambientali globali. Sociologia urbana e rurale, Franco Angeli, Milano

Sennett R (2018) Building and Dwelling: Ethics for the City. Farrar

Spagnuolo S (2017) Le alluvioni un disastro ambientale annunciato: il caso di Benevento. In: Mela A, Mugnano S, Olori D (eds) Territori vulnerabili. Verso una nuova sociologia dei disastri italiana; Franco Angeli s.r.l, Milano

Chapter 21
Pre-disaster Examination as Post-disaster Managerial Thinking Ahead for Hoi An, Vietnam

Fatemeh Farnaz Arefian

Abstract Post-disaster reconstruction in historic cities brings together disaster management, urban development and heritage maintenance systems, representing a great level of multi-faceted complexities to address multiple objectives in the field. Yet, in practice, adding a historical layer to the already complex reconstruction systems adds further expectations from reconstruction activities on "*what*" to achieve, while the main issue of "*how*" to achieve remains problematic. Addressing the *how*, this chapter employs a multi-perspective model for understanding and examining organised activities for post-disaster reconstruction. It offers a pre-disaster examination of practicalities from a managerial point of view for a probable disaster in Hoi An in the case it entails large scale post-disaster reconstruction and retrofit. The historic city centre in Hoi An, a world heritage urban core, is located in the disaster-prone central coast of Vietnam which is at the risk of floods, typhoon and other environmental challenges.

Keywords Disaster · Historic cities · Reconstruction management · Organisational architecture · Tourism economy · Pre-disaster planning · Hoi An · Vietnam · Multi-perspective model · Bam

21.1 Introduction

Reconstruction and recovery scenarios in cities are greatly complex and uncertain. Urban reconstruction overlaps with urban development planning, regulations and processes and has to deal with existing concurrent urban systems and networks, such as social, economic, physical and governance and administration and their relations to each other that might or might not be directly apparent (e.g. Johnson and Blackburn 2014; Shilderman 2010; Chelleri 2012; Comfort and Kapucu 2006). In the case of historic cities, theoretically, reconstruction activities must consider how to deal

F. F. Arefian (✉)
Silk Cities, The Bartlett Development Planning Unit, UCL, London, UK
e-mail: f.arefian@silkcities.org; f.arefian@ucl.ac.uk

© The Author(s), under exclusive license to Springer Nature Switzerland AG 2021

F. F. Arefian et al. (eds.), *Historic Cities in the Face of Disasters*, The Urban Book Series,
https://doi.org/10.1007/978-3-030-77356-4_21

with the historic built environment, while dealing with the urgent needs of a disaster-affected population and the complexities of reconstruction that are widely acknowledged. In practice, adding a historical layer to the already complex reconstruction systems adds further expectations from reconstruction activities on "*what*" to achieve, while there are also other pressing fundamental expectations from reconstruction.

Those growing expectations are a result of advancements interconnections of disaster and development studies and practices, and the critical stage of reconstruction activities which links post-disaster situation to future urban development. Although since 1990s the entry point to dealing with disasters shifted from re-active disaster management to pro-active disaster risk reduction, over 90% of all funds for disaster risk management in recent decades went to disaster relief and post-disaster reconstruction according to the World Bank. Vulnerability to environmental hazards, development failures and disaster risk are inter-dependent and the importance of facilitative post-disaster reconstruction and recovery for improving future development is now recognised. The current state of the art and global agenda expressed in the priorities for action in the Sendai Framework for Disaster Risk Reduction, 2015–2030, focuses on *"enhancing disaster preparedness for effective response" and to "Build Back Better" in recovery* (e.g. UNISDR 2015; GFDRR 2014; Bosher 2008). Reconstruction is therefore a strategic physical agent for a multi-disciplinary recovery, and this has raised expectations on what reconstruction of the built environment should be (Arefian 2018). Among them are: addressing the escalating risks and destruction of historic cities, for example destruction of urban centres, significant or monumental buildings, landscapes and archaeological sites.

At present, there is an emerging attention to the urban-related cultural heritage during reconstruction. Given the escalating number of historic cities damaged and destroyed by natural hazards or human-induced conflicts, for example those in the Middle East during the last decade, the issue of post-crisis reconstruction has become an urgent and pressing issue for International Council for Monuments and Sites (ICOMOS), an international NGO dedicated to the protection and management of cultural heritage, with an initial remit of archaeological monuments and sites (ICOMOS 2016). While ICOMOS and its International Scientific Committee on Risk Preparedness (ICORP) traditionally focus on monuments it acknowledges that beyond monuments, the challenge exists for traditional buildings and historic city centres which had specific characteristics from morphology to architectural typology (e.g. Jigyasu 2016; ICOMOS 2016; UNESCO 2015).

Yet, despite the wide acknowledgement and debates on "*what*" needs to be achieved the question of "*how*" remains underexplored. Addressing the *"how"*, a recent multi-perspective framework for reconstruction process to achieve multiple strategic objectives, Arefian (2018) brings together the multiple insights which are connected in practice. The overarching perspectives are as follows:

- Organisational architecture
- Social aspects
- Strategic aspects
- Multi-organisational aspects.

Building on the above model, this chapter addresses overlapping areas of reconstruction processes and heritage processes and practical opportunities and areas for consideration for pre-disaster planning for post-disaster activities in the case of historic Hoi An in Vietnam.

The ancient city of Hoi An is in Viet Nam's central coast, Quang Nam Province, on the north bank near the mouth of the Thu Bon River. This historic city was inscribed as the World Heritage in December 1999 (UNESCO 1999), and since then, it has been enjoying the international recognition of its historic heritage translated to the tourism industry. Yet Hoi An is located in a disaster-prone area as the central coast is heavily exposed to storms and floods and other environmental challenges. The city experiences floods on a yearly basis during the rainy season and suffers from coastal erosion. The region has been among disaster-affected areas in Vietnam for some recent major incidents, for example in 2017 and 2016. The recent flooding in Hoi An, although it was not at a disastrous level, exceeded emergency level one. In October 2020, heavy rains for three days in the central coast flooded many areas and caused erosion in mountainous districts. Flooding in the historic centre of Hoi An started on 8 October 2020, and an interim report by the local flood and storm prevention agency indicated that the Hoài River had risen 1.29–0.29 m higher than emergency level one. Flood levels were forecast to continue to rise in the coming days with heavier rainfall and added water released from reservoirs that were nearing capacity (Vietnam News 2020). Figure 21.1 shows flood levels in the Hoi An city centre.

This chapter offers a pre-disaster examination of practicalities from a managerial point of view for a probable disaster in Hoi An if it entails large scale post-disaster retrofit and reconstruction. It also offers some reflections and examples from another World Heritage urban fabric which was struck by a disaster: The case of Bam, a historic city in Iran after the 2003 earthquake.

21.2 Tourism-Oriented Urban Heritage Maintenance System in Hoi An

As the UNESCO (1999) briefly describes, Hoi An constitutes an exceptionally well-preserved example of a Southeast Asian trading port from the period of the fifteenth to the nineteenth centuries. Its buildings and its street pattern reflect the influences, both indigenous and foreign, that combined to produce this unique heritage site (Fig. 21.2). Since 2009, Hoi An is one of the two Special National Cultural Heritage sites in Vietnam (Thinh et al. 2019).

In recent years, Hoi An has become a prominent heritage tourist attraction in Southeast Asia. From having 195,000 tourist visitors in 2000, according to the People's Committee it had 400,000 in 2004, and in the first four months of 2005, according to the town's tourism authority, Hội An received 559,500 visitors (Nguy˜ên Ngọc Tu´ân 2005; Thị Thu Hà 2008, p. 48). As the steady increase of the number of tourists continued, according to the Quang Nam province's Department of Culture, Sports and Tourism, 3.2 million tourists visited Hoi An in 2017 while the city population was 121,000, unofficial statistics report of 5 million tourists in 2018.

Fig. 21.1 Flood levels in historic urban fabric of Hoi An (author 2019)

Not surprisingly, the economy of Hoi An and livelihood of its residents are heavily reliant on heritage tourism, and at the same time, funds for the protection and maintenance of the World Heritage Site come primarily from tourism. Over 75% of the local households are involved in tourism, and in 2016, the total tourism revenue was about 8 million USD, the most important section of the local economy (Thinh et al.

Fig. 21.2 Typical roads in the historic core of Hoi An (author 2019)

2019). It has, however, been criticised that such sites become over-commercialised and protection has been replaced by commercial exploitation (Lask and Herold 2004). From the environmental sustainability perspective, Hoi An is also facing new challenges like rapid urbanisation, economic growth and vulnerability to climate change. Responsibility for dealing with the ancient town's heritage and its preservation is part of the Ministry of Tourism which is also working with UNESCO on dealing with sustainability challenges.

The large urban scale of the heritage and variety of building typologies in Hoi An led to the development of an urban process and mechanism across the city centre of dealing with the maintenance of various properties which are part of the ancient city, including the privately owned ones. As part of the comprehensive planning for tourism development in Vietnam's central coast, it has determined the core zone, buffer zone and protection zone for the ancient city of Hoi An. With a focus on the core historical zone, detailed mappings determine various criteria for each property located there. Examples are lot boundary of historical buildings by street names, detailed usage of each building, rank of historical building importance, stories of historic buildings in the core zone, roof material, building style and historical typology, ownership, year of the construction of property (starting from 1697), building condition (good, moderate, bad). As the UNESCO Regional Unit for Culture in Asia and the Pacific (2008) explains, the following factors are detailed to establish how the built heritage is to be dealt with:

- Heritage value which is grouped in five categories. The extent of original elements, unique architectural and artistic structures, maintained in an integrated manner, is factors to establish the heritage value. Special category and category one represent the highest historical, cultural and scientific value.
- Building typology and structure such as communal houses, family chapels, shrines, residential houses, bridges, pagodas, temples, assembly halls, wells and other cultural works.
- Ownership, privately owned or government owned.
- Location, which depends on whether the building is located on the main road or in small lanes and alleys.

For privately owned properties, such as houses, family chapels and shop houses, the operational urban heritage maintenance process is based on owner-driven approach accompanied by financial support from the government. Governmental support for funding considers the above factors as criteria. Financial support to repair and maintain special category buildings located in small lanes and alleys is 75% of the total cost. For buildings in categories three and four located on the main road, the financial support is 40% of the total repair cost.

There are also guidelines for the public to assist them with the formal process of repairing private buildings. For example, The Homeowner's Manual is aimed to inform and communicate a toolkit used by historic property owners and tenants, conservation architects, contractors, urban planners and local government officials active in the historic–touristic core of the city. The manual starts with the explanations of key concepts and an understanding of buildings from an architectural and heritage perspective and of their significance, naming stakeholders and step-by-step guidance for the administrative procedure (the process) (UNESCO Bangkok et al. 2008).

21.3 Disaster Management and Post-disaster Reconstruction in Vietnam

Of the countries in the East Asia and Pacific region, Vietnam is among those at highest risk from natural hazards and disasters, with floods, droughts, severe storms, landslides, and forest fires having substantial economic and human impacts annually. In recent years, Vietnam has experienced an upsurge in intensity of these natural disasters. For example, during the last three months of 2016, heavy rain caused five consecutive flooding events, affecting 18 provinces in central Vietnam (GFDRR 2017a). In 2017 also, Typhoon Damrey made landfall in Vietnam in November with winds of up to 135 km per hour impacting 15 provinces.

Yet interestingly, it can be observed that in Hoi An typhoon and floods are not perceived as a real threat to the city and its urban fabric heritage, not in the view of heritage stakeholders nor tourists. The reason from the local stakeholders' side might be the pressing issue of coastal erosion during the last decade that also affects tourism revenue from resorts on the beach side (Thinh et al. 2019). In fact, floods

Fig. 21.3 Overcrowded tourism in city centre (author 2019)

the city experiences during the rainy season are considered routine events. As local heritage stakeholders during the field work indicated, tourists have even started to see the flooding of the city as the adventurous part of their trip during the wet season. A simple Internet search brings out selfies and photographs of tourists happily riding evacuation boats and posing for their photograph. In parallel, local heritage stakeholders see fire and overcrowded tourism as the real disaster risk (Fig. 21.3).

In Viet Nam, the legislation and regulation systems and institutional arrangements for disaster risk management focus mainly on disaster risk prevention and reduction. According to United Nations Development (UNDP), Vietnam has been progressing in disaster risk reduction through collaborating with international agencies that promote the application of the Sendai framework within the umbrella of Sustainable Development Goals (SGD). However, disasters do not wait for cities and communities to get prepared. The need for strengthening recovery processes in Vietnam has recently been highlighted in a specialist workshop organised by the Vietnam Disaster Management Authority (VDMA) and the UNDP in Ha Noi (UNDP 2018).

In Vietnam's post-disaster reconstruction and recovery international agencies and organisations play an important role. In the continuity of the Sendai Framework, the international support for Vietnam to deal with post-disaster reconstruction and recovery starts with providing information on desired expectations advocating a multi-faceted "*Building Back Better*". For example, in response to the need for reconstruction and recovery in post-Typhoon Damrey in 2017, according to the World Bank (2017b), the bank advocated for "Building Back Better" in the most-affected province and provided essential information on principles to assist all relevant stakeholders in formulating a cross-sectoral framework to address institutional, social and financial aspects of recovery and reconstruction in immediate and medium- to long-term post-disaster recovery. This information laid down the expected considerations and principles of reconstruction and recovery programmes that must be led by the provincial institution backed by the central government:

- Institutional arrangements to have an entity for core functions of planning and oversight to meet recovery objectives
- Capacity building during the recovery process
- Enhancing the participatory approaches and decentralised planning and programming
- Gender sensitivity
- Financing mechanism and mobilisation of funds in a timely manner
- Contributing to the economic revitalisation of the affected communities
- Integrating the resilient infrastructure design in all sectors
- Monitoring, evaluating and learning and having participatory monitoring and evaluation mechanisms
- Mainstreaming DRR in the recovery and reconstruction process so that appropriate measures to manage and reduce risks are included (GFDRR 2017b).

The above principles are transferable from one post-disaster situation to another in Vietnam as international organisations like the World Bank adopt the international agenda of the Sendai Framework and country profile. It is foreseeable that similar expectations will be promoted by international stakeholders in any new case of post-disaster reconstruction in the country. In the case of a potential disaster in Hoi An specifically, it can be expected that UNESCO will also be a key international stakeholder.

The economy in Hoi An is highly reliant on tourism, which is the first industry to be damaged by a real disaster. Therefore, the city will face severe challenges for the economic revitalisation of the affected communities and will be intensifying its reliance on national and international financial support for post-disaster reconstruction and recovery.

21.4 A Multi-perspective and Multidisciplinary Model

The Hoi An case presents not only a lack of preparation and planning for post-disaster reconstruction and recovery but also a lack of recognising the potential scale of the disaster risk. This issue possibly resonates in other historic cities too. In parallel, there is a sectoral fragmentation among the disaster management system and heritage maintenance, and lack of strategic approach connecting the well-established tourism-oriented urban process with the newly established disaster management perspective.

Yet in an unfortunate disastrous incident in Hoi An, these two aspects have to be closely effective and influential, as happened in other cases in Kathmandu, Nepal 2015; L'Aquila, Italy 2009; and Bam Iran 2003. They must interact in post-disaster reconstruction and retrofit for all and every property in the city centre and its buffer zone. In practice, such post-disaster activities in historic cities are ultimately multi-objectives, but their delivery systems are new to the locals. The reason is that from the disaster management side they must include measures for future risk reduction, addressing previous developmental failures and potentially new stakeholders (new system elements). The scale of the disaster and destruction informs the need for human resources and division of labour in various professional and administrative roles and responsibilities across the delivery system.

Integrating the multiple organisational perspectives, namely organisational architecture, strategic, social and multi-organisational aspects of reconstruction activities, a conceptual framework, shown in Fig. 21.3, offers an integrated tool of insights for the understanding of organised activities of post-disaster reconstruction for such cases (Arefian 2018). From the managerial point of view, it helps to bring together disaster management and heritage systems in practice to be implemented in the field. The model has been applied in the case of post-disaster reconstruction in Bam, Iran, another world heritage area after the 2003 destructive earthquake. It provides a real-life example to inform the Hoi An case. The model shown in Fig. 21.4 builds on the following aspects.

Regardless of whether it is publicly declared or not, reconstructions must be strategic to serve multidimensional developmental recovery through multi-objective programmes and projects, such as future risk reduction and safeguarding urban heritage as it will be the case in Hoi An and was in Bam and other historic cities. It is associated with vertical cascading from higher levels of strategies to the operational levels of practical considerations, harmonising each level of activities and addressing dependencies and interconnections and priorities.

Similarly, while the participation of disaster-affected people is essential, disaster-affected people are just one albeit important element of the programme delivery system. It has to be kept in mind that the social dynamism of a society in a participatory reconstruction blurs the boundaries between organisational society and society as a whole. Social dynamics of a participatory reconstruction programme influences the system formation and implementation. Dynamics of society—as a whole (or its representatives)—mixes with the dynamics of interactions within the system and influences the way of doing things. By the same token, decisions made for organisational architecture and configuration of participatory programmes can facilitate or hinder social learning.

Post-disaster reconstruction activities are multi-actor and multi-organisation unlike single organisation or disciplinary activities in normal situations. This adds

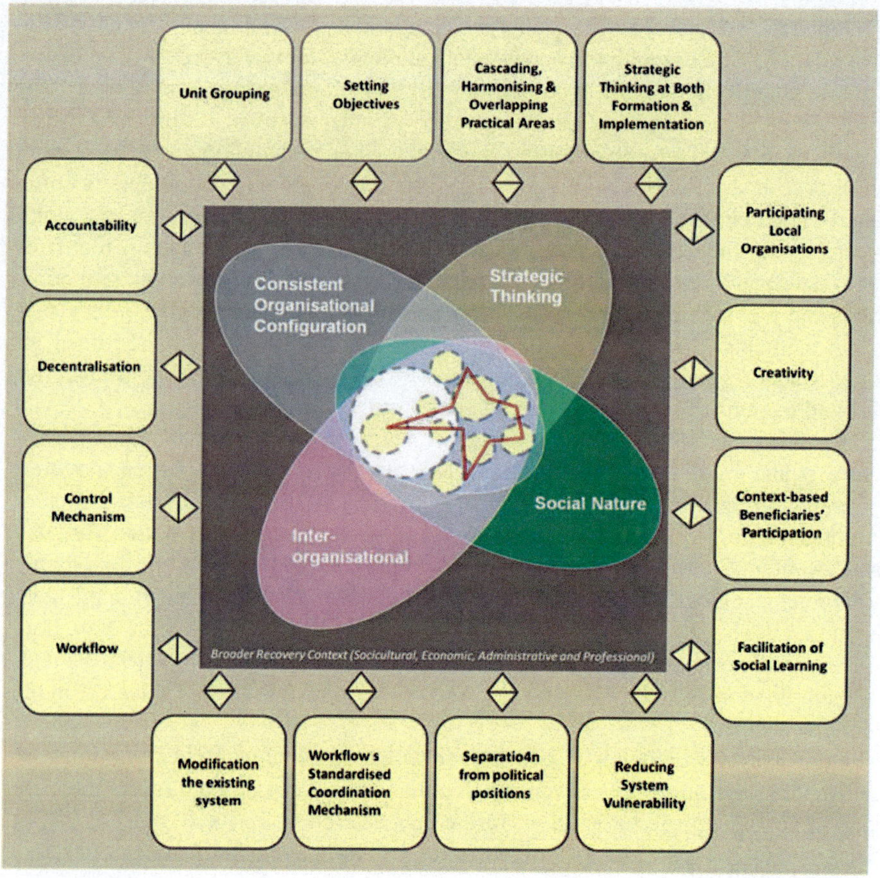

Fig. 21.4 Conceptual model for organisation design and management for post-disaster reconstruction (Arefian 2018)

to the complexity, as the reconstruction delivery system will indeed be a mega-system, within which each component is a system. This makes it even more difficult. In practice, all major dynamics of the multi-organisational nature of a reconstruction programme interrelates with other organisational configurations, and the strategic nature and social nature of the programme, and influences the programme performance through its system formation and implementation.

For the organisational design and formation of such organised activities, the model identifies reconstruction projects and programmes as innovative socio-technical systems and highlights the advantages of modifying existing systems if there are any in place. The crucial point is the strategic direction, alignments and consistency among various organisational attributes in their organisational architecture. It then lists key practical considerations and organisational attributes of organisational architecture that their consistencies or inconsistencies influence the project and programme performance. Organisational attributes and decisions rest on practical considerations, such as operational unit grouping, ways of delegation of accountability, decentralisation, allocation of human resources and workflow process which is a production chain.

21.5 What if? A Managerial Point of View for Hoi An

Looking at the case of Hoi An employing the multi-perspective model offers interesting results and points of attention in the case of a possible disaster. An important feature is that there is an existing urban process that addresses heritage maintenance in Hoi An, although the way it gets to work is not what is expected at the local level. At present, local heritage stakeholders state that this systemic mechanism is also used for repair, retrofit and reconstruction of affected properties in case of potential future disaster damage. Yet, this is a single-disciplinary statement and the preparation and establishment of this mechanism is for normal situations, linking up to the tourism industry, but not linking with the country's disaster management system. The above statement can be over-optimistic because the city's experiences of floods, fire and/or other destructive incidents and their damages have not reached the "*disastrous*" level so far. Nor has it been tested against a traumatic complex scenario of disaster which is beyond the local capacity and entails sudden flux of demands and needs, extra levels of funding and the engagement of other stakeholders such as international agencies, besides other disaster risk reduction (DRR) expectations. In fact, according to the World Bank, in Vietnam even for a non-historic urban context despite significant investment in better planning, the government continues to face a funding gap after disasters (GFDRR 2017b). Fortunately, there are still opportunities for both streams to inform each other in principle and in practice.

The existing detailed documentation of the core ancient city is a valuable asset for post-disaster needs assessment. Having active local organisations with great knowledge on Hoi An is strategically impactful and advantageous for both short-term reconstruction and rehabilitation and long-term recovery. Post-disaster reconstruction in historic cities brings urban development, heritage and disaster systems together. This is an advantage that many of other historic cities may not have. In the Bam case, the detailed documentation was only existed for the city's largest monument, its citadel, but the documentation of the other historic buildings and typologies was compiled in post-disaster.

The urban and heritage systems are already integrated in Hoi An. From the organisational architecture perspective shown in the model, an existing urban process is a positive attribute as it will be beneficial for formulating the post-disaster reconstruction process in Hoi An. In the Bam case, the reconstruction programme was based on a modified system of the existing housing development process, but there was a disconnection between this system and heritage maintenance in normal situations before the earthquake. Those too required to be integrated during post-disaster reconstruction which intensified reconstruction complexities in Bam.

The existing system in Hoi An can be modified to integrate post-disaster requirements and necessary organisational elements and steps. While this mechanism cannot be used as is if a disaster would strike, it can provide a basis to work on, adjust and improve to include other post-disaster considerations linked with the nature and scale of disaster. As other chapters in this book are discussing after a destructive disaster, such principles become of utmost importance.

It is crucial to look at the existing system of Hoi An's heritage maintenance process and examine it from the perspective of contemporary post-disaster principles. For example, mainstreaming DRR in recovery and reconstruction in practice means that technical measures need to be added to the post-disaster repair and reconstruction of properties. First is necessary to identify the space in the existing system of areas for improvements. The modification of the delivery system to implement those improvements may require other stakeholders and organisations, new operational unit groupings, the allocation of new tasks and responsibilities to different organisations and for different system elements, delegations of accountability, control mechanisms and other organisational attributes. They are decided during the formation of the modified integrated system which brings disaster management and heritage preservation together and contemplates the scale of the required reconstruction, retrofit, funds and human resources. The system process in turn acts as a workflow process connecting all stakeholders, organisations and disaster-affected beneficiaries to create a production chain and a standardised coordination mechanism among them. In the Bam case, new organisations were added for the modification of the existing housing development process to address new measures to safeguard historic architecture and reduce disaster risk reduction. A new operational unit grouping for technical and administrative matters was then creatively formed on an end-focus and area basis that also broke down the sale of the operation and decentralised the operations, although it did not apply to contractors. The control mechanisms were centralised and accountability was mostly aligned with the control mechanisms.

Additionally, in all post-disaster recovery principles participation of local people in the reconstruction and retrofit of their houses is a key component, especially the owner-driven approach. This is what already exists in the urban process for heritage maintenance in Hoi An. As the case of Bam also shows this approach frames people's participation in a realistic way within the broader urban governance context. The familiarity of Hoi An residents of the core city with the process and local organisations is hugely advantageous and time saving. The current existing guidelines and criteria for the public will present familiar references to assist them with the modified procedure. Yet, the guidance requires an update and inclusion of additional measures which are necessary for the post-disaster recovery process.

The above points are connected to the first principle stated by the World Bank for post-Typhoon Damrey that is about institutional arrangements to have an entity for core functions of planning and oversight to meet recovery objectives. These are important organisational advantages in place that support managerial decisions for formulating post-disaster reconstruction and recovery processes. Yet, they are necessary but not sufficient. As reflected in the model and as the experience of post-disaster reconstruction in Bam showed there are questions to be answered and points to be considered so that any potential post-disaster Hoi An can fully benefit from the likelihood of a destructive disaster. They include:

- Are the vision and strategic objectives clearly defined?
- Are objectives prioritised? What is the priority when it comes to preserving the current heritage as it exists against future disaster risk reduction? Do participant organisations have shared understanding of the priorities?
- As the formal coordinating mechanism and process for the modified integrated system, is the new or modified workflow process fluent? Are organisational architecture attributes, such as operational unit grouping, division of labour and control mechanism and number of technical staff and funding stages consistent, contingent and aligned with the objectives?
- How is accountability delegated and monitored?
- How does decentralisation reflect the strategic direction and the bigger picture?
- Are lower-level policies and practical considerations cascaded from higher level strategies and objectives? Are the interdependencies of objectives clear to stakeholders? Are practical considerations harmonised towards the strategic directions provided by the objectives?
- Is auto-adaptation among organisational stakeholders monitored against the greater good of the reconstruction objectives?

From the managerial perspective, decisions on the above form attributes of the organisational architecture of reconstruction projects. The fluency of the workflow process hugely depends on the consistency (or inconsistency) among those organisational attributes and understandings of priorities. For example, in the Bam case, operational unit grouping, decentralisation and administrative zoning were consistent, but there were inconsistencies with unit grouping for contractors, the control mechanisms and their human resource allocations and lack of shared understanding of priorities affected the fluency of the workflow. They caused delays in implementing the reconstruction programme in its early stages that had to be solved creatively and ultimately affected the programme performance. For example, although in practice objectives were clearly defined and translated into lower-level practical considerations, but their practical independencies were not considered, and objectives were not prioritised. Thus, in practice, the objective of maintaining the historic architecture was hindered in solving the unforeseen problems. However, an examination of the organisational design from a multi-perspective could have helped with reducing the problems by exposing bottlenecks and identifying areas requiring attention, so that the reconstruction management could polish and reformulate the programme and its workflow process before the implementation. Figure 21.5 graphically presents the organisational design and configuration in the Bam case.[1]

Many of those principal decisions and operational policies for bringing post-disaster reconstruction, retrofit and urban heritage maintenance systems together in

[1] For detailed review of the Bam case and further explanation of the model, refer to the book below: Arefian, F.F., 2018, Organising Post-Disaster Reconstruction Processes, Springer. https://doi.org/10.1007/978-3-319-70911-6, link: https://www.springer.com/gp/book/9783319709109.

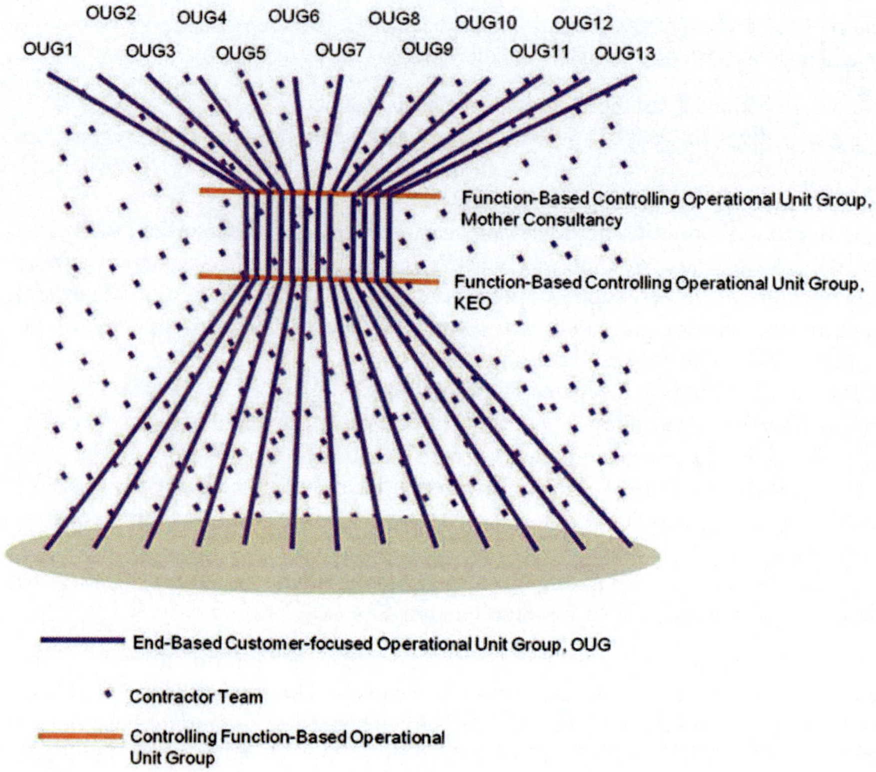

Fig. 21.5 Organisational configuration for the Bam case (Arefian 2018)

Hoi An can be decided pre-disaster when the disaster urgency is absent. Such advance blueprint multi-perspective examinations are needed in advance in the Hoi An case—and indeed other disaster-prone historic cities—to make post-disaster complexities more manageable and reduce the likelihood of unforeseen surprises in the field. The scale of the probable disaster affects the preparatory scenarios, and some factors are only unfolded after the disaster occurs, for example the extent of the damage which determines the overall extent of required financial, technical and administrative supports. Nevertheless, operational polices for breaking down the scale of operations are closely related to the above questions, such as on unit grouping, decentralisation, accountability and so on and how together they achieve consistency and fluency for achieving multiple objectives and priorities. Preparatory activities can be done regardless of scale. For example, it is possible to examine the current urban system and to anticipate the entry points of improving the system and the different stages of the heritage maintenance process, to develop scenarios for intervention and to initiate cross-sectoral dialogues for establishing priorities in a post-disaster case.

21.6 Conclusion

This chapter offered a prospective analytical examination of opportunities and areas of considerations for pre-disaster planning towards post-disaster reconstruction and interventions in Hoi An, Vietnam. The case exemplifies a historic city which is heavily tourism-oriented and has an urban governance for which heritage serves economic purposes. Tourism in fact became a driving force in establishing a detailed system and urban process for maintaining the historic core of the city built on people's participation, active engagement of local organisations, documentations and relevant guidance for the public and professionals. Yet, there is a fragmentation between the heritage system and the disaster management system. While the city has not yet experienced adverse incidents at a disastrous level, it is located in the disaster-prone region and suffers from heavy rains and erosion.

A multi-perspective and multidisciplinary understanding and overview is required to examine and anticipate how urban process, heritage process and reconstruction can potentially interact in a post-disaster scenario. In Hoi An, there exists a close tie between urban development process and heritage in the city centre. This is not the case in many other historic cities where detailed heritage documentation, people's engagement in the maintenance of the historic city are ignored. Yet, in Hoi An, bringing disaster management system into the equation and forming a multi-objective post-disaster reconstruction remains a challenge. As the multi-perspective model for strategic management and organisational design for post-disaster projects and programmes shows the existing system can offer an entry point to think ahead about post-disaster planning for Hoi An. Nevertheless, this does not guarantee success per se, as various organisational attributes of a strategic organisational architecture and the way it addresses social and multi-organisational aspects must be consistent, aligned and harmonised around the objectives, their priorities and interconnections.

Two crucial questions remain for Hoi An and other historic cities. What prevents them from working on potential post-disaster planning before a disastrous incident happens and, more importantly, how can local stakeholders be persuaded of the need and urgency of doing that in a pre-disaster situation?

References

Arefian FF (2018) Organising post-disaster reconstruction processes. Springer, ISBN 978-3-319-70911-6

Bosher L (2008) Introduction: the need for built-in resilience. In: Bosher L (ed) Hazards and the built environment: attaining built-in resilience. Routledge, London, pp 3–20

Chelleri L (2012) From the "Resilient City" to urban resilience. A review essay on understanding and integrating the resilience perspective for urban systems. Documents d'Anàlisi Geogràfica 58(2):287–306

Comfort LK, Kapucu N (2006) Inter-organizational coordination in extreme events: the world trade center attacks, September 11, 2001. Nat Hazards 39:309–327. https://doi.org/10.1007/s11069-006-0030-x

Global Facility for Disaster Reduction and Recovery, GFDRR (2014) Guide to developing disaster recovery framework (consultation report). Washington DC, USA. Available from https://www.gfdrr.org/sites/gfdrr/files/publication/DRF-Guide_FINAL_small_REVISED_FULL-disclaimer.pdf

ICOMOS (2016) Post-trauma reconstruction colloquium proceedings, 4 March 2016, Charenton-le-Pont—France, Available from: http://openarchive.icomos.org/1707/1/ICOMOS%2DPost%2DTrauma_Reconstruction_Proceedings%2DVOL1%2DENGok.pdf. Accessed on 9 April 2016

Jigyasu R (2016) Because culture cannot wait! Post-earthquake recovery of Nepalese cultural heritages. In: Post-trauma reconstruction. Proceedings of the 1-Day Colloquium at Icomos Headquarters, 4 March 2016. Presented at the ost-Trauma Reconstruction, France, pp 7–8

Johnson C, Blackburn S (2014) Advocacy for urban resilience: UNISDR's making cities resilient campaign. Environ Urban 26(1):29–52.

Lask T, Herold S (2004) An observation station for culture and tourism in Vietnam: a forum for world heritage and public participation. Current Issues Tourism 7(4–5):399–411. https://doi.org/10.1080/13683500408667993

Ngọc Tu´ân (2005) New Developments Lure Visitors to Hoi An. Vietnam News, 13 June 2005 http://vietnamnews.vn/travel/143626/new-developmentslure-visitors-to-hoi-an.html. Accessed on 1 Dec 2014

Shilderman T (2010) Putting people at the centre of reconstruction. In: Lyons M, Schilderman T, Boano C (eds) Building back better: delivering people-centred housing reconstruction at scale. Practical Action Pub, Warwickshire UK, pp 7–33

The World Bank, Global Facility for Disaster Reduction and Recovery, GFDRR (2017a) Vietnam Post-Typhoon Damrey Rapid Damage and Needs Assessment. Available from https://www.gfdrr.org/sites/default/files/publication/vietnam-damrey-rapid-assessment-report-en.pdf. Accessed on 11 July 2019

The World Bank, Global Facility for Disaster Reduction and Recovery, GFDRR (2017b) Vietnam Rapid Damage and Needs Assessment. Available from https://www.gfdrr.org/sites/default/files/publication/Vietnam%20Rapid%20Damage_FinalWebv3.pdf. Accessed on 11 July 2019

Thinh NA, Thanh NN, Tuyen LT et al (2019) Tourism and beach erosion: valuing the damage of beach erosion for tourism in the Hoi An World Heritage site, Vietnam. Environ Dev Sustain 21:2113–2124. https://doi.org/10.1007/s10668-018-0126-y

Thi Thu Hà N (2008) Tourism development and the sustainable management of cultural heritage—a case study of Hoi An an ancient town in Vietnam [Thesis], KDI School of Public Policy and Management, South Korea

UNDP (2018) Project Capacity Building on Disaster Recovery in Vietnam. UNDP in Viet Nam. Available from https://www.vn.undp.org/content/vietnam/en/home/operations/projects/environment_climatechange/capacity-building-on-disaster-recovery-in-vietnam.html. Accessed 11 Jan 2019

UNESCO (1999) Hoi An ancient town (Viet Nam), world heritage convention, available from Hoi An Ancient Town (Viet Nam). Available from https://whc.unesco.org/en/decisions/2637. Accessed 1 Nov 2020

UNESCO Office Bangkok and Regional Bureau for Education in Asia and the Pacific, UNESCO Office in Hanoi, Hoi An Centre for Monuments Management and Preservation, Showa Women's University. Institute of International Culture (2008) Hoi An world heritage site, Viet Nam: heritage homeowner's preservation manual. Available from https://unesdoc.unesco.org/ark:/48223/pf0000193014. Accessed 20 Sep 2020

UNESCO (2015) Background document, Post conflict reconstruction in the middle east context, and in the old city of Aleppo in particular [meeting] 18–19 June 2015, Paris, France. Available from https://whc.unesco.org/en/events/1286/. Accessed 4 July 2021

UNISDR (2015) Sendai framework for disaster risk reduction 2015–2030 (No. UNISDR/GE/2015-ICLUX EN5000 1st edition). Geneva, Switzerland. Available from http://www.preventionweb.net/files/43291_sendaiframeworkfordrren.pdf. Accessed on 20 Nov 2016

Vietnam News (2020) https://vietnamnews.vn/environment/793257/flood-hit-areas-evacuated-in-hoi-an.html. Accessed on 15 Oct 2020

Chapter 22
Play Street: Experimenting Tactical Urbanism for Urban Resilience in Iran

Soheila Sadeghzadeh and Azadeh Lak

Abstract In recent years, a new approach of intervention in urban spaces, called Tactical Urbanism (TU), has been introduced to solve current problems of urban areas. The main purpose of this approach is to meet the needs of citizens in urban spaces. This chapter investigates the experience of a TU project in form of "play street" in a disadvantaged area in District 10, Tehran. Held in 2017, the project called "Street in Hands of Children" and attempted to bring a relatively new approach in urban governance that emphasised on greater interactions of various community groups with urban spaces and improve the relations between the public and urban authority. The project was accompanied with a qualitative case study research with real-time interviews with participants and officials to analyse their feedback and understand potentials for enhanced presence of children in urban spaces and needs of community. This chapter reflects on the formation of the project and examines the event from community resilience perspective. It discusses how such an approach can potentially enhance social resilience by improving local knowledge and learning capacity, community networks and relationships and organisational capacity and communal assets. It proposes an approach comprising small-scale provisional measures to respond to the social issues and improving social capital among children and adolescents and offer points of consideration in future TU events, which can be seen as a systematic process for improving the quality of life and community resilience.

Keywords Tactical Urbanism · Play street · Urban resilience · Community resilience · Tehran · Iran

S. Sadeghzadeh
Tehran Municipality, Tehran, Iran

A. Lak (✉)
Shahid Beheshti University (SBU), Tehran, Iran
e-mail: a_lak@sbu.ac.ir

© The Author(s), under exclusive license to Springer Nature Switzerland AG 2021

F. F. Arefian et al. (eds.), *Historic Cities in the Face of Disasters*, The Urban Book Series,
https://doi.org/10.1007/978-3-030-77356-4_22

22.1 Introduction

Recently, cities face various socio-cultural and economic problems, and because of their complex nature, they need different solutions. In this regard, to improve the quality of life and urban resilience, local authorities have created new trends in public administrations. Economic constraints have multiplied sensitivity in decision making; But still, the dominant approach to solving these issues is large scale, costly interventions that, in addition to imposing heavy costs on the city's economic resources, have not been able to operate efficiently and improve the people's daily lives (Lydon and Garcia 2015). Nevertheless, sometimes a simple, low-cost, incremental and informal intervention can create new energy in the city and has a more positive impact on the lives of citizens, especially children and adolescents. Those interventions are called Tactical Urbanism (TU) (Lydon and Garcia 2015).

TU projects can be considered as an example of a bottom-up process, a "spontaneous" and informal activity that has integrated local and informal initiatives with managerial strategies in a coordinated, but an unplanned and unexpected process. The specific needs of a play street project will be briefly examined, especially in terms of building resilience and social capital (Barber 2013). This chapter examines a recent experience of running a play street in District 10, Tehran that applied the principles of TU in practice. District 10 is one of the smallest districts in the capital of Tehran, yet it is one of the most densely populated and socio-economically disadvantaged districts. In fact, the District, which has been formed over a century ago, is one of the poorest and oldest parts in the central Tehran. Its high population density is one of the most prominent features of the district (Asl and Lak 2017; Sadeghzadeh et al. 2018).

The municipality of Tehran planned a relatively new approach inspired by international urban design experiences and organised an event called "*Street in hands of children*" as a Play Street event in December 2017. The event aimed to boost socially interactive urban spaces in District 10 with the presence of children and different social groups for three hours, followed by the high presence of the people to seek solutions to improve the quality of life for residents in District 10. This event aimed to increase interactions and social trust among residents, between residents and local authorities, and eventually enhance urban resilience. Local groups participated in this event included children, adolescents, parents and seniors. At the same time, individuals and groups from outside the eighbourhood were invited for consultation on the programmes arranged for the event day.

This chapter reports on a qualitative case study research which used this case to examine impacts of social events in urban spaces from community resilience perspective. One of the most significant features of such research is the contemplation of a certain example of a phenomenon, referred to as the case (Merriam 1998).

The research employed participatory observation, semi-structured interviews during the event and qualitative data analysis. Researchers attended the event which informally called the *Street Festival*. Additionally, during th event, 13 experts from various departments in the District municipality used open-ended questionnaires to engage directly with event paricipants and conducted semi-structured interviews. They aimed to understand participants' needs and opinion on social affairs, traffic, urban services, environment, traffic police, fire department, the Red Crescent, neighborhood councils and emergency centre. Questions were focused on the improvement of local knowledge and learning capacity, social networks, organisation capacity and social structures of local communities in their neighbourhoods. Face-to-face interviews of 10–15 min each were conducted used to complete the data. As well, the results of semi-structured interviews were used for triangulation and validation of qualitative research. The reports were subsequently compiled following the process of the event.

22.2 Dimensions of Community Resilience

The literature defines that resilience is the ability of a social-ecological system to cope with and adapt to external social, political, or environmental disturbances (Cinner et al. 2009). The concept of "community resilience" is invariably viewed as positive, being associated with greater local capacity (Kennedy et al. 2013), social support (Aldrich and Meyer 2015) and resources (Patel et al. 2017), and decreasing risks (Turnbull et al. 2013), miscommunication (Chandra et al. 2011) and trauma (Bonanno 2004).

The notion of community is strongly tied with social capital, the social glue that holds a community together and the interactions which build it can be either formal or informal, from the structures of a regular meeting to a casual wave to the guy down the street. Social capital is seen as "connections among individuals—social networks and the norms of reciprocity and trustworthiness that arise from them" (Putnam 2000). The perceived benefits of strong social capital do not end with community resilience. Strong connections between people are linked to more happiness and better mental health. Trust between individuals in a community reduces crime, improves the functionality of public institutions and improves the community's self-image, as well as the outward image, with flow-on effects in regards to visitors and economic growth (Barber 2013).

Community resilience consists of three main elements which are local knowledge and capacity for learning, community networks and relationships and organisational capacity and communal assets, as shown in Fig. 22.1 (Patel et al. 2017). Local knowledge could be affected to mitigate community vulnerabilities. The first dimension of knowledge is the *factual knowledge base* of the community. Defined as the information, education and experience acquired concerning a learning capacity in a society (Patel et al. 2017). *Training and education* is the second dimension based on community education. Additionally, Moore and colleagues (2013) proposed a practice to be an element of community resilience with activities such as community training and exercises proposed to build local knowledge and capacity (Moore et al. 2013). Effective training and education should lead to learning (Cutter et al. 2008).

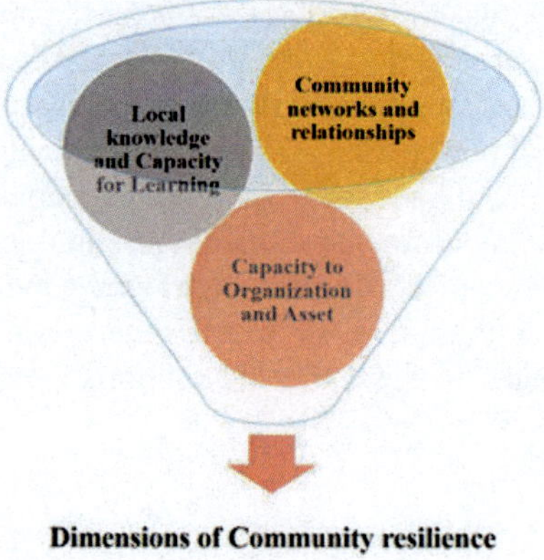

Fig. 22.1 Components of community resilience (graphic by authors, adopted from Patel et al. 2017)

Community networks and relationships are claimed to impact positively on a community and its members during a crisis. The connectedness of a community sometimes called "social network," as defined by the linkages within a community. Creating links among community members based on social relationships and/or between communities was examples of connectedness (Patel et al. 2017). The cohesion of a community is based on the nature of these links, typically described as weak or strong ties. Several factors that determine the strength of a tie, including trust and shared values might be relevant to enhance community resilience. The connectedness of the networks and their cohesion were also discussed as important aspects of social capital, which conceptually focuses on bonding, bridging, and linking (Aldrich and Meyer 2015; Patel et al. 2017).

Finally, improving organisational capacity and communal asset included quantitative data on involvement in community organisations, migration, community decision making, and population (Cinner et al. 2009). The indicators of assets include community-level infrastructure to make wealth or social status within a community such as public services (Cinner et al. 2009).

22.3 The Experiment

District 10 in Central Tehran with 817 ha area is one of the smallest districts out of the 22 ones in Tehran. It has three divisions, ten neighbourhoods and ten community centres. The population of the district is about 320,000 and has a population density of 420 people per hectare. In this regard, it is one of the most densely populated districts, and its population is four times the standard and twice the density in Tehran. About 100,000 households live in this area, and women make 49.15% and men make up 50.85% of the population. Youth of between 19 and 24 years old include 23.86% of the population, and the district has the literacy rate of 86.12% (Asl and Lak 2017; Sadeghzadeh et al. 2018).

In addition to the socio-economic context, main roads in this district provide city-scale business and trade services; they always have heavy traffic and do not provide space for social engagement of the wider local community. The envisioned experiment therefore aimed to showcase and enhance the social engagement, especially to respond to residents' social need for greater interaction in urban public spaces that in turn and ultimately could address social justice for the marginalised groups of residents in the district.

The project idea evolved around creating a car-free road with various stalls and activities for children in one of the main high streets in the district that in normal days, because of the traffic and business activities, are not socially engaging with diverse groups of local residents and families.

The proposed location was Roodaki Street, one of the busiest streets and most important public area in the district, both in terms of business (retail) and religious spaces. It appeared that the date should be a public holiday so the traffic disruption to the surrounding area is lowest, and businesses were not affected. Also a public holiday related to a religious occasion seemed an appropriate time for marking the event for locals and broader parties.

To realise the idea, all steps and coordination took place very quickly. First, a local decision-making committee was formed. The committee members were the mayor of District 10, deputies of urban and environmental services, social and cultural and transportation and traffic, as well as mayors of the three divisions and representative of the secretary of the local councils. Each member proposed ideas and took on responsibilities for the event day. One of the most important advocacy efforts was from the police to stop the traffic and redirect it so the Roodaki street to be a car-free urban space for three hours.

Next was to identify the target audience from marginalised groups which the event could create a platform for their engagement. Considering the great number of children in the district and their need for playing games, they were given the priority. Additionally, in plans, stands for table tennis, table football and citizenship education were added to increase the mobility of the adults and adolescents as well. Other stands measured and controlled blood pressure and blood sugar aimed for senior citizens. All services were free of charge. The event was called "Street in hands of children".

Considering the risks involved in the implementation of formal events and the lengthy beurocratic process of obtaining required permits and to improve relationship between locals and urban officials, the urban governance took the initiative to organise and run the event as an informal one. The District 10 municipality took responsibility for all possible problems and dissatisfaction of various urban management departments. Managing the festival event engaged various departments at both district and local levels, as well as sectoral organisations. During the festival, secretaries of district councils and district administrators played a significant role in controlling the population. The local authorities had significantly participated in providing games for children and were also in charge of the Red Crescent, Emergency and Police Department in the field. Traffic police was in charge of the safety and traffic. The Regional Emergency Center, Red Crescent, local police and the fire department were in charge of security, health, the public discipline and first aids in case of accidents and injuries at the larger geographical scale.

This small-scale experiment of Play Street—an example of TU—became extremely successful in bringing local residents to the street and children to reclaim the space (Fig. 22.2). The turnout for this buzzing event was great. In fact, the success of the project called for more small-scale measures targeting the children, women and senior citizens as marginalised groups (Asl and Lak 2017; Sadeghzadeh et al. 2018). Since the experience was well-received by children, women and the senior residents, the district-level urban governance decided to organise a series of small-scale and informal measures with sustainable effects on the lives of the residents within the framework of the TU approach to enhance community resilience.

The real-time research conducted during the event confirmed children and senior citizens as the ones should be prioritised in such TU-oriented initiatives. Findings were later communicated and discussed with a relatively wide and varied group of local authorities, e.g. members of the Tehran City Council and community centres, middle managers of the district-level municipality, heads of neighborhood councils, secretaries of the neighborhood councils in three meetings and relevant professional institutes such as The Institute of Urban Designers and other local NGOs concerned

Fig. 22.2 First "Street in Hands of Children" Roodaki St, District 10, Tehran (authors 2017)

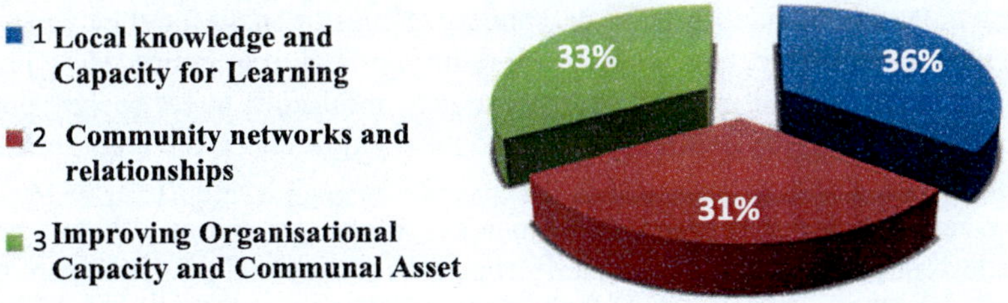

Fig. 22.3 Result of the open-ended questionnaire

with the well-being of marginalised groups such as women and children. The street as the event venue was prepared by municipality staff.

22.4 The Experiment from Community Resilience Perspective

As Fig. 22.3 presents, the research showed that over 36% of respondents saw it as the best possible venue for citizenship education both for children and others. To over 33% of participant locals and urban officials, the event provided great opportunities for the interaction between urban officials and the public. The increased satisfaction of urban space can lead to expanded collective capital and increased public welfare. Over 30% believed that the event and its activities could help the formation of a new type of neighbouring ties and interactions and could eventually bring further liveliness, satisfaction, knowledge of the neighbourhood, happiness and joy, as well as improving children's relationship with families and entertaining kids and adolescents. They would altogether help increase social capital and social interactions.

22.5 Local Knowledge and Capacity for Learning

For a purposeful TU event, the target audience had to be defined. In this experiment children and their parents were chosen as the target groups in need of improving their urban knowledge for example on environment and citizenship. Especially, children were chosen as the first target of this TU project. Next was to identify the needs and requirements of their age range to outline the objectives and activities of the event. Several meetings with child psychologists helped with the identification of children's needs and suitable games. The event therefore included a varied and wide range of games. Taking into consideration the appropriate games, some child psychologists also attended the event to facilitate the target group engagement in citizenship education and to increase the social interaction among participating children. Stalls were made for painting banners, games of snakes and ladders, citizenship education, music, educational storytelling, drama, trampoline, Holy poem, kids' train, table football, martial arts, hopscotch, mini-golf, street soccer, poetry recital and storytelling by children and a healthy stand to check the blood pressure and pulse of the participants.

Those activities attracted many participants of other age groups although being initially designed for children. This clearly exposed a broader need and the scarcity of open spaces in District 10 as such even courtyards and backyards in the disadvantage neighbourhood, which eventually rendered it difficult for children to access such games and adults to enjoy urban space activities.

Interestingly, although the event was initiated and planned by District 10 municipality for neighbourhoods around Roodaki St, but on the event day, numerous children from all neighbourhoods of the district attended the event. It became a district-scale event. Evaluating the scope of influence of this event showed the positive impact the event had on locals understanding of various issues the event activities addressed. It also showed that people of the adjacent divisions and other neighbourhoods also took part in the event.

The provision of such small-scale activities for children to attend and engage with citizenship education and learning through games requires a safe and healthy environment for them. Despite children's wish, in order to prevent causing trouble for local trades, the games were removed after three hours. This initiative verified the dire absence of and the need for safe and enabling urban spaces to support children excersicing their rights to the city in the southern districts of Tehran.

22.6 Community Networks and Relationships

As the research showed, participants believed that "Play Street" could enhance social relationships during the event among residents. Besides questionnaires, interviews with participants on the event day showed that such events could meet the needs of all population group such as children, adolescents, parents and street vendors who took part in the festival. A parent puts it in words:

> I am so excited about the event held in our neighbourhood. My daughter loved to participate. It was advertised in their school. All her classmates were here, and she played a lot. She normally does not get out. I bought stuff from the peddlers. I ran into a lot of friends and neighbours, and we had a chat.
>
> (A 45-year mother with her daughter)

Local residents are mostly long rooted in this neighbourhood. Thus they are keen on the activities and traditions of both religious and social natures. The TU project was a response to the local residents' requests to provide safe urban places for children in their neighbourhoods. It proved that such platforms are effective measures with a wider scope of influence.

Suitable games and activities, street vendors and shops that provide the everyday needs of the population can prepare the infrastructures for the realisation of more events from the TU projects. One of the attractions of such an event is various performers and especially music bands. The research showed the importance of allocating spaces to peddlers, games, pop-up exhibits and art performances for more social interactions and strengthening community relationships. It is about providing better mingling conditions for all social groups including women, seniors and children to learn about local culture.

Since TU in Iran is a new approach, the process of training, evaluation and satisfaction of such projects still requires calls and consultation sessions with urban managers and experts, representatives of NGOs and general public so that the future events are better contemplated.

22.7 Improving Organisational Capacity and Communal Asset

Considering the location, physical form and density in District 10, in practice finding the suitable space to be the event venue was relatively difficult. The experiment showed that any possible initiative for any given social group—in particular for neglected marginalised groups—can engage with large number of local people. Thus, the identification of potential properties and spaces with required features and assessment of the needs of social groups like children, the elder and women can increase the chances of success. A member of local community puts it in words:

> As the secretary of the neighborhood council, I did not expect such huge participation. There are a lot of problems in our neighbourhood, but still, the participation rate was high (Secretary of the neighborhood council- Roudaki neighbourhood).

Since the event was run as an unofficial and informal event, despite the previous attempts and communications, the existing regulations which could have supported the event as an official one, in reality would not provide the municipality with the necessary executive power if anything had gone wrong as a result of the event. However, the enthusiastic reception of the event by the public and participation of local officials paved the way for further interaction between the two groups and building trust.

> The event was so well-received, and participants asked for more similar ones in their neighbourhoods. We made promises to do so in other neighbourhoods as well.
>
> (The former mayor of District 10)

The feedback extracted from interviews with participants and organisers also showed that children, adolescents, the youth, parents and the local authorities, e.g. social services, traffic police, urban services, fire station departments, the Red Crescent, the Emergency Centre, neighborhood councils and chiefs of neighbourhoods were actively present in the event, but what brought all these together was also the local retail shops. Interestingly, although car access was restricted during the festival hours, the local shop owners suggested Roodaki St to be the venue for the next event again, since their sales were highly satisfactory. This should have logically trigger other ways of thinking in related organisations about roads, cars and business vitality (Fig. 22.4).

To most participants, having such events on a regular monthly basis would improve locals' satisfaction about the performance of the district municipality, increased liveliness and participation on the side of the citizens, increased vitality and mobility among children and increased knowledge of citizenship rights, which would in turn lead to further social resilience.

Many local officials see organising such events in other districts as a way forward to increase the citizens' vitality. Similar events can continue to take place and even become a regular plan in various parts of Tehran to further promote pedestrian-oriented activities for residents, especially for children, women and senior citizens.

Fig. 22.4 Street in Hands of the Children, November 2017, organised by the municipality (authors 2017)

22.8 Tactical Urbanism as an Enabling Tool

The importance of this study is the presence and cooperation of locals and the enhancement of social resources in the form of events in urban spaces. Given the success of this experiment and its learning opportunities, it can be used as an example of a small-scale initiative to pursue various objectives. Examples of objectives are: to enhance social resilience as it was the case in District 10 and to improve environmental quality and the quality of urban life as an eventual objective of urban design and development.

It appears that citizen-centred approaches to tactical urbanism can be put into practice more in Iran on a local scale. They can enhance community resilience and social resources in some ways. Firstly, the learning capacity can be enhanced by providing informal educational programmes in events for different groups of people. For example, conducting book reading competitions and educational shows. Secondly, social relationships can be enhanced in tactical events through joint programmes and facilitating public participation. Thirdly, organisational capacity is developed by involving general and governmental organisations.

Therefore, altogether the factors influencing success or failure of tactical urbanism will depend on the following main groups: people and citizens, experts and urban developers, responsible institutions and organisations such as municipalities and so on. Each of those groups, their perceptions and participation are influential and can be a barrier or facilitator to the implementation of TU projects (Fig. 22.5). Those small-scale TU initiatives can act as enabling platforms for bringing together

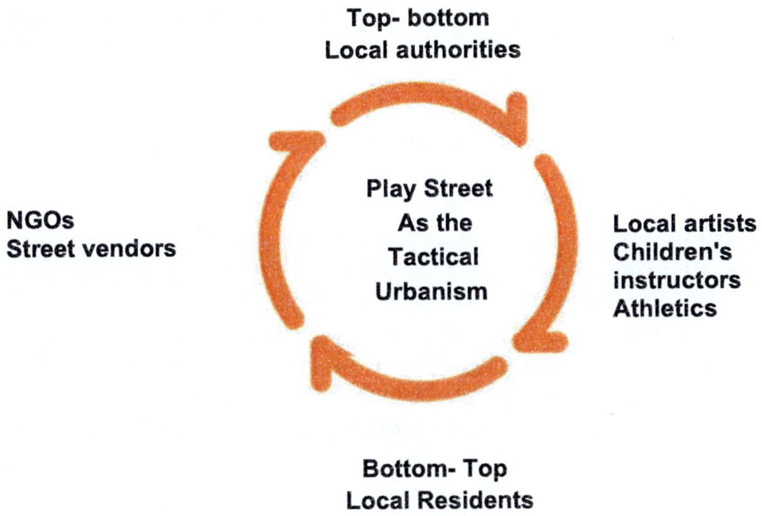

Fig. 22.5 Diagram of participation in the TU experimental event Street in Hands of Children, District 10 Tehran

various stakeholders. They can become a meeting point for top-down and bottom-up approaches.

Events of similar nature can improve resilience through building trust among residents and bringing the public and urban officials to a direct interaction with locals. Such interactions can enhance the legitimacy regarding the governance rules and entities. They also increase the accountability of the officials that would bring clarity in urban governance processes.

The results also suggest that future TU projects require better coordination between local authorities, familiarity of municipalities with tactical initiatives, managing public transport and access to the venue, flexibility in dealing with participants' demands, providing sufficient resources and in general, and all-out support by local authorities via a strategic plan to enhance urban resilience through small-scale initiatives.

22.9 Conclusion

Employing Tactical Urbanism as a tool to enhance resilience calls for reflection on the TU experiment "Street in Hands of Children" in District 10 of Tehran, 2017. Directly and indirectly, the experiment addressed local knowledge, community networks and organisational capacity and communal assets, the pilars of what considered important for community resilience. The experiment was built on the mission to improve the relationship between locals and urban authority. It proved to be a success and can be used as an example of a small-scale initiative to pursue various objectives. Importantly, further evaluation and research on how the TU performed against its initial vision for its target audience and more is thoroughly important. It can help the evolving nature of such events in the process of place-making and resilience towards better initiatives and more effective measures on the urban fabric in future.

To this, planning such events based on the existing social capitals is necessary along with the improvement of the required infrastructures for engaging with and entertaining local general public. To improve the learning capacities in the community, it is necessary to invest in both formal and informal tools and initiatives for various community groups such as children, women and senior citizens in the form of social, cultural, or religious events with educational activities.

Among the limitations of this study, the small number of projects conducted in Iran can be mentioned. The implementation of different TU events and the qualitative and quantitative studies of their effects social resilience can help policymakers, urban authorities, planners and developers to maximise the potentials of tactical urbanism.

References

Aldrich DP, Meyer MA (2015) Social capital and community resilience. Am Behav Sci 59(2):254–269. 10.1177/0002764214550299 (SAGE PublicationsSage CA: Los Angeles, CA)

Asl SS, Lak A (2017) How safe is your neighborhood? Iranian women's perception of safety and security. Mediterr J Soc Sci 8(1):419–430. https://doi.org/10.5901/mjss.2017.v8n1p419

Barber R (2013) Making do: tactical urbanism and creative placemaking in transitional Christchurch, New Zealand. Available at: https://researchrepository.murdoch.edu.au/id/eprint/21075/. Accessed: 19 June 2019

Bonanno GA (2004) Loss, Trauma, and human resilience: have we underestimated the human capacity to thrive after extremely aversive events? Am Psychol 59(1):20–28. https://doi.org/10.1037/0003-066X.59.1.20

Chandra A et al (2011) Building community resilience to disasters: a way forward to enhance national health security. Rand Health Q 1(1):6. Available at: http://www.ncbi.nlm.nih.gov/pubmed/28083162. Accessed: 19 June 2019 (The RAND Corporation)

Cinner J, Fuentes MMPB, Randriamahazo H (2009) Exploring social resilience in Madagascar's marine protected areas. Ecol Soc 14(1). https://doi.org/10.5751/es-02881-140141

Cutter SL et al (2008) A place-based model for understanding community resilience to natural disasters. Glob Environ Change 18(4):598–606. https://doi.org/10.1016/j.gloenvcha.2008.07.013 (Pergamon)

Kennedy G et al (2013) Disaster mitigation: initial response. South Med J 106(1):13–16. https://doi.org/10.1097/smj.0b013e31827cb037

Lydon M, Garcia A (2015) A tactical urbanism how-to. In: Tactical urbanism. Island Press/Center for Resource Economics, Washington, DC, pp 171–208. https://doi.org/10.5822/978-1-61091-567-0_5

Merriam S (1998) Qualitative research and case study applications in education. Revised and Expanded from" Case Study Research in Education.&quot. Available at: https://eric.ed.gov/?id=ED415771. Accessed: 19 June 2019

Moore M, Chandra A, Feeney KC (2013) Building community resilience: what can the united states learn from experiences in other countries? Disaster Med Public Health Preparedness 7(3):292–301. 10.1001/dmp.2012.15

Patel S et al (2017) 'What do we mean by "community resilience"? A systematic literature review of how it is defined in the literature. PLOS Currents Disasters 1–32. https://doi.org/10.1371/currents.dis.db775aff25efc5ac4f0660ad9c9f7db2

Putnam RD (2000) Bowling alone: America's declining social capital. In: Culture and politics, pp 223–234. New York: Palgrave Macmillthe n US. https://doi.org/10.1007/978-1-349-62397-6_12

Sadeghzadeh S, Golkar K, Ghaffari A (2018) Tools for the Governance of urban design: the Tehran experience. Armanshahr 11(22):69–80

Turnbull M, Sterrett C, Hilleboe A (2013) Toward resilience: a guide to disaster risk reduction and climate change adaptation. Available at: https://oxfamilibrary.openrepository.com/bitstream/handle/10546/297422/bk-ecb-toward-resilience-drr-climate-change-adaptation-guide-030113-es.pdf?sequence=3. Accessed: 19 June 2019

Chapter 23
The Preservation of Rural Landscapes for Building Resilience in Small Towns: Insights from North Italy

Paola Branduini and Fabio Carnelli

Abstract In small towns, rural landscape historically has been closely linked to its urban dimension in shaping relationships, local history, local identities and senses of belonging. Furthermore, landscape can be fully considered as heritage, to the extent to which it refers to the tangible and intangible heritage of rural areas. Nonetheless, rural landscapes hitherto have been considered less as a driver of resilience to recover from or react to disturbances, although the values and relationships incorporated in them can contribute to addressing local communities' needs in different terms. For these reasons, in this chapter, the potential of rural landscape in terms of promoting resilience against urban and subsequent landscape changes has been tackled. The study site chosen was the area of the Ticino Park (Parco del Ticino) in north Italy, where water systems and rural landscapes still play an important socio-cultural and economic role and where, at the same time, a traditional agricultural production technique, the marcita meadows, is now at risk of disappearing. Over-industrialisation, overbuilding and a recent project for the construction of a new highway are at present the latest threats. A qualitative research method was adopted to analyse how institutions and local actors' represent and engage in the construction and preservation of the historical rural landscape, in face of the potential loss of tangible and intangible heritage linked to the highway project. The chapter demonstrates how the rural landscape, when acknowledged as component of a landscape system, can be a resilience resource for the local population if understood and valued in terms of local knowledge, as part of the historical and social system.

Keywords Rural landscape · Heritage · Resilience · Small towns · Marcita meadows · Ticino · Italy

P. Branduini (✉)
Politecnico di Milano, Milan, Italy
e-mail: paola.branduini@polimi.it

F. Carnelli
Eurac Research, Bolzano, Politecnico di Milano, Milan, Italy
e-mail: fabio.carnelli@eurac.edu

© The Author(s), under exclusive license to Springer Nature Switzerland AG 2021
F. F. Arefian et al. (eds.), *Historic Cities in the Face of Disasters*, The Urban Book Series,
https://doi.org/10.1007/978-3-030-77356-4_23

23.1 Introduction

The 2012 United Nations Conference on Sustainable Development (Rio + 20) significantly acknowledged how the well-being of local communities is strictly dependent on their cultural heritage, especially for the most vulnerable actors. Furthermore, cultural diversity, like biodiversity for natural systems, has the potential to act on the resilience of social-ecological systems, whose heritage can play a key role in facing or adapting to new and previous disturbances, such as a disaster or a long process of urbanisation. Heritage thus:

> can contribute to sustaining and increasing the adaptation and resilience of rural landscapes by supporting rural and urban inhabitants, local communities, governments, industries and corporations as integral aspect to managing the dynamic nature, threats, risks, strengths and potentialities of such areas (*ICOMOS-IFLA Principles Concerning Rural Landscapes As Heritage*, adopted by the 19th ICOMOS General Assembly, New Delhi, India, 15th December, 2017).[1]

The aforementioned key concepts inspired the research presented here. It is part of a bigger project: the *Reach Culture H2020* (REACH, Horizon 2020) project which, for the Italian pilot actions, is led by the Italian Ministry of Economic Development (MiSE), in collaboration with the Department of Architecture, Built Environment and Construction Engineering (DABC) of Milan Polytechnic University.[2] The aim is to help some communities in Italy where their cultural identity either has been put at risk by disruptive events or affected by a progressive downturn due to changes in the social and economic fabric. For these purposes, two communities threatened by two different disturbances were identified, because they could have potentially similar effects on their rural heritage. The first community examined is a network of small towns in Central Italy (in the areas of Norcia and Amatrice) hit by the 2009 and 2016–2017 earthquakes. This research aims at contributing to the recovery and maintenance of the rich cultural identity and values linked to tangible and intangible heritage, namely the various local handcraft activities and rural production processes that characterise this area. The other community addressed is located in the area of the Ticino Park (*Parco del Ticino*) in north Italy.[3] In this area, a traditional agricultural production technique, the *marcita meadows*,[4] was developed in the Middle Ages by Cistercian monks and local farmers: it is now at risk of disappearing due to over-industrialisation and to the construction of a new highway. Over the past centuries, the *marcita meadows* technique, originating from this area, is supposed to have been passed down to Benedictine monks in Norcia.

---

[1] International Council on Monuments and Sites; International Federation of Landscape Architects.

[2] *Ministero dello Sviluppo Economico.*

[3] *Parco del Ticino.*

[4] The meaning of the name "marcita" has two attributions, following the interpretation of Domenico Berra in "Dei prati del basso milanese detti a marcita" (1822): first interpretation is referred to the last harvest of grass in autumn that was let rot on the meadow; the second one is referred to the swampy aspect of the meadow before being well drained and becoming *marcita meadows.*

This research focusses on the Ticino Park area to analyse how institutions and local actors' engagement in the construction and preservation of historical rural landscapes can play an active role in adapting to or resisting the potential loss of tangible and intangible heritage. It is hoped that discussions and recommendations in this chapter assist institutions and politicians to better involve communities in preserving and enhancing their rural heritage in order to create awareness within heritage communities.

23.1.1 Methodology

To achieve comparable results, the research methodology adopted was that of a pilot action for the aforementioned H2020 project, the "small town" one, adapting the data collection to the study's specific case study and goals. A qualitative research method was adopted, by undertaking focussed ethnographical research, mainly through document analysis (using policy documents, reports, local newspapers, websites and social media) and qualitative interviews. Initiatives promoted by institutions, stakeholders and the public were collected and analysed, and 14 semi-structured classic and walking interviews with the main actors were undertaken. To this extent, geographers and sociologists (Kusenbach 2003; Evans and Jones 2011) have shown the power of walking interviews in highlighting how the relationships between human actors and the surrounding environment can emerge strongly.

Field work was undertaken mainly in Abbiategrasso and Albairate, which are both in the metropolitan area of Milan and would be crossed by the highway: the former is the biggest "small town" of the Milanese Ticino Park area (around 32,000 inhabitants), the latter one of the affected small towns which are populated by between 1,500 and 6,000 inhabitants and are located in the rural area inside or just next to the Ticino Park area.

The questions for our interviews were based on the research questions developed in the Horizon 2020 (H2020) pilot action but adapted to our case study, as follows:

- What is heritage and what do you consider to be protection worth heritage?
- Which (and what kind of) initiatives are organised for heritage promotion/conservation?
- How rural landscape and *marcita meadows* are considered in heritage-making processes? What kind of participation?
- How the highway project is seen and what kind of impact it will have on heritage and rural landscapes?
- How heritage and participation in heritage-making are used to adapt/resist to the disturbances brought by the highway project?

The interviews undertaken were transcribed and through empirical thematic analysis of the collected data it was possible to identify three main issues, as follows: what people consider as heritage; how they perceive their landscape; how they consider the highway project.

23.2 Between Heritage and Resilience: The Rural Landscape

As mentioned, the specific focus was on the potentialities of rural landscape in terms of promoting resilience against urban- and subsequent landscape changes, linking cultural heritage, landscape and resilience. Cultural heritage (CH) should be considered something dynamic, it being the product of cultural aspects which are produced and reproduced by communities over time. As communities change, so does their heritage. The REACH Culture project uses pilot case studies to examine some emerging communities more in depth, to engage with them in participatory activities, to discover and promote best practices and to create the conditions to encourage their commitment beyond the project's end. The objective of project is to advocate socio-economic values of civic participation in preservation, use and management of CH, by exemplifying best practices in the development of resilient policies in community and territorial preservation, education, data management and protection of intellectual rights.

"Resilience" is referred to as the capacity to deal with disturbances or changes without altering the main characteristics of social-ecological systems (Plieninger and Bieling 2012). Consequently, from this perspective, landscape changes are analysed as:

> consequences of a complex interplay between nature and society, following patterns and processes of reciprocal effects, feedback loops, non-linearity, thresholds, surprises, legacy effects, time lags, different degrees of resilience (Plieninger and Bieling 2012, p. 16).

Indeed, in small towns, rural landscape historically is strictly linked to an urban dimension, in shaping relationships, local history and local identities and senses of belonging: it forms a system of built and open spaces that gives to the landscape its uniqueness (Scazzosi 2018).

Furthermore, landscape can be fully considered as heritage, to the extent to which it refers to the tangible and intangible heritage of rural areas. Rural landscape as heritage encompasses physical attributes, as described by the ICOMOS-IFLA *Principles Concerning Rural Landscapes as Heritage*, adopted in New Delhi 15/12/2017:

> the productive land itself, morphology, water, infrastructure, vegetation, settlements, rural buildings and centres, vernacular architecture, transport, trade networks, etc.—as well as wider physical, cultural and environmental linkages and settings. It also includes associated cultural knowledge, traditions, practices, expressions of local human communities' identity and belonging and the cultural values and meanings attributed to those landscapes by past and contemporary people and communities. Rural landscapes as heritage encompass technical, scientific and practical knowledge, related to human-nature relationships.

Nonetheless, rural landscapes are considered less in recovery and reconstruction planning and practices, although the values and relationships incorporated can contribute to addressing local communities' needs in terms of recovering those place-based social-cultural aspects that create a feeling of "being safe at home" again (Carnelli 2012). Moreover, the importance of people's engagement in safeguarding cultural heritage is universally acknowledged: it goes from "the right of everyone to partake freely in the cultural life of the community" (defined in ICOMOS Stockholm Declaration 1998) to the encouragement to be part of heritage community as "people who value specific aspects of cultural heritage which they wish, within the framework of public action, to sustain and transmit to future generations?" (defined in the Faro Convention, Council of Europe 2005). Art. 12 of the Convention highlights the importance of sharing responsibility for cultural heritage and public participation, so it encourages everyone first to "participate in the process of identification, study, interpretation, protection, conservation and presentation of the cultural heritage"; in that sense, "public reflection and debate on the opportunities and challenges which the cultural heritage represents" are also stimulated. These principles stimulate the approach of "participatory governance of cultural heritage", promoted in 2018 by a working group of member states' experts on the EU Commission, as an innovative and creative process, not finalised at quantitative results but focussing on the continuous and ongoing procedure of involving stakeholders in processes usually reserved for and run by experts, officials and politicians.

23.2.1 The Marcita Meadows Landscape: A Case Study

As mentioned before, *marcita meadows* is an ancient practice based on a thin layer of underground and surface water flowing over the meadows (Fig. 23.1); this flow prevents the grass freezing in winter (Fig. 23.1) and guarantees up to seven harvests per year, instead of the usual four or five. It is based on the availability of water in the Po valley plain and especially in the Ticino valley. The availability of water has made it possible to create a fertile and varied landscape, with a mix of cereal fields and meadows. Water control is managed by the work of the *camparo*, a traditional figure that constantly monitors the amount of water and the canal banks. This can be considered a landscape system than interweaves tangible and intangible heritage. Nowadays, it is at risk of disappearing due to changes in cattle feeding, lack of ability to manage it and also availability of winter water. Since the creation of the Ticino Park, this landscape system is under the protection of the Italian law for cultural heritage preservation.[5] The *Piano Territoriale di Coordinamento* (PTC—Territorial Plan) of the Ticino Park (approved by *Deliberazione Giunta Regionale* D.G.R. 5983/2001) protects agriculture both for its multifunctional role and for its use in entrepreneurial activity, which plays an irreplaceable role in the preservation, management and conservation of the territory. Within the area of the park, there are Natura 2000 sites as well as ecological corridors and priority areas for biodiversity within the Rete ecologica nazionale (Regional Ecological Network). The Ticino Park has also been recognised since 2002 as a Biosphere Reserve as part of the UNESCO

[5] *Codice dei Beni Culturali e del paesaggio* (Law 490/99).

Fig. 23.1 Framework of *the marcita meadows*' landscape in the Ticino Valley: underground water, which is warmer than surface water, supplies the canals and irrigates continuously water meadows (in winter and summer) (authors)

MAB Program. Ticino Park promotes many initiatives in order to spread knowledge and the value of the *marcita* meadows landscape, using initiatives such as a travelling exhibition, an itinerary-based bicycle path, conferences and a taught course for new *campari*: and for this latest remarkable action, in 2018, it was awarded a prize by the Ministry of Cultural Heritage in the framework of the Landscape Prize established by the Council of Europe. Therefore, this area is essentially characterised by a rural landscape with a high cultural and historical significance, which is endangered by the construction of the proposed highway.

23.2.2 *The Highway Project*

The highway project is called "*Tangenziale Vigevano-Malpensa*" and stems from a development idea related to the potential growth of the Malpensa International Airport designed in the 1990s. This neoliberal idea of development was mainly linked to heavy urbanisation and industrialisation and was devoted to the concept of high movement of people and goods and delocalisation, with very low consideration of any local environmental or social impacts, and this is one of the main criticisms put forward by local stakeholders, NGO's and some institutions who proposed changes to the project. Initially envisioned in the 1990s, included in the 1999 *Piano Territoriale d'Area Malpensa* (Malpensa Area Territorial Plan) and finally approved in 2008 with the Strategic Law (*Legge Obiettivo,* 443/2001). In parallel, the project has been contested through an appeal at the *Tribunale Amministrativo Regionale* (TAR—Regional Administrative Court) and a petition to the EU Parliament, also based on threats to the UNESCO status of part of this recognised area. Following a lack of funding by the state, the project has been revised and consistently reduced: the route connecting Vigevano and Abbiategrasso (which are both intense commuting areas with Milan) was eliminated, resulting in a new illogical route for the highway, which would connect the Malpensa International Airport to a rural area of extremely important landscape and significant agricultural value in the Ticino park (Fig. 23.2). In 2015, an alternative proposal to reduce the environmental impact of the project (by reducing the length size of the roadway and renovating existing roads and changing the route) was rejected by the Lombardy Regional administration, although substantially supported by local NGO's and institutions, stakeholders and the *Città Metropolitana di Milano* (Metropolitan Council of Milan) together with the Ticino park. Indeed, several criticisms concerning environmental, landscape, social and economic aspects have been consistently put forward by a series of different stakeholders at different levels of governance, as will be discussed in the analysis of our results. Although only some municipalities support the project, over the last 30 years, the Lombardy Region has repeatedly proposed the project to different National Governments and, in theory, work should have finally begun in 2020. Figure 23.3 presents the timeline of the project.

Fig. 23.2 Extract of the highway project made by ANAS close to our field research area with Abbiategrasso on the left side (illustration by authors on Google maps. The highway layout was taken from SILVIA, territorial system for environmental impact evaluation, Lombardy Region)

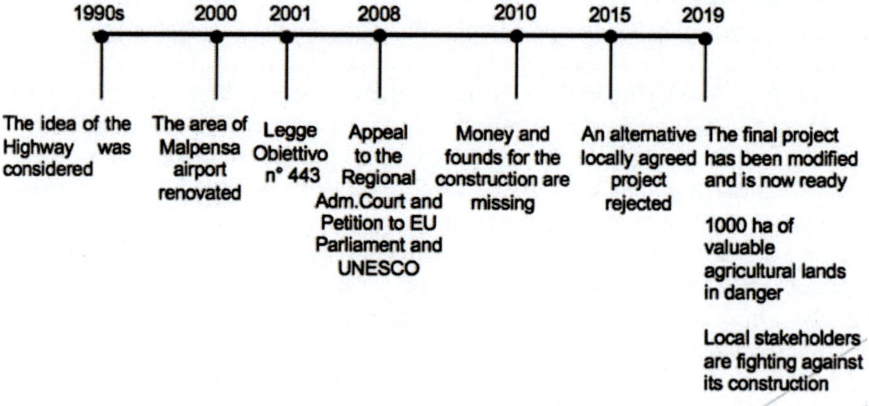

Fig. 23.3 Timeline of the highway project (authors' elaboration)

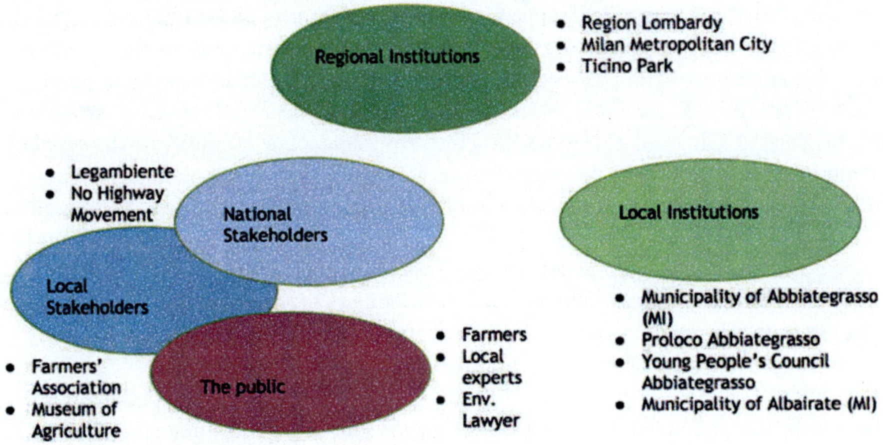

Fig. 23.4 Actors' map of the study area (authors' elaboration)

23.3 Results and Discussion

Following the empirically identified themes that emerged from our data collection and interviews, it was decided to focus on the three most significant: the construction of the concept of heritage, the representation and consideration of rural landscape and *marcita meadows*, the impact of the highway project on these aspects and the relationships between these factors in terms of resilience. Interviewees were also clustered into three governance levels: institutions (at local and regional level[6]), stakeholders (at local, regional and national level) and the public (at local level). Figure 23.4 shows the actors' map that was defined by accessing the field and tracing the relationships between different actors and governance levels: the three different colours show the type of interviewee, while the bullet list identifies the specific interviewee (names were made anonymous due to privacy reasons). In this section, some excerpts from the interviews were reported, which were undertaken in Italian and then translated into English.[7]

Among local institutional actors, the mayor of Abbiategrasso was interviewed (since June 2017 and Municipal councillor 2007–2012) and also Albairate (May 2014 to May 2019), the deputy-chair of the Abbiategrasso Proloco, which is the official agency for the promotion of culture, heritage and tourism in the town since 1999, and a member of the Abbiategrasso Youth Council, an association of young people (18–30 years old) nominated by the Municipal Council to manage cultural activities in the

[6] It also was planned to interview a representative of Region Lombardy—as one of the main institutional actor to have designed this project—but no reply to requests were received.

[7] Around half of the interviews were undertaken by the authors, while the other half were collected with the students of the course "Sociology of the Environment" (M.A. *Sustainable Architecture and Landscape Design*, A.Y. 2018/2019) under the supervision of the professor of the course, one of the authors.

town since 2014.[8] At regional level a member of the Ticino Park management board was interviewed, together with the deputy-chair for the Landscape Commission of the metropolitan city of Milan (2006–2011 and 2014–2015), the deputy-chair (deputy head of mobility, infrastructures and security) of the Metropolitan City of Milan and the current Chair of the Italian Society for Landscape Ecology.[9] Among the stakeholders, a local officer of Legambiente, the major environmentalist association in Italy, was interviewed as well as of a local member of the No-Highway Movement, which was founded 12 years ago.[10] They have collected 15,000 signatures against this project, promoted different petitions and court appeals and also built up a national network on different issues. At local level, the members of the association running the Agricultural Museum of Albairate were interviewed as well as a farmer, whose farm in Albairate would be crossed by the highway and who is also the local representative of a women' association of farmers.[11] Among the public, a farmer in Abbiategrasso who is highly engaged in the promotion of *marcita meadows* was interviewed, a local expert and consultant, who has been commissioned by the No-Highway Movement to do an agronomic evaluation of the project. Last but not least, we also interviewed one of the lawyers who is supporting the appeal to the Regional Administrative Court (TAR).[12]

23.3.1 The Construction of Heritage

Heritage is built in different ways by different actors. Except for the Ticino Park management and the ex-mayor of the smallest town, Albairate, institutional actors mainly see heritage as something linked to the town, referring to the historical built environment of a supposed mediaeval origin. An acknowledgement of rural landscape as heritage was identified, but usually when it was linked to the town, to the extent to which local products, local food initiatives and farmers' markets are considered to be local initiatives of heritage preservation (Fig. 23.5). Though mainly linked to the urban historic centre, the Abbiategrasso Proloco reported a very interesting idea of heritage, which anticipates what the local stakeholders put forward:

> Heritage is the integration between the environment and the people who live there, so it includes almost everything, because nowadays also intangible culture is considered "heritage" (I am thinking of UNESCO, for example, which is definitely going in this direction) and I understand it perfectly because of the lifestyles around here, it is clear that it is very difficult to replicate them in other area. Therefore, from this point of view, "heritage" was born

[8] *Proloco di Abbiategrasso*; *Consulta Giovani di Abbiategrasso.*

[9] *Società Italiana di Ecologia del Paesaggio.*

[10] *No-Tangenziale.*

[11] *Donne in Campo.*

[12] *Tribunale amministrativo regionale.* It is quite evident how the majority of our interviewees are against the project. Since it was quite difficult to find local stakeholders for the project, this reflects reality and has been critical considered in our data collection and analysis, by using a reflexive approach.

Fig. 23.5 Historical centre of Abbiategrasso and its rural landscape with a farmhouse (authors)

right as a product of human relationships with the space that surrounds them (Abbiategrasso Proloco, 6th May 2019).

Local relationships established between inhabitants and their surrounding environments turn out to be fundamental in framing the concept of heritage for whoever has an active relationship with it. Indeed, the Ticino Park considers heritage as "almost everything connected to the landscape as heritage" (Ticino Park, 6th May 2019): the rural landscape with *marcita meadows* (as UNESCO heritage), but also the rural landscape including degraded landscape and ruined rural buildings like *Cascinas*, the old farmhouses.

All stakeholders interviewed mainly identified heritage as everything connected to the rural landscape, mainly derived from a direct (local) knowledge of it: from a general "rural culture" (farmer of Albairate, 6th May 2019), to the "daily life of farmers as rural heritage" (Agricultural Museum of Albairate, 6th May 2019) to a "historical rural vocation of the area as a fundamental element of making people aware of their territory" (No-Highway Movement, 6th May 2019).

A deep local knowledge of the area and local practices turns out to be what establishes the link between heritage and rural landscape. Indeed, among the public we found an original interpretation of the concept of heritage. The farmer of Abbiategrasso we interviewed considered "water" (with its huge water network and different uses) the heritage of this area, to the extent to which water is the wealth and the historical architect of this landscape and its cultural identity: "Water is a fundamental asset, if there was no more water how could we do" (Farmer of Abbiategrasso, 6th May

2019). In this regard, the local agronomist, who did the agronomic evaluation, defined heritage as:

> a combination of factors that act in an ecosystemic way for the metropolitan area and the local area itself: quality, production, work opportunities (out of 30 hectares, 5-10 people work, while in a single-crop company e.g. in Lomellina [a close rural area, NoAs] an area of 200 hectares has barely 1 person working there, having a ratio 1/100 of employment intensity and employment of a certain quality (Local agronomist, 26th July 2019).

Heritage is here acknowledged as a mix of ecosystem services (Braat and De Groot 2012), which can have positive benefits also on local economy and urban areas.

An exception among local institutions was the opinion of the ex-Mayor of Albairate, who considered soil and water to be the heritage of an area historically based on agriculture. On the contrary to Abbiategrasso, here the rural landscape is considered a vivid present value to be preserved; he also mentioned a strong effort to stop land consumption since the 1980s and limit the built urbanised area up to 17% of the administrative area of the town.

23.3.2 The Rural Landscape and Marcita Meadows

Focussing on the *marcita meadows*, again with the exception of the Ticino Park, institutional actors mainly see them as something dead, "a relic of the past" (Abbiategrasso Proloco, 6th May 2019), again linked to a supposed mediaeval identity, a part of a past identity; that is an "old tradition", with an historical and cultural value, somehow lost and not linked to a (potentially) current productive value. A lack of local knowledge both of tangible and intangible aspects of this traditional technique and landscape emerged. For some institutional actors, such as the Metropolitan Council of Milan, this is clearly part of a collective history which designed the rural landscape, but has not got an active role as heritage in the present:

> More than being a type, let us say, of agricultural production, it is the story of the collective intelligence of a territory and the story of the quality and value of the water system in our territory, because then marcita meadows - together with all the canals and the springs - was what generated the design of the landscape. But now I don't think we should duplicate marcita meadows (Metropolitan Council of Milan, 26th July 2019).

On the contrary, the Ticino Park representative highlighted current potentialities of the *marcita meadows*, considering them both a unique, internationally acknowledged value to be preserved for their ecosystemic continuity and a potential future model for a sustainable circular economy for the town:

> Alongside landscapes of this kind, we have rural landscapes of great interest because they are traditional landscapes. In particular, the marcita meadows, which are a remnant of the Medieval period on the one hand but, on the other, open the door to the model of circular economy. Marcita meadows are emblematic for the production of hay for grazing, which in turn gives dairy products, and there is no waste (Ticino Park, 6th May 2019).

The link between past and present is considered to be active in terms of economic practices through traditions. In one famer's opinion, *marcita meadows* as heritage were important for the identity of the place (which is made of hard rural work), and "future generations understand "certain" values: the value of water" (Farmer of Abbiategrasso, 6th May 2019). Again, water becomes an identity-making element through past and present rural work, building relationships through practices.

With the exception of the Museum of Agriculture, all other actors see *marcita meadows* as heritage that can play an active role in the future:

- considered similar to heritage of the town to raise awareness on local key characteristics (No-Highway Movement);
- as both symbol and practice of a place's identity (already recognised as an internationally acknowledged value for ecosystem continuity) which can have a potential role in terms of circular economy and sustainable rural practices for the future (Farmer, Albairate);
- as a symbol of past relationships with a rural past evoking also shame (evoking poverty, bad living conditions, extensive agriculture and hard work) therefore representing a conflict between this rural identity and associated benefits for present and future quality of life—mainly compared to Milan and its Metropolitan Area) (Legambiente).

Again, local knowledge seems to act here as differential knowledge (Carnelli et al. 2020), in modelling the landscape as heritage. Heritage not as a noun but as a verb is a process of remembering and forgetting, which can also act on socio-cultural and economic models as cultural practices (Smith and Akagawa 2018). Indeed, the Museum of Agriculture representative considers *marcita meadows* as "something disappeared", to the extent to which they are part of a past rural life which disappeared. No longer practised, and not turned into a symbol, *marcita meadows* stop being heritage as part of an experienced rural landscape . For these reasons, rural landscapes can be seen in a transformative way, as a trigger of economic development and socio-cultural aspects:

> Heritage is the result, in a reality like ours, of the design of the territory, of the ability of agricultural companies to continue to produce, to have a quality irrigation system, to have ecological connections, ecological passages... the heritage is all these things here, along with the cultural heritage [...] the wealth of the Park of Ticino, together with the South Agricultural Park of Milan (Metropolitan Council of Milan, 26th July 2019).

Therefore, the use of the landscape in terms of heritage initiatives reflects this view. Indeed, in the biggest town, Abbiategrasso, initiatives promoting heritage are mainly linked to either mediaeval folklore festivals or to symbolically "bringing the countryside into the town": rural markets in the historic centre, food festivals with celebrity chefs (here the result is odd); or the landscape is simply used for recreational purposes: a bike ride, a walk along the Naviglio (the big channel providing water to Milan). On the contrary, in Albairate some rural folklore festivals still celebrate this culture: e.g. St. George's feast day is linked to traditional milk production, or a religious procession with the cattle and other animals.

23.3.3 The Highway Project's and Its Impact on the Rural Landscape

Whether or not the rural landscape is practised as heritage, the highway project is mainly seen (by every interviewee, with the exception of the mayor of Abbiategrasso) as a "useless project", "a nonsense". The mayor of Abbiategrasso claims that it would have a bad impact on the environment, but it would benefit citizens (due to the significant lack of existing infrastructure), local industries and traffic in the area. In contrast, the Ticino Park representative claims there is no real need for this highway, that it would not be useful for commuters, but only for private interests; it would have various negative impacts on soil, water and air quality. It would have a disruptive impact on drainage of the water grid and water resurgence and springs; therefore, it would divide the landscape. He clearly stated.

> The project dates to 1993, made when it was a different world that took nothing into account, and we have never been given the opportunity to rethink it significantly with an approach that was different from that of the engineers of those years. It must be reconsidered because there are terrible and immense junctions, there are very high viaducts, the whole thing is very impactful (Ticino Park, 6th May 2019).

The deputy-chair of the Metropolitan Council of Milan was even more critical:

> I trust the Vigevano-Malpensa won't be built as planned, because it is a pointless work […]. It is not necessary to build that type of infrastructure. The real goal is to connect Vigevano and Milan, not Vigevano and Malpensa, or rather that was a goal of 20 years ago. Intelligence says you can't persevere just because you have to spend money. I would like there to be an ex post evaluation, in 10 years I will be right, no industrial development, etc [...] it's just a matter of parties and politics, it's right wing against left wing […]. It'll be an environmental and naturalistic disaster, and for farmers very substantial, needlessly [...] harmful - this is really stupid (Metropolitan Council of Milan, 26th July 2019).

The farmers, who have a practical local knowledge of the area, confirm this and explain very clearly the problems: the extent to which the highway would decrease the level of the aquifer and springs, that it would disrupt the groundwater system and hence (rural) identity and local economy. Indeed, on the one hand, it would disrupt the rural landscape by literally dividing and fragmenting farms, fields, local streets and local relationships between local farmers (the Farmer of Albairate claimed). On the other hand, it will affect a place identity. Technically speaking, as the Ticino Park representative clearly explained:

> It will fragment, divide the landscape, you have areas on either side of the road where there is a problem of soil and water pollution and this being an agricultural landscape this is a major problem. It will provoke air pollution, there also will be visual impacts because this road won't be all at ground level but largely elevated, so it will have a disruptive impact on the water grid. We are in an area of water springs (risorgive) formed because there are underground flows that sometimes meet a layer of clay and rise to the surface. These fragile and sensitive areas will find themselves with an infrastructure that in some points digs in and in others loads huge weights from the road, embankments, pylons and vehicles. This weight crushes the land and the territory risks behaving like a dam towards the water and deviates from the water grid, draining all the resurgences and springs (Ticino Park, 6th May 2019).

So, given that the No-Highway movement also claims that it would disrupt rural and social landscapes (social relationships), it would have an impact on agricultural production and on the water system; therefore, it would accelerate an environmental disaster. From this point of view, which is shared among the farmers, the local experts, stakeholders and some institutional actors, the disruption of the rural landscape would affect social relationships and the local economy; the degradation of this landscape is seen as a potential disaster. On the other hand, as the Proloco representative explained, the project would inevitably change the local identity by changing the rural landscape. It would thus affect traditions and daily life of a society particularly tied to its landscape.

23.3.4 Rural Landscape Enhancing Resilience

First of all, when something is uncertain, little known or unpredictable, people often prefer to look for familiar behaviours or even rituals, which make the "unknown" something finally "known", so that they can evaluate and manage it (Mugnaini 2015). Disasters usually appear exactly "at the point of connection between society, technology and the environment, at the intersection of human practices and environmental materiality" (Gugg 2017), when some social vulnerabilities are at stake (Blaikie et al. 1994). As noted previously, the highway becomes a potential disaster, as it configures itself as the threat of losing something at the intersection between human practices and environmental materiality. Uncertainty is given both by long-term planning (which fragments expectations) and by the breaking of a daily routine and some symbols with an identity connotation, which are both based on this rural landscape. Indeed, as the lawyer supporting the TAR appeal confirms:

> The route passes through an area of environmental value, and this is all Park. There are agricultural areas, the last, at present, agricultural areas that we have got in Lombardy. In addition, there are areas recognised as UNESCO World Heritage Sites, buffering urban interventions. Residual agricultural environment is the most important - and environment, green and agricultural areas have to be protected to ensure health and quality of life of citizens (Environmental lawyer, 10th June 2019).

The value of the "environmental materiality" that survives, identified in terms of rural landscape, increases with the loss of land, so strategies of resilience to preserve this enhances its value in terms of heritage. In terms of potential damage to be resisted, the agronomist interviewed reported to the TAR (Molina 2018) a combination of direct and indirect damages which highlights the strong interrelations of socio-cultural and economic factors incorporated into the rural landscape. The role of the rural landscape as heritage, with *marcita meadows* as a key element (historical and heritage), is acknowledged by the stakeholders as a tool enhancing resilience when it:

- transmits shared values and local knowledge;
- promotes sustainability and raises awareness;

- promotes urban regeneration and stronger local networks for a different economic and social system;
- is part of the mechanism of a circular economy.

Furthermore, water is the element that has been shaped at physical, economic and social level, the rural landscape acknowledged as heritage with its socio-hydrological relationships. Indeed:

> The historical evolution that has led to the current landscape is very much linked to the elements of irrigation designed by Leonardo and then developed in the period prior to the Industrial Revolution at the end of the 19th century (from the Naviglio Grande to the Canale Villoresi), so much so as to indicate it as "Terre dei Navigli" (Navigli´s Areas). The landscape is still varied and harmonious, with important elements indicative of balance between man and nature present in the contexts of traditional agricultural landscapes built over centuries of coexistence: woods, hedgerows, irrigation channels with their vegetation, weaving of the fields intertwined without dominant monoculture elements, size of plots adequate for a fair relationship between cultivation and elements of ecological balance, presence of permanent meadows, in some cases roosters with winter irrigation is still active (Local agronomist, 26th July 2019).

The point is to acknowledge the existence and development of this supposed balance in terms of creation, maintenance and development of local social-ecological relationships, which are incorporated into the rural landscape, as being identified as heritage. A heritage which is conceived as the result of a process:

> through which cultural and natural elements in time receive the status of heritage, elements worthy of being preserved, enhanced and transmitted [...] a construction process inducing both a devaluation, through selection and delimitation, as well as a revaluation involving changes and redefinition, reflecting the needs of the present rather than past configurations (Le Mentec and Zhang 2017, p. 351).

This is exactly the kind of heritage incorporated in this rural landscape: a network made of cultivated fields, farmers, products and relationships shaped by a water network, which historically fuelled economic development, since the seventeenth century, at least. The same water network could drive future sustainable rural and economic practices, if and only if it is locally acknowledged and culturally practised (Smith and Akagawa 2018) by actors having a strong local knowledge of the area, such as the farmers and stakeholders who were interviewed.

23.4 Conclusion

This chapter discussed the potential of the rural landscape to be considered as both heritage and a tool to enhance resilience in facing disrupting events. The results of the interviews undertaken in the Ticino area have highlighted different positions concerning the rural landscape, that ranges from a mere corollary of the urban built heritage (e.g. some of the institutional actors: Abbiategrasso Proloco, Mayor of Abbiategrasso) to a source of production and life (farmers, local agronomist, Ticino Park), to a fundamental part of an historical identity (Mayor of Albairate, partially Proloco, Museum of Agriculture, Metropolitan City of Milan) and a potential resource for a sustainable future society (Ticino Park, No-Highway Movement, Environmentalist lawyer, Legambiente). In terms of reacting to a disturbance, the rural landscape resulted in being not just "the surroundings" of a "small town", but rather an essential component of its landscape system. The landscape system can be a resource of resilience for local people if it is understood and evaluated in terms of local knowledge as a part of the historical and social system. Furthermore, this could embody and transmit tangible and intangible aspects which potentially encapsulate a sense of identity and place and which are essential in order to recover from disturbances. The rural landscape could work as a key factor in resisting potentially disruptive disturbances. It could connect people at local level—it both shapes and is shaped by a social network (which is *inherently* embedded into socio-natural elements such as water). It could connect (and visualise) social-ecological relationships, thereby promoting sustainable urban regeneration and raising local knowledge and awareness. It could also work as source and place of alternative economic approaches, turning the rural landscape as heritage into an active and present element of continuity between a diversified past and a challenging future.

References

ANAS (2008) Progetto definitivo 1° stralcio funzionale. Relazione generale descrittiva, Available at: http://www.legambienteabbiategrasso.org/wp-content/uploads/2015/06/A.1.1-Relazione-generale-descrittiva.pdf. Accessed 6 Sept 2019

Blaikie P, Cannon T, Davis I, Wisner B (1994) At risk. Natural hazards, people's vulnerability and disasters. Routledge, New York

Berkes F, Folke C (eds) (1998) Linking social and ecological systems: management practices and social mechanisms for building Resilience. Cambridge University Press, New York

Berkes F, Colding J, Folke C (2000) Rediscovery of traditional ecological knowledge as adaptive management. Ecol Appl 10(5):1251–1262

Braat LC, De Groot R (2012) The ecosystem services agenda: bridging the worlds of natural science and economics, conservation and development, and public and private policy. Ecosystem Serv 1(1):4–15

Carnelli F (2012) Istantanee etnografiche postsismiche da Paganica (L'Aquila). In: Calandra L (ed) Territorio e Democrazia. Un laboratorio di geografia sociale nel doposisma aquilano. L'UNA, L'Aquila

Carnelli F, Short C, Mugnano S (2020) Local knowledge as key factor for implementing effective nature-based solutions for flood risk mitigation. Rassegna Italiana di Sociologia, LVI (2): 381–406, 2020. https://doi.org/10.1423/97838

Council of Europe (2005) Convention on the value of cultural heritage for society. Faro, Portugal.

Creswell JW, Poth CN (2018) Qualitative inquiry and research design: choosing among five approaches. Sage publications, London

Cutter SL, Ash KD, Emrich CT (2016) Urban–rural differences in disaster resilience. Ann Am Assoc Geogr 106(6):1236–1252

Evans J, Jones P (2011) The walking interview: methodology, mobility and place. Appl Geogr 31:849–858

Gadgil M, Olsson P, Berkes F, Folke C (2004) Exploring the role of local ecological knowledge in ecosystem management: three case studies. In: Berkes F, Colding J, Folke C (eds) Navigating social-ecological systems building resilience for complexity and change. Cambridge University Press, Cambridge

Gugg G (2017) La cima della montagna ha fatto un fumo molto caliginoso (seconda parte). il lavoro culturale (online). Available at: http://www.lavoroculturale.org/vesuvio-incendi-2017/. Accessed 15 Jun 2020

ICOMOS (1998) The Stockholm declaration. Available at: https://www.icomos.org/charters/Stockholm-e.pdf. Accessed 15 Sept 2019

ICOMOS (2017) ICOMOS-IFLA principles concerning rural landscapes as heritage, adopted by the 19th ICOMOS general assembly, New Delhi, India, 15th December, 2017

Klusáková L, Parafianowicz H, Brzozowska H (2019) The strategic use of heritage representations: the small towns of Podłasie province. Urbanities 9(1):54–73

Kusenbach M (2003) Street phenomenology. The Go-along as Ethnographic Tool. Ethnography 4:455–485

Le Mentec K, Zhang Q (2017) Heritagization of disaster ruins and ethnic culture in China: recovery plans after the 2008 Wenchuan earthquake. China Inf 31(3):349–370

McEwen L, Garde-Hansen J, Holmes A, Jones O, Krause F (2017) Sustainable flood memories, lay knowledge and the development of community resilience to future floods. Trans Roy Geog Soc 42:14–28

Molina G (2018) Controdeduzioni agroecologiche, juridical doc, not published

Mugnaini F (2015) A flagello terraemotus, libera nos homo. Dal presagio al calcolo: verso una consapevolezza politica del rischio sismico. In Carnelli F, Ventura S (eds) Oltre il rischio sismico. Valutare, comunicare e decidere oggi. Carocci, Roma

Plieninger T, Bieling C (eds) (2012) Resilience and the cultural landscape understanding and managing change in human-shaped environments. Cambridge University Press, Cambridge

Scazzosi L (2013) Lombardy. In: Agnoletti M (ed) Italian historical rural landscapes, environmental history 1. Springer, Dordrecht, 221–245

Scazzosi L (2018) Landscapes as systems of tangible and intangible relationships. Small theoretical and methodological introduction to read and evaluate rural landscape as heritage. In: Rosina E, Scazzosi L (eds) The conservation and enhancement of built and landscape heritage. A new life for the ghost village of Mondonico on Lake Como. Poliprint, Milano

Smith L, Akagawa N (2018) Introduction. In: Smith L, Akagawa N (eds) Intangible heritage. Key issues in cultural heritage. Routledge, London

Ticino Park (2001) Territorial Plan (PTC), approved by DGR 5983/2001

Wisner B (2009) Local knowledge and disaster risk reduction. Keynote speech delivered at the side meeting on indigenous knowledge global platform for disaster risk reduction, Geneva, 25th June 2009, Available at: www.radixonline.org/resources/WisnerLocalKnowledgeDRR_25-6-09.doc. Accessed 15 Sept 2019

Chapter 24
Antigua Guatemala, from History of Disasters to Resilient Future

Barbara Minguez Garcia

Abstract Disasters have marked the history of Antigua Guatemala since its foundation. Actually, the current location was the third establishment of Santiago de los Caballeros de Guatemala: the cultural, economic, religious, political and educational centre for the entire region during almost three centuries. The capital lasted until the Santa Marta earthquakes in 1773, when it was transferred again and for good to current Guatemala City. However, despite the destruction and partial abandonment of the city in ruins, some inhabitants remained, rebuilt it and started to refer to it as Antigua (Ancient) Guatemala. The regulations on reconstruction preserved the ruins and the city's sixteenth-century Renaissance grid pattern as signs of identity of Antigua, being included in the UNESCO World Heritage list in 1979. Meanwhile, the action of natural hazards didn't end in historic times, and the city kept suffering several earthquakes, hurricanes and mudslides. Nowadays, the area is even more vulnerable due to urban growth and tourism development. A new call for attention was launched due to the Volcán de Fuego eruption on 3 June 2018. Tonnes of ash reached Antigua Guatemala, gathering over the historic cobblestoned streets, squares and buildings. The authorities realised the need to integrate cultural heritage into the recovery plan. The Executive Secretariat of the National Coordinating Office for Disaster Reduction (SE-CONRED) (Secretaría Ejecutiva de la Coordinadora Nacional para la Reducción de Desastres) launched an initiative with the support of the World Bank and the Global Facility for Disaster Reduction and Recovery (GFDRR), to work together with the Ministry of Culture on developing a strategy to integrate disaster risk management (DRM) for cultural heritage. From the international perspective, while there has been significant progress on both disciplines in recent years, in most countries DRM and cultural heritage are not connected, and there is a lack of communication between specialists. This chapter analyses the benefits of integrating cultural heritage into DRM strategies, and vice-versa, developing DRM plans for heritage sites, and the role of international institutions to support this, through the example of Antigua Guatemala in connexion with other international initiatives.

B. M. Garcia (✉)
World Bank/Global Facility for Disaster Reduction and Recovery (GFDRR), Washington D.C., USA
e-mail: Bminguezgarcia@worldbank.org

© The Author(s), under exclusive license to Springer Nature Switzerland AG 2021

F. F. Arefian et al. (eds.), *Historic Cities in the Face of Disasters*, The Urban Book Series,
https://doi.org/10.1007/978-3-030-77356-4_24

Keywords Cultural heritage · Disaster risk management · Resilience · Guatemala

Acronyms

DECORBIC	Department of Conservation and Restoration of Cultural Property
DRM	Disaster risk management
GFDRR	Global Facility for Disaster Reduction and Recovery
GIRRD	Directorate of Integral Management for Disaster Risk Reduction
ICCROM	International Centre for the Study of the Preservation and Restoration of Cultural Property
IDAEH	Institute of Anthropology and History
INSIVUMEH	National Institute of Seismology, Volcanology, Meteorology and Hydrology
MCD	Ministry of Culture and Sports
SE-CONRED	Executive Secretariat of the National Coordinator for Disaster Reduction
SEGEPLAN	Secretariat for Planning and Programming of the Presidency
UCEE	State Building Construction Unit
UNESCO	United Nations Educational, Scientific and Cultural Organization

24.1 Introduction

While hazards are natural, disasters are not. Natural events such as earthquakes or volcanic eruptions cannot be predicted, but they are recurrent. The concept and practice of disaster risk management (DRM)[i] have evolved considerably in the last decades, from a reactive to a preventive approach, recently including the study of climate change[ii] as an aggravating factor in the effects of hydrometeorological events. At the same time, although cultural heritage[iii] is particularly vulnerable to hazards due to its specific characteristics, such as ageing and previous restorations, it is not usually considered as a priority sector, nor taken into account in the DRM plans of most countries.

Guatemala, located in Central America—one of the most hazard-prone regions of the planet—has around 2200 archaeological sites dating from pre-Hispanic times (from around 2000 BC to 1524 AD), in addition to the numerous monuments, buildings and churches from the colonial (1524–1821), republican (1821–1898) and contemporary (1898–1944) eras. Three of them are United Nations Educational, Scientific and Cultural Organization (UNESCO) World Heritage Sites[iv]: the city of Antigua Guatemala, the Archaeological Park and Ruins of Quirigua and Tikal National Park (Fig. 24.1). These sites are exposed to frequent earthquakes, volcanic eruptions, landslides, fires, hurricanes and floods (Crasborn and Navarro 2011; Garcia Acosta 1996).

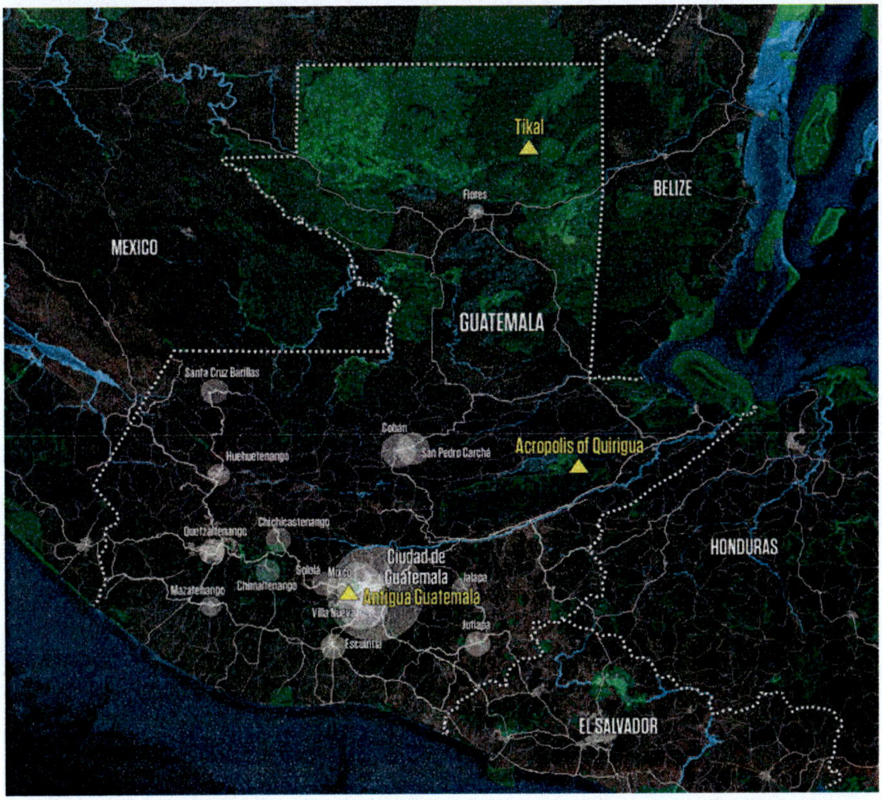

Fig. 24.1 Location of the current capital, Guatemala city, and the three World Heritage Sites: Antigua Guatemala, Ruins of Quirigua and Tikal National Park (map courtesy of Naiara Vegara and Marie Diez, Fundacion Metropoli)

The case of La Antigua Guatemala is a particularly interesting example of the interaction between disasters and heritage. The city has been marked by catastrophes throughout its history, becoming the important legacy that it is nowadays. It was recognised as a World Heritage Site by UNESCO in 1979, in part due to the action of an earthquake.

As this chapter aims to show, in order to face the effects of future hazards and new challenges, such as the pressure of increasing urbanisation, the incidence of tourism and the action of climate change, the integration of the practices of DRM and cultural heritage conservation is fundamental to protect these historic legacies. From the international perspective, while there has been significant progress on both disciplines in recent years, in most countries DRM and cultural heritage are not connected, and there is a lack of communication between specialists. However, international organisations such as UNESCO, the International Centre for the Study of the Preservation and Restoration of Cultural Property (ICCROM) and the World

Bank are promoting and emphasising the benefits of integrating cultural heritage into DRM strategies, and vice-versa, developing DRM plans for heritage sites.[v]

This chapter presents: (i) an overview of the history and disasters that made Antigua Guatemala become World Heritage; (ii) a review of the DRM for cultural heritage methodology based on international examples, applied to the Antigua case; (iii) a reflection on the current situation after the eruption of Volcán de Fuego and the proposal to strengthen resilience through heritage in Guatemala.

24.2 History and Disasters in Antigua

Disasters have characterised the history of Antigua Guatemala since its foundation. The current location was indeed the third settlement of the original capital, Santiago de los Caballeros de Guatemala. The historic evolution is reflected in Table 24.1.

Despite the destruction, transfer of the capital and partial abandonment of the city in ruins, some of the inhabitants remained, were progressively rebuilding their city and began to refer to it as Antigua Guatemala. From the mid-1800s, the area began to be repopulated because of agricultural production, particularly coffee.

Table 24.1 Historic evolution of Antigua Guatemala

Date	Event
1524	**First foundation**: Pedro de Alvarado founded the first capital occupying the Mayan city of Iximche, capital of the Kaqchikel people, renaming it as Santiago de los Caballeros de Goathemalan. However, only a few years later it was abandoned after an indigenous uprising caused some fires in the city
1527	**Second foundation**: After the uprising, Santiago de los Caballeros moved to the Almolonga Valley, to what is Ciudad Vieja today. This new foundation of the capital lasted only fourteen years
1541	**Lahar from the Volcán de Agua**: the lake at the crater overflowed after heavy rains, and a lahar of mud, stones and trees devastated the city and caused the death of 600 people, according to existing sources
1543	**Third foundation**: The new location for the capital was transferred to the Panchoy Valley, where the capital Santiago de los Caballeros was once again founded in what currently is La Antigua Guatemala. The urban plan followed the pattern of straight lines established by the grid of north–south and east–west streets and inspired by the Italian Renaissance. It is one of the best examples of a sixteenth-century city in Latin America
1717	**The "1717 Crisis"**: an eruption of the Volcán de Fuego on 27 August 1717, and a series of earthquakes, including the San Miguel earthquake which took place on the day dedicated to the saint on 29 September destroyed churches, convents and houses, and caused also numerous landslides and lahars on the slopes of the Volcán de Agua
1773	**The Santa Marta earthquake**: comprised a series of earthquakes that caused an important destruction of the city. The authorities decided to move the capital to another area, and it was established in the Hermitage Valley in 1776 of what is the current City of Guatemala today

It is important to highlight that Antigua Guatemala was not just the capital of a country. It was particularly relevant in colonial times, since it played a key role in the historical formation of the territoriality of the modern American continent. Also known as the Kingdom of Guatemala, it encompassed and had jurisdiction over most of Central America through southern Mexico. During more than two and a half centuries, its capital was one of the most important political, economic, religious, educational and cultural centres of the continent. Thanks to the partial abandonment of the city and the regulations prohibiting the repair and construction of new buildings, the ruins and city's sixteenth-century Renaissance grid pattern have been preserved as signs of identity of Antigua.

At present, Antigua Guatemala has numerous cultural attributes, both tangible and intangible,[vi] with different and important associated values, ranging from the ruins, historical buildings, museums and cultural centres, to intangible aspects such as the sixteenth-century Renaissance grid pattern and the cultural landscape, characterised by the volcanoes that surround the city. In addition, one of the unique characteristics of the historic architecture of Antigua is the *Barroco Antigüeño*, which consists of a regional adaptation of the Baroque style, designed to withstand the seismicity in the region.

Antigua was also a centre for the exportation of religious images and statues to the rest of the American continent and to Spain during the seventeenth and eighteenth centuries. This tradition remains present in the celebrations during *Semana Santa* (Easter). The intangible culture of Antigua, especially during these religious festivities, is, therefore, another fundamental sign of identity and attracts an increasing number of visitors every year. These include the traditional *alfombras*—carpets over the cobblestone streets made by dyed sawdust, flowers and vegetables—and the *procesiones*—processions with images representing the Passion of Christ carried on huge wooden platforms.

The recognition of Antigua Guatemala was increasing. It was listed as National Monument in 1944, denominated Monumental City of America in 1965 and included in the UNESCO World Heritage list in 1979 (Fig. 24.2). Likewise, the list of disasters affecting the city kept growing. It suffered the M8.3 and M7.5 earthquakes in 1942 and 1976, respectively, hurricanes Mitch in 1998, Stan in 2005, tropical storm Agatha, mudslides from Volcán de Agua in 2010 and earthquakes and eruptions of Volcán de Fuego in 2017 and 2018.

24.3 Disaster Risk Management for Cultural Heritage

Antigua Guatemala is an example of historical resilience[vii] whose characteristic cultural heritage is partly the result of the continuous history of disasters. However, natural and man-made hazards continue to put the historic legacies at risk. Connecting the practices of DRM and cultural heritage conservation is, therefore, fundamental to protect these unique sites that are also part of the identity of their people.

Fig. 24.2 Three criteria of the outstanding universal value of Antigua Guatemala: *Criterion (ii)* Baroque style of the eighteenth century preserved today as ruins; *Criterion (iii)* city basic grid plan from 1543 has been maintained; *Criterion (iv)* Barroco antigueño: regional adaptation of Baroque designed to withstand the earthquakes. *Source* UNESCO[1]; (author 2015)

A key action for the integration of these two disciplines lies in the necessary collaboration between the respective specialists from both sectors, at the local and national levels. Top-down and bottom-up approaches need to be simultaneous. The integration of cultural heritage in the national DRM plans is as important as the integration of DRM measures into the management plans of the heritage sites.

Cooperation between national and local DRM and cultural heritage agencies and institutions is fundamental, and international organisations play also an important role in fostering this connextion and providing technical support and knowledge exchange.

Basic principles to establish interagency collaborations and develop DRM for cultural heritage encompass:

- **Understanding the risk, as the fundamental first step**. This includes the risk to the physical monuments and the integrity of the site, the cultural and economic activity in and around the site and the well-being of local people and communities. In order to be effective and thorough, the local community[viii] living and working nearby, must be part of the process to identify the risks and then protect and preserve this site, its character and its integrity. Figure 24.3 summarises the equation to calculate risk: Hazards × Exposure × Vulnerability = Risk; where exposure would be the cultural heritage.

---

[1] whc.unesco.org/en/list/65/.

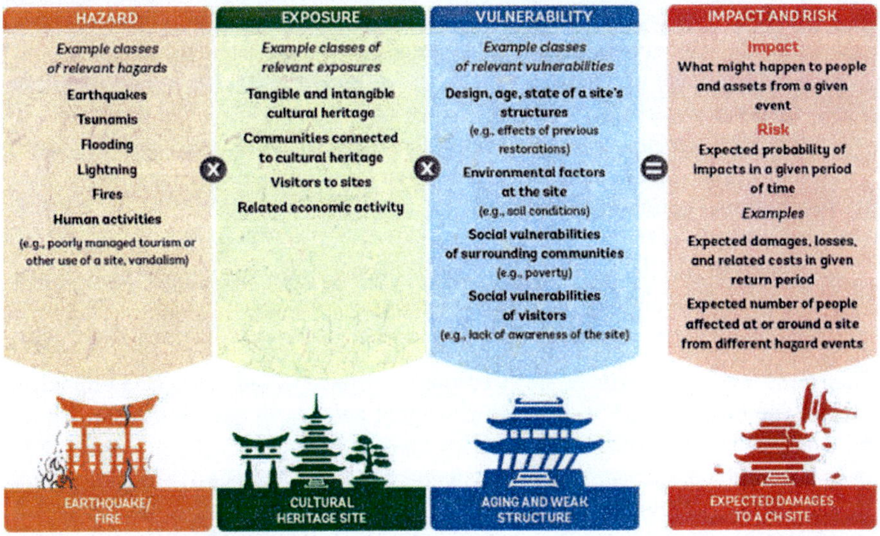

Fig. 24.3 Risk calculation for cultural heritage assets (World Bank 2020)

- **Considering all the components related to the site's culture**. Cultural heritage is not limited to architectural structures; hence, they are not the only ones at risk of being affected by disasters. In addition to tangible (monuments, museums), there is the intangible culture (traditions, celebrations). For example, in the case of Antigua Guatemala, the risk analysis cannot be limited to the ruins of the old city or the heritage buildings, but needs to include other aspects such as vernacular architecture, the characteristic *Barroco Antigüeño*, movable heritage, etc., at the same time, it needs to consider intangible aspects such as traditions and celebrations. For instance, during *Semana Santa* (Easter), the number of visitors unfamiliar with the area increases notably, and they need to be taken into account in emergency response plans.
- **Integrating an assessment of cultural values and their relationship with communities**. Firstly, targeting local, but also national and international communities should be considered, since many cases require the need to prioritise actions. For example, some heritage sites with less architectural value may have greater religious value for the local community. Likewise, some places are considered World Heritage Sites by UNESCO, which provides international recognition. However, cultural heritage is about people and it is critical to understand its meanings and values in order to protect it properly.
- **Following all the phases of disaster management**, including:

  (1) *pre-disaster*: identification of specific risks for heritage including secondary hazards, such as the risk of fire after earthquake—crucial, for example, in old wooden churches with candles; prevention and mitigation measures,

emergency preparedness, evacuation routes, etc., being important to evaluate if the local community is prepared, through drills and training, to assist in the heritage protection;

(2) *during disaster*: the priority is always saving lives; in a second moment, it is essential to properly act on cultural heritage to avoid irreplaceable losses; the action of emergency experts without knowledge of heritage may cause additional unintended damage;

(3) *post-disaster*: cultural heritage specialists and professionals should be in charge of assessing damages and proposing and carrying out restoration treatments, including measures to strengthen resilience in order to avoid or at least mitigate the effects of future disasters.

- **Including an evaluation of both response capacity and technical capacity within the city**. Cultural heritage sites cannot be seen in isolation; the need to be conceived as part of the whole urban context is to ensure that the necessary decisions and measures are taken to safeguard them. This is easier to understand in historic cities such as Antigua Guatemala, where heritage is visible in streets, buildings, monuments, etc., across the whole urban fabric. DRM authorities need to ensure capacity to respond properly to hazardous events in heritage areas in order to preserve their values.

24.3.1 Lessons from Japan

International knowledge exchange and adaptation of lessons learned to the own context may help integrate these principles into actual actions. The case of Japan is a key example and reference to learn about the integration of DRM and cultural heritage disciplines:

- Japan is a very hazard-prone country—an archipelago with over 100 active volcanos and 2000 active fault lines, threatened by earthquakes, volcanic eruptions, tsunamis, typhoons, floods, landslides and fire—and at the same time it has an extensive and important cultural heritage, which includes temples, shrines, castles and many other notable monuments, mainly built in wood. These conditions have motivated actors across Japanese society to develop an extensive experience in DRM for cultural heritage. The World Bank has been documenting this expertise and facilitating knowledge exchange to help some developing countries learn from it. Some key ideas to inspire other countries include the Japanese systems for protection of cultural heritage (institutions and regulations), and DRM strategies and techniques applied to heritage sites, particularly for prevention and mitigation, emergency preparedness, emergency response and reconstruction and recovery (Newman and Minguez Garcia 2017; World Bank 2020).
  In the case of Antigua Guatemala, despite having a very different kind of heritage, some parallelisms can be found, and therefore, some Japanese solutions could be studied and adapted to the local context. For example, the city of Kyoto has

developed very strong firefighting systems due to the great number of wooden constructions. Churches in Antigua, and several other Spanish colonial cities, have important wooden movable heritage to be protected from fire and, therefore, could adapt and apply some Japanese solutions.

- One of the best Japanese examples to illustrate this is the Kiyomizu-dera area in Kyoto, a UNESCO World Heritage Site. The temples have a traditional seismic-resistant architectural design, firefighting and lightning systems are distributed throughout the area and camouflaged to avoid visual impact, and there are stabilisation measures for landslides and monitoring. In addition, innovative programmes are put into place to involve and train the local community in disaster preparedness and heritage protection, which is critical since local people may be the first respondents in case of disaster (Fig. 24.4).

  A learning exchange between Kyoto and Antigua could be very beneficial for the Guatemalan city to improve the integration of DRM and cultural heritage in its own context.

Fig. 24.4 DRM measures in Kiyomizu-dera area in Kyoto (World Bank 2020)

An example of cross-collaboration, building on the Japanese expertise, is the case of Bhutan. This small country in the Himalayas has been working with the World Bank to develop guidelines for integrating DRM in the conservation of their cultural heritage sites. Their vernacular architecture, *lhakhangs* (temples) and *dzongs* (fortresses), as well as *nangtens* (interior cultural heritage assets such as paintings, sculptures and carvings), are integral to daily life and particularly vulnerable to fire. Bhutan is also an earthquake and windstorm prone country.

- The Bhutanese Department of Culture at the Ministry of Home and Cultural Affairs (MoHCA) organised a workshop including participants from different disciplines, including the Department of Disaster Management, Royal Bhutan Police, Royal Bhutan Army, DeSuung, Ministry of Agriculture and Forests, National Centre for Hydrology and Meteorology (NHCM) and District Administrations, to learn from Japan and from each other, working together to find the best solutions for Bhutan's cultural heritage protection.
- Among the key actions they (i) shared current and planned efforts to safeguard cultural heritage and to manage and reduce disaster risk, (ii) worked to improve operational coordination and cooperation and (iii) developed principles and practices to draught guidelines in accordance with the above, focussing on: pre-disaster risk identification, reduction, monitoring and preparedness; post-disaster response and recovery mechanisms for cultural heritage sites.

A similar experience has been developed in Antigua Guatemala, where some findings from this Bhutanese experience were shared, setting the basis to strengthen the resilience of cultural heritage in the Central American country.

24.4 Resilient Cultural Heritage in Guatemala

Despite the damages caused by the eruption of Volcán de Fuego on 3 June 2018, they were not destructive in Antigua Guatemala, but they acted as a call of attention and revealed the need to consider cultural heritage, not only in Antigua but also at the national level, within the sphere of DRM in Guatemala.

The Executive Secretariat of the National Coordinator for Disaster Reduction (SE-CONRED)[2] organised a joint workshop with the Ministry of Culture and Sports (MCD)[3] and the support of the World Bank and GFDRR, inviting other relevant actors from the DRM and cultural heritage sectors in Guatemala. It was held in March 2019, at the Spanish Cooperation Training Centre in La Antigua Guatemala, and attended by representatives of the SE-CONRED, including the Directorate of Recovery and the Directorate of Integral Management for Disaster Risk Reduction (GIRRD),[4] among others; the MCD, including the Directorate of Cultural and Natural

[2] Secretaría Ejecutiva de la Coordinadora Nacional para la Reducción de Desastres.

[3] Ministerio de Cultura y Deportes.

[4] Dirección de Gestión Integral de la Reducción del Riesgo a Desastres.

Heritage,[5] the Department of Conservation and Restoration of Cultural Property (DECORBIC),[6] and the Institute of Anthropology and History (IDAEH)[7]; the Secretariat for Planning and Programming of the Presidency (SEGEPLAN)[8]; the Municipality of Antigua Guatemala; the State Building Construction Unit (UCEE),[9] part of the Ministry of Communications, Infrastructure and Housing,[10] and the UNESCO Office in Guatemala. The participants also identified other institutions related to the management and conservation of cultural heritage and DRM, to be contacted and involved in the next initiatives. The first principle to develop DRM for cultural heritage presented in this paper was, therefore, greatly achieved, by establishing the connexion between multidisciplinary actors.

The main objective of the collaboration established between institutions at this workshop was to ensure the protection of Guatemala's cultural heritage from natural hazards, such as earthquakes, hurricanes, floods and volcanic eruptions, as well as from anthropic hazards, such as fires, vandalism or erroneous post-disaster evaluations, among others. In order to do so, the workshop participants established a roadmap to integrate and strengthen DRM in the cultural heritage of Guatemala.

The roadmap was developed around six main areas of action:

1. Inter-institutional collaboration: It was clearly stated that DRM for cultural heritage is a multidisciplinary process that is not confined to a single institution and must be developed through collaboration and cooperation between different actors. The need to provide technical advice and support between DRM and cultural heritage specialists was noted, and in order to do so, the workshop participants recommended to sign an inter-institutional agreement. This was proposed within the framework of the current legislation and the Guatemala DRM National Plan 2018–2022, whose fifth section (*Eje 5*) integrates the MCD actions. Likewise, the necessity was clear to foster capacity building for both technical staff and also local communities to ensure that there is awareness of the risk affecting cultural heritage, as well as people commitment to protect and conserve it.
2. Generation of information to build databases: During the workshop, participants highlighted the lack of information related to certain aspects required to carry out risk analysis for cultural heritage. For example, it would be necessary to prepare inventories of heritage assets, as well as compilations of historical documentation and previous studies on the one hand, and to develop studies about the natural hazards, the exposure of those heritage assets and their vulnerability on the other hand. In sum, there was a need to develop databases combining both types of information related to heritage and to disaster risk and to keep them updated regularly. Moreover, it would be useful to develop a classification

[5] Dirección General del Patrimonio Cultural y Natural.

[6] Departamento de Conservación y Restauración de Bienes Culturales.

[7] Instituto de Antropología e Historia.

[8] Secretaría de Planificación y Programación de la Presidencia.

[9] Unidad de Construcción de Edificios del Estado.

[10] Ministerio de Comunicaciones, Infraestructura y Vivienda.

or categorisation of the cultural heritage to help decision making on prioritising interventions, together with some guidelines or maintenance manuals for heritage management and conservation. In this regard, Japan might be a good reference with the classification of its cultural heritage into six cultural property categories, which is the foundation for their protection system (World Bank 2020).

3. Risk communication for cultural heritage: Another fundamental point highlighted during the workshop was the need to establish clear communication channels to share the information related to cultural heritage at risk between practitioners and also with local populations. Some proposed measures were to integrate risk and heritage issues in schools and education programmes; include risk management measures in tourism plans, emphasising communication strategies and evacuation plans; to conduct campaigns in local communities located in risk areas, with a focus on those in isolated zones.
4. Pre-disaster technical actions: Regarding technical aspects, the workshop participants divided the specific DRM measures into ex-ante and ex-post to the occurrence of a disaster. Among the ex-ante actions—to establish prevention, mitigation and preparedness measures—the workshop participants proposed the establishment of specific risk identification programmes for cultural heritage; for example, through the selection and prioritisation of heritage sites or buildings to carry out preventive restorations. A key action would be to update emergency response plans by including the advice of cultural heritage experts. Another ex-ante key action would be the development of technical guidelines for the protection of heritage buildings before interventions, as well as training to follow those guidelines for institutional technical staff, and also for the local population.
5. Post-disaster technical actions: Among the ex-post actions—focussed on emergency response and, particularly, to enhance recovery after a disaster—the participants proposed the development of mechanisms for damages and losses assessment of cultural heritage, such as standard forms for rapid evaluation to collect key data and information from movable and immovable heritage assets. In order to do so, it is crucial to maintain communication channels between national and municipal agencies and to develop and execute recovery plans in coordination with the institutions responsible for cultural heritage at both the national and local levels. A good reference for this, might be the *First Aid and Resilience to Cultural Heritage in Times of Crisis* programme and materials developed by the ICCROM.[11]
   Another of the recommended ex-post actions, in line with the ex-ante, would be the development of technical guidelines on the recovery of heritage buildings, and also to organise training to capacitate institutional technical staff, and the local population.
6. Administration and logistics: Finally, the participants identified a series of supporting measures, important both to simplify and facilitate the administrative processes and to improve inter-institutional communication. These actions

[11] iccrom.org/publication/first-aid-cultural-heritage-times-crisis.

include reducing the procedures of providing intervention permissions while ensuring that they are endorsed by cultural heritage conservation professionals; establishing focal points in each relevant institution to ensure constant inter-institutional communication and to coordinate the DRM for cultural heritage activities.

The workshop participants identified a large number of potential actions to be organised in the short, medium and long term. However, they proposed to initially focus on a small number of more specific actions to establish the roadmap that would lead progressively to the integration of DRM and cultural heritage (Minguez Garcia 2019).

This has not been the only measure taken in Guatemala to connect DRM and cultural heritage. In 2013, the Ministry of Culture and Sports, through the Vice Ministry of Cultural and Natural Heritage, had developed a risk management methodology for the effects and impacts of climate change, to be used on archaeological sites and parks (Rojas 2013a). It was developed with the objective to contribute to the improvement of heritage sites management, using the Archaeological Park of Quiriguá, as a pilot case to test it (Rojas 2013b). Considering that the workshop participants were not aware of this initiative it should have been made more widely known to attract follow-up activities and promoted to share experiences and lessons learned to ensure continuity and the possibility to be applied to other sites.

24.5 Conclusion

Guatemala faces the challenge and the opportunity to strengthen the protection of its cultural heritage against the natural hazards that threaten its territory. Earthquakes, volcanic eruptions, landslides, hurricanes and floods—these last two increased by the action of climate change—present a potential danger to a rich cultural heritage conserved and evolved over millennia of history. Mayan archaeological sites, vestiges of great pre-Columbian civilisations, Hispanic colonial cities and among others, the development of a unique architecture which adapted European styles to the characteristics of the American territory, are part of the great Guatemalan cultural heritage to be preserved in the face of the twenty first century new challenges.

DRM for cultural heritage has emerged and improved during the last decades demonstrating effectiveness to protect historical legacies around the world. Central American Guatemala has already started to connect both disciplines, but one of the main challenges remains to keep developing this and moving forward.

Based on experiences and good practices internationally shared to establish DRM for cultural heritage, the initial measures are taken by Guatemala; a series of key actions to solve the remaining gaps can be identified to be adapted to the national context. These actions should be considered as part of a continuous process that may integrate different phases and be based on the collaboration and inter-institutional cooperation already established through the workshop that connected DRM and cultural heritage authorities and professionals from national and local levels. Some of these potential actions to design the next steps would include:

- consolidate a specific working group on DRM for cultural heritage with members from the institutions involved, and invite other relevant actors and agencies to join, such as the National Institute of Seismology, Volcanology, Meteorology and Hydrology (INSIVUMEH).[12] Agree on the basic objectives and common commitment.
- Identify the gaps and information needed to develop a solid DRM strategy for cultural heritage, including: exhaustive inventory of the country's heritage sites/assets classified according to previously established categories; maps and analysis of the natural hazards affecting those heritage sites; vulnerability studies and risk calculations based on the above.
- Develop and implement the joint roadmap with short, medium and long-term priorities. This roadmap should include: (i) the identification of priority actions at the national level, based on the analysis of the current situation, to establish risk reduction strategies for cultural heritage; (ii) the development of the necessary documents to contribute to the process of risk reduction (for example: risk assessment for heritage, guidelines for preparedness, priority actions in case of emergency in heritage sites, etc.); (iii) the establishment of protocols for emergency response and resilient recovery at the national level (including the development of a recovery strategy); (iv) the analysis of specific cases to establish risk reduction measures at the site level (starting with pilot cases) and studies to test the implementation of specific actions; (v) the establishment of a timeline of action; (vi) the collection of lessons learned throughout the process, to be applied in successive phases.
- Ensure that local communities are involved during the whole process, since they are the key to protect their cultural heritage, and often the first available to act in case of emergency. It is critical to consider the traditional knowledge from these communities, about their environment, natural hazards and signs from nature that might alert them on threats or risks. This is indeed part of the intangible heritage that proves very useful in creating resilience in local communities and their cultural heritage sites. It is also important to develop communication campaigns including understanding risk, actions to reduce it and measures to protect cultural heritage sites, taking into account that they are usually touristic areas with people unfamiliar with the area, the hazards and sometimes the language.
- Establish the necessary financing measures to develop and maintain the DRM plans for cultural heritage, including a budget for emergency situations. The investment in protection and promotion of cultural heritage has proven to be profitable. The regeneration of historic centres and cities, including measures to increase resilience, significantly improves the living conditions for both inhabitants and visitors. At the same time, it makes cities become more attractive and

[12] Instituto Nacional de Sismología, Vulcanología, Meteorología e Hidrología.

competitive, which increases the possibilities of attracting private investment and foster job creation.

Natural hazards only become disasters when they affect human life negatively. Identifying and reducing risk, and setting emergency preparedness and response measures makes a big difference, particularly in historic areas. Connecting people, professionals, specialists and local communities to establish effective collaborations is the first step to strengthen the resilience of cultural heritage.

Ultimately, cultural heritage is the reflection of the people who identify with that culture, and protecting it against disasters is to protect the community and preserve its legacy for future generations. The role of the local community is, therefore, essential and needs to be integrated into DRM action plans.

Endnotes—Glossary

i. **Disaster risk management**—the application of disaster risk reduction policies and strategies to prevent new disaster risk, reduce existing disaster risk and manage residual risk, contributing to the strengthening of resilience and reduction of disaster losses—undrr.org/terminology/disaster-risk-management.

ii. **Climate change**—change in the state of the climate that can be identified (e.g., by using statistical tests) by changes in the mean and/or the variability of its properties and that persists for an extended period, typically decades or longer. Climate change may be due to natural internal processes or external forcings, or to persistent anthropogenic changes in the composition of the atmosphere or in land use—archive.ipcc.ch/pdf/special-reports/srex/SREX-Annex_Glossary.pdf.

iii. **Cultural heritage**—relates to both tangible and intangible legacies, including artefacts, monuments, group of buildings and sites—archaeological and underwater—that have a diversity of values including symbolic, historic, artistic, aesthetic, ethnological or anthropological, scientific and social significance, as well as practices, representations, expressions, knowledge, skills and instruments and cultural spaces associated therewith that communities recognise as part of their identity—whc.unesco.org/en/convention/

iv. **UNESCO World Heritage**—UNESCO seeks to encourage the identification, protection and preservation of cultural and natural heritage around the world considered to be of outstanding value to humanity. This is embodied in an international treaty called the Convention concerning the Protection of the World Cultural and Natural Heritage, adopted by UNESCO in 1972. The World Heritage Committee is responsible for the implementation of the World Heritage Convention, defines the use of the World Heritage Fund and allocates financial assistance upon requests from States Parties. It has the final say on whether a property is inscribed on the World Heritage List. It examines reports on the state of conservation of inscribed properties and asks States Parties to take action when properties are not being properly managed. It also decides on the inscription or deletion of properties on the List of World Heritage in Danger—whc.unesco.org/en/about/

v. **International cooperation**—cultural heritage is increasingly exposed to disasters caused by natural and human-induced hazards such as earthquakes, floods, fires, typhoons, theft, terrorism, etc. Recent examples include Earthquakes in Central Mexico in 2017, Kumamoto in Japan, Central Italy and Myanmar in 2016, Nepal earthquake in 2015, UK floods in 2015, Balkan floods in 2014 and ongoing conflicts in Syria and Yemen. In 2005, UNESCO, ICCROM and ACA held a thematic session entitled Cultural Heritage Risk Management at the 2nd World Conference on Disaster Risk Reduction, in cooperation with the Ristumeikan University (Japan) and other partners. This would be the first time cultural heritage was discussed within the larger field of disaster risk reduction policy and programmes—iccrom.org/news/cultural-heritage-and-disaster-resilient-communities. Since then, many other organisations are developing programmes to support countries, cities and local communities.

vi. **Intangible cultural heritage**—includes traditions or living expressions inherited from our ancestors and passed on to our descendants, such as oral traditions, performing arts, social practices, rituals, festive events, knowledge and practices concerning nature and the universe or the knowledge and skills to produce traditional crafts—ich.unesco.org/en/what-is-intangible-heritage-00003.

vii. **Resilience**—The ability of a system, community or society exposed to hazards to resist, absorb, accommodate, adapt to, transform and recover from the effects of a hazard in a timely and efficient manner, including through the preservation and restoration of its essential basic structures and functions through risk management—undrr.org/terminology/resilience.

viii. **Local communities**—Communities are fundamental to developing disaster risk management (DRM) for cultural heritage for two major reasons. First, communities are often the main users and custodians of cultural properties; they interact with and have an interest in it and thus play a crucial role in its conservation and management. Secondly, communities can also respond quickly to disasters, since they know the environment and can reach affected places before emergency teams and authorities can. In short, developing DRM activities at the community level at cultural heritage sites is essential in ensuring preparedness and proper emergency response in case of an event, for the ultimate protection of the cultural heritage assets—gfdrr.org/en/publication/resilient-cultural-heritage-learning-japanese-experience.

References

Crasborn J, Navarro H (2011) Natural hazards and the cultural heritage of guatemala: an overview from the vantage of the Quirigua archaeological park. Mesoweb Articles. www.mesoweb.com/articles/Quirigua/Crasborn-Navarro-2011.pdf. Accessed 1 July 2019

Garcia Acosta V (ed) (1996) Historia y Desastres en América latina. Volumen I. Red de Estudios Sociales en Prevención de Desastres en América Latina. www.desenredando.org/public/libros/1996/hydv1/HistoriaYDesastresVol_I-1.0.0.pdf. Accessed 1 July 2019

Minguez Garcia B (2019) Guatemala—study on disaster risk management of cultural heritage (English). World Bank Group, Washington, DC. documents.worldbank.org/curated/en/516071560920320974/Guatemala-Study-on-Disaster-Risk-Management-of-Cultural-Heritage. Accessed 1 July 2019

Newman J, Minguez Garcia B (2017) Technical deep dive on resilient cultural heritage and tourism: summary report. World Bank, Washington, DC. http://documents.worldbank.org/curated/en/843101528951984265/pdf/127173-WP-P162816-PUBLIC-13-6-2018-16-0-47-ResilientCHTTDDSummaryReport.pdf. Accessed 1 July 2019

Rojas OE (2013a) Metodología de gestión de riesgo climático para Sitios y Parques Arqueológicos. Guatemala, Viceministerio del Patrimonio Cultural y Natural del Ministerio de Cultura y Deportes. https://www.wmf.org/sites/default/files/article/pdfs/WMF%20Micude%20Metodologia%20de%20gestion%20de%20riesco%20climatico%20para%20sitios%20y%20parques%20arqueologicos.pdf. Accessed 1 July 2019

Rojas OE (2013b) Análisis y plan de gestión de riesgo-adaptación ante el impacto del cambio climático del Parque Arqueológico Quiriguá. Guatemala, Viceministerio del Patrimonio Cultural y Natural del Ministerio de Cultura y Deportes. https://www.wmf.org/sites/default/files/article/pdfs/WMF%20Micude%20Analisis%20y%20plan%20de%20gestion%20de%20riesgo-adptacion%20an%20el%20impacto%20del%20cambio%20climatico.pdf. Accessed 1 July 2019

World Bank (2020) Resilient cultural heritage: learning from the japanese experience. World Bank, Washington, DC. https://www.gfdrr.org/en/publication/resilient-cultural-heritage-learning-japanese-experience. Accessed 10 October 2020

Chapter 25
Emergency Management for the Built Heritage Post-earthquake: Emilia-Romagna and Lombardy, Italy

Enrica Brusa

Abstract When an earthquake occurs, after operations to rescue human lives have been completed, it is crucial to implement prompt interventions to secure damaged buildings to avoid further collapses caused by aftershocks. This urgency is particularly true for built cultural heritage since it comprises often old and vulnerable constructions in many cases. The 2012 earthquake in Italy affected several regions with various levels of damage for buildings including the historic ones. The most damaged region was Emilia-Romagna, which was the nearest to the epicentre, but also Lombardy suffered severe damages. Just a few months before the 2012 earthquake, a new organisation in Italy had been set up that for the first time offered the opportunity to test the procedure from various perspectives. Therefore, soon after the earthquake, a new operative process with specific units for the built heritage was activated during the emergency phase. This chapter revisits new procedures defined to deal with earthquake-damaged built heritage during the emergency phase in Lombardy and Emilia-Romagna. It discusses organisations, procedures and competences developed, aiming to identify advantages and limitations which can be useful for other cases.

Keywords Earthquake · Safety countermeasures · Built cultural heritage · Emergency management · Emilia-Romagna · Mirandola (MO) · Church of S. Francesco · Lombardy · Mantova · Church of S. Barbara; Italy

Acronyms

C.C.T.	Territorial Coordination Centre
C.O.T.	Municipal Operative Centre
Di.Coma.C.	Central Command Direction
M.I.B.A.C.	Ministry of Cultural Heritage and Activities (Ministero per i Beni e le Attività Culturali)

E. Brusa (✉)
Department of Architecture and Urban Planning (D.A.St.U.), Politecnico di Milano, Milan, Italy
e-mail: enrica.brusa@polimi.it

© The Author(s), under exclusive license to Springer Nature Switzerland AG 2021

F. F. Arefian et al. (eds.), *Historic Cities in the Face of Disasters*, The Urban Book Series,
https://doi.org/10.1007/978-3-030-77356-4_25

U.C.C.N.	National Crisis Unit (Unità di Crisi a Coordinamento Nazionale)
U.C.C.R.	Crisis Units for Regional Coordination (Unità di Crisi a Coordinamento Regionale)
T.P.C. unit	"Protection of the Cultural Heritage" team of the *Carabinieri* (Nucleo Tutela Patrimonio Culturale)
S.A.F. team	Rescue team of the Italian Fire Brigades (Speleo Alpin and Fluvial)

25.1 Introduction

It is widely known that damage generally caused by an earthquake depends not only on the magnitude of the seismic event, but also on the vulnerability of the buildings. Moreover, the repetition of strong aftershocks can significantly increase the extent of damages (Podestà and Scandolo 2017). Awareness of this issue has grown in Italy after the last earthquakes which struck Emilia-Romagna and Lombardy in 2012 and the Central regions of Italy in 2016, causing the irreversible loss of many historic buildings. Indeed, the 2012 earthquake had several aftershocks. Two of them had a magnitude of about 6.0 and caused the greatest damage to buildings.

The link between vulnerability and aftershocks is particularly dangerous for historic buildings, which are usually the most vulnerable due to both their age and their construction characteristics—i.e. their shape and materials. Hence, it is important to rapidly defining safety countermeasures for historic buildings to limit the damage caused by aftershocks and reduce potential permanent losses.

In recent years, various procedures have been developed in Italy to achieve efficient management of both rescue activities and building safety countermeasures based on knowledge acquired from recent emergencies (Dolce et al. 2015). Since earthquakes are not predictable, it became necessary to establish an efficient strategy designed to allow a prompt reaction to reduce damages to our heritage. In fact, a few months before the 2012 earthquake, a new organisation had been set up that for the first time offered the opportunity to test the procedure from various perspectives.

This chapter revisits how damaged built heritage was dealt with during the first months that follow the 2012 earthquake—the "emergency phase". It focuses on the procedures, selection criteria, intervention methodologies and actors/stakeholders involved in organising the emergency activities and provides examples of specific interventions on historic buildings. Aiming to extract important criteria to be useful for other interventional procedures on built heritage, this study identifies advantages, criticalities and limitations that still need to be improved.

After a summary of the main features of the 2012 earthquake, the chapter presents the operative procedure used by the Italian Government for the emergency. An analysis of the management of the interventions on damaged built heritage is then given, highlighting main actors and available operational tools for technicians. Practical examples of the procedure for built heritage for two prominent churches in Emilia-Romagna and Lombardy will be offered, followed by reflections on important points of the 2012 seismic emergency management.

25.1.1 The 2012 Earthquake

In May 2012, two big earthquakes occurred in Northern Italy with a short interval. The first happened on twentieth May with magnitude 6.1. Its epicentre was in the city of Finale Emilia in the province of Modena in the Emilia-Romagna region. A few days later, on 29th May, a second strong earthquake was recorded with magnitude 5.9 with epicentre always in the province of Modena in the city of Medolla (National Institute of Geophysics and Volcanology 2015).

The most serious effects of the earthquakes were observed in three Italian regions located near the epicentres, precisely Emilia-Romagna, Lombardy and Veneto. Six provinces of these regions were affected by earthquake: Modena, Ferrara, Bologna and Reggio Emilia in Emilia-Romagna; the province of Mantova—and, partially, of Cremona in Lombardy; and the province of Rovigo, which is in Veneto.[1] In the Emilia-Romagna region, 91 municipalities suffered damages caused by the earthquake (Segretariato Regionale Emilia-Romagna 2016), while in the province of Mantova, the damaged municipalities were 31. All these towns registered damages also to their own built heritage (Lattanzi 2014). A summary of the affected municipalities in the provinces discussed by this chapter is shown in Table 25.1.

The greatest damage was recorded in Emilia-Romagna where there were also 27 fatalities although there were few serious collapses among the residential buildings. Industrial buildings also suffered considerable economic damages, mainly due to underestimation of the seismicity of the areas during their construction (Stucchi et al. 2012).[2] Historic buildings too faced severe damages. Two main causes contributed to this: firstly, the vulnerability of old buildings themselves due both to their age and to the consequent physical decay of the construction materials (Podestà et al. 2013); secondly, the repeated strong quakes that hit the buildings a few times. The repetition of a second high magnitude strain compromised the resistance of the historic buildings as they had been already weakened by the presence of the previous damages.[3] The 2012 earthquake affected several regions, with various levels of damage for buildings including the historic ones.

Figure 25.1 shows the epicentres of the seismic events that were registered during the first two months after the shake of 20th May 2012. Yellow circles, squares and stars refer to days 1–30, while the red ones refer to days 31–60.

25.2 Emergency Management in Italy

The National Government in Italy establishes the state of emergency for areas affected by a disastrous event. This authorises to start interventions in derogation of the ordinary law. Generally, the state of emergency has a duration of 60 days. During this period, the Ministry of Internal Affairs, which operates in the field through the Civil Defence System, coordinates first rescue operations. The Civil Defence System is responsible for organising and performing the rescue operations for the population,

Table 25.1 Municipalities where the 2012 earthquakes caused damages to the local built heritage (Author)

Region	Province	No. of municipalities	Municipalities
Emilia-Romagna	Reggio Emilia	14	Boretto, Brescello, Campagnola Emilia, Correggio, Fabbrico, Gualtieri, Guastalla, Luzzara, Novellara, Reggio Emilia, Reggiolo, Rio Saliceto, Rolo, San Martino in Rio
	Modena	19	Bastiglia, Bomporto, Campogalliano, Camposanto, Carpi, Castelfranco Emilia, Cavezzo, Concordia sulla Secchia, Finale Emilia, Medolla, Mirandola, Modena, Nonantola, Novi di Modena, Ravarino, San Felice sul Panaro, San Possidonio, San Prospero, Soliera
	Bologna	17	Argellato, Baricella, Bentivoglio, Bologna, Castel Maggiore, Castello d'Argile, Crevalcore, Gallera, Malalbergo, Minerbio, Molinella, Pieve di Cento, Sala Bolognese, San Giovanni di Piano, San Giovanni in Persiceto, San Pietro in Casale, Sant'Agata Bolognese
	Ferrara	8	Argenta, Bondeno, Cento, Ferrara, Mirabello, Poggio Renatico, Sant'Agostino, Vigarano Mainarda
Lombardy	Mantova	31	Bagnolo San Vito, Borgoforte, Borgofranco sul Po, Castellucchio, Curtatone, Felonica, Gonzaga, Mantova, Marcaria, Moglia, Ostiglia, Pegognaga, Pieve di Coriano, Poggio Rusco, Pomponesco, Quingentole, Quistello, Revere, Roncoferraro, Sabbioneta, San Benedetto Po, San Giacomo delle Segnate, San Giovanni del Dosso, Schivenoglia, Sermide, Serravalle Po, Sustinente, Suzzara, Viadana, Villa Poma, Virgilio

guaranteeing assistance and supporting logistics. It also operates by partnering other National Security Forces, such as the police, army and fire brigades, who are involved in rescue and safety interventions. An active and important role is then offered by voluntary organisations, which are mainly concerned with assisting disaster-affected population.

Once essential medical and logistics services are provided, the public sector staff and technicians conduct the first accessibility verifications on buildings. This is,

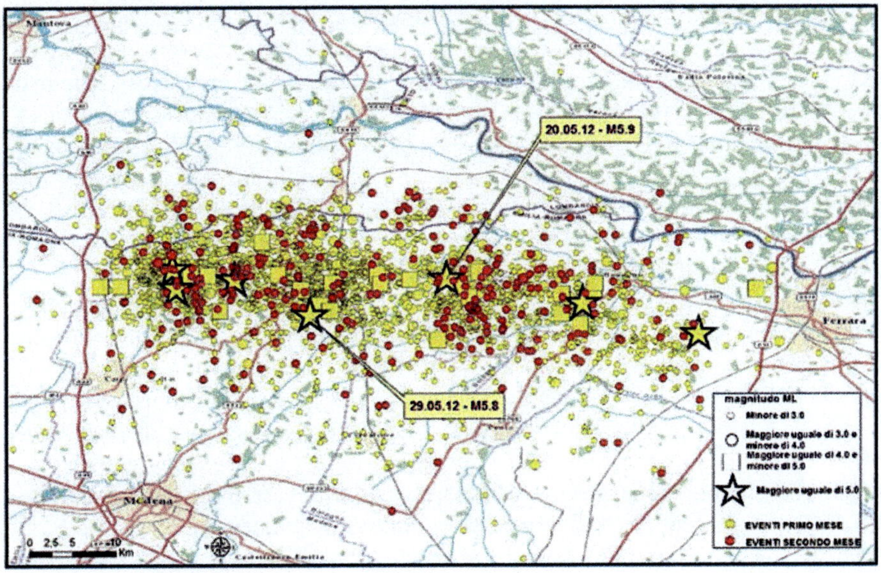

Fig. 25.1 Localisation of the epicentres in the period: 20 May 2012–20 July 2012 (photo by Istituto Nazionale di Geofisica e Vulcanologia—I.N.G.V. Available at: https://ingvterremoti.wordpress.com/2013/03/29/i-terremoti-in-pianura-padana-emiliana-del-maggio-2012-e-la-pericolosita-sismica-dellarea-che-cosa-e-stato-sottostimato/#more-3422)[4]

above all, designed to guarantee access to key public buildings—i.e. hospitals, municipalities, schools—and to inform residents if their own homes are considered safe. In the urgent cases, they issue authorisations necessary to start securing damaged buildings. They are often supported by other experts, such as:

- The army and the law enforcement corps—i.e. the police, the "Protection of the Cultural Heritage" team of the *Carabinieri* (Nucleo Tutela Patrimonio Culturale), *also called "T.P.C." unit*, etc.—for the additional supply of both vehicles and human resource;
- Municipality staff, who support assistance operations and interventions in disaster-affected towns;
- Staff from the "Ministry of Cultural Heritage and Activities", who support the securing damaged cultural heritage, assisted by universities, trained engineers and architects.

To facilitate communications and ensure effective coordination of the interventions, some logistic centres are activated in the affected areas too. The centres end their activity after the 60-day state of emergency has elapsed. Their functions are then transferred to the officials of administrative agencies that are normally present in the territory—i.e. municipalities, prefectures, provinces—and management of those still on-going operations—i.e. viability checks, securing of buildings, reconstruction—is transferred to the presidents of involved regions.

After the 2012 earthquake, the "state of emergency" was declared at two different times, each after the two main quakes. The first one was on 22nd May for the provinces of Modena, Ferrara and Bologna in Emilia-Romagna and for the province of Mantova in Lombardy. The second one was on 30th May for the provinces of Reggio Emilia in Emilia-Romagna and for the Rovigo province in Veneto. Figure 25.2 shows main entities, organisations and structures that were involved in the management of the emergency activities after the earthquakes of May 2012. The structures named as "Di.Coma.C.", "C.C.T.", "C.O.C." are some emergency operative centres, all manged by the Civil Defence System.

During the emergency phase, intervention plans to secure roads and buildings, including both listed and historic ones, are generally drawn up only if they constitute imminent dangers for the success of rescue operations or for public safety. Hence, in 2012, the interventions performed on historic buildings during the first two months mainly concerned constructions that had either been classified as "dangerous" for residents and/or for rescue operations—e.g. buildings that were threatening to collapse on public streets—or when the necessary economic resources were available to implement the countermeasures almost immediately.

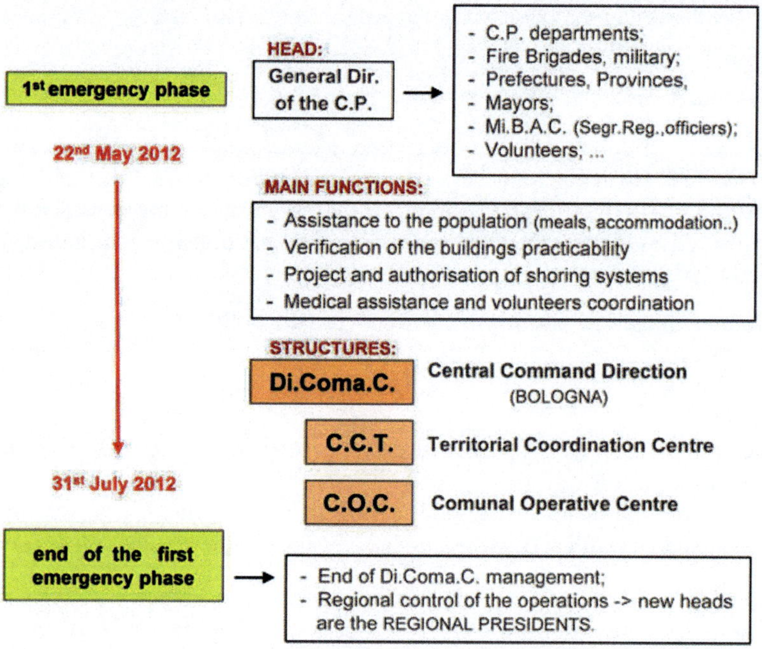

Fig. 25.2 Emergency management diagram (illustration by author)

25.3 Emergency Management for Cultural Heritage: New Approach

Coincidently, in February 2012 just a few months before the earthquakes, a new operating model concerning interventions on cultural heritage was experimented for the first time. This procedure had been developed by the "Ministry of Cultural Heritage and Activities" (Ministero per i Beni e le Attività Culturali—Mi.B.A.C). Conversely to previous practices, this procedure involved the immediate activation of a new Mi.B.A.C. called the national crisis unit (Unità di Crisi a Coordinamento Nazionale—U.C.C.N.) whose purpose was to enhance the efficiency of communications between the Ministry and the other institutions involved. This efficiency enhancement also covered the evaluation of management of interventions and protection of damaged cultural heritage. In so doing, waiting times were also reduced.

Tasks assigned to the U.C.C.N. included monitoring all activities of securing and restoring earthquake-affected heritage in the emergency phase and afterwards until the end of recovery and reconstruction operations. The U.C.C.N took charge of checking the compliance of operating procedures and dealt with the supervision of surveys and cataloguing activities. They were responsible for installing provisional systems and had to manage the subsequent restoration and/or seismic improvement interventions performed on the listed buildings.

In addition to U.C.C.N., it was also decided to create some special regional crisis units to be activated for each disaster-affected region to improve the presence of rescue services in the area. The local units, called "*Crisis Units for Regional Coordination*" (Unità di Crisi a Coordinamento Regionale—U.C.C.R.), count local officers of the Mi.B.A.C. (Ministero per i Beni e le Attività Culturali). They are expected to regulate organisational relations with the local authorities for securing cultural heritage—i.e. members of the civil defence system, fire brigades, T.P.C. unit of the *Carabinieri*, dioceses, local offices of the Mi.B.A.C., etc. The U.C.C.R also aimed to ensure the safety of cultural heritage and to train personnel required to carry out the operations. Furthermore, officers are responsible for communicating with the provincial superintendence.

To manage all the operations, the regional units consist of three different operative sections: "*Operating Unit for damage surveys*", "*Operating Unit for technical coordination*" and "*Operating Unit for temporary warehouses*". Their specific tasks are:

- *Operating unit for damage survey* to organise surveys; catalogue and file damage survey records;
- *Operating unit for technical coordination* to train workforces involved in the surveys (i.e. officers of the Mi.B.A.C. and external technicians and experts) and plan provisional interventions on buildings;
- *Operating unit for temporary warehouses* to choose suitable locations for warehouses; catalogue and restore the movable heritage.

Figure 25.3 presents temporary warehouses that were opened for recovering the

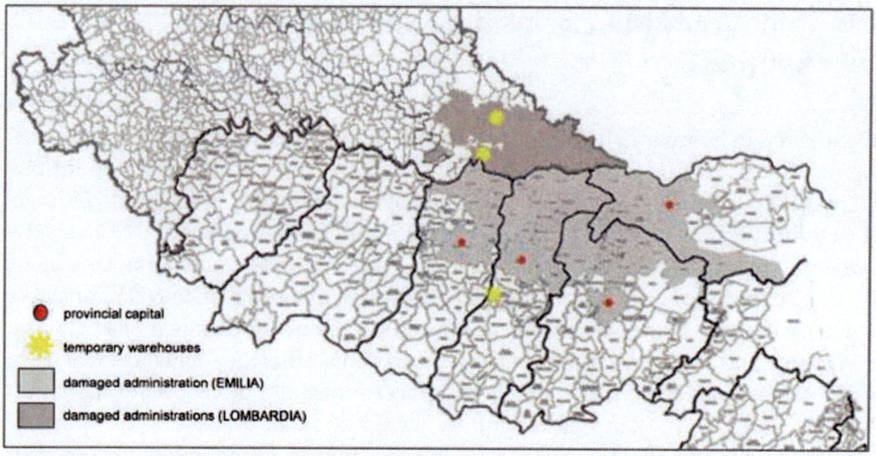

Fig. 25.3 Location of temporary warehouses for the movable heritage damaged in 2012 (illustration by author)

damaged movable heritage are shown. Two warehouses were opened in Lombardy, in the city of Mantova and in the town of Suzzara. While one warehouse was opened in Emilia-Romagna, in the town of Sassuolo. These warehouses are indicated with yellow circles in the map.

Hence, the U.C.C.R. units coordinated activities designed to ensure the safety of damaged heritage—both movable and immovable. They managed both internal personnel training and communications with external experts—i.e. technicians and professionals from universities. They also planned transfer operations of movable heritage into temporary deposits. To carry out all these tasks, each U.C.C.R. has three operative units. A simple diagram of the organisation of the U.C.C.R., together with the tasks of every operative unit, is shown in Fig. 25.4. The same picture, in the two black-framed white boxes, also shows main public entities which collaborate with the U.C.C.N. and U.C.C.R.

Damage surveys on the built heritage were carried out by teams of technicians from other entities: there were one or more specialised technicians from the fire brigade,[5] operators of the civil defence system and at least one staff from the Mi.B.A.C. that could be accompanied by a restorer if required.

These teams were assigned the task of verifying access to buildings and their locations to certify the absence of potential hazards for people, to understand the activated kinematic movements affecting each building, to estimate damages and thus design the shoring systems required to ensure the safety of the buildings.

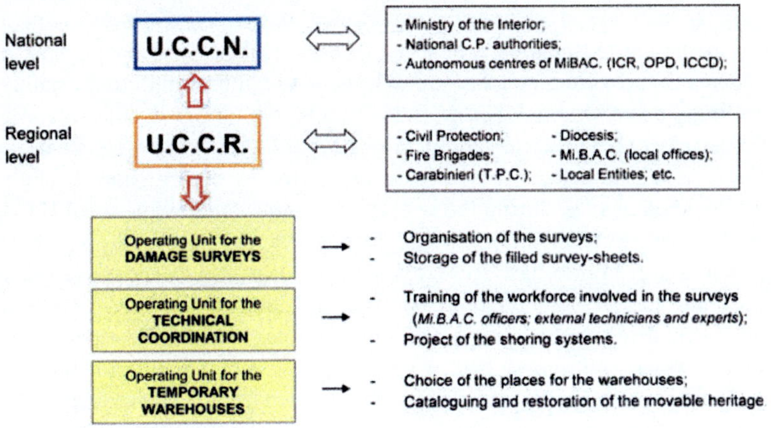

Fig. 25.4 Diagram of the Italian Crisis Units instituted by the Ministry of Cultural Heritage and Activities (Mi.B.A.C.) (illustration by author)

25.4 Surveillance and Cataloguing Tools

According to the new procedures, technicians in regional crisis units (Unità di Crisi a Coordinamento Regionale) should use special cataloguing forms to record kinematic movements activated and the damage to the built heritage. These forms were especially aimed at identifying and classifying both characteristics of buildings and the damage mechanisms that had been activated (Mi.B.A.C.T. 2015). Based on damages witnessed during previous Italian earthquakes, three types of forms were developed (Dip. Protezione Civile and Ministero per i Beni e le Attività Culturali 2013). These forms, which were used also during the 2012 emergency, concern the typical damages that occur in three different building typologies: churches, palaces and movable heritage.[6] The forms concern main built typologies to whom the built heritage could be referred to. Otherwise, specific forms called "AeDES forms" or "forms for the Agibility and the Damage in the Seismic Emergency" (Scheda di Agibilità e Danno nell'Emergenza Sismica) have to be filled when surveying residential buildings that are not listed as built heritage (Dip. Protezione Civile 2008).

Depending on the building typology, these forms record information about construction features—such as location, geometry, materials—and about the damages on the building—i.e. the location of damage, its extent and its possible reasons, etc. A further section of the forms also includes a space to suggest necessary countermeasures to keep the building safe.

Other important operative tools were then the S.T.O.P. manuals (Grimaz et al. 2010a, b). The S.T.O.P. stands for "Technical Forms for the Provisional Systems" (Schede Tecniche per le Opere Provvisionali—S.T.O.P.). This series of manuals have been created by the Italian fire brigades in collaboration with the University of Udine in Italy after the experience of the earthquake that affected L'Aquila in 2009. They concern provisional interventions that can be implemented by the fire brigades. The

S.T.O.P. manuals are based on practical experiences of fire brigade units that worked directly on the field after the most recent Italian earthquakes.[7] These technical forms and manuals help the technicians of the fire brigade to put promptly under safety the listed buildings, protecting them from further collapses and protecting the officers during the surveys (Dolce et al. 2006). So, the S.T.O.P. (Schede Tecniche per le Opere Provvisionali) manuals are a very useful tool for the fire brigades that help the officers of the Mi.B.A.C.T. during the operations coordinated by the local U.C.C.R.

Hence, they constitute an important reference to design the necessary countermeasures for the damaged buildings. These manuals are drawn up based on geometrical and physical parameters—i.e. the dimensions of the damaged building, the shape of the façade, the quality and the materials of the walls, etc.,—and they help technicians identify appropriate solutions to deal with the activated kinematic movements and damages, dimensioning the elements of the provisional systems on the basis of both the dimensions of the damaged walls and the external conditions—i.e. the presence of other buildings, the width and/or the slope of the street and so on.

25.5 Two Examples in Lombardy and Emilia-Romagna

Two interventions carried out during the emergency phase in Lombardy and Emilia-Romagna that shed light on the procedures governing interventions to secure the historic built heritage. One is the Church of Santa Barbara in Mantova, a town in Lombardy that is listed by UNESCO since 2008 for its important monuments and historic centre, built by the Gonzaga dynasty. The second is the Church of San Francesco in Mirandola, a town in Emilia-Romagna situated in the province of Modena, very close to the epicentre of the 2012 earthquakes and known in Italy as the heat of a duchy during the Renaissance and as birth place of the famous humanist Pico Della Mirandola (MO). Those churches exemplify a type of the built heritage that is particularly vulnerable to earthquakes, due to both morphological and construction features. Major collapses occur in such buildings often cause irreparable losses to Italian heritage as churches are greatly exist in all Italian regions and have strong association with common perception on Italian identity.

The Church of Santa Barbara, built between 1562 and 1572, is situated in the historic centre of Mantova, inside the building complex of the Palazzo Ducale. Building cracks opened after the 20th May earthquake, especially on the top of the bell tower. On that very spot, the second quake of 29th May caused the "lantern" above the tambour to collapse. This element had been badly affected by damages of the first earthquake and its supporting structures were compromised. The church is located at a square, which provides access to both a museum and to the local offices of the Mi.B.A.C.. For this reason, adequate interventions and conservation of its bell tower were strongly recommended by authorities. The scope of intervention was to prevent any further collapse in the event of a subsequent earthquake by restoring the structural elements and box-like behaviour. Considering the height of the bell tower and the risk of a collapse of remainder elements, along with the need to safeguard the firemen assigned to perform the task, it was decided to use a long-arm crane to instal the countermeasures, which were pre-mounted on the ground by the fire brigades. The intervention was planned by an interdisciplinary team with members from the Lombardy U.C.C.R., the Diocese of Mantova, the Civil Defence System and the S.A.F.—Speleo Alpin and Fluvial' team (unità Speleo Alpino Fluviale) of fire brigades. The intervention was completed on 14th July 2012, in just over a month (Lattanzi 2014).

The Church of San Francesco from fifteenth century had its floor plan revisited several times—the last intervention was between 1927 and 1938. The church suffered several damages when the earthquakes on 20th and 29th May occurred. In particular, the second quake caused the collapse of its bell tower. In a domino-effect phenomenon, this caused both the roof and part of the nave walls also to collapse. Then the building façade remained standing without the full support of the nave walls. It was at the high risk of collapsing in the churchyard square and blocking access. Therefore, a provisional intervention was required to ensure the accessibility of the square, which is situated along a road of primary importance for the town. The operation was carried out by a specialised fire brigade unit, which pre-mounted the tube-joint countermeasures on the ground, away from the church's façade. Firemen then installed the pre-mounted structures on the façade, using a long-arm crane. The intervention began on 22nd June 2012 and ended in the first half of July, making the square accessible again within two months after the earthquake (Il Resto del Carlino 2012). The internal parts of the church were later secured by installing shoring systems for both the counter-façade and for interior nave wall (Nikravan 2014).

The above examples highlight some criteria, such as the accessibility to urban squares near the damaged buildings, availability of specialised operators and technicians in order to pre-instal the provisional systems surely, timing the securing intervention without losing time—that helped the operators choose suitable solutions to secure the built heritage, underscoring the importance of safety for both the public and the workers. They also show that collaborations between technicians and experts from different institutions—Mi.B.A.C., Civil Defence System, S.A.F. unit of the fire brigades, etc.—permitted to achieve good efficiency standards by working rapidly, using existing certain procedures that had already been experienced previously. Figures 25.5 and 25.6 present the provisional systems placed by the fire brigades in those churches.

25.6 Discussion: Advantages and Limitations

As the gap between the first and second earthquake was only 9 days, in practice the possibility of effective intervention on the built heritage was limited. Yet, the seismic sequence of May 2012 engendered awareness about the need to ensure a timely intervention for protecting cultural heritage after an earthquake. As observed, attention to aftershocks and the ability to face them is also important. The examples

Fig. 25.5 Church of Santa Barbara in Mantova (photo by Geobia available from https://commons.wikimedia.org/wiki/File:3520MantovaSBarbara.jpg)

Fig. 25.6 Church of San Francesco in Mirandola (photo by Ministero dell'Interno available at https://commons.wikimedia.org/wiki/File:Chiesa_di_San_Francesco_ingabbiata_dopo_il_terremoto.jpg)

show how an effective reaction to the damage caused by the first quake can make the difference in the chances of saving a historic building.

From operational perspective, several advantages can be extracted from the 2012 earthquake and the exemplary cases concerning the management of emergency operations for the built heritage. First is the importance of establishing a dedicated office for dealing with the management of a disaster-induced damages suffered by the built heritage, as it is likely to enhance the efficiency of field operations. The creation of this office ensures having a precise informative channel and a targeted decision-making system, without burdening the workload of other offices of the Ministry. Nevertheless, it is important to have clear and consistent rules and criteria during this phase, to obtain an effective standard of coordination between the various forces involved in relief.

Another is the usefulness of informative and diagnostic operative tools such as technical manuals and survey forms derived from the previous post-seismic experiences. This also highlights the importance of learning from past events (Dolce et al. 2015). Moreover, available information about the existing built cultural heritage in an earthquake-stricken area is also important for the emergency management phase. Available catalogues of the local heritage after the 2012 earthquake were essential to this end. For instance, they were referred to during the operations of securing the movable heritage, which were extensively recorded during the first two months after the earthquakes. In those churches, the fire brigade and Civil Defence System teams—together with their volunteers and Ministerial officers—handled not only the removal of statues, paintings and decorations that were present inside the damaged buildings, but also their subsequent transfer into temporary warehouses, such as the museum of the Palazzo Ducale in Mantova, the museum of Suzzara (MN), the Palazzo Ducale of Sassuolo (MO) and in other local museums.

In parallel, the 2012 experiences exposed limitations too. Despite the above awareness, the main operative limitation was that during the emergency phase, interventions on historic buildings were usually implemented only based on safety requirement after a public order criterion. Additionally, in most cases, interventions to secure the built heritage were completed only after the end of the emergency phase, being strictly governed by safety requirements, together with the availability of material means and economic resources.

Another limitation was the imbalance in the provision of qualified human resources with required specialisms. For example, the number of firemen involved in dealing with the built heritage during the emergency was limited to a few dozen technicians, all specialised in the implementation of structural countermeasures. They were of course employed according to the safety priority criterion based on the extent of the damages. The largest number of teams was thus committed to solving the critical issues in the Emilia region, especially in the industrial sector. Subsequently, to avoid scarcity of firemen available for surveys, the Regional Directorate for the cultural heritage of Lombardy—which is one of the Regional headquarters of the Mi.B.A.C.—started a collaboration with structural engineers from universities (Lattanzi 2014). In fact, engaging qualified professionals and specialised workers compensated the low number of firemen, allowing to carry out surveys and to plan the work with an adequate standard of expertise.

Since 2012, a lot has been done in this regard, although the number of available resources for the management of timely interventions to secure damaged built heritage is still not satisfactory, as the recent earthquake of 2016 in central Italy has shown once again (Podestà and Scandolo 2017). This is the case for the special units of the fire brigades and ministry staff assigned to cataloguing and training activities. However, results achieved at that time show, once again, that the path undertaken had several advantages. Hence, it is logical to continue it until an adequate number of both collected data and trained personnel is reached. Furthermore, there is still no "independent" and separate operative path to serve only those securing activities on the listed historic buildings that can be carried out *in parallel to* the existing post-disaster emergency procedures for securing unlisted historic buildings, without either interfering with them. This lack seriously limits the number of promptly implemented countermeasures to deal with damages to the built heritage, as the preservation of a building's cultural value was often considered secondary, if compared to the one of logistic-productive functions around the priority criterion (Capriotti 2014). The operational procedures aimed at securing listed buildings still depend on the availability of means and technicians at both the Civil Protection Department and the fire brigades that as things stand, their resources must be used for both unlisted and listed buildings. This could be a disadvantage for built heritage, since it may provoke delays to staring interventions on the built heritage. However, it is important to remember that the identity of places does not only depend on the economic and administrative frameworks of communities. Indeed, local identity is also significantly characterised by the presence of cultural heritage, which is unique for every community (Vecco 2010). It is, therefore, desirable to progress further, considering the possible solutions that can already be implemented during a "time of normality", without forgetting, as often happens, what learned from previous experiences. For example, some methods of independent financing could be defined to provide for the costs of provisional or restoration works—i.e. through dedicated insurance, maintenance and/or improvement works of their seismic strength capacity, and so on.

25.7 Conclusion

The analysis of procedures developed and practiced for the built heritage during the 2012 emergency phase offered some reflections. One is that the earthquake that affected the two Italian regions of Emilia-Romagna and Lombardy in May 2012 was an important event with regard to the management of interventions for securing and stabilising the built heritage during the seismic emergency phase. This earthquake, characterised by two main shakes of great magnitude close to each other, has raised the awareness that the occurrence of big aftershocks is possible even after a short time and that they could seriously worsen the damage already caused by the first earthquake. It has clearly shown the need to timely secure historic buildings, early before the damage caused by a possible second earthquake becomes irreparable.

Given that the operating model had been developed only a few months earlier by the Ministry of Cultural Heritage and Activities (Ministero per i Beni e le Attività Culturali), it was the first important Italian earthquake during which a new model could be concretely experimented. The creation of the regional office and the operating model was a novelty at that time. According to this model, the emergency management of both the damage surveys and the securing activities on the cultural heritage is entrusted to the U.C.C.R., with the collaboration of other public entities, like Civil Protection and fire brigades.

The examples of churches in Mantova and Mirandola showed, the application of this model could be carried out efficiently and quickly if it is possible to provide required human resources by participant organisations, i.e. the Ministry of Cultural Heritage and Activities (Ministero per i Beni e le Attività Culturali) and fire brigades. They showed an effective intervention strategy can be realised quickly, namely within the first two months after the earthquake. They showed the presence of adequately trained personnel, knowledge of historical cultural heritage that was present in the area, availability of the technical means for the securing interventions and the sharing of the operative criteria by all involved technicians and personnel are crucial to carry out securing interventions on historic buildings.

This model of emergency management for the cultural heritage is still valid in Italy and has been replicated for other earthquakes after 2012. Some critical issues during the emergency management of the earthquake of Emilia-Romagna and Lombardy and in subsequent ones essentially relate the lacking or insufficiency of the above characteristics. Introducing an independent and separate path for the built heritage will potentially assist better planning for interventions to secure damaged built heritage by calling designated means and human resources. The adoption of procedures, developing information and training solutions during the time of normality allow for rapid and timely interventions to prevent any further damages that might be caused by an aftershock or another earthquake. This ultimately contributes to effectively limit the loss of unique and unrepeatable built heritages as representatives of local identities.

Notes

i. Damages amounting to more than 13.2 billion euros were estimated two months after the earthquake. The most damaged region was obviously Emilia-Romagna, which was closer to the epicentres. This area suffered economic losses totalling 11.5 billion, mostly in the provinces of Modena and Ferrara: 2.7 billion due to the losses of the cultural heritage and about 5 billion for damages to the industrial sector. Concerning the other regions, a loss of 980 million was estimated for the Lombardy Region—considering only the province of Mantova—while another loss of 51 million was calculated for the Veneto Region—precisely the province of Rovigo (Regione Emilia-Romagna 2012).

ii. Although there have been several high intensity earthquakes in the past, such as the well-known earthquake that struck the city of Ferrara in 1570, the area between the Via Emilia and the Po river was not classified as an area of high seismicity at least until 2003. This meant that even the industrial buildings—very often characterised by large lights—were not always built according to all the anti-seismic criteria, consequently showing numerous collapses of the sheds and damaging both the goods and the machinery.

iii. The National Institute of Geophysics and Volcanology. (I.N.G.V.) recorded about 2,000 seismic events during the 30 days after the first earthquake. On 20 and 29 May, when the major shaking happened, there were 7 quakes with a magnitude equal to or greater than 5.0: three of them occurred on 20 May, while four were recorded on 29 May (INGV terremoti 2012).

iv. This picture is licensed under the Creative Commons Attribution-No Derivatives 4.0 International License. To view a copy of this license, visit http://creativecommons.org/licenses/by-nd/4.0/ or send a letter to Creative Commons, PO Box 1866, Mountain View, CA 94,042, USA.

v. Fire brigades have different specialised operating units for interventions in many emergency situations. There are at least two specialised units for seismic emergencies, precisely the "Speleo-Alpino-Fluviale" rescue team and a team expert in the "treatment system for structural criticality".

vi. The three forms—"A-DC", "B-DP", "C-BM"—are available online at the following URL: https://www.beniculturali.it/mibac/multimedia/MiBAC/documents/1437986369344_Allegato1_strumenti_schedografici.pdf.

vii. The most recent Italian earthquakes where the Italian fire brigade units worked directly on the field, developing the intervention techniques that are illustrated in the "S.T.O.P." (Schede Tecniche per le Opere Provvisionali) manual are: the earthquake of L'Aquila (2009), the earthquake of Emilia-Romagna and Lombardy (2012), the earthquake of Central Italy (2016).

References

Il Resto del Carlino (2012) Mirandola, al via i recuperi nelle chiese S.Francesco e Gesù. https://www.ilrestodelcarlino.it/modena/cronaca/2012/06/22/733175-mirandola-viaggio-nella-zona-rossa-vigili-del-fuoco.shtml. Accessed 6 July 2019

Capriotti P (ed) (2014) Ricostruire l'emergenza. Cronologia della gestione istituzionale del sisma e sintesi tematica. Regione Emilia-Romagna

Dip. Protezione Civile (2008) Scheda di 1° livello di rilevamento danno, pronto intervento e agibilità per edifici ordinari nell'emergenza post-sismica. http://www.protezionecivile.gov.it/resources/cms/documents/Scheda_AEDES.pdf. Accessed 9 July 2019

Dip. Protezione Civile and Ministero per i Beni e le Attività Culturali (2013). Strumenti schedografici. https://www.beniculturali.it/mibac/multimedia/MiBAC/documents/1437986369344_Allegato1_strumenti_schedografici.pdf. Accessed 15 April 2019

Dolce M, Speranza E, Dalla Negra R, Zuppiroli M, Bocchi F (2015) Constructive features and seismic vulnerability of historic centres through the rapid assessment of historic building stocks. The experience of ferrara, Italy. In: Toniolo L, Boriani M, Guidi G (eds) Built heritage: monitoring conservation management. Springer, London, pp 165–175

Dolce M, Liberatore D, Moroni C, Perillo G, Spera G, Cacosso A (2006) Manuale per le opere provvisionali urgenti post-sisma (OPUS), Roma

Grimaz S, Barazza F, Bellizzi M, Bolognese C, Cavrian M, Caciolai M, D'Odorico A, Maiolo A, Malisan P, Mannino E, Moretti A, Munaro L, Ponticelli L (2010a) Manuale Opere Provvisionali. L'intervento tecnico urgente in emergenza sismica, Roma

Grimaz S, Cavriani M, Mannino E, Munaro L, Bellizzi M, Bolognese C, Caciolai M, D'Odorico A, Maiolo A, Ponticelli L, Barazza F, Malisan P, Moretti A (2010b) Vademecum STOP. Schede tecniche delle opere provvisionali per la messa in sicurezza post-sisma da parte dei Vigili del Fuoco, Roma

INGV terremoti (2012) Terremoto in Pianura Padana Emiliana: alcuni dati dei due mesi di attività sismica. https://ingvterremoti.wordpress.com/2012/07/23/terremoto-in-pianura-padana-emiliana-alcuni-dati-dei-due-mesi-di-attivita-sismica/. Accessed 7 July 2019

Istituto Nazionale di Geofisica e Vulcanologia (2015) CPTI15—parametric catalogue of Italian earthquakes 2015. https://emidius.mi.ingv.it/CPTI15-DBMI15/query_eq/. Accessed 7 July 2019

Lattanzi D (2014) Gestione dell'emergenza sismica e tutela del patrimonio culturale in Lombardia. Attività dopo il terremoto che ha colpito Emilia-Romagna, Lombardia e Veneto il 20 e 29 maggio 2012 (Tesi di laurea). Rel: Farina, P.M., Politecnico di Milano—Scuola di Specializzazione in Beni Architettonici e del Paesaggio, Milano, a.a. 2013–2014

Mi B.A.C.T. (2015) Allegato 1—strumenti schedografici. In: Procedure per la gestione delle attività di messa in sicurezza e salvaguardia del patrimonio culturale in caso di emergenze derivanti da calamità naturali. National law 'Direttiva Mi.B.A.C.T. 23/04/2015'

Nikravan N (2014) Mir.R.O.R.—Progetto di reintegrazione innovativa della chiesa di S.Francesco a Mirandola (MO) (Tesi di laurea). Rell: Belardi P, Gusella V, Ansaloni E, Università degli studi di Perugia, Perugia, a.a. 2013–2014, 77–84

Podestà S, Scandolo L, Moro L (2013) Sulla vulnerabilità sismica, quando l'uomo non ricorda, ci (ri)pensa il terremoto. ANANKE 68:42–51

Podestà S, Scandolo L (2017) L'assenza di opere di pronto intervento: l'evoluzione del danno nelle chiese a seguito degli eventi sismici del Centro Italia. In: Anidis 2017. L'ingegneria sismica in Italia—proceedings of the XVII national conference, Pistoia, pp SS03–24–SS03–33

Segretariato Regionale Emilia-Romagna (2016) Il terremoto in Emilia-Romagna. http://www.emiliaromagna.beniculturali.it/index.php?it/325/il-terremoto-in-emilia-romagna. Accessed 10 July 2019

Stucchi M, Meletti C, Bazzurro P, Camassi R, Crowley H, Pagani M, Pinho R, Calvi GM (2012) The earthquakes of May 2012 and the seismic hazard of the area: anything underestimated? Progettazione Sismica 4(3):63–73

Vecco M (2010) A definition of cultural heritage: from the tangible to the intangible. J Cult Herit 11:321–324

Regione Emilia-Romagna (2012) Terremoto: stimati danni per circa 11,5 miliardi. http://www.regione.emilia-romagna.it/notizie/2012/luglio/terremoto-stimati-danni-per-oltre-11-5-miliardi. Accessed 6 July 2019

Chapter 26
Factors of Educational Poverty and Resilience Responses in L'Aquila's Young Population

Alessandro Vaccarelli, Silvia Nanni, and Nicoletta Di Genova

Abstract Disasters destabilise the lives of both individuals and communities, starting from trauma situations and stressful conditions that do not only characterise the first months of an emergency. Following the 2009 earthquake in L'Aquila, the problem of not only material but also cultural reconstruction became evident, considering the new social and economic dynamics that have greatly destabilised the quality of life of individuals, groups (families) and educational institutions (schools) in the post-emergency. Over the last few years, situations of educational poverty have evidently increased, especially among young people. In this context, several studies showed problems of social and individual fragility. Some educational projects and paths have been activated, aimed at responding to these critical issues and supporting both social and individual resilience. After over a decade from the seismic event, it is therefore possible to begin to reflect, especially with those individuals affected by the earthquake when they were children or teenagers. This chapter presents a study reflecting on the above. The qualitative study included interviews with fifteen girls and boys with the aim of reconstructing their "educational biographies" in the scenario of the reorganisation of urban, family and school life. The intention of the interviews was to search for possible causes of educational successes or failures and understand which forms of resilience—both individual and institutional resilience—and which vulnerability factors to conditions of educational poverty are at stake.

Keywords Education poverty · Resilience · Educational development · Educational biographies · L'Aquila · Italy

A. Vaccarelli (✉) · S. Nanni · N. Di Genova
Department of Human Sciences, University of L'Aquila, L'Aquila, Italy
e-mail: alessandro.vaccarelli@cc.univaq.it

S. Nanni
e-mail: silvia.nannii@cc.univaq.it

N. Di Genova
e-mail: nicoletta.digenova@uniroma1.it

© The Author(s), under exclusive license to Springer Nature Switzerland AG 2021

F. F. Arefian et al. (eds.), *Historic Cities in the Face of Disasters*, The Urban Book Series,
https://doi.org/10.1007/978-3-030-77356-4_26

26.1 Introduction

In recent years, situations of educational poverty have increased evidently, especially among young people, in particular among school-age children. In 2014, *Save the Children* developed, for the first time in Italy, an index to measure Educational Poverty (IPE), which consists of 14 indicators concerning the educational offering and the use of recreational and cultural activities, with the aim of monitoring how much the Italian regions favour, or not, the educational development of children. The Abruzzo region, the focus of this study, is 13th out of a total 18 regions assessed (Save the Children 2014, p. 20). In the same report, educational poverty is defined as the impossibility for children and teenagers to learn, experiment, develop and freely foster their capacities, talents and aspirations.

A fundamental discriminating factor is, therefore, the access to essential services and the quality of the services themselves. With the same household income, it makes a huge difference to have a good network of public services in the territory. Fighting poverty among young people means to offer concretely all children and teenagers, regardless of their parents' income, equal educational opportunities. For that reason, a great investment in education, in a broader sense, from school to services for minors, is unavoidable. The recommendations of the 2018 OECD Report on social mobility also move towards this direction, indicating as a priority for Italy the access to quality education, from kindergarten to tertiary education, for disadvantaged children and young people. The description so far of the construct of educational poverty makes it clear which kind of impact emergency situations have on this phenomenon and how much the occurrence of educational poverty is likely to increase in emergency contexts. The consequences of catastrophes often reflect on issues that are strictly related to equality, or rather equity, in citizens' access to quality educational experiences. Both the earthquakes that struck the city of L'Aquila in 2009 and the one that struck central Italy in 2016 have provided the possibility to circumscribe, through numerous research studies carried out thereon, some areas of vulnerability that may present risk factors of educational poverty closely related to emergency situations.

26.2 Post-emergency and Educational Risk

In the post-earthquake context in L'Aquila and Amatrice in 2016, the idea of emergency pedagogy took shape, today recognised by the scientific community as a specific area that reflects, studies and intervenes in situations of individual and collective criticality framed in catastrophic scenarios.

Several studies have shown cases of social and individual fragility, and both educational projects and pathways have been activated with the aim of responding to situations of educational poverty, in support of social relationships and individual resilience. With regard to post-emergency, the earthquake in L'Aquila has in fact opened up various scenarios. These refer to the relationships between people and urban space, to weaken and unstructure social networks, to the relationship, in short, with a social geography of the post-earthquake that has deeply altered the way of life of the community. The shadow of the *shock economy* and *disaster capitalism* (Klein 2007), fallen on emergency and post-emergency policies, is still producing, ten years after the earthquake, an extremely complex and problematic situation: the reconstruction of the historic centre still far from being completed; the housing displacement as a living condition for many citizens; the reconstruction of schools not yet started and several problems regarding the relocation of schools in usable buildings; economic and working conditions marked by weakness.

In "Children of Katrina", Fothergill and Peek (2015) affirm that in the context of disasters, all aspects and spheres of children's lives go into crisis: family, housing situation, school, peer relationships, health, leisure and extracurricular activities. We therefore understand that the whole system of needs (primary and secondary) is pervasively affected by the extraordinary situation, with a more general system that cannot normally guarantee the answers to the needs themselves: conditions of severe stress or trauma affect both the single subject considered in itself and the other formal and informal actors, activating further elements and variables that add criticality to criticality. In this regard, as an example, a family may go into crisis and adults, who are in turn victims of the event, tend to assume dysfunctional behaviours that can temporarily compromise the normal parenting skills connected to the needs of security and the feeling of a secure base (Langher 2018).

Along with the spheres identified by Fothergill and Peek, we mention the sense of community, which must be considered as a background-concept for all the dimensions previously considered. The whole community enters crisis mode and risks, as often happens, the dispersion on the territory, the redefinition of its identity traits that tend to "flatten out" on the event (in this sense, one becomes an "earthquake victim", a "flood victim", a "displaced person"…), the loss of decision-making capacity and the weakening of its participation in choices, as an effect of the emergency policies marked by the intervention of the *shock economy*. The emergency pedagogic intervention can never be characterised outside the idea of community (cf. Catarci 2018; Nanni 2018a; Tramma 2009), even when it is intended (always and only illusorily) for the single individual.

In emergency situations, a critical factor is represented by the difficulty of the primary group and the extra-familial reference group (e.g. at school, with classmates or friends) to functionally guarantee a secure base, and this happens because catastrophes affect communities in their entirety, "widening" the stress and traumas that they cause on all the community members. In this regard, we often observe:

(a) the alteration of parenting skills: parents experience strong feelings of anxiety that can generate controlling and overprotective behaviours towards their children and/or alternate moments of lax control linked to depression and post-traumatic stress disorder. Moreover, the most vulnerable subjects in catastrophes are those who have scarce social and economic resources, which do not allow them to prepare themselves adequately, respond and face extreme events. It has been widely demonstrated that subjects living in the margins (ethnic minorities, low-income families, etc.) suffer significantly more for the negative consequences of catastrophes (Fothergill and Peek 2015);

(b) the alteration of relational skills by teachers: as well as for parents, even the educational activity of teachers, when they are primary victims and are therefore part of the affected community, might be influenced by emotional states and stressful situations, although some research shows that teachers' responses sometimes tend to be "meliorative" with regard to the exercise of their profession (Mariantoni and Vaccarelli 2018; Salerni and Vaccarelli 2019).

In this perspective, some studies have investigated the living conditions and the adaptation to environmental stress of male and female children in L'Aquila, the development of specific representations of the urban space and the emotions connected to it. Babies born just a short time before or after the earthquake—the "babies of the earthquake" as A. Caporale (2010) defined them referring to the situation in Irpinia, "the generation without a city", as defined by Vaccarelli (2012)—find themselves living in a strongly altered reality, without being aware of it (*Ibidem*). Children are not able to identify their roots, since they lack the necessary memory to build a new representation of the city. We run the risk of letting the children's idea of city be limited to the idea of earthquake as an identifying characteristic and that the sense of belonging may be built within a context generated by the emergency and post-emergency policies. As it happened elsewhere, in Basilicata for example, where, more than 30 years after the disastrous earthquake, today's youth still pay for the unfulfilled promises of development and still perceive a condition intrinsically characterised by a past that they have only heard about (Fondazione MIdA 2012). Hence, the need to move from resilience to resistance (Contini 2009).

Considering children who barely conceptualise (compared with children of other places and contexts) the more general idea of city and lose the spatial and social referencing (Vaccarelli 2015), the research group coordinated by A. Vaccarelli has given attention to various action research experiences, aimed at promoting resilience, sociality and reappropriation of the urban space through a series of activities based on outdoor education methodologies that re-established contact with the idea of city and the urban space itself, promoting resilience and socialisation (Aja González et al. 2016, 2017).

After over ten years from the seismic event that struck the city of L'Aquila and 60 neighbouring municipalities, it is therefore possible to begin to reflect, especially with those individuals affected by the earthquake when they were children or teenagers, and try to understand which paths, difficulties, institutional responses or *resilient energies* may have helped or hindered their educational growth in a city under reconstruction.

This chapter presents a qualitative research that has interviewed 15 girls and boys aged between 17 and 30, in an attempt to reconstruct their educational biographies in the scenario of the reorganisation of urban, family and school life. The intention of the interviews was to search for possible causes of educational successes or failures and understand which forms of resilience (individual or institutional resilience) or which vulnerability factors (internal and external vulnerability) are at stake.

It is about individuals who preserve a memory of the city before the earthquake and who have a clear understanding of the relationship between before and after, but also about individuals who have not had enough time to fully develop their socialisation in a context of life not altered by the disaster and the subsequent policies.

26.3 Earthquake and Educational Risk: Some Qualitative Data

In the stories collected, educational poverty has been investigated on the basis of different existential events, as a condition that directly concerns the interviewees, in different ways. Some stories highlight vulnerability factors linked to the private and individual sphere (mourning, disease, disability, family socioeconomic status), other stories, instead, highlight a great common factor that turns, in fact, into educational risk and structural vulnerability: the earthquake and all its consequences in terms of reorganisation of the individual and social life. By focusing on education, we tried to trace the risk and vulnerability factors, but also the resilience resources in various areas, including the school, the extra-school, the sociality and the look at the future. Given the scope of the study, the family and the university were not focused because they would deserve a separate discussion.

We drew transversal lines of interpretation in all the interviews, outlining at least four macro-areas, from which some excerpts have been taken and subsequently reported. The names of the interviewees are fictitious to protect their privacy.

26.3.1 The School

Immediately after the earthquake, the closing of schools and their reopening in tent cities, or the attendance in schools far away from L'Aquila were often marked by experiences of disorientation, also in view of the educational institution, which did its utmost (especially in the person of managers and teachers) to guarantee the service for students (cf. Di Genova and Vaccarelli 2019).

Maria (24 years old)[1] tells us about her school experience on the Adriatic coast, in classes formed in an emergency situation and composed entirely of students and teachers affected by the earthquake:

> I went to another school in Roseto and... but... it was the... I mean... not the worst experience, but... almost... because I did not have my schoolmates, many teachers were not my school teachers. For example, I was attending the classical high school, and there was no Latin and Greek teacher... there were teachers who were anyway available to... they had also been moved from L'Aquila to the coast... they gave their availability to come to us and then

[1] In accordance with research ethics protocols the names of the interviewees have been changed to protect their identity.

arrived a law teacher that, interesting, EVERYTHING FINE, but still, it was quite heavy because… what do you want to talk about? You could not have normal classes, we basically used to talk about the earthquake.

As well as the frequent relocation over the years in temporary buildings (M.U.S.P.[2]: Temporary modular school buildings) or in buildings "on loan" (sometimes with the need for double shifts), on which political responsibilities have impinged and are still impinging, responsibilities that to this day, after ten years from the earthquake, have not provided yet both for the reconstruction of schools and the full safety of buildings that were stated as usable after the earthquake, but that still present structural problems, which make them, in effect, unusable. The Safe Schools Committee continually denounces[3] this situation, while the political choices for reconstruction to this day see only one project started (the Mariele Ventre School), the fragmentation of an entire school (the Cotugno High School) in buildings that are very distant from each other. Of the 11,500 students, 4,900 adolescents and children still attend lessons in 32 M.U.S.P. Federico, 17 years old, tells us about his experience as a high school student over the last years:

Well, there would be so much to say about that, especially about a couple of things, many of which I am not particularly aware of, because I do not know what led to this decision, surely politics is involved. […]. However, the fact that we are inside that M.U.S.P. is a good thing, but the fact that there's only us of the language high school, isolated, that means a lot. Because, before all this mess happened, we were all together. We were... in the same building […] And yes, four school floors. We have moved from four unusable school floors—we knew it at the end——to a M.U.S.P. with mixed classes, where you cannot even stand a metre away that you feel right inside a can. That is the fact... we are really missing the union of before. By the way, a word went out from the school magazine: "The stew of Cotugno".

26.3.2 Society

Also society is strongly affected by the earthquake, not only because of the logistical and housing problems, the continuous intra- and extra-urban transfers that weaken the previous friendships and social networks but also because the city itself takes another form, it becomes the altered space of social relations: The city centre loses its function in terms of aggregation and sociality often orients itself elsewhere, to commercial galleries, or to friendly relationships cultivated at home.

Immediately after the earthquake, in what Giusti and Montari (2000) define *the honeymoon phase*, *warm* forms of sociality are also experienced, which gradually leave room, however, for relationships marked by spatial anonimity and disorientation that risk becoming social disorientation (cf. Iorio 2018, p. 122). Antonella, 20 years old, tells:

---

[2] Translator's note: M.U.S.P. = Modulo ad Uso Scolastico Provvisorio.

[3] https://news-town.it/cronaca/21190-le-parole-non-bastano-pi%C3%B9-lettera-del-comitato-scuole-sicure-alla-provincia.html.

> No... actually, about the earthquake, since I have been in the in the tent cities of my village... there has been... I had the chance to have even more friends, to be all sticked together, maybe there is also the neighbour who you just used to say: "Hello, how are you?" and that is enough, as we were every day together with them, also the circle of friends was growing, I was talking to guys that I did not talk to before, anyway I felt that being from L'Aquila was, it was taken as a pride thing, everybody felt to be more "aquilan", and all together we were looking for a solution, that is, we were expanding, that is, just like relationships, we were more united, we saw some people who used to argue, but now they were more united in mourning, in city mourning...

Carla, 17 years old, tells instead of a sociality that takes shape in a frayed, decentralised, in some way repelling city:

> Now, we miss a meeting point by the way. I think the meeting point now is the shopping centre, the Aquilone, the Globo or places like that anyway. We are missing points of reference; we are a bit disoriented. For example, if I tell my friends "I will see you in Via Sallustio", maybe they do not even know where it is. Or the main street, they do not know that it is Corso Vittorio Emanuele that reaches the Portici and then Piazza Duomo, they do not know what it is called, so I would say, it is a bit... not embarrassing, but bad that they do not know their city. And this is exactly because they are disoriented. I think it will take a long time before L'Aquila starts to liven up again. However, students are here because there are some good universities, but even at school the girls who do the fifth year already think they want to leave to study elsewhere because here there are few possibilities, and this is an unpleasant image. But we will see what happens.

26.3.3 Extracurricular Training Opportunities

After the earthquake, the city reacts in an extremely diversified way with respect to extracurricular training opportunities, which represent an excellent chance to prevent forms of educational poverty and which can also represent an alternative to the loneliness of those who live in the *Progetto C.A.S.E.*[4] or far from consolidated social networks. The educational planning of associations and organisations, often in synergy with the school and sometimes with the university (Nanni 2018a), has produced significant experiences, which can be summarised as follows:

- opening of schools in the afternoon, with laboratory activities (art, sports, music, etc.);
- opening of the "Centro Punto Luce di Sassa" (by *Save the Children*);
- good reaction from sports clubs;
- projects for the reappropriation of urban spaces (Outdoor activities);
- projects directly aimed at fighting educational poverty, including the "Ony standing places—

The project "Educating beyond the desks" is supported by the Foundation WITH CHILDREN, with whom we work in partnership. This project is part of the sector

[4] Translator's note: Progetto C.A.S.E. = Sustainable and Environmentally friendly Anti-Seismic Complexes Project.

involved in fighting the child educational poverty and is conceived and carried out by the Leading Association "Brucaliffo" and other 15 partners, 7 Associations, 1 Cultural enterprise, 4 Schools, the University and the Municipality of L'Aquila, 1 Foundation (for impact assessment), involving around 1200 children (1160 aged between 9 and 11 and 40 children aged between 11 and 14); about 130 teachers on all schools and at least 1200 families (parents of the children participating in the project).

Most of these activities mainly concerned the children aged between 6 and 13, while for younger children and especially for the adolescents, for the young people and for the adults, the extracurricular training activities were lower numerically and still seem to diminish; placing the so-called young adults and adults in front of a city that has no moments of education mediation or alternatives to a sociality hit by the earthquake. Luca, 30 years old, tells in this regard:

> Well [...] I think there are few recreational activities unfortunately. We do not know how to address the time we could invest in recreational activities and most of the time it is lost, spent on pointless things. Well, we have one of the most beautiful universities in Italy, and students do not know what to do next. I myself... um... then wanting to do something... um... I am powerless because I do not have access to all the activities that are carried out here in L'Aquila, so maybe I do not know many of them.
>
> First of all, the services that the city offers are no longer the same, for example the houses, I am still in the Progetto C.A.S.E. after ten years, once I used to live in my house in Bagno and the house that I had there before was quite large... indeed, it was very large. The house where I am now, is very small, maybe that is why I am always around, because anyway...what can you do? It is very small, it is a hole! I mean, in the end I used to do the same things before, but I was just more at home, of course because it was larger and more pleasant to be there. Now, instead, life meant as a private life, um, that is, when I am at home... I mean, now when I am around, for example, my relationship with the others, that has not changed!

26.3.4 A Look at the Future

The educational poverty seems to be in strong relation with the post-earthquake conditions of the city and is translated not only in the loss of individual potential, but also in a look towards the future that in almost all cases does not bet on the rebirth of the city, "presentizing" both the past and the future: In short, we are not looking at tomorrow. Mario, 30 years old:

> Let us say that L'Aquila is emptied... houses may be ready, I mean, people could live there easily, but they do not go there, out of fear. I know a lot of people who have the possibility to go back to their homes in the centre, but maybe they prefer to rent it to students, rather than go and live it. Well... maybe that is what... I mean, in the sense that people now develop around L'Aquila. They do not return to L'Aquila anymore. That is, L'Aquila is now just a name... that is the centre, you know that there is the centre, but it is not inhabited, that is, it is not lived daily... maybe it is lived because they open new shops, they open pubs, they open anything, but it is not lived as families... I mean... everyday life... in the sense that you live there and then you park there to go home and not go to a shop. And nothing... before it was full of families, now instead you see empty houses, with a sign saying "FOR RENT" at most, or "FOR SALE".

All this gives rise to doubt whether individuals, especially the younger ones, should stay or go and live elsewhere. Federico, again:

> I just tell you this: I see myself in a normal city, because this is not a normal city, I see that I am continuing my studies and I do not know anything about work. I have a clear view of my university pathway, about the job, we will see. But I see myself continuing my studies outside L'Aquila, even though I will have to leave all my friends, but I do not think I am the only one with these ideas.

26.4 Concluding Discussion

With the notion of fighting educational poverty, the focus is not only on resilience skills of individuals but also of institutions, associations and anyone who is publicly involved in education. It means, therefore, to give this concept a political interpretation as well, where it is too often assumed that resilience is an attitude strictly connected to the resources of individuals. This is the reason why the study of vulnerability and resilience factors can lead us to focus on the points that we need to rely upon in the educational design, in search of systemic approaches, by putting at the centre well contextualised goals in relation to the local realities, the social and institutional networks, the actors and the vision for the future.

In this direction, pedagogy can act as a key science on three fronts: (1) the prevention and, therefore, the risk education, the acquisition of knowledge and attitudes that are not only related to the implementation of behavioural procedures ("What to do if ...") but which also raise a wider scope of problems and concern environmental, local and political issues and values closely related to the ethical dimension. (2) the emergency management: On the one hand, the psychological experiences that require also an *educational care* (educating to resilience, redesigning existence, reformulating the traumatic event, etc.), on the other hand, social and territorial issues (reorganising school and educational services, avoiding phenomena like the social marginalisation and the crumbling of communities and social relations, educating to social resistance in the presence of possible political and economic forces suspected of speculation, etc.); (3) the post-disaster management, which often risks assuming the aspect of chronic emergency, when the effects of educational poverty can be more visible.

Dealing with education within a community hit by a disaster (such as geophysical, environmental disasters or wars) means taking an interest in the sense of the community's perspective itself, in working on social identity and, at the same time, on a project of resistance, a "change of scene" that is meliorative and not subtractive compared to before, at least as far as values are concerned (Cf. Vaccarelli 2015).

The concept of—and the invitation to move from resilience to—resistance (Mantegazza 2000; Contini 2009) brings the social, ethical and not least, pedagogical sense of self out, as a direction, path or tool to assert ourselves as citizens, therefore as individuals and as a community and to assert the right to life of wounded territories and communities, as a past, present and future space able to give identity and sense of belonging. From a pedagogical point of view, it deals with an institutional resistance and the need to define and/or enhance good and innovative educational practices, oriented towards the present and the future, which allow to promote a sense of conscious and active citizenship. Resistance also in respect of upstream choices (settlement choices, unreliable construction techniques that even involve school buildings, lack of prevention, etc.), which can represent the "hindsight" for the victims, but that may be, for the rest of a country at risk, the warning and the spur to finally take other directions. "Before" the emergency we find prevention, risk education, ethical sense, without which we cannot go too far from a *status quo* that contemplates—after every seismic event—the catastrophe that could be contained, the dead that could keep on living, the destruction that could be avoided (Vaccarelli 2012).

The risk prevention and education stimulate pedagogy to expand to an idea of space (school, life, urban and working spaces) to be wholly frequented and explored, in which the same idea of care expands from the intimate and immaterial dimension of the relationship to the material and tangible dimension of life and educational context. As Ulivieri claims, it is necessary to develop active and responsible participation in the resolution of problems and relationships, thus favouring the development of a new fundamental, relevant, necessary knowledge, an environmental, both local and global, ethics of mankind for today and for the future, in which man recognises himself daily in his individuality, in his community and in his society (Nanni 2018b; Ulivieri 2015). This is a warning that gives sense, meaning and perspectives back to education in the time of nihilism, a nihilism that can cross, even more severely than usual, contexts affected by disasters.

References

Aja González T, Calandra LM, Vaccarelli A (2016) L'educazione outdoor. Territorio, cittadinanza, identità plurali fuori dalle aule scolastiche. Pensa Multimedia, Lecce

Aja González T, Pardo R, Vaccarelli A, García-Arjona N (2017) L'écologie corporelle dans un contexte urbain de post urgence: une expérience éducative à L'Aquila (Italie). CORPS 15

Caporale A (2010) Terremoti Spa. Dall'Irpinia all'Aquila. Così i politici sfruttano le disgrazie e dividono il Paese. Rizzoli, Milano

Catarci M (2018) Educazione e comunità competente. In: Mariantoni S, Vaccarelli A, di Cura A (eds) Individui, comunità e istituzioni in emergenza. Intervento psico-socio-pedagogico e lavoro di rete nelle situazioni di catastrofe. Franco Angeli, Milano

Contini M (2009) Elogio dello scarto e della resistenza. Pensieri ed emozioni di filosofia dell'educazione. CLUEB, Bologna

Di Genova N, Vaccarelli A (2019) Il contrasto alla povertà educativa nelle emergenze: il ruolo dell'educazione alla resilienza. In: Isidori MV, di Cura A (eds) La formazione dell'insegnante inclusivo. Superare i rischi vecchi e nuovi di povertà educativa. Franco Angeli, Milano

Fondazione MIdA—Osservatorio permanente suk dopo sisma (2012) Lucantropi. Tra il dito e la luna, scelgo la Luna. Salerno

Fothergill A, Peek L (2015) Children of Katrina. University of Texas Press, Texas

Giusti E, Montari C (2000) Trattamenti psicologici in emergenza con EMDR per profughi, rifugiati e vittime di traumi. Sovera, Roma

Iorio C (2018) La relazione d'aiuto in emergenza. In: Mariantoni S, Vaccarelli A, di Cura A (2018) Individui, comunità e istituzioni in emergenza. Intervento psico-socio-pedagogico e lavoro di rete nelle situazioni di catastrofe. Franco Angeli, Milano

Isidori MV, Vaccarelli A (2012) Formazione e apprendimento in situazioni di emergenza e post-emergenza. Armando, Roma

Isidori MV, Vaccarelli A (2013) Pedagogia dell'emergenza/Didattica nell'emergenza. I processi formativi nelle situazioni di criticità individuali e collettive. Franco Angeli, Milano

Klein N (2007) Shock Economy. L'ascesa del capitalismo dei disastri. Rizzoli, Milano

Langher V (2018) Quando i bambini hanno paura e gli adulti sono spaventati: teoria dell'attaccamento e risorse della comunità. In: Mariantoni S, Vaccarelli A, di Cura A (eds) Individui, comunità e istituzioni in emergenza. Intervento psico-socio-pedagogico e lavoro di rete nelle situazioni di catastrofe. Franco Angeli, Milano

Mantegazza R (2000) Filosofia dell'educazione. Bruno Mondadori, Milano

Mariantoni S, Vaccarelli A, di Cura A (2018) Individui, comunità e istituzioni in emergenza. Intervento psico-socio-pedagogico e lavoro di rete nelle situazioni di catastrofe. Franco Angeli, Milano

Nanni S (2018a) Educazione degli adulti, sviluppo di comunità, pedagogia critica. Angela Zucconi e il Progetto Pilota Abruzzo. Franco Angeli, Milano

Nanni S (2018b) Il rischio come possibilità educativa: dallo "spazio interstiziale" alla comunità. In: Mariantoni S, Vaccarelli A, di Cura A (eds) Individui, comunità e istituzioni in emergenza. Intervento psico-socio-pedagogico e lavoro di rete nelle situazioni di catastrofe. Franco Angeli, Milano

Save the Children (2014) La Lampada di Aladino - L'Indice di Save the Children per misurare le povertà educative e illuminare il futuro dei bambini in Italia

Salerni A, Vaccarelli A (2019) Supporting school resilience: a study on a sample of teachers after the 2016/2017 seismic events in Central Italy. J Educ Cult Psychol Stud (ECPS) 19

Tramma S (2009) Pedagogia della comunità. Criticità e prospettive educative. Franco Angeli, Milano

Ulivieri S (2015) La *mission* sociale dell'educazione e della scuola. Pedagogia oggi 2

Vaccarelli A (2012) La generazione dei senza-città. I bambini dell'Aquila dopo il terremoto. In Corsi M, Ulivieri S (eds) Progetto generazioni. Bambini e anziani: due stagioni della vita a confronto. ETS, Pisa

Vaccarelli A (2015) Emotions and representations of "the city" after the 2009 earthquake in L'Aquila: children, education and social reconstruction in a post-catastrophe context. Ricerche di Pedagogia e Didattica—J Theories Res Educ 10:3. Open access: https://rpd.unibo.it/article/view/5913

Vaccarelli A (2017b) Pedagogisti ed educatori in emergenza: riflessioni, stimoli ed esperienze per una professionalità declinata nelle situazioni di catastrofe. Pedagogia oggi 2

Vaccarelli A, Mariantoni S (2018) Children after a natural disaster. Materials for educators and teachers. Franco Angeli, Milano

Chapter 27
Dropout, Resilience and Cultural Heritage: A Focus of the ACCESS Project in a Highly Fragile Area

Antonella Nuzzaci and Iole Marcozzi

Abstract This chapter presents main results of a survey research carried out on the territory of L'Aquila as a highly fragile area because of the 2009 earthquake, to deepen the understanding of the relationship between traumatic event and early school leaving and university leaving. The question is whether and to what extent the dropout is related to the catastrophe situation, and how it relates to the lack of certainty caused by the existential displacement linked to it. Based on the emerging evidences of a European scale project (the Erasmus+ project ACCESS-KA2), it explores issues related to the risk of dropout by university students and examines the reasons relating to their perseverance to attend school or not, to their ability to be able to graduate and to not leave formal education. It emerges of the results as for individuals residing in a highly fragile territory the risk of dropout becomes more pronounced. The territorial positioning of these subjects seems to reinforce the obstacles linked both to the realisation of personal goals and to the ability to work with others to make significant long-term changes. Linking educational dropout risk factors with territorial variables, the chapter offers new perspectives on innovative strategies. These strategies concern the cultural heritage as an instrument of methodological innovation and disciplinary intersection, capable of contributing to removing some obstacles deriving from "disasters" that become "individual and social catastrophes".

Keywords Early school leaving · University leaving · Risk factors · Resilience · Cultural heritage · Earthquake · L'Aquila · Italy

A. Nuzzaci (✉) · I. Marcozzi
University of L'Aquila, L'Aquila, Italy
e-mail: antonella.nuzzaci@univaq.it

I. Marcozzi
e-mail: imarcozzi@ilmiofuturo.it

© The Author(s), under exclusive license to Springer Nature Switzerland AG 2021
F. F. Arefian et al. (eds.), *Historic Cities in the Face of Disasters*, The Urban Book Series,
https://doi.org/10.1007/978-3-030-77356-4_27

27.1 Introduction: Dropout, Earthquake and Resilience

Dropping out of school and university is a multidimensional problem which gets worse in people who have low resilience, accentuating their difficulties such as (1) the inability to achieve goals and the incapacity to ask for help if needed; (2) the failure to establish positive relationships with teachers, peers and friends; (3) the inability to plan, to make choices and to make decisions; (4) the inability to promote independent and responsible learning practices.

Multi-factoriality exists in the sense that the causes derive from the interaction between a set of personal, scholastic, family and contextual factors (Tanner-Smith and Wilson 2013). This interaction explains the importance of validating integration models that facilitate the creation of early diagnostic tools, the identification of risk groups (Bowers 2010) and the design of intervention proposals suitable for different situations. The literature (Gairín et al. 2010; Chen 2012; Dale 2009) has highlighted the lack of a standardised system for measuring dropout and determining causes that lead a student at some time to decide not to continue school (Rumberger 2011) or with a university course. Although specific experiences have been made in many countries, dropout is a clear indicator of a worrying institutional failure. However, there is a lack of systematic knowledge, data and indicators on educational success in Europe, which indeed implies that there is a "successful outcome" of the "training path", in terms of the development of academic skills which are indispensable in life.

Furthermore, the data available throughout Europe on the success of the study are different in terms of availability, methods of data collection, definitions and use, both related to completion rates, and to other indicators of success of the study such as dropout, average time for complete the acquisition of a university degree, etc.

In particular, reports by the Organisation for Economic Cooperation and Development (OECD)—*Education at a Glance Reports, published every year*—must be looked at carefully for differences in the definition and choice of indicators in the contexts of different European countries. In the past, Italy had one of the highest dropout rates in universities in Europe. Its completion rate of 45% of tertiary type A programmes (ISCED 5A) was substantially lower than the OECD average of 69%. However, the latest OECD report (2019) indicates a significant improvement in Italian completion rates, and the success of the study is increasingly becoming an integral part of quality assurance through the integration of completion, conservation and dropout by the different institutions (European Commission 2017, 2014, 2013; European Commission/EACEA/Eurydice/Cedefop 2014). The fact remains that even today the dropout represents a problem, especially in areas of high territorial fragility and exposed to potential disasters that require further research (Benvenuto 2011). In addition to the individual characteristics, the institutional context in which students live can also foresee the future dropout (Chen 2012), as well as the student's family background and schools attended, while the family background is generally made operational by the socio-economic status of the student. As far as the latter is concerned, the level of education of one of the parents and their income today remains an important contribution to academic success. Even when checking other student characteristics, students with low income backgrounds are more likely to dropout (Alexander et al. 2001). With reference instead to the institutional variables, the research is much more limited since only some studies have included the scholastic characteristics in multilevel models; which led to results of different nature (Goldschmidt and Wang 1999; Lee and Burkam 2003; Luyten et al. 2003; Lamote et al. 2013).

The high dropout rates in the Italian university system are often due to a multiplicity of factors (Cingano and Cipollone 2007). A study conducted by Di Pietro (2004, 2006) seemed to hypothesise how university students with low incomes hardly maintain an acceptable standard of living in Italy and are tempted to interrupt the university and leave it when they find it extremely difficult to complete their studies. These considerations suggested that an important cause of university leaving was to be found both in some features of Italian university demand and of the educational services, thus contributing to the attrition of students. In line with this last approach, dropout rates could be due both to the inadequacy of the services provided by the university in order to meet the needs of students and to guarantee the quality of education (methods, approaches, teaching skills, extension of curricula, etc.), and to contextual factors (as environment conditions). Although there are many studies focused on factors that may have an impact on individual students' success, almost no research took care of better understand about risk factors not frequent, as those found in specific areas with high territorial fragility.

This chapter examines a focused survey, carried out on the territory of L'Aquila as a highly fragile area because of the 2009 earthquake, which represents a case study to deepen the understanding of relationship between traumatic event and early school or university leaving. This survey arose from the Erasmus+ project ACCESS (*Active Cross-sectoral Cooperation for Educational and Social Success*—Development of Innovation—KA2),[1] which is aimed to tackle, in preventative way, the early school dropout caused by school-related risk factors, such as the negative school climate, which includes conflicts with teachers and peers, poor school results, lack of motivation, misalignment between teaching methods and learning styles (Nuzzaci and Marcozzi 2019). The rationale of the project refers to the recommendation of the ET 2020[2] Working Group on Schools Policy (2015) which has stressed the need for a whole school approach to tackling early school leaving (ESL). It has underlined the importance of adopting strategies aimed at: (1) creating supportive and quality learning environments; (2) improving teachers' skills; (3) developing professional communities among schools; and (4) opening to wider multidisciplinary teams and agencies. The framework of this study is given by the ACCESS Project, which involved Italy, Portugal, Lithuania and Romania which were in the building of the strategic project partnership, which involved their universities, upper schools, research and training institutes. It used a specific criterion to group the countries and

[1] ACCESS project, funded by European Erasmus+ programme, aims to face early school leaving by reinforcing pedagogical quality and innovation through the improvement of teachers' skills and the creation of a positive learning environment.

[2] ET 2020 is the strategic framework for European cooperation in education and training.

create three different clusters as indicated by European Educational and Training Monitor Report (European Commission 2017).

Italy along Portugal belonged to a cluster with a decrease in ESL rates from 2010 to 2016 (from 18.6% in 2010 to 13.4% in 2016 in Italy; and from 28.3% to 14.0% in Portugal respectively). The ESL rates, however, remained higher than the European benchmark which is fixed at less than 10%.

As the literature highlights (Boon 2008; Larsen et al. 2013), more research is needed to increase the understanding of the processes leading both to school dropout and university leaving decisions, above all with regard to the specific conditions-related factors in areas at risk or affected by disaster. The educational literature and the ACCESS Project acted as a background to this survey research, which involved University of L'Aquila students and explored the relationship between dropout, earthquake experience, resilience skills. In the prospective introducing innovative strategies by teachers in order to help students at risk of dropout to face the uncertainty that governs their action or non-action, the survey has also explored the teaching strategies linked to the cultural heritage as a tool of methodological innovation and disciplinary intersection. The survey's objectives were as follows: (a) to identify the relationship between risk factors of dropping out (both school and/or university) and the earthquake experience; and (b) to identify the correlation among dropout, the earthquake experience and resilience skills.

27.2 The Survey

The survey had both prospective and retrospective dimensions. As a prospective study, it explored the relationship between risk factors of dropout both of school and of university, taken also in consideration a retrospective point of view of university students. It means, participants were asked also about past matters, when they attended the upper school. The study sample of 148 participants included students selected from the degree programmes in the fields of education at the University of L'Aquila,[3] including Course Degree in Social work (28.8%), Master's Degree in Planning and Management of social and educational interventions (17.3%), Course Degree in Primary teacher education (54.7%), all for the academic year of 2018–2019. Participants consisted of 91.89% females 8.11% males, reflecting the fact that university courses in educational fields involve significantly more female than male. Participants represented two different groups of university students: those who experienced 2009 earthquake of L'Aquila and those who did not. The research hypotheses were as follows:

H1 The earthquake experience increases the risk factors of dropping out from school and from university.

[3] The survey involved with university students who were asked about their intention to leave the school (in a retrospective dimension, as specified). Instead ACCESS project, that is the background to this research, involved with secondary school students.

This hypothesis was linked to a sub-hypothesis (*H1.2*) that the earthquake-related conditions affect students than other risk factors.

But, at the same time,

H2 Resilience skills (related to both dropout and traumatic events) increase in the students who did not leave school after the earthquake and they contribute to the reduction of dropout risk factors for the future (for example during the university) that identify the conditions that significantly increase the likelihood of students.

The questionnaire included six sections, corresponding to macro-dimensions, as follows:

- The risk factors about early leaving both school and/or university;
- Underlying motivations linked to pulling out/pushing out/falling out factors;
- Opinions about the potential causes of why students leave school/university (from those who have never thought about doing it);
- Dimensions related to the school/university well-being;
- Teaching methods used and their importance for engaging students;
- Resilience skills level.

The format was composed of closed-ended questions, Likert scales and Resilience scale (Connor and Davidson Resilience Scale-25) (Connor and Davidson 2003). Data analysis was accomplished through correlation method, which examines the relation and co-variation of many variables. Correlation indicates the tendency that the variables have to vary together, that is, to hatch. The research examined the co-variation between traditional risk factors, several variables linked to the earthquake (as wish to live/to study in another place; and to move in another place; negative emotions), and resilience skills. Statistical techniques (SPSS Statistics 22.0) were as follows: Analysis of variance (ANOVA); multivariate analysis of variance; regression measurement in order to predict, to support the hypothesis, and to measure factors reliability. Through correlational analysis the risk factors of the educational literature were compared with the risk factors related to the earthquake conditions. And subsequently, again through the correlational analysis, the risk factors related to the earthquake conditions were analysed with each of the 25 related five-dimensional items of the Connor and Davidson resilience scale. In the analysis, the afore mentioned correlations were also carried out in the light of the two groups of students: those who experienced the 2009 L'Aquila earthquake and those who did not.

27.3 Main Results

The analysis between the features considered as risk factors of dropping out, on the basis of the literature and the specific elements connected to the territorial context affected by an earthquake, which in the first research hypothesis (H1) represented additional risk factors, has highlighted several significant correlations. Listed below are the correlations (Pearson correlation coefficient)[4] between commonly used risk factors and the so-called earthquake experience-related risk factors:

1. *Quality of education*: I do not have a good relationship with university professors correlates with:

    a. Discomfort related to the conditions of the earthquake zone (.482**)[5];
    b. Move to a different place following the earthquake (.36**);
    c. I do not feel a sense of belonging to my university (.486);
    d. I am always anxious because I think there may be new shocks (.237); teachers use teaching strategies suitable for students (−.258**) and non-engaging teaching methods (.312**).

2. *Discomfort towards the University experience*: I do not feel at ease with my university correlates with:

    a. I do not feel a belonging to the university (.740);
    b. Move to a different place following the earthquake (.278);
    c. I am always anxious because I think there may be new shocks (.252).

3. *University failure/bad grades:* I have failed in some tests correlates with I do not feel a belonging to the university (.304*).
4. *Learning difficulties*, correlate with:

    a. I do not feel a belonging to the university (.312*);
    b. Move to a different place following the earthquake (.271).

5. *Lack of sense of belonging/low sense of belonging to the university*, correlate with:

    a. I am always anxious because I think there may be new shocks (.229), I would like to study in a non-earthquake city (.258).

In addition, earthquake-related risk factors in the second research hypothesis (H2) potentially were configured as additional risk factors or factors which strengthen the "traditional" factors of dropping out of University. Addressing this, the survey results demonstrated strong correlation with resilience skills. In the analysis phase, each item connected to each of five dimensions of resilience skills based on Connor-Davidson Resilience Scale 25-CD-RISC-25. It is a self-administered test that measures resilience or how well one is equipped to bounce back after stressful events, tragedy or trauma. It contains 25 items referring to five factors: (1) personal competence and tenacity; (2) self-confidence and tolerance of negative affect; (3) positive acceptance of change and secure relationships; (4) control; and (5) spiritual influence.

They are as below.

[4] Pearson correlation coefficient has value between +1 and −1; +1 is total positive linear correlation and −1 is total negative linear correlation, 0 is no linear correlation.

[5] Consolidated international scientific "code" is valid, where very significant correlations are indicated/marked with two asterisks (**); those a little less significant with one asterisk (*).

Dimension 1: "Positive acceptance of change/secure relationship" correlate to the earthquake-related risk factors as follows:

- I have at least one close and secure relationship that helps me when I am stressed:
  - I do not feel comfortable at university (−.475**);
  - Desire to live in another place (.218);
  - Low sense of belonging/no sense of belonging to university (.257)
- During times of stress/crisis, I know where to turn for help:
  - I do not feel comfortable at university (−.297);
  - Desire to live in another place (.246);
  - Discomfort linked to living and study conditions in a place hit by the earthquake (−.300*).
- I can make unpopular or difficult decisions that affect other people, if it is necessary:
  - Desire to live in another place (.447);
  - Move to another place following the earthquake (.269).
- I take pride in my achievements:
  - Intention of leaving the university (.329);
  - Move to another place following the earthquake (−.190).

In summary, with increasing the positive acceptance of change/secure relationship, firstly increased the sense of well-being at University; and secondly decreased the discomfort linked to living and study conditions in a place hit by the earthquake.

Dimension 2 "Personal competence/tenacity"

- I try to see the humourous side of things when I am faced with problems:
  - I would like to study in a non-earthquake-stricken university (-.206);
  - Living in a city where there was an earthquake causes me anxiety (−.229**);
  - Studying in a city where there was an earthquake make me feel at quiet (−.301).
- Having to cope with stress can make me stronger:
  - Low sense of belonging/no sense of belonging to university (−.327);
  - Studying in an earthquake city doesn't make me feel comfortable (−.176).
- I tend to bounce back after illness, injury, or other hardships:
  - I am always anxious because there may be new shocks (−, 327**);
  - Studying in a city where there was an earthquake does not make me feel at quiet (−.298**);
  - Living in a city where the earthquake occurred give me anxiety (−.261**);
  - I would like to study in a different city, not affected by the earthquake (−.225**).
- I believe I can achieve my goals, even if there are obstacles:

- Earthquake experience (I experienced the 2009 earthquake) (.222**).

With increasing the personal competence/tenacity, those anxiety and fear related to living in a place hit by the earthquake decreased, and at the same time, the sense of belonging to the university and to the city increased.

Dimension 3 "Self-control"

- Under pressure, I stay focused and think clearly:
  - I experienced the 2009 earthquake (.221**).
- I can handle unpleasant or painful feelings like sadness:
  - Transfer to another place following the earthquake (−.276).
- I can handle unpleasant or painful feelings like fear:
  - I'm always anxious because there may be new shocks (−.226**).
- I can handle unpleasant or painful feelings like anger:
  - I would like to study in a different city, not affected by the earthquake (−.216).
- In dealing with life's problems, sometimes you have to act on a hunch without knowing why:
  - Desire to live in another place (.236).

It showed that a greater capacity for self-control reduced the risk factors related to the specific conditions of the earthquake such as anxiety, the desire to move to another place, the desire to study in a city not affected by the earthquake. Self-control was associated with a greater willingness to deal face problem and to take risks.

Dimension 4 "Self-confidence and management of negative emotions"

- I Prefer to take the lead in solving problems rather than letting others make all the decisions:
  - Desire to live in another place (.293*).
- I am not easily discouraged by failure:
  - I am always anxious because there may be new shocks (-.221**);
  - Desire to live in another place (.256*).

Those who have greater self-confidence and ability to manage negative emotions had no anxiety about other earthquakes, despite having the desire to live in another place.

Dimension 5 "Fatalism"

- When there are no clear solutions to my problems, sometimes fate or god can help:
  - Intention school leaving (.219**).
- Good or bad, I believe that most things happen for a reason:

- Desire to live in another place (.353).

With increasing fatalism, the intention to leave school early increased but not the one to leave the university. In fact, the fatalistic view was generally linked to a greater propensity to drop out of school and decreased with age. The literature (e.g. fall and roberts 2012; Sahin et al. 2016) also shows that youth who have this fatalistic view at high risk for dropping out of school. In this direction, in fact, the same fear of failure can be interpreted as an example of fatalism at the level of the student's experience of school (Downes 2006).

In summary, the high correlation among factors and their high statistical significance level(**) showed that the relation among earthquake, resilience and dropout, is very close, especially in terms of the environmental well-being of students and in reference to the need to introduce innovative strategies by teachers to help students at risk of dropout.

27.4 New Perspectives to Counter the Dropout: Cultural Heritage as an Intervention Strategy

As stated previously, the educational risk factors of the dropout can be identified at community, family, individual or school levels and can also be cumulatively correlated with other factors (de Witte et al. 2013), for example, territorial ones (poverty, systems family, urban areas, rural residence, etc.). Territorial factors cover a broad spectrum of variables, whose to be interpreted require a phenomenological approach which enriches the understanding of the problem and the risks by mapping them (Boudesseul et al. 2016; Broccolichi et al. 2007). Hence, there is a need to examine these territorial factors in terms of specific local and national contexts. The fight against dropout in fact mobilises the territory as a space for concrete actions. Therefore, educational institutions are urged to develop a clear political and strategic framework (MIUR 2018), favouring at the territorial level action plans to combat dropout. This brings together heterogeneous situations and can refer to different orientations and support practices (Knesting 2008). It is not easy to define the territory as an operational level aimed at adopting necessary measures and strategies in order to prevent and resolve the early dropout of university studies. This approach derives from the importance of defining precise measures against early dropout of university studies through: (1) the way in which people appropriate the resources available in a specific area; (2) training provision, available jobs, continuing education opportunities, etc. (Bernard and Bell 2016); and (3) mostly, educational governance at local level.

Teaching strategies based on the use of cultural heritage, as museums, libraries, etc., go in this direction and seem capable of contributing to removing some obstacles deriving from "natural hazard induced disasters", which become "individual and social disasters" and those undermine identity. The negative effects due to territorial fragility on the perception of individuals could be stemmed focusing on an education in the reconstruction of memory and of individual, cultural and social identity.

However, the question is whether and to what extent the dropout is related to a disastrous incident such as the 2009 earthquake, and how it relates to the lack of certainty caused by the existential displacement linked to it (Nuzzaci 2018).

The survey also explored students' perceptions, as "key witnesses", those who experienced the 2009 earthquake and those who did not. This unique opportunity made it possible to better focus on the link between the ESL and school welfare and the use of possible approaches, strategies or tools. Students reported a variety of reasons that surge them to for dropping out of school and University; therefore, the solutions too must be multidimensional. They identified effective strategies related to cultural heritage that have a most positive impact on reducing school dropout. The strategies can be categorised as the following:

- School and community perspective that refers to school-community collaboration at different levels of the education territorial system, in a logic of local governance;
- Early interventions that refers to early childhood education, early literacy and multi-literacies, child development, family engagement and similar;
- Basic core strategies which refers to opportunities of use of cultural goods, after-school/out of school activities, alternative schooling, mentoring/tutoring and so on;
- Educational interventions that refers to the ability to support active learning, using technologies and resources in a strategically integrated way and also leveraging personalised education and the like.

Although those strategies appear to be independent, but they work well together and complete each other. Indeed, they can be implemented as stand-alone strategies but optimum positive outcomes will be achieved when schools and universities or other agencies develop quality improvement plans and instructional design forms that encompass more than one of these strategies. From the general point of view, however, they must give rise to a continuity of actions.

Given the results of this research, it is hoped that schools and universities will design and implement strategies to enable students to use and enjoy cultural heritage in order to cope with difficulties related to existential displacement, discomfort and inadequate teaching methods. The latter is one of the major risk factors for early school leaving appears in the literature and in the focus of the ACCESS project on university students. The following is a list of recommendations:

- Promote the use of local cultural heritage as tool of literacy to create positive attitudes among students towards instruction (multi-literacies), school, university, school subjects, disciplines, students, teachers and the school community in its entirety (Nuzzaci, 2011);
- Design curricula and programmes to educate the school and university communities to understand the role that the local cultural heritage plays in education and to use it in a didactically effective way, so as to allow constructive reflection on the culture of the school and the territory;

- Elaborate interventions educating students to use cultural heritage it in a positive and critical way (as a resource), in order to identify innovative teaching strategies and materials to counteract the forms of dropout;
- Increase the empirical and evidence-based research about the use of the local cultural heritage in education with the aim of containing the phenomena of school dropout in areas of high fragility and existential displacement;
- Increase researches on the relationship between territorial variables linked to the educational use of local cultural heritage and variables linked to the quality of education, for example, the relationship between the student's grade point average and negative effects due to territorial fragility and so on;
- Increase researches on the use of the local cultural heritage to feed the methodological innovation in school and university teaching.

Advocating the above, it must be noted that schools and universities continually seek to develop strategies on innovative teaching and teaching-learning processes, since the traditional learning methods no longer meet students' needs. The literature on the importance of incorporating local cultural heritage in education is now widely used (Nuzzaci 2011). Such assets can contribute only if properly exploited, to the development of individual repertoires and of local schools and university communities. The can potentially create a continuity inside and outside the school-university that strengthens the disciplinary and generic competencies of the students. However, it is equally true that schools like universities should be able to create adequate teaching conditions and flexibility for using local assets to develop precise skills in students. Education must provide real opportunities for the use of local resources to ensure the acquisition of high quality learning, as well as to be able to innovate teaching, even from a methodological point of view. The quality of education is influenced by inadequate context in which the school/University is inserted. In this sense, it is appropriate to identify precise approaches, strategies (imitative, heuristic and creative) and innovative didactic tools which teachers can use to help weaker students or those at the risk of dropout to achieve educational success, leveraging on local cultural heritage as central dimension of culture, identity and literacy.

27.5 Conclusion

The survey on students from the L'Aquila university aimed to identify the relationship between risk factors of university and school dropout and the earthquake experience and to identify the correlation among dropout, the earthquake experience and resilience skills. As it emerges, for individuals residing in a highly fragile territory the risk of dropout becomes more pronounced. The territorial characteristics or variables seem to reinforce the obstacles linked both to the realisation of personal goals and to the ability to work with others to make significant long-term changes. It has become a priority to investigate how teachers can identify suitable forms of didactic support to help the students a greater risk of dropout to achieve educational success, using research data to support the decisions of teachers in context. Teachers can provide training support and motivate students in difficulty with incisive pedagogical approaches, employing the local cultural and environmental goods of the territory for educational purposes that better meet individual needs of students and direct independent and responsible learning practices (Brackenreed 2010) and positive relationships (Condly 2006). In any case, the Italian educational system at the regional level is facing important challenges. Educational researches relevant to the didactic use of the cultural heritage will contribute to improve the learning contexts that will transform into environments rich of resources and opportunities for any child, adolescent or adult. This will permit to enhance the quality of education, as well as making formal, informal and non-formal learning contexts interacting.

The possibility to use the cultural heritage to counter the dropout phenomenon is, by all means, a challenge for policy makers, researchers and teachers. The future research perspectives must consider the development and the validation of strategies to face increasingly subtle forms of dropout, based on the idea that the territorial resources are value assets within an entire community that become more and more "educating".

A better comprehension of how and why the multiple territorial factors may be linked to the dropout still represents an important challenge for the future.

References

Alexander K, Entwisle D, Kabbani N (2001) The dropout process in life course perspective: early risk factors at home and school. Teachers College Record 103(5):760–822

Benvenuto G (ed) (2011) La scuola diseguale. Dispersione ed equità nel sistema di istruzione e formazione. Anicia, Roma

Bernard PY, Bell L (2016) Territoires, offre de formation et expériences du décrochage scolaire: une étude de cas», Espaces et Sociétés 166(3):95–112

Boon HJ (2008) Risk or resilience? What makes a difference? Aust Edu Res 35(1):81–102

Boudesseul G, Caro P, Grelet Y, Minassian L, Monso O, Vivent C (2016) Atlas des risques sociaux d'échec scolaire: l'exemple du décrochage. CEREQ, DEPP, Marseille-Paris

Bowers A (2010) Grades and graduation: a longitudinal risk perspective to identify student dropouts. J Edu Res 103(3):191–207

Brackenreed D (2010) Resilience and risk. Int Edu Stud 3(3):111–121

Broccolichi S, Ben Ayed C, Mathey-Pierre C, Trancart D (2007) Fragmentations territoriales et inégalités scolaires: des relations complexes entre la distribution spatiale, les conditions de scolarisation et la réussite des élèves. Édu Formations 74:31–48

Chen R (2012) Institutional characteristics and college student dropout risks: a multilevel event history analysis. Res High Educ 53(5):487–505. https://doi.org/10.1007/s11162-011-9241-4

Cingano F, Cipollone P (2007) University dropout: the case of Italy. Temi di Discussione del Servizio Studi 626, from https://www.bancaditalia.it/pubblicazioni/temi-discussione/2007/20070626/en_tema_626.pdf

Condly S (2006) Resilience in children: a review of literature with implications for education. Urban Edu 41(3):211–236

Connor KM, Davidson JRT (2003) Development of a new resilience scale: the Connor-Davidson resilience scale (CD-RISC). Depress Anxiety 18(2):76–82. https://doi.org/10.1002/da.10113

Dale R (2009) Early school leaving: lessons from research for policy makers. Report written on behalf of NESSE (Network of experts in social science of education and training), from http://www.nesse.fr/nesse/activities/reports/activities/reports/early-schoolleaving-report

De Witte K, Cabus S, Thyssen G, Groot W, Van Den Brink H (2013) A critical review of literature on school dropout. Tier working paper series, Tier WP 14/14

Di Pietro G (2004) The determinants of university dropout in Italy: A bivariate probit model with sample selection. Appl Economic Lett 11:187–191

Di Pietro G (2006) Regional labour market conditions and university dropout rates: evidence from Italy. Region Studies 40(6):617–630

Downes P (2006) Quality development of out of school services. QDOSS, Dublin

European Commission (2013) Reducing early school leaving: key messages and policy support. Final report of the thematic working group on early school leaving, November 2013, from http://ec.europa.eu/education/policy/strategic-framework/doc/esl-group-report_en.pdf

European Commission (2014) Thematic working group on early school leaving. Early warning systems in Europe: practice, methods and lessons. Publications Office of the European Union, Luxembourg

European Commission/EACEA/Eurydice/Cedefop (2014) Tackling early leaving from education and training in Europe: strategies, policies and measures. Publications Office of the European Union, Luxembourg

European Commission (2017) European Educational and Training Monitor Report. Publications Office of the European Union, Luxembourg

Fall A, Roberts G (2012) High school dropouts: interactions between social context, self-perceptions, school engagement, and student dropout. J Adolesc 35(4):787–798. https://doi.org/10.1016/j.adolescence.2011.11.004

Gairín J, Figuera P, Triadó X et al (2010) L'abandonament dels estudiants a les universitats catalanes. AQU Catalunya, Barcelona

Goldschmidt P, Wang J (1999) When can schools affect dropout behavior? A longitudinal multilevel analysis. Am Educ Res J 36(4):715–738

Knesting K (2008) Students at risk for school dropout: supporting their persistence. Preventing Sch Fail 52(4):3–10

Lamote C, Van Damme J, Van Den Noortgate W, Speybroeck S, Boonen T, de Bilde J (2013) Dropout in secondary education: an application of a multilevel discrete-time hazard model accounting for school changes. Qual Quant 47(5):2425–2446

Larsen MS, Kornbeck KP, Kristensen RM, Larsen MR, Sommersel HB (2013) Dropout phenomena at Universities: What is dropout? Why does dropout occur? What can be done by the Universities to prevent or reduce it? A systematic review. Danish Clearinghouse for Educational Research, Department of Education, Aarhus University, Copenhagen. https://edu.au.dk/fileadmin/edu/Udgivelser/Clearinghouse/Review/Evidence_on_dropout_from_universities_technical_report_May_2013.pdf

Lee V, Burkam D (2003) Dropping out of high school: the role of school organization and structure. Am Educ Res J 40(2):353–393

Luyten H, Bosker R, Dekkers H, Derks A (2003) Dropout in the lower tracks of dutch secondary education: predictor variables and variation among schools. Sch Effectiveness Sch Improv 14(4):373–411

MIUR (2018) Una politica nazionale di contrasto del fallimento formativo e della povertà educativa. Cabina di regia per la lotta alla dispersione scolastica e alla povertà educativa, gennaio 2018, from http://www.miur.gov.it/documents/20182/0/Rapporto+sul+contrasto+del+fallimento+formativo/7575f155-63f9-479a-a77f-1da743492e92?version=1.0

Nuzzaci A (ed) (2011) Patrimoni culturali, educazioni, territori: verso un'idea di multiliteracy. Pensa MultiMedia Editore s.r.l. I, Lecce-Brescia

Nuzzaci A (2018) Patrimoni locali, identità e linguaggi: educare "*ai* e *con i*" beni culturali e ambientali in aree ad elevata fragilità. In: Mariantoni S, Vaccarelli A (eds) Individui, comunità e istituzioni in emergenza. Intervento psico-socio-pedagogico e lavoro di rete nelle situazioni di catastrofe. FrancoAngeli, Milano, pp 213–229

Nuzzaci A, Marcozzi I (2019) Fattori di rischio scolastici e dropout nella percezione degli studenti: il progetto internazionale ERASMUS KA2 ACCESS—School risk factors and dropout in students' perception: the international ERASMUS KA2 ACCESS project. Giornale Italiano della Ricerca Educativa—Italian J Edu Res 12(23):48–68. https://doi.org/10.7346/sird-022019-P48

OECD (2019) Education at a glance: OECD indicators. OECD Publishing, Paris. https://doi.org/10.1787/f8d7880d-en

Rumberger R (2011) Dropping out. Why students drop-out of high school and what can be done about it. Harvard University Press, Cambridge

Sahin S, Arseven Z, Kılıç A (2016) Causes of student absenteeism and school dropouts. Int J Instr 9(1):195–210

Tanner-Smith EE, Wilson SJ (2013) A meta-analysis of the effects of dropout prevention programs on school absenteeism. Prev Sci 14(5):468–478. https://doi.org/10.1007/s11121-012-0330-1

Chapter 28
How Can Teachers Promote Resilience in Schools?

Sara Gabrielli and Cesare Fregola

Abstract This chapter addresses building and enhancing resilience in children experiencing a crisis or facing adversity in the twenty-first century. It reflects on an experimental research project for developing an educational and didactical path in order to promote resilience which could be introduced into teaching practices and to verify the effectiveness of this formative path. The chapter examines the possibility of influencing the connection between learning and the affective, cognitive, and psychological variables, especially when promoting resilience. The teaching contents for the project were chosen in a way to directly connect theory and practice. Resilience was connected to scientific thinking, history, geography, sport and storytelling, and the didactical path was structured into learning units in which different methods and instruments were used. The project focused on enhancing social competences, problem-solving, creativity and humour to recognise resources at three levels: personal, classroom and environmental. The result makes a case for the need to realise the full potential of school teachers as resilience tutors.

Keywords Resilience · Tutor · School children · Earthquake · Dedactical path · Italy

28.1 Introduction

In Italy, the earthquake in Umbria, Marche, Lazio and Abruzzo occurred on the night of 24 August 2016. It was a 6.0 magnitude earthquake whose epicentre was located between Accumoli, Amatrice and Arquata del Tronto. Subsequent earthquakes occurred twice in October 2016 and once in January 2017, affecting 131 villages (Save the Children 2017a, b).

S. Gabrielli (✉)
Sapienza University of Rome, Rome, Italy
e-mail: sara.gabrielli@uniroma1.it

C. Fregola
Università degli studi Roma Tre, Rome, Italy
e-mail: cesare.fregola@uniroma3.it

© The Author(s), under exclusive license to Springer Nature Switzerland AG 2021

F. F. Arefian et al. (eds.), *Historic Cities in the Face of Disasters*, The Urban Book Series,
https://doi.org/10.1007/978-3-030-77356-4_28

The earthquake also damaged 669 schools in total and concerned 84,399 pupils in kindergarten, primary and secondary schools, as shown in the ISTAT (National Statistical Institute, Italy) Report. In this context, children happen to be exposed to serious vulnerability, mainly because they risk losing contact with their origins. In fact, children and families are often forced to move from mountains to coast or to near villages not affected by the earthquake. In some cases, families will not be able to come back to their native village for a long period.

An earthquake like the one that occurred in Italy in 2016—involving Umbria, Marche and Lazio—could create a sense of disaffection towards historical and collective spaces (Vaccarelli et al. 2016). It underlines how urgent an educative response is to reinforce or to rebuild the feelings of connection between communities and territories. Thanks to this affective bond, people can perceive history as being built collectively. According to this perception, Calandra (2018) recognises how people feel a connection between defining their actions and their *topophilia*: the love for their own land, literally "love for place" from Ancient Greek.

Fragmentation is a prevailing trait of the places affected by the earthquake. The population is divided into small towns and villages, especially in the mountains. Moreover, parents commute every day to get to work and sometimes children also attend schools in other villages, moving there by bus. The disaster also caused a fracture among people, reducing the population density—which was already lower than the national average, for example, in 2016, it equalled 74,5 people per square km, while as the average in Italy there are 200 people per square km, according to ISTAT (2016).

Villages contain the environmental and cultural legacy of traditions that need to be rediscovered and valued, above all after an earthquake and, most importantly, by children who spent less time in their environment and shared less tradition with their community. Earthquakes are traumatic events for the human psyche and have the power to cause strong destabilisation towards the affected communities. As a process of coping with trauma, resilience can be an answer after an earthquake (Salerni and Vaccarelli 2019).

Regarding resilience in communities after the earthquake, it is important to consider two different levels: the need of promoting resilience in the whole community and the role of the community in enhancing children's resilience by communicating memories and territory-related narratives. A third consideration could be that the use of the community in increasing resilience in children could itself represent a way of promoting environmental resilience. Based on these considerations, this chapter presents a didactical path designed to promote resilience in children who are experiencing such a crisis, explored also by other authors (e.g. Cyrulnik and Malaguti 2005; Inguglia and Lo Coco 2013), and who are facing adversity and complexity referred to as "complexity construct" by Bocchi and Ceruti (2007). More generally, resilience has been defined as a dynamic process that can be learnt at any given point in life by an individual (Mash and Barkley 2003; Masten 2007 in Isidori 2013).

In general terms, a didactical path refers the design and experimentation of teaching sequences followed in the pedagogical intervention. In this project the didactical path to promote resilience was undertaken in three contexts. One context was in one of the villages involved in the earthquake and considered as highly damaged in the

ISTAT Report (2016) focused on parents' and grandparents' narrations as resources. The same path was also conducted in two other contexts: an Italian school in Spain and a school in Rome, where resilience capability is considered important to succeed in everyday challenges and to face future emergencies in a preventive way. There, the role of the school system and teachers as resilience tutors is emerging—not only after a trauma, but in everyday life.

Malaguti (2005) describes resilience tutors as resilient persons who have met, in their own lives, some helpful people (resilience tutors) in auspicious environments. She underlines the capability of resilience tutors of seeing the best side in children, of providing support, of being a protective influence.[1] Reference could be made to teachers, who must focus on strengths not weaknesses in pupils to improve their self-reliance and self-esteem, akin to the protective shield described by Garmezy (1991). Indeed, "the ability of a child to react positively depends on internal factors, but also on how much the adult world is able to accompany them in the stressful or traumatic experience and the extent to which this experience is managed and elaborated by the adults themselves" (Vaccarelli 2018, p. 28).

28.2 Why Talking About Resilience in Schools?

Different international studies introduced resilience in schools (e.g. Rutter 1987; Werner and Smith 1982), recognising the significant role of schools and teachers in developing protective factors in students. Benard (1991, 1995) acknowledges the role of schools in influencing children's development, self-esteem and self-efficacy, although teachers and educators still underestimate their potential in supporting children—as demonstrated by different researchers (e.g. Oswald et al. 1999, 2003; Russo and Boman 2007). Gagné (1973) indicated that "people make choices of experiences that enable them to feel self-efficient, and choices of environments that challenge their abilities and stimulate their development of competence. Students and/or teachers need to mediate, support and provide feedback so that self-efficacy can lead to shifts in cognitive, metacognitive and affective variables" (Fregola 2015, p. 20). Further studies on relational dimensions and on affective dimensions intend to examine in-depth connections to transactional analysis, a theory focused on phenomena based on behaviours, according to which it is possible to act on resilience during the learning and teaching process (Fregola 2016a; Montuschi 1993).

Resilience has been introduced into schools by international, European and Italian legislation, defining resilience as a competence to be promoted inside classrooms, by schools and teachers. Three documents are considered (Fregola 2016b):

- At the international level, the World Health Organization (WHO) first referred to life skills in 1993 as "abilities for adaptive and positive behaviour, that enable individuals to deal effectively with the demands and challenges of everyday life". They were listed as: "decision making, problem solving, creative thinking,

[1] Translation by the author.

critical thinking, effective communication, interpersonal relationship skills, self-awareness, empathy, coping with emotions, coping with stress" (p. 5). Life skills referred to personal and relational abilities to handle relationships in the "knowledge society".

- In 2006, the European Parliament and the European Council adopted a "Recommendation on Key Competences for Lifelong Learning". It defined the competences each European citizen needs "for personal fulfilment and development, employment, social inclusion and active citizenship", in a society non-static and in change. "Competence-oriented education is regarded as advantageous in a time when the knowledge base of our societies is developing at an immense speed and when the skills required need to be transferred to and developed in many different societal contexts, including those unforeseen in the future" (CE, 962/2006).
- In 2015, the World Economic Forum published a report entitled "New Vision for Education: Unlocking the Potential of Technology". Its premise is: "to thrive in a rapidly evolving, technology-mediated world, students must not only possess strong skills in areas such as language, arts, mathematics and science, but they must also be adept at skills such as critical thinking, problem-solving, persistence, collaboration and curiosity" (WEF 2015, p. 1).

Problem-solving, creativity, cooperation and the ability to find solutions are considered as the way to succeed in a rapidly changing world, and all those skills form strictly part of the resilience construct. Recently, in Italy the Ministry of Education, University and Research (MIUR)[2] enacted a "Guideline to a certificate of competences at the end of the first cycle of education" (MIUR, DM 742/2017), effectively used in schools to evaluate students in their final years. In the guideline, resilience relates to flexibility and creativity, well applied by pupils when they "react to unpredictable situations or needs with functional and divergent solutions, through creative means".

Educational research (e.g. Bellhouse et al. 2005; Cefai and Cooper 2008) shows a successful connection between good relationships among teachers and peers, self-esteem, the ability of working in teams to achieve goals and resilience. Research by Hawkins et al. (1999) underlines that interventions have a better lasting effect on resilience when connected with school satisfaction and self-esteem perceptions. Moreover, Isidori (2013) also analyses how studying and learning during an emergency or in a post-emergency phase can be challenging for children because the learning process is conditioned by the trauma the individual is exposed to.

28.3 Schools and Promoting Resilience

Studying represents a way towards a better future, it studying could make up for students' damage and loss. Through their studies their will to survive and not to be overcome by the adversity the earthquake had caused in their lives (Isidori 2013). The role of schools in promoting resilience is recognised in that "it is possible to

[2] MIUR—Ministero dell'Istruzione, dell'Università e della Ricerca.

use tools of education (music, theatre, play, art, active education, didactic paths and integrated disciplines" (Malaguti 2005, p. 175). According to the literature, there exist various useful intervention strategies to promote socio-emotional, cognitive and physical resilience in children.

Some areas of attention, intervention strategies and how they are addressed are involved in the project's didactical path is project are described below.

Social competences such as empathy and communication skills are important in creating and strengthening the bonds in classrooms and to stimulate the perception of schools as "safe places" (Iorio 2018). "The authentic expression of emotions requires a free, welcoming, empathetic, respectful, non-judgmental and constructive context" (Di Genova 2018, p. 62).

Given the above, in the didactical path used as intervention in this project, social competences have been enhanced through activities focused on peer support, for example, by asking children to indicate five positive aspects of one of their classmates. The idea of the project is that schools—especially in the post-emergency phase—can be regarded as the place to be supportive and to share, due to its "ordinary magic" (Masten 2001), making it possible to recognise each other inside the community.

Humour is regarded in the literature as a key competence to promote resilience (e.g. Vanistendael, 2005; Isidori and Vaccarelli 2013). The ability to smile after a trauma activates a process to rethink emotions connected to the traumatic event and to find a new meaning. In Cyrulnik's words (2001), "humour is anything but a joke" and it is useful to create a distance from ourselves and our problems, as also recognised by Frankl (1998).

Therefore, during the project, humour was used in different activities, trying to keep smiling and to establish a good relationship between teachers and children.

Resilience is often associated with ideas of flexibility, creativity and problem-solving. In this way, Antonietti and Pizzingrilli (2013) regard resilience as the ability to manage and handle difficult situations, finding uncommon and different ways to succeed and changing perspective.

Given the above, the didactical path asked children to find alternative solutions in different ways, encouraging them to think out of the box. In some cases, pupils were asked to assist a book or movie character by using role playing. Another method to implement problem-solving is through sport. According to Boerchi et al. (2013), sport focuses on perseverance and persistence in obtaining results and improves self-efficacy and self-esteem. Body training also trains resilience, challenging children to find innovative strategies to solve problems (Vaccarelli 2016) and to establish that "optimal challenge" mentioned by Olmetti Peja (2016).

Regarding identity and self-efficacy, self-knowledge and self-understanding are often linked to narration and self-narration as a privileged way to process traumas and experiences (Padovani 2018). Through narration, experiences acquire new meanings for us and others and are building value to events (Bianchi 2020). Finding a space for narrations is essential and vital for both affective and cognitive personal growth (Vaccarelli 2016). From the didactical point of view, "the most difficult situations that people find themselves living very often find a space of recognition in self-narration—be it in the form of diary, or in other literary genres—to recompose fragments of existence and to re-attribute sense and meaning to what was experienced" (Padovani 2018, p. 53).

Given the above importance of narration, in the project, a diary has been used during the didactical path to promote self-reflection and self-narration. Furthermore, children wrote letters to exchange with pupils from other schools, describing themselves, their strengths and weaknesses. Recognising personal abilities is important to understand each other and to build identity. Narrations have been used with reference to picture books, in particular Lorenzo's Dipper's (Carrier 2011), "The Warm-Off Tale" (Steiner 2009) and the Eli's "Whisper Series" (Bellini 2015a, b), arousing a range of feelings and emotions (wonder, enchantment, curiosity, laughter) (Zizioli and Franchi 2018).

Regarding community and culture, it is acknowledged that through "topophilia", people could be active citizens, conscious of problems and resources of their own environment and analyse possibilities of growth. There is a connection between society and territory. The latter builds on the first, not only reflecting culture, but also sustaining functions (Turco 2019). Territoriality, as a reflective elaboration, is represented by collective narrations, built on the relationship between community and territory. The sense of belonging and identity are developed thanks to those narrative constructions (Calandra 2013). A territory could be regarded as the place where subjects and the whole community can express themselves (Berdoulay and Entrikin, 1998). Catarci (2018) observed that involving children into educational programmes based on territorial awareness could be significant in a perspective of social fairness (Jackson 2011). In reality, the environment defines pedagogical dimensions, activating processes of knowledge through social and active experiences (Catarci 2018).

Given the above, during the project those school classes which were affected by the earthquake (in the village) participated in analysing their environmental resources. The project consisted of different steps aimed to assist students to recognise the territory as valuable, as described below.

First, children produced a mental map of the places they considered most important in the district or in the village: schools, the church, the home, sport societies, best friends' houses, etc. Children reconstructed the history of their environment by conducting interviews and research at home, but most of all by involving families and—when possible—the whole community. The project required the participation of the parents in doing homework, developing a dialogue between schools and the community and promoting an historical approach using oral, written or visual sources. Grandparents were also considered as resources. They had a story to tell, and they shared pictures, photographs, objects and tales of their past with children. Moreover, participants acquired knowledge about social and ecological aspects, encouraging them to participate and cooperate to achieve common goals.

28.4 Research Design of the Didactical Path

In accordance with the existing literature and by examining the school context involved in the earthquake, the didactical path was designed to promote resilience and self-esteem in children and to enhance their ability of identifying internal and external resources to face change.

The didactical path was structured into four phases (Gabrielli 2019):

1. The observation of school classes design the didactical path to respond as much as possible to the needs of children, teachers and schools, according to their lesson plans.
2. The Child and Youth Resilience Measure (CYRM) (developed by Ungar and Liebenberg 2011) was applied as pretest.
3. Each school to include 60 hours of interventions.
4. The CYRM was applied as post-test, to measure differences of resilience in children and to evaluate the efficacy of the intervention.

The didactical path was structured as interdisciplinary, involving Italian literature, physical education, geography, history, geometry and technology. It was formed by three learning units: the first focused on personal resources, the second on school class resources and the third on community resources. In particular, the first unit explored self-knowledge using picture books, whereby children could empathise with characters who faced challenges. Physical activities, above all with barriers and obstacles, were used as metaphor of everyday problems. Resilience was essentially connected with citizenship, using geography and history, involving the whole community.

28.5 Method and Results

The research involved 79 students (41 males and 38 females—aged 8–9) in three different contexts:

1. A Spanish school with an experimental group of 15 children: 3 males and 12 females, and a control group of 16 children: 10 males and 6 females. The school was bilingual, there were children from both Italy and Spain, and lessons were in both Italian and Spanish.
2. A school involved in the 2016 Italian earthquake with an experimental group of 19 children: 10 males and 9 females. It was not possible to have a control group because of the critical situation that followed the earthquake, as the school was attended by few students.
3. A school in Rome with an experimental group of 17 children: 10 males and 7 females and a control group of 11 children: 8 males and 3 females. The school faced challenges by including children from different countries (some of which did not speak Italian) and with different abilities.

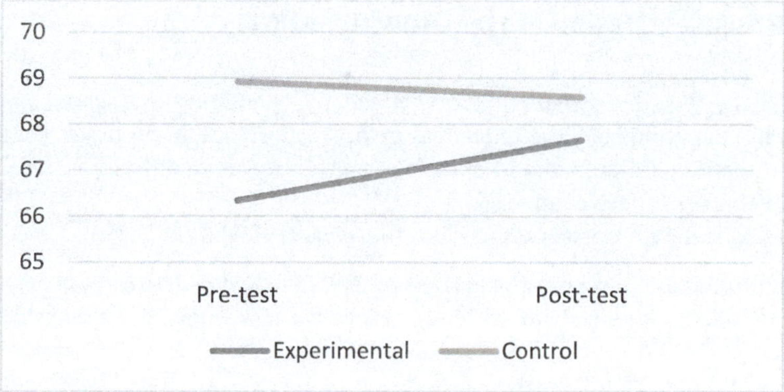

Fig. 28.1 ANOVA results of CYRM

The CYRM was adapted to Italian for research purposes and distributed as pre- and post-test, in both the experimental and control groups. The experimental groups participated also in the didactical path.

CYRM is a questionnaire composed of 26 items and eight subscales. Each item requires expressing frequency on a 3-point Likert scale (1 = no, 2 = sometimes, 3 = yes). Its objective is to evaluate different resources (individual, relational, environmental, cultural) inherent in every person and able to improve or affect resilience. The questionnaire was first developed in the International Resilience Project (IRP) from the Resilience Research Centre (RRC).

Data analyses were conducted using SPSS software, ver. 25, to verify the effectiveness of the didactical path in enhancing resilience in children. An ANOVA was conducted, using it twice: administration of instruments pre- and post-intervention within factors, and control versus experimental group between factors. The results showed a significant ($p = .04$) increase of resilience in the experimental groups, but not in the control groups. According to these results, the didactical path experimented in the project seems effective in promoting resilience in children who participated (Fig. 28.1).

28.6 Conclusion

Regarding the idea of schools and teachers as resilience tutors, the chapter describes an experimental didactical path to promote resilience in children in different contexts: a village affected by the 2016 earthquake, a school in Rome and an Italian school in Spain.

The project and experimental didactical path incorporated the principles and considerations suggested in literature for promoting resilience in schools, such as enhancing social competences, humour, creativity and problem-solving, identity and self-knowledge, community and environment. Thanks to its focus on relationships, environments and communities, the proposed didactical path contributed significantly to children's resilience, as shown by the CYRM results. It also had important outcomes regarding the resilience of communities by creating cooperation and, most importantly, by bringing the community involved in the earthquake to implement the project in all the classrooms of the school.

This relatively small scale experiment clearly makes a case for further research on the subject matter, focusing on the capabilities of teachers as resilient tutors. If resilience is to be explored further as a process, not as a static ability, it would be important this experimental project to be followed up with children involved in the project.

In a future perspective, the research could be extended by involving more children and schools from different countries and contexts, also to validate the questionnaire adapted in the Italian language.

In practical terms, a formative programme could be realised to highlight and further explore the role of teachers as protective factors and resilience tutor for children, by involving in-service or pre-service teachers to measure their resilience. Moreover, teachers who participated in the formative programme could undertake educational and didactical programmes by themselves to enhance resilience in children. Resilience is clearly important on a global level—above all during the COVID-19 emergency—for all communities and individuals, and didactical paths to promote this ability appear to be essential.

References

Antonietti A, Pizzingrilli P (2013) Resilienza e creatività: nessi possibili. In: Castelli C (ed) Tutori di resilienza. Guida orientativa per interventi psico-educativi. EDUCatt, Milano

Bellhouse B, Fuller A, Johnson G, Taylor N (2005) Managing the difficult emotions. Paul Chapman Publishing, London

Bellini L (2015a) Eli sottovoce… l'uovo. Valentina Editore, Milano

Bellini L (2015b) Eli sottovoce… le carte. Valentina Editore, Milano

Benard B (1991) Fostering resiliency in kids: protective factors in family, school and community. Northwest Regional Educational Centre, Portland

Benard B (1995) Fostering resilience in children. ERIC Clearinghouse on Elementary and Early Childhood Education, Urbana

Berdoulay V, Entrikin JN (1998) Lieu et sujet. Perspectives théoriques. L'Espace géographique 2:111–121

Bianchi L (2020) La narrazione nella relazione educativa: le parole dell'intercultura. Edu Sci Soc 1:270–279

Bocchi G, Ceruti M (2007) La sfida della complessità. Mondadori, Milano

Boerchi D, Castelli C, Rivolta M (2013) Costruire resilienza attraverso lo sport. In: Castelli C (ed) Sport e resilienza. Vita e Pensiero, Milano

Calandra L (2013) Cultura e territorialità: quando l'abitare diventa multitopico. Esempi da L'Aquila post sisma. In: Pedrana M (ed) 2013. Multiculturalità e territorializzazione. Casi di studio. IF press srl, Roma

Calandra L (2018) Comunità e post-emergenze: il territorio dei miei sogni per ricostruire il futuro. In: Mariantoni S, Vaccarelli A (ed) 2013. Individui, Comunità e Istituzioni in emergenza. Intervento psico-socio-pedagogico e lavoro di rete nelle situazioni di catastrofe. Franco Angeli, Milano
Carrier I (2011) Il pentolino di Antonino. Kite, Padova
Catarci M (2018) Educazione e comunità competente. In: Mariantoni S, Vaccarelli A (eds) 2018. Individui, comunità e istituzioni in emergenza. Franco Angeli, Milano
Cefai C, Cooper P (2008) Promoting resilience in the classroom: a guide to developing pupils' emotional and cognitive skills. Jessica Kingsley Publisher, London
Cyrulnik B (2001) Les vilains petits canards. Editions Odile Jacob
Cyrulnik B, Malaguti E (2005) Costruire la resilienza. La riorganizzazione positiva della vita e la creazione di legami significativi. Erickson, Gardolo
Di Genova N (2018) Contact with one's own emotion and those of others: playing Emotica. In: Vaccarelli A, Mariantoni S (eds) Children after a natural disaster. Materials for educators and teachers. Franco Angeli, Milano
European Parliament and Council (2006) Raccomandazioni del parlamento europeo e del consiglio. Available on https://eur-lex.europa.eu/legal-content/IT/TXT/?uri=celex%3A32006H0962. Accessed 19 July 2019
Frankl VE (1998) Senso e valori per l'esistenza la risposta della logoterapia. Città Nuova Editrice, Roma
Fregola C (2015) Two empirical research projects into the impact of teaching the concept of drivers to preschool children in Italy. IJTAR 6(2):19–30
Fregola C (2016a) Insegnamento, Formazione e AT del campo educativo nella prospettiva del lifelong learning. NEOPSICHE 21:138–165
Fregola C (2016b) La relazione di apprendimento e le Skills del XXI Secolo. IAT J 1:47–60
Gabrielli S (2019) La scuola come tutrice di resilienza. Una sperimentazione condotta in Italia e in Spagna. Franco Angeli, Milano
Gagné RM (1973) Le condizioni dell'apprendimento. Armando, Roma
Garmezy N (1991) Resilience in children's adaptation to negative life events and stressed environments. Pediatric Ann 20:459
Hawkins JD, Kosterman R, Catalano RE, Hill KG, Abbot R (1999) Preventing health-risk behavior by strengthening protection during childhood. Arch Pediatric Adolescent Med 153:226–234
Inguglia C, Lo Coco A (2013) Resilienza e vulnerabilità psicologica nel corso dello sviluppo. Il Mulino, Bologna
Iorio C (2018) How to manage mourning children and adolescents: what teachers and educators need to kwon. In: Vaccarelli A, Mariantoni S (eds) Children after a natural disaster. Materials for educators and teachers. Franco Angeli, Milano
Isidori MV (2013) Emergency and post-emergency situations. Educational and teaching model. J Theor Res Edu 8(2):1–8
Isidori MV, Vaccarelli A (2013) Pedagogia dell'emergenza/Didattica nell'emergenza. I processi formativi nelle situazioni di criticità individuali e collettive. Franco Angeli, Milano
ISTAT—Istituto Nazionale di Statistica (2016) Caratteristiche dei territori colpiti dal sisma del 24 agosto, 26 ottobre e 30 ottobre 2016. Available on https://www.istat.it/it/archivio/194346. Accessed 19 July 2019
Jackson S (2011) Sustaining communities: conclusion. In: Jackson S (ed) Lifelong learning and social justice. Communities, work and identities in a globalized work. National Institute of Adult Continuing Education, Leicester
Malaguti E (2005) Educarsi alla resilienza. Come affrontare crisi e difficoltà e migliorarsi. Erickson, Trento
Mash EJ, Barkley RA (2003) Child psychopathology. The Guilford Press, New York
Masten AS (2001) Ordinary magic: resilience processes in development. Am Psychol 56(3):227–238
Masten AS (2007) Resiliency in developing systems: progress and promise as the fourth wave rises. Dev Psychopathol 19:921–930

MIUR—Ministero dell'Istruzione dell'Università e della Ricerca (2018) Linee guida per la certificazione delle competenze nel primo ciclo d'istruzione. Available on www.miur.gov.it/-/linee-guidacertificazione-delle-competenze. Accessed 19 July 2019

Montuschi F (1993) Competenza affettiva e apprendimento. La Scuola, Brescia

Olmetti Peja D (2016) Sfida ottimale, resilienza, autoefficacia: il ruolo della motivazione nella costruzione del campione. In: Ruggiero M (ed) Vince chi si alza prima. Talento, valori, successo, pp 153–166. Aliberti Compagnia Editoriale, Reggio Emilia.

Oswald M, Johnson B, Howard S (1999) Quantifying and prioritizing resilience factors: Teachers' views. Available on website www.aare.edu.au/99pap/osw99728.html

Oswald M, Johnson B, Howard S (2003) Quantifying and evaluating resilience promoting factors teachers' beliefs and perceived roles. Res Edu 70:50–64

Padovani L (2018) Narratives and resilience. In: Vaccarelli A, Mariantoni S (eds) Children after a natural disaster. Materials for educators and teachers. Franco Angeli, Milano

Russo R, Boman P (2007) Primary school teachers' ability to recognise resilience in their students. Aust Edu Res 34(1):17–32

Rutter M (1987) Psychosocial resilience and protective factors. Am J Orthopsychiatry 57:316–331

Salerni A, Vaccarelli A (2019) Supporting school resilience: a study on a sample of teachers after the 2016/2017 seismic events in Central Italy. EPCS J 19:83–103

Save the Children (2017a) Terremoto in Italia centrale: l'intervento di Save the Children. Available on https://www.savethechildren.it/sites/default/files/files/rapporto%20terremoto%202017.pdf. Accessed 19 July 2019

Save the Children (2017b) L'impatto degli eventi sismici dell'Italia centrale sui diritti dei bambini e degli adolescenti. Available on https://www.savethechildren.it/sites/default/files/files/uploads/pubblicazioni/limpatto-degli-eventi-sismici-italia-centrale-sui-diritti-dei-bambini-e-degli-adolescenti.pdf. Accessed 19 July 2019)

Steiner C (2009) La favola dei caldomorbidi. Edizione illustrata. Artebambini

Turco A (2010) Configurazioni della territorialità. Franco Angeli, Milano

Ungar M, Liebenberg L (2011) Assessing resilience across cultures using mixed-methods: construction of the child and youth resilience measure-28. J Mixed Methods Res 5(2):126–149

Vaccarelli A (2016) Le prove della vita. Franco Angeli, Milano

Vaccarelli A (2018) Promoting resilience after a catastrophe: a pedagogical approach. In: Vaccarelli A, Mariantoni S (eds) Children after a natural disaster. Materials for educators and teachers. Franco Angeli, Milano

Vaccarelli A, Ciccozzi C, Fiorenza A (2016) Resilienza, socialità e intervento pedagogico a cinque anni dal sisma dell'Aquila del 2009: una ricerca-azione nelle scuole primarie. Epidemiol Prev 40(2):98–103

Vanistendael S (2005) Umorismo e resilienza: il sorriso che fa vivere. In: Cyrulnik B, Malaguti E (eds) Costruire la resilienza. La riorganizzazione positiva della vita e la creazione di legami significative. Erickson, Gardolo

WEF (2015) New vision for education: unlocking the potential of technology. Available on www3.weforum.org/docs/WEFUSA_NewVisionforEducation_Report2015.pdf. Accessed 19 July 2019

Werner EE, Smith RS (1982) Vulnerable but invincible: a longitudinal study of resilient children and youth. McGraw-Hill, New York

WHO—World Health Organization (1993) Life skills education for children and adolescent in schools. Available on http://apps.who.int/iris/handle/10665/63552. Accessed 19 Sept 2019

Zizioli E, Franchi G (2018) Silent educators for group readings in emergencies. In: Vaccarelli A, Mariantoni S (eds) Children after a natural disaster. Materials for educators and teachers. Franco Angeli, Milano

Part IV
Bringing the 21st Century into Reconstruction

Chapter 29
Cities in Transformation: Smarter Reconstruction in Historic City Centres

Donato Di Ludovico

Abstract This chapter analyses potential effects of highly innovative projects on historic centres from the technological point of view. It particularly aims to analyse the effects of the strong acceleration of the urban transformation of reconstructed historic cities after a disaster such as L'Aquila, an instable historic centre of Central Italy hit by an earthquake in 2009. In this historic centre, there are many innovation-oriented initiatives, that refer to the principles of the smart city, but the local context is incapable to transfer technologies and innovative solutions to create new effective services for citizens. The analysis of the case study of L'Aquila and scientific literature, which presents clear gaps on the subject, made it possible to identify two technologies macro-categories: tangible technologies and intangible technologies. In the context of these two macro-categories and in relation to the general theme of the application of these to changes in the historic centre, the research has highlighted a significant impact on urban design and on society, that it calls into question two concepts that the chapter seeks to reinterpret in a smart key: the conservation and protection of historical values, that direct or constrain the transformations of our cultural heritage (tangible), and new citizenship rights (intangible). It is a remarkable field of experimentation in which, however, scientific urban planning and design insufficiencies are compared, and where above all experimentation clashes with the acceptance of new technologies by local communities. The chapter stresses the need to re-create the concepts of conservation and citizenship rights, activating deep cognitive and mapping processes of local communities.

Keywords Historic centre · Smart city · Reconstruction · Technologies · Big data · Resilient city · L'Aquila · Italy

---

D. Di Ludovico (✉)
Dept. of Civil, Construction-Architectural and Environmental Engineering, University of L'Aquila, L'Aquila, Italy
e-mail: donato.diludovico@univaq.it

© The Author(s), under exclusive license to Springer Nature Switzerland AG 2021

F. F. Arefian et al. (eds.), *Historic Cities in the Face of Disasters*, The Urban Book Series,
https://doi.org/10.1007/978-3-030-77356-4_29

29.1 Introduction

The research concerns the impact of new technologies on historic centres and consequently on cultural heritage, especially the smaller, fragile and subject to abandonment, a topic that recalls the principles of the smart city and involves first of all the resilience, but also the quality of life and new citizenship rights.

Resilience is a broad concept and has been analysed by many authors in different scientific fields such as social sciences, environmental sciences, engineering, land-use and spatial planning, urban design, business management and more (Fleming and Ledogar 2018; Bhamra et al. 2011; Haimes 2009). This research connects to the concept related to urban design, concerning the shape of the city and its fragility (Borsekova et al. 2018; Zhang and Lid 2018; Meerow et al. 2016; Patel and Nosal 2016), a city composed of parts, centre (sometimes historic centre) and suburbs. In this sense, the definition of resilience by the organisation for Economic Co-operation and Development (OECD) can be used: "resilient cities are cities that can absorb, recover and prepare for future shocks (economic, environmental, social & institutional). Resilient cities promote sustainable development, well-being and inclusive growth" (OECD 2019). Therefore, a resilient city must also deal with both general topics such as quality of life and environmental sustainability and specific topics such as "*Minimal human vulnerability*", "*Social security*", "*Reduced physical exposure*", "*Continuity of critical services*", "*Reliable communications and mobility*", "*Integrated development planning*" and others (Patel and Nosal 2016). Many of these topics are closely linked to the evolution of technologies that are significantly changing urban life, whether to be in city centres or suburbs.

This chapter discusses primary results of a research which reviewed international scientific literature on historic city centres and their urban transformations following the application of new technologies. The research then examined the case study of the historic centre of L'Aquila in Abruzzo region, Italy. L'Aquila presents a special case for such investigation because the rapid transformation of its historic centre is the result of a response to a disaster. L'Aquila is a city of medieval origin (1254), which has a remarkable historic centre, both from the point of view of the extension (160 ha inside the medieval walls) and from the point of view of the concentration of cultural heritage (Properzi 2009). About 8,000 residents and about 7,000 students lived in the historic centre before the earthquake (Di Ludovico and Di Lodovico 2019). The city centre hosted several institutional offices such as municipality, provincial and regional offices and other public services. In 2009, the city was hit by a destructive earthquake (6.3 MW) with its epicentre within the city and very close to the historic centre (Di Ludovico and Dominici 2019). After the earthquake, almost the entire historic city centre was declared as "red zone", meaning it was inaccessible due to extensive damage to the buildings and urban fabric. It is currently under reconstruction.

Section 29.2 describes projects that use innovative technologies and smart principles in the post-earthquake reconstruction of the historic centre of L'Aquila. Section 29.3 deals with these topics, for the historic centres, based on a survey of the scientific literature, differentiating innovative technologies in tangible and intangible. Sections 29.4 and 29.5 analyse the impact of these technologies on two major themes typical of historic centres, conservation and new citizenship rights. The conclusion (Sect. 29.6) summarises the first results of the research and outline the next steps.

29.2 The Post-earthquake Reconstruction of the Historic Centre of L'Aquila and Innovative Technologies

The post-earthquake reconstruction of L'Aquila, as reported by many scholars (Di Ludovico et al. 2019; Di Ludovico and Di Lodovico 2019; D'Ovidio et al. 2015) is not based on a vision for the future of the city or short and long-term development strategies, but on a mere financial approach, and the reconstruction plan post 2009 earthquake is essentially a financial programme. Faced with the lack of an overall vision for the city and its territory, the huge injections of financial resources by the EU, State and private sector led to a significant number of experiments with innovative technologies during reconstruction that also concern its historic city centre. Those independent experimental interventions are rapidly changing the way L'Aquila and its historic centre are used and partially changing their morphology by pursuing the principles of smart cities (Mora et al. 2018). University of L'Aquila is also involved in many of those experiments as listed below:

- Smart tunnel (Fig. 29.1) is a 12.5 km tunnel in which the underground networks are housed in a single site under the road surface of the historic centre. Inside it is being tested. It includes a multi-core optical fibre produced by Japanese

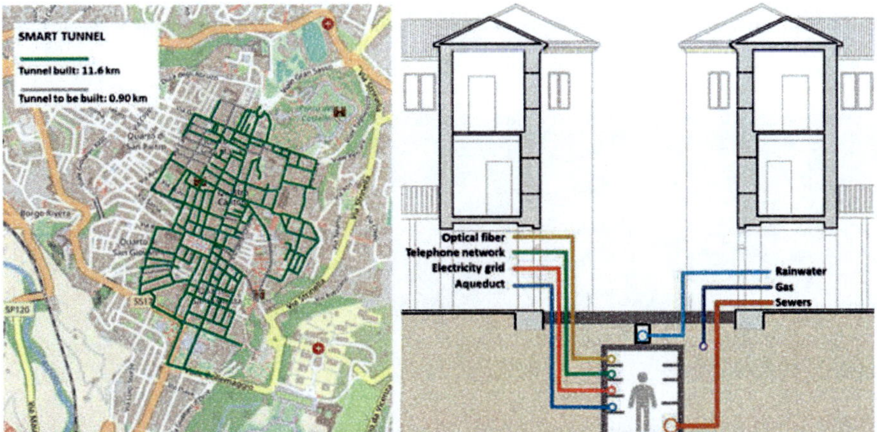

Fig. 29.1 Smart tunnel. The areas of the historic centre subject to intervention and the inside of the tunnel (http://www.sottoserviziaq.it/)

Sumitomo, capable of transporting data 10–15 times faster than the optical fibres currently in use.

- L'Aquila smart grids (Fig. 29.2) enable technologies and services required for becoming a smart city. It is the result of the memorandum of understanding between the National Electricity Board (Ente Nazionale per l'Energia Elettrica—ENEL) and the Municipality of L'Aquila in 2013. This is a series of interventions on the electricity distribution network aimed at implementing the functions of intelligent energy networks capable of distributing energy and communicating with customers. Smart grids are equipped with automation, remote control systems and devices to inform customers about what and how they are consuming. They help customers to be more aware of their electricity consumption (Urbanelli 2016).
- Structural monitoring (Fig. 29.3) uses sensors for assessing the vulnerability of structures and for planning maintenance activities (Alaggio et al. 2019).

Fig. 29.2 Smart grids. The intervention area, which includes the historic centre of L'Aquila (Urbanelli 2016)

Fig. 29.3 Structural monitoring of cultural heritage. The Basilica of Collemaggio in L'Aquila, positioning of the accelerometers on the east wall (left) and on the top of the facade (right) (Alaggio et al. 2020, 2019)

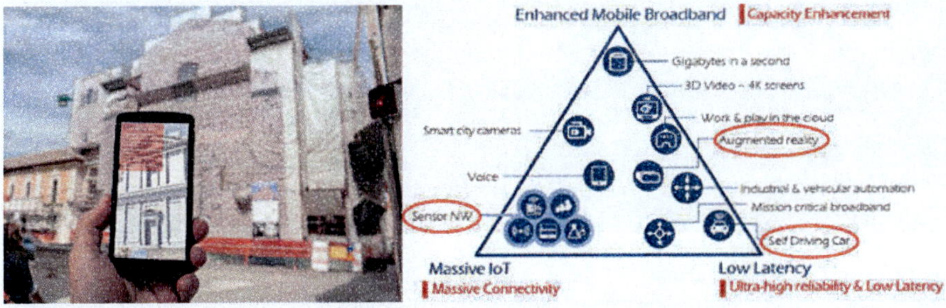

Fig. 29.4 (left) The experimentation of augmented reality on the monumental heritage of L'Aquila in the framework of the INCIPICT project of the University of L'Aquila (Brusaporci et al. 2017). (right) The 5G usage scenarios envisaged for the International Mobile Telecommunications—IMT 2020 (IEEE 2017)

- Augmented reality (Fig. 29.4, left) to represent new levels of information and preserve the memory of the reconstruction phases of the city and its history, also to experiment with new modes of urban use (Brusaporci et al. 2017).
- The 5th generation of mobile technology—5G—to bring advanced technology to the city (Fig. 29.4 right). L'Aquila is one of the cities of experimentation and which will significantly improve the capabilities of mobile broadband and address the new needs of the network society. The University of L'Aquila will develop research into the potential of 5G connected to the self driving car, augmented reality and sensor networking (University of L'Aquila/ZTE agreement; IEEE 2017).

Alongside these experiments, various reconstruction activities (design and construction) produce an enormous amount of information (Big Data) that represent a considerable advancement of the knowledge on the historic centre (Di Ludovico 2019) and therefore of its resilience capacity. A large amount of data, already transferred into a GIS and digital database, covers the characteristics of the buildings before and after the reconstruction, economic and financial analysis of the reconstruction, data on mobility and smart city projects. This data can help build an urban performance verification tool and understand how much the reconstruction has increased the resilience of the city. Specifically, this knowledge concerns:

- Architecture of buildings that are analysed in historical terms and typologies. For example, there has been an extensive use of the 3D laser scanner survey (Fig. 29.5, left), considerably deepening the knowledge level of buildings. Produced 3D scans contain a great extent of information such as geometric, structural, materials and more (Dominici et al. 2013).
- Structure of buildings that has been subjected to extensive tests and investigations, but to which innovative construction techniques and technologies have also been applied, such as the use of seismic isolators. The WebGIS of the Special Office for the Reconstruction of L'Aquila—USRA has published information on

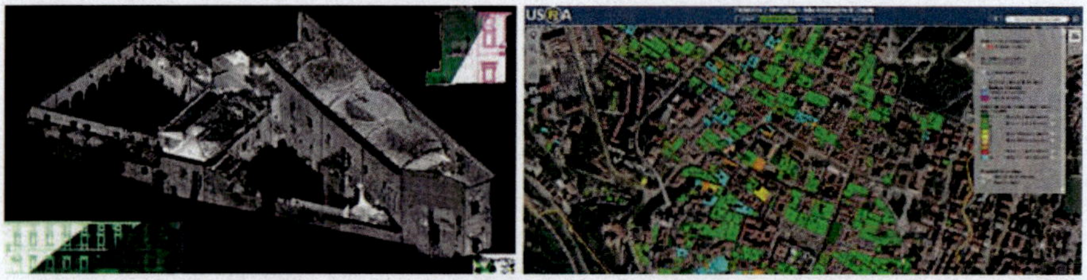

Fig. 29.5 (left) The 3D survey with laser scanner of the monastery of San Basilio (L'Aquila) (Dominici et al. 2013). (right) The webgis of the special office for the reconstruction of L'Aquila—USRA (screenshot of the site http://webgis.comuneaq.usra.it/mappa_def.php, accessed 04 August 2019)

buildings including the seismic vulnerability through a pre- and post-intervention risk indicator for each building (Fig. 29.5, right).

- Soil characteristics are locally identified at their seismic response level through extensive tests and investigations. Just after the earthquake, the National Department of Civil Protection (DPCN) prepared maps of the 3rd level seismic microzonation. Maps characterised soils by quantitative parameters of seismically homogeneous areas.

The above technical innovative projects characterise the reconstruction of L'Aquila's historic city centre and present an important step towards a safer and more resilient city. But the lack of a long-term vision for the future and a medium-term strategy that refers to the principles of the smart city to position L'Aquila among the most advanced European cities could potentially make those substantial invested resources useless (Di Ludovico 2017, 2019; Di Ludovico and Di Lodovico 2019). These innovative projects can represent the new drivers for the growth of cities, with the aim of relaunching pieces of cities, fragile and problematic as the historical centres in which today the quality of life has several critical elements. However, there are still some open questions about the rigid conservative purpose of historic buildings that in L'Aquila's case (as an example) did not allow the installation of advanced technical systems, which could have produced renewable energy (SBAP 2011). Remembering that the 160 hectares of the historic city centre can accommodate up to 20,000 inhabitants; therefore, the issue of clean energy is fundamental and must necessarily be addressed to meet the current needs of the contemporary society. Also remains the issue of extensive data generation without tapping on their potentials, and that even today (as of 2020), the big data are crammed into the archives of some public institutions. Yet, the generated big data are not analysed nor made available to the community to structure new services or articulate new urban management tools but are used by private individuals for their business purposes reducing access by citizens to new citizenship rights.

The rapid post-earthquake transformation of L'Aquila and its historic centre, generated by the implementation of innovative projects adhering to the principles of the smart city, therefore brings out some innovative themes. These themes are in fact related to the use of new smart technologies in the transformation processes of contemporary cities, with a significant impact on urban design and society. For example, two fundamental themes that emerge are the conservation of historical values and the new citizenship rights that will have to be analysed with reference to the principles of the smart city.

29.3 Innovative Tangible and Intangible Technologies for Historic City Centres

In the scientific literature, the intersection of smart technologies and historic city centres has not been adequately investigated. In general, the themes identified in that intersection can be distinguished in two main macro-categories: (a) those of "*tangible*" technologies and (b) those of "*intangible*" technologies. The latter are mostly connected to the theme "big data", their production and prediction-oriented analysis.

There are numerous publications on tangible technologies (Table 29.1), however, the greatest number concerns energy retrofit (Franco and Magrini 2017), much less exist on smart materials (Kumawat et al. 2017) and even more less on urban utility tunnels (Valdenebro and Ginema 2018). On intangible technologies (Table 29.1), there are numerous references to "urban information systems" (Amato et al. 2017) aimed at tourist and communicating identity and cultural values (Borda and Bowen 2017), a theme that also involves augmented reality (Barile et al. 2014). Additionally, various literature concern smart communities (Lillevold and Haarstad 2019) and sensors, in particular about structural monitoring (Zonta et al. 2010). There is much less literature on autonomous driving (Franke et al. 1998), smart grid (Marzal Diaz 2017) and remote sensing (Parrinello and Picchio 2019) that address historic centres.

On intangible technologies, there is also the heritage building information modelling—HBIM (Megahed 2015), that recognises cultural heritage aspects of a building and therefore indirectly addresses historic city centres, but this experimentation of BIM is not yet extended from building scales to urban fabrics and blocks in which the cultural heritage forms a context. This issue also is a common characteristic of tangible technologies, for which scientific sources rather than referring directly to historic centres as urban sites in transformation, refer to the restoration and conservation of cultural heritage of individual buildings, leaving out the urban scale, shape and design. Furthermore, some literature rather than referring to the historic city centre, refer to the old town or downtown (Franke et al. 1998).

As Table 29.1 and the case of L'Aquila present new tangible and intangible technologies plus big data represent emerging themes of changes in historic city centres. However, their application is not neutral; it requires smart redefinition of the fundamental concepts of cultural heritage conservation and citizenship rights whose theoretical development of paradigms are significantly influenced by the principles of the smart city.

Table 29.1 Tangible and intangible technologies in smart-historic city centres

Technology	Location	References
"Tangible" technologies		
1. Urban utility tunnels	Pamplona (Spain)	Valdenebro and Ginema (2018)
2. Smart materials/adaptive and active materials	Cultural heritage (Italy and other Europe countries)	Kumawat et al. (2017); Ausiello (2018)
3. Energy retrofit	Cultural heritage (Italy and other Europe countries)	Franco and Magrini (2017); AiCARR (2014); Broström and Nilsen (2012); Kilian et al. (2010)
"Intangible" technologies		
5. Urban information systems/atlas of knowledge/big and open data	Paestum (Italy); Chicago (USA)	Amato et al. (2017); Longley and Duxbury (2016); Mone (2015)
6. Virtual and augmented reality applications	Naple (Italy)	Barile et al. (2014)
7. HBIM approach in historic preservation and management	Teorical	Megahed (2015)
8. Autonomous driving	Teorical	Franke et al (1998)
9. Smart communities/communities networks/identity	Røros (Norway)	Lillevold and Haarstad (2019)
10. Smart IoT-based building/bluetooth sensors network/sensors for sites protection—structural monitoring	Valencia (Spain); Trento, Verona, Portogruaro, Novara, Vicoforte, Modena, Orvieto, Rome (Italy)	Plumè et al. (2019); De Stefano et al. (2016); Zonta et al. (2010)
11. Smart grid, smart urban lighting	Sutri, Spoleto (Italy)	Marzal Diaz (2017); EnelX (2018)
12. Remote sensing/drone	Bethlehem (Palestine)	Parrinello and Picchio (2019)

29.4 Evolution of the Concept of "Conservation" of Cultural Heritage

The concept of conservation is connected to two fields of research: First is architecture in which conservation is through the restoration of buildings as cultural heritage. Second is urban development and planning in which conservation is expressed through the rehabilitation of historic urban fabrics. This research addresses the second field, which is inextricably linked to the notion of valorisation in Italy. In its contemporary meaning, valorisation was introduced in 2005 by the Italian Ministry of Heritage and Cultural Activities in the "Project for the definition of a model for the realisation of the Management Plans of UNESCO sites" (Progetto di definizione di un modello per la realizzazione dei Piani di Gestione dei siti UNESCO). It presents an innovative methodology that refers to the concept of "integrated approach", ideally capable of integrating two sets of needs to conserve sites and enhance the socio-economic development of their territories (Mibac 2005). Thus, on one hand, there is the protection and conservation of cultural heritage (understood as buildings but also as contexts) and on the other, development and transformation (valorisation).

Another influential document is the European Commission (EC) report "Getting cultural heritage to work for Europe", which considers cultural heritage a factor of economic production, cohesion and inclusion. The report identifies cultural heritage as an element to be used in an innovative and sustainable way for enhancing and valorising the environment and cultural landscape. It refers to European trends (but goes also beyond Europe) that aim to reach the "balance between preservation/conservation and adaptive reuse and upgrade" (EC 2015, p. III), introducing the need for a reuse adaptive aimed at valorisation, alongside the conservation. UNESCO also emphasises on this balance in its "recommendation on the historic urban landscape" indicating that the "conservation has become a strategy to achieve a balance between urban growth and quality of life on a sustainable basis" (UNESCO 2011, p. 2). This does not happen in the "Faro" Convention (a framework convention on the value of cultural heritage for society), which emphasises the sustainable use of cultural heritage with the goal of quality of life (CoE 2005).

In Italy, urban development, adaptive reuse of cultural heritage and specifically of historic centres are conflicting themes with those of conservation and restoration. Today, these are two fields which reject the use of some new technologies such as those to produce renewable sources energy (photovoltaic, small wind) or wireless/mobile networks installations or more generally ICT networks installations. That of new technologies is a very wide and rapidly evolving field that the conservation discipline, which instead evolves slowly, is not capable of to check. An attempt to overcome this conflict by pursuing an "integrated" approach and an adaptive reuse is by Salvador Muñoz Viñas, a researcher trying to go beyond the traditional concept of conservation. He states that in the contemporary conception of conservation, the notion of "truth" has been replaced with the notions of "function, use or value", thus questioning the crucial notions, deriving from classical theories, of "reversibility, universality and objectivity" (Muñoz Viñas 2002). This new approach introduces a new element in the evaluation of conservation, the desires and role of communities and stakeholders, which identify the new functions of the protected cultural heritage in urban system by adapting their meaning to places and their identities. Thus, it will also be possible to introduce the concept of "flexible conservation", more adapted to the needs of contemporary "digital" and "technological" society, without, however, overshadowing all the traditional disciplinary research on conservation, cantered on the object and materials.

Despite the above, in the case of L'Aquila and its city centre with considerable historical value, the concept of conservation has been applied in a restrictive object-oriented way, focused on single buildings. Instead of articulating actions on the urban system, restoration actions in post-earthquake L'Aquila were merely organised around a single element of the system. It can be predicted that the result of such an approach will be a reproduced Historic Centre at the end of reconstruction period that can be considered a museified centre in which the cultural heritage is not valorised as a whole.

29.5 Citizenship Rights and Technological Innovations

The concept of "right to the city" was formulated at the end of the 1960s by Lefebvre who understood it as a superior form of rights concerning first of all freedom, socialisation and living, but also fruition of spaces (the latter differs from the right to property). These new aspects of living in city implied a democratic change of public decision-making methods which characterised the rise of the city of that period strongly marked capitalism (Lefebvre 2014). The above issues have been deeply analysed by David Harvey. In one of his last essays, Harvey analyses the new demand for citizenship rights that is emerging in the world and what happened from the 1970s to the present. This is a question generated by fragmentated claims on what citizenship rights are (Blokland et al. 2015). According to Harvey, "'the right to the city' is far more than a right of individual or group access to the resources that the city embodies: it is a right to change ourselves by changing the city more after our heart's desire" (Harvey 2012, p. 4). It is, therefore, a collective right rather than an individual one because reinventing the city inevitably depends upon the exercise of a collective power over the processes of urbanisation. Based on these general premises, Harvey states that the freedom to make a remake ourselves and our cities is one of the most precious yet most neglected of our human right, because today (but also at the time of Lefebvre's conceptualisation) is the capitalism to guide urbanisation, as capitalism needs urbanisation to absorb the surplus of products it continually produces (Harvey 2012).

The new demand for rights to the city is now concentrated on some major issues such as the right to housing, inequalities and socio-spatial injustices (Busquet 2013) that are not very different to those identified by Lefebrve in the 1970s. But this new demand also refers to aspects of equality more appropriate to contemporary society, for example, those linked to globalisation and multiculturalism (Božilović 2012), contrasting with the widespread individualism of our society (Bauman 2001).

Rights to the contemporary city also concern innovative technologies and their accessibility that in turn involve the quality of life (Rights to the city → Quality of life). The question is whether those technologies are or can be a component of the capitalism-urbanisation connection, and how they impact citizenship rights and urban shape. As observed in L'Aquila's case, the reconstruction process is producing a huge amount of data mostly related to the building and structural factors of the buildings, or, for example, the soils and their characteristics. They are less concerned with social factors and their descriptors or, e.g. public and private mobility. Unfortunately, these data, let it be clear in the hands of public bodies, are not made available and therefore cannot be processed and used to structure new services to the population and expand the boundaries of citizenship rights, i.e. the rights to live in an up-to-date city with "smart" principles.

It can be argued that the category of tangible technologies previously introduced has a positive effect on the quality of life, and their impact on the right to the city is connected to the community's ability to access these technologies in terms of costs and ease of use. The category of intangible technologies is much more complex, because it includes one of the most important new intangible technological subjects in recent times: the generation and use of mass data. It involves the so-called big data and their related infrastructures, new forms of social connection called digital connection, new polarisations but also new exclusions, exacerbating even the exclusions of marginalised groups or those who cannot access the digital network (Willis 2019).

"Big data, artificial intelligence, cybernetics and behavioural economics are shaping our society" (Helbing 2018, p. 81). According to literature, big data therefore can be used to better understand social change, to assist groups and organisations who work for social change, to support and facilitate public discussion and public awareness on social issues, to support and develop research, to study systems and solutions for the inclusive society and to address the risks and confidentiality of data (Tseng and Harmon 2018). Additionally, such data analyses, which are based on data extrapolated from new technologies (e.g. several sensor networks), can be also used to help citizens and the city. For example, they can be used to create new services for citizens, to improve existing ones (e.g. technological infrastructure management), to feed the algorithms of new models of urban simulation (Artificial Intelligence) whose development is accelerating considerably (Townsend 2013). The above examples are a few of various ways big data-related technologies, and innovations can facilitate life in fragile and inaccessible places such as historic city centres, thus adding new meanings and value to citizenship rights. However, these new rights are strictly limited by how big data is used. Since they are almost exclusively private property, these are used for "private" purposes, aimed at the growth of companies rather than social and economic development. It is therefore necessary to recognise and protect the new citizenship rights deriving from the use of big data, to experiment with technologies that leave the ownership of data to the communities who produce them, to allow the same communities to self-determine through their use (Kitchin et al. 2019).

Additionally, attention needs to be made to different demographic and community groups when talking about right to the city and innovative technologies. For example, for a community that inhabits a small and fragile historic centre or a remote rural location, perhaps even in a phase of abandonment, those above mentioned are even more complex. The reason is the demographic characteristics of the historic centre's population presumably with more elderly with less physical accessibility to services and infrastructure. In such cases, the role of the community should be considered truly central, as well as that of one's own data, to prevent the fragility of the community from becoming chronic. Since small and fragile historic centres produce small amounts of data, it would be necessary to set up networks of small historic centres capable of producing significant masses of data. The produced data can be analysed and managed through innovative services aimed at reducing travel time, increasing physical and digital accessibility (digital divide), facilitating new working methods (for example remote working), enhancing tourism, strengthening services and urban management (mobility, community services, etc.) and developing digital systems dedicated to agriculture (Smart Rural Grid, Farmers' Markets) (Di Ludovico et al. 2014).

29.6 Conclusion

Understanding the impact of the application of innovative technologies such as digital ones on historic urban areas opens a new path of forward looking research domain. This includes those technologies applied in post-disaster reconstruction of historic city centres, as it was in the case of L'Aquila after the 2009 earthquake. This chapter contributed to the subject matter by examining scientific literature on the smart-historic centre binomial and using the L'Aquila's case to focus on a special category of small and more fragile historic centres. The research identified two major directions and perspectives related to the smart-historic centres and the needs of contemporary society, namely the concept of "conservation" and that of "new citizenship rights". They characterise potential capabilities and impacts of the application of tangible and intangible innovative technologies and provide analytical directions to understand such innovative technologies. From conservation perspective, it is crucial to manage the impact of innovative technologies on restoration, rehabilitation and management of cultural heritage and on historic centres. Yet, a balance must be sought between conservation and protection of physicality of buildings and the socio-economic development and facilitating the adaptive reuse of historic cities. A community-oriented functionalist approach is needed to give local communities, stakeholders and their desires a primary role while thinking of innovative technologies in those contexts. The attribution of value and function to cultural heritage can help to achieve this balance. From the perspective of "new citizenship rights" in fragile urban contexts, the focus should be made on the potential of innovative networks in small historic centres and their communities. Communities must be seen as producers of information and data, which can feed new types and levels of services capable of generating and enhancing social and economic development of those fragile contexts.

As an emerging and evolving research domain, further examinations of various aspects of linking conservation of built heritage, innovative technologies and citizenship rights are needed. For example, the next step of this research will address the design of the platform that will integrate the "community mapping system" with the "knowledge system" and the implementation of urban simulation models supporting data-driven planning.

References

AiCARR (eds) (2014) Historical and existing buildings: designing the retrofit. In: Proceedings of the 49° AiCARR international conference, Milan

Alaggio R et al (2020) Two-years static and dynamic monitoring of the Santa Maria di Collemaggio basilica. Constr Build Mater 121069. https://doi.org/10.1016/j.conbuildmat.2020.121069

Alaggio R, Antonacci E, Galeota D (2019) Il monitoraggio della Basilica di Collemaggio a L'Aquila. https://www.ingenio-web.it/20239-limportanza-del-monitoraggio-strutturale-nella-conservazione-degli-edifici-monumentali-un-caso-studio. Accessed 4 Aug 2019

Amato F et al (2017) Big data meets digital cultural heritage: design and implementation of SCRABS, a smart context-aware browsing assistant for cultural environments. ACM J Comput Cult Heritage 10(1):1–23. https://doi.org/10.1145/3012286

Ausiello G (2018) Materiali innovativi e conservazione. In: D'Agostino S (ed) Ingegneria e Beni culturali. Il Mulino, Bologna

Barile F et al (2014) ICT solutions for the OR.C.HE.S.T.R.A. project: from personalized selection to enhanced fruition of cultural heritage data. In: 2014 tenth international conference on signal-image technology and internet-based systems. Marrakech, pp 501–507. http://dx.doi.org/10.1109/SITIS.2014.12

Bauman Z (2001) The individualized society. Polity Press, Malden

Bhamra R, Dania S, Burnard K (2011) Resilience: the concept, a literature review and future directions. Int J Prod Res 49(18):5375–5393. https://doi.org/10.1080/00207543.2011.563826

Blokland T et al (2015) Urban citizenship and right to the city: the fragmentation of claims. Int J Urban Regional Res 39(4):655–665. http://dx.doi.org/10.1111/1468-2427.12259

Borda A, Bowen JP (2017) Smart cities and cultural heritage—a review of developments and future opportunities. In: Proceedings of EVA London 2017. http://dx.doi.org/10.14236/ewic/EVA2017.2

Borsekova K, Nijkamp P, Guevara P (2018) Urban resilience patterns after an external shock: an exploratory study. Int J Disaster Risk Reduction 31:381–392. https://doi.org/10.1016/j.ijdrr.2018.05.012

Božilović J (2012) The global city and changes in the concept of citizenship. Facta Univ Series Philos Sociol Psychol History 11(1):103–114

Broström T, Nilsen L (eds) (2012) Energy efficiency in historic buildings, postprints from the conference, Visby, 9–11 Feb 2011, Gotland University Press, Visby. http://uu.diva-portal.org/smash/get/diva2:507200/COVER02.pdf. Accessed 13 Aug 2019

Brusaporci S et al (2017) Augmented reality for historical storytelling. The INCIPICT project for the reconstruction of tangible and intangible image of L'Aquila historical centre. Proceedings 1(1083):1–20. https://doi.org/10.3390/proceedings1091083

Busquet G (2013) Question urbaine et droit à la ville. Mouvements 74:113–122. https://doi.org/10.3917/mouv.074.0113

CoE (2005) Council of Europe framework convention on the value of cultural heritage for society. Faro Convention. Council of Europe. https://rm.coe.int/1680083746. Accessed 14 Aug 2019

D'Ovidio G, Di Ludovico D, La Rocca GL (2015) Urban planning and mobility critical issues in post-earthquake configuration: L'Aquila city case study. Proc Eng 161:1815–1819. https://doi.org/10.1016/j.proeng.2016.08.670

De Stefano A, Matta E, Clemente P (2016) Structural health monitoring of historical heritage in Italy: some relevant experiences. J Civil Struct Health Monit 6:83–106. https://doi.org/10.1007/s13349-016-0154-y

Di Ludovico D (2017) Il Progetto Urbanistico, prove di innovazione per la città del futuro. Aracne Editrice, Canterano

Di Ludovico D (2019) L'Aquila città della conoscenza e dell'innovazione: le sperimentazioni in atto. Urbanistica Informazioni 280–281:57–58

Di Ludovico D, Di Lodovico L (2019) L'Aquila a 10 anni dal sisma. In: Pierantoni I, Salvi D, Sargolini M (eds) Nuovi sentieri di sviluppo per l'appennino marchigiano dopo il sisma del 2016. Quaderni del Consiglio Regionale delle Marche, Ancona

Di Ludovico D, Dominici D (2019) How to combine the smart city and the historic centre: suggestions from a case study. In: Bologna S (ed) A set of good practices and recommendations for smart city resilience engineering and evaluation. AIIC Publisher, Rome, pp 36–45

Di Ludovico D, Properzi P, Graziosi F (2014) From a smart city to a smart up-country. the new city-territory of L'Aquila. Tema, J Land Use Mobility Environ (Special issue):353–364 (June 2014). http://dx.doi.org/10.6092/1970-9870/2482

Di Ludovico D, D'Ovidio G, Santilli D (2019) Post-earthquake reconstruction as an opportunity for a sustainable reorganisation of transport and urban structure. Cities 96: https://doi.org/10.1016/j.cities.2019.102447

Dominici D et al (2013) Cultural heritage documentation using geomatic techniques, case study: San Basilio's monastery—L'Aquila. In: Proceedings of the 2013 digital heritage international congress, 28 Oct–1 Nov, Marseille (Fr), vol 1, pp 211–214

EC (2015) Getting cultural heritage to work for Europe. Report of the Horizon 2020 Expert Group on cultural heritage. Directorate-General for Research and Innovation. Publications Office of the European Union, Luxembourg. https://ec.europa.eu/programmes/horizon2020/en/news/getting-cultural-heritage-work-europe. Accessed 14 Aug 2019

EnelX (2018) Enel Sole illuminates the streets of Spoleto and Pescara with sustainable light. https://www.enelx.com/it/en/news-media/notizie/2018/02/energy-efficiency-enelsole-spoleto-pescara. Accessed 13 Aug 2019

Fleming J, Ledogar RJ (2018) Resilience, an evolving concept: a review of literature relevant to aboriginal research. Pimatisiwin 6(2):7–23

Franco G, Magrini A (eds) (2017) Historical Buildings and Energy. Springer International Publishing, Cham, Switzerland

Franke U et al (1998) Autonomous driving goes downtown. IEEE Intell Syst Appl 13(6):40–48. https://doi.org/10.1109/5254.736001

Haimes YY (2009) On the definition of resilience in systems. Risk Anal 29(4). http://dx.doi.org/10.1111/j.1539-6924.2009.01216.x

Harvey D (2012) Rebel cities: from the right to the city to the urban revolution. Verso, London, New York

Helbing D et al (2018) Will democracy survive big data and artificial intelligence?. In: Helbing D (eds) Towards digital enlightenment. Springer. Cham. https://doi.org/10.1007/978-3-319-90869-4_7

IEEE (2017) ITU-R agrees on key performance requirements for IMT-2020="5G". http://techblog.comsoc.org/2017/03/02/itu-r-agrees-on-key-performance-requirements-for-imt-20205g/. Accessed 25 Jan 2019

Kilian R, Vyhlidal T, Broström T (eds) (2010) Developments in climate control of historic buildings. In: Proceedings from the international conference "climatization of historic buildings, state of the art", Fraunhofer IRB, Verlag

Kitchin R, Cardullo P, Di Feliciantonio C (2019) Citizenship, justice and the right to the smart city. In: Cardullo P, Di Feliciantonio C, Kitchin R (eds) The right to the smart city. Emerald publishing, Bingley

Kumawat A et al (2017) Review of recent advances in smart materials. Int J Innov Res Sci (Special Issue) 15:189–192

Lefebvre H (2014) Il diritto alla città. In: Morosato G (eds). Ombre Corte editore, Verona

Lillevold K, Haarstad H (2019) The deep city: cultural heritage as a resource for sustainable local transformation. Local Environ 24(4):329–341. https://doi.org/10.1080/13549839.2019.1567481

Longley A, Duxbury N (2016) Mapping cultural intangibles. City Cult Soc 7(1):1–7. https://doi.org/10.1016/j.ccs.2015.12.006

Marzal Díaz F (2017) Un proyecto "Smart" para el casco histórico de Sutri. In: ACE: architecture, city and environment = Arquitectura. Ciudad y Entorno 11(33):197–212. http://dx.doi.org/10.5821/ace.11.33.5156

Meerow S, Newell JP, Stults M (2016) Defining urban resilience: a review. Landscape Urban Plann 147:38–49. https://doi.org/10.1016/j.landurbplan.2015.11.011

Megahed NA (2015) Towards a theoretical framework for hbim approach in historic preservation and management. Archnet-IJAR 9(3):130–147

Mibac (2005) Progetto di definizione di un modello per la realizzazione dei Piani di Gestione dei siti UNESCO. Ministero per i Beni e le Attività Culturali and Ernst & Young Financial Business Advisor S.p.A. http://www.sardegna.beniculturali.it/getFile.php?id=10391. Accessed 14 Aug 2019

Mone G (2015) The new smart cities. Commun ACM 58(7):20–21. https://doi.org/10.1145/2771297

Mora L, Deakin M, Reid A (2018) Strategic principles for smart city development: a multiple case study analysis of European best practices. Technol Forecast Soc Chang. https://doi.org/10.1016/j.techfore.2018.07.035

Muñoz Viñas S (2002) Contemporary theory of conservation. Rev Conserv 3

OECD (2019) Resilient cities. https://www.oecd.org/regional/resilient-cities.htm#:~:text=Resilient%20cities%20are%20cities%20that,cities%20can%20increase%20their%20resilience. Accessed 6 Oct 2020

Parrinello S, Picchio F (2019) Integration and comparison of close-range SFM methodologies for the analysis and the development of the historical city center of Bethlehem. Int Arch Photogramm Remote Sens Spatial Inf Sci XLII-2/W9:589–595. https://doi.org/10.5194/isprs-archives-XLII-2-W9-589-2019

Patel R, Nosal L (2016) Defining the Resilient City. United Nations University Centre for Policy Research. Working paper 6. https://pdfs.semanticscholar.org/aaec/cdb4b59824958c0442be5af3116003da73fe.pdf. Accessed 22 Jan 2019

Plumé JM et al (2019) Evaluation of the use of a city center through the use of bluetooth sensors network. Sustainability 11:1002. https://doi.org/10.3390/su11041002

Properzi P (2009) 20 Città a confronto, L'Aquila. In: Benevolo L, Piroddi E (eds) Il Nuovo Manuale di Urbanistica, vol 3. Gruppo Mancosu Editore, Roma, pp 280–325

SBAP (2011) Prescrizioni per gli interventi nei centri storici di L'Aquila e frazioni. Comune dell'Aquila, Soprintendenza per i Beni Architettonici e Paesaggistici per l'Abruzzo. http://www.comune.laquila.it/moduli/output_immagine.php?id=1877. Accessed 25 Jan 2019

Townsend AM (2013) Smart cities. Big data, civic hackers and the quest of a new Utopia. Norton & Company, New York, London

Tseng FM, Harmon R (2018) The impact of big data analytics on the dynamics of social change. Technol Forecast Soc Chang 130:56. https://doi.org/10.1016/j.techfore.2018.02.010

UNESCO (2011) Recommendation on the historic urban landscape. United Nations, Educational, Scientific and Cultural Organization. https://whc.unesco.org/uploads/activities/documents/activity-638-98.pdf. Accessed 14 Aug 2019

Urbanelli G (2016) Il cunicolo tecnologico (parte della Smart City). Enel Distribuzione SpA. http://www.serviziarete.it/wp-content/uploads/2016/04/Gianfranco.URBANELLI.enel.pdf. Accessed 14 Sept 2019

Valdenebro JV, Gimena FN (2018) Urban utility tunnels as a long-term solution for the sustainable revitalization of historic centres: the case study of Pamplona-Spain. Tunn Undergr Space Technol 81:228–236. https://doi.org/10.1016/j.tust.2018.07.024

Willis KS (2019) Whose right to the smart city? In: Cardullo P, Di Feliciantonio C, Kitchin R (eds) The right to the smart city. Emerald publishing, Bingley

Zhang X, Lid H (2018) Urban resilience and urban sustainability: what we know and what do not know? Cities 72:141–148. https://doi.org/10.1016/j.cities.2017.08.009

Zonta D et al (2010) Wireless sensor networks for permanent health monitoring of historic buildings. Smart Struct Syst 6(5–6):595–618

Chapter 30
Evaluating Visitors' Experiences at St Augustine's Abbey, Canterbury

Ayda Majd Ardekani, Sophia Labadi, and Rocio von Jungenfeld

Abstract This chapter discusses an ongoing research which evaluates visitors' experiences at St Augustine's Abbey before and after in situ projections of reconstructed imageries of non-existent Abbey artefacts. The research is based on the contrasting opinions of Viollet-Le-Duc and William Morris on reconstruction and how it should be wrought. It introduces a case study on visitors' experiences at St Augustine's Abbey focused on visitors' views on themes associated with digital reconstruction such as authenticity and realism. The results indicate that a considerable number of respondents thought in situ digital reconstructions of the Abbey artefacts would have a positive impact on their visitation experience. Visitors associate authenticity with accuracy and factuality and inauthenticity with not being original. Respondents stated that digital reconstructions are more hyperreal than real as they create an illusionary vision of reality. The case study also analyses visitors' perception of the sixteenth-century Virtual Reality (VR) of the Abbey with emphasis on immersion and quality of the information provided. Lastly, the chapter introduces methods for digital reconstruction of non-existent artefacts for future workflows of the research.

Keywords Digital reconstruction · St. Augustine's Abbey · Visitor experience · Authenticity · Virtual Reality (VR) · Augmented Reality (AR) · Realism · Canterbury · UK

A. M. Ardekani (✉) · S. Labadi · R. von Jungenfeld
University of Kent, Canterbury, UK
e-mail: am2324@kent.ac.uk

S. Labadi
e-mail: s.labadi@kent.ac.uk

R. von Jungenfeld
e-mail: r.von-jungenfeld@kent.ac.u

© The Author(s), under exclusive license to Springer Nature Switzerland AG 2021

F. F. Arefian et al. (eds.), *Historic Cities in the Face of Disasters*, The Urban Book Series,
https://doi.org/10.1007/978-3-030-77356-4_30

30.1 Introduction

This chapter introduces themes around digital reconstruction in the context of cultural heritage artefacts. It presents two philosophical concepts of authenticity and realism that are interwoven with the subject of reconstruction. It refers to pro and anti-restoration manifestos about reconstruction and restoration as well as which values must be acknowledged in the act of recovery of a cultural heritage object.

St Augustine's Abbey is part of a UNESCO World Heritage Site designated in 1988. Over time, the Abbey has been significantly demolished and reduced to its foundations. Based on a qualitative case study with 65 participants at St Augustine's Abbey in Canterbury the research evaluates the perception of visitors regarding three notions: firstly, the public acceptance of having digital reconstruction in heritage sites, including users' opinions on already available Virtual Reality (VR) headsets and their thoughts on having digital installations in heritage sites; secondly, visitors' views on authenticity and how they perceive it; thirdly, the public's understanding of the concepts of the real and the hyperreal compared to how the terms of representation and simulation are defined in philosophical discussions.

The chapter also presents workflows of digital reconstruction and exemplifies the reconstruction of heritage sites using such methodologies.

30.2 St Augustine's Abbey, What Remains and Whether to Restore Non-existing Artefacts?

St Augustine's Abbey is a Benedictine monastery in Canterbury (Kent), in the south-east of the UK (Fig. 30.1). It was founded by St Augustine in 588 who was sent on a mission by Pope Gregory the Great to reintroduce Christianity into the south of England. It was a centre of learning and spirituality for almost a thousand years until suppressed by order of King Henry VIII as part of the dissolution of the monasteries (Roebuck 1997). The Abbey is part of the UNESCO World Heritage Site "Canterbury Cathedral, St Augustine's Abbey and St Martin's church" designated in 1988. It is of outstanding universal value according to the selection criteria (ii) of the "Operational Guidelines for the Implementation of the World Heritage Convention" due to its exhibition of "an important interchange of human values, over a span of time and within a cultural area of the world, on developments in architecture or technology, monumental arts, town planning and landscape design" (UNESCO 2017, p. 25). The influence that St Augustine's Abbey had in the Middle Ages extended far beyond Kent and Northumbria in Anglo-Saxons kingdoms (ICOMOS 1988). The Abbey, as a consequence of a series of developments, demolitions, natural hazards and indeed the dissolution, has been reduced to its foundations. At present, the remains of the Abbey are ruins of stone foundations (Fig. 30.2) on the site and a collection of excavated and preserved artefacts housed in the Abbey museum as well as at the English Heritage collections.

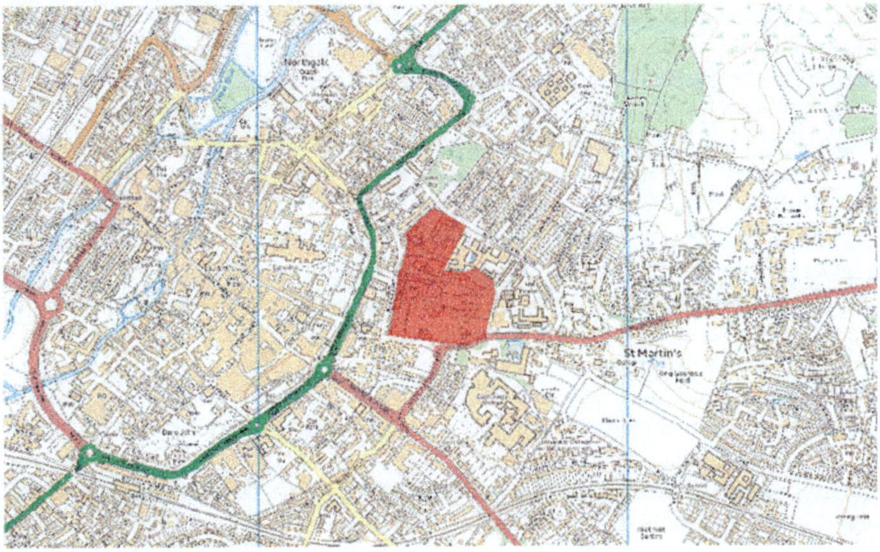

Fig. 30.1 Ordnance Survey, St Augustine's Abbey, 1:10000, Historic England. © Crown copyright and database rights 2018 ordnance survey (Ordnance Survey 2018)

Fig. 30.2 Remaining foundations of the Nave area at St Augustine's Abbey, Canterbury (author)

Historically, two very opposing points of views have been expressed on the subject of restoration. At one end of the spectrum, there are pro-restoration manifestos that welcome and encourage restoration of cultural heritage artefacts, and at the other end, there are anti-restoration theories which prohibit the recovery of an object created in the past. To Viollet-le-Duc, restoration meant an act of re-establishing an edifice in a finished form, which actually may have never existed. Two questions arise within the context of restoration from Viollet-le-Duc's point of view: whether to restore an edifice according to its original state or to restore it taking into account later developments and modifications that occurred to the structure. What matters in restoring an artefact is to retain its integrity (Viollet-le-Duc 1854).

Contemporary to Viollet-le-Duc, John Ruskin and William Morris expressed their strong anti-restoration opinions. For example, in the manifesto of "The Society for the Protection of Ancient Buildings" (1887), Morris draws attention to the need to apply as little intervention as possible and to prevent further decadence by carrying out everyday care. Ruskin considered "restoration" as the worst destruction a building can undergo, involving false descriptions of the ruined monument. He believed that each piece of work of art is unique and cannot be redone without creating a fake, also that a restored monument would not be a genuine manifestation of the original and that we do not have the right to intervene with the creations from people of the past unless unconditionally necessary (Ruskin 1849).

In his manifesto, Morris points at the opposed approaches towards restoration. What distinguishes the two is how restoration is wrought and what the outcome is. Firstly, the act of restoration which was applied in the fashion of the time of restoration. For example, a historical building of the twelfth century would have been restored in the sixteenth century or later styles. Restoration in this context would have included added to or altered features. Although, "whatever history it destroyed, left history in the gap, and was alive with the spirit of the deeds done midst its fashioning, though harsh and visible enough, were by their very contrast, interesting and instructive and could by no possibility mislead" (Morris 1877, p. 1). Secondly, the act of restoration applied contemporary to Morris's time. It attempted to restore a historical artefact to its best time by imagining what former builders ought or may have done to restore it (Morris 1877). Having said that, the manifesto is concerned with the protection of ancient buildings and the need to pass them on to the next generations with respect and integrity.

In the modern conservation theory, Alois Riegl describes a monument, in its oldest and most unique sense, as a work of human created for the particular principle of "keeping human deeds or destinies (or a complex accumulation thereof) alive and present in the consciousness of future generations" (Riegl 1982, p. 21). There are numerous associated values to every monument, and these depend on how different people observe its characteristics. Historical values are much more commonly comprehended and are affiliated with what once existed but no longer does. In a modern and more inclusive interpretation of this, the historical value is associated with the belief that "everything that once was can never be again, and that everything that once was forms an irreplaceable and inextricable link in a chain of development" (Riegl 1982, p. 21). Three concerns arise in conserving any object of value: (1) what is considered to be the whole of the object; (2) what is the context of the object; and (3) what is the history of the object. The whole of the object requires considerable attention because due to a habit of classification we tend to scatter the monuments and display them in divisions according to mutual characteristics such as the techniques incorporated in their creation. Thus, a historical monument can be easily pulled into isolated pieces of sculptures, paintings, etc., throughout the museum grounds or gallery displays. The second significant concern is context, which refers to the immediate surrounding of the object which might lead to the truthful interpretation of the object. The context consequently lies in the traditional surroundings of the object and is crucial to the scale, significance and social circumstances when the object was originally in use. If possible, taking the whole of the object and context into account is the best approach to conserve objects in situ and maintaining their full values (Philippot 1972).

30.3 Themes Around Digital Reconstruction

30.3.1 Technology for Heritage Context

The question here is how we employ the digital reconstructions of heritage or archaeological sites in today's context. One adoption of digital reconstruction is in the museum context, which can easily expand the diversity of visitors' groups and their engagement. Digital reconstruction allows cultural heritage to be displayed in museums in different forms through virtual technologies. The procedure of recording, preserving and presenting cultural and heritage significances through "electric manipulations of time and space" forms Virtual Heritage (Stone and Ojika 2000, p. 73). Once digital reconstruction is completed and virtual heritage formed, the question becomes how to incorporate them in a museum context. Virtual media and subsequently, virtual heritage can be displayed in different mediums. Of course, VR headsets or other wearable technologies are broad-ranged possibilities for virtual display and have been extensively applied in the context of heritage tourism. An alternative to VR technologies is visual projections. Projection units embody features which relate to the technological approaches in this research project. The units can be installed in situ to represent the virtual heritage content on-site rather than in remote or isolated locations. In comparison with VR wearables, this project aims to investigate how encountering digital interventions displayed in some form of projection directly on site may influence visitors' experiences.

The study of visitors' experiences at St Augustine's Abbey aimed to observe how participants understood and learnt from the site and the museum offerings before encountering digital and creative installations on the Abbey ruins. The evaluation of the study is based on primary data gathered from Abbey visitors which includes individuals who observed the site themselves or attended a guided tour offered by the Abbey site manager. It is based on four predominant subjects. Firstly, the demographic features of the visitors including their interest in and frequency of visits to heritage sites. Secondly, the comparison between visitors' expectations prior to their visit and the experience itself; whether their visit is in line with their expectations of what to see and learn about the site or not and what can differentiate the two. Thirdly, the efficiency levels and usability of the available technological devices for better and more comprehensive learning about the site. Finally, what are the visitors' views on authenticity and realism and by what means do they relate these concepts to reconstructed imageries.

This study examined visitation experiences of the general public visiting the Abbey as opposed to specialists in heritage, archaeology and affiliated subjects. Hence, participants of the study were recruited from an extensive range of visitors to the Abbey. In terms of the age range, participants of the study fall into extensive classifications. Amongst all, 26.15% of the participants were aged between 18 and 24 years old, 18.64% were aged between 25 and 44 years old, 18.64% visitors were 45–65 years old and 36.92% were seniors, aged above 65.

One criterion measured in the study captured whether installing digital and creative installation in the Abbey ruins would influence visitors' engagement, and whether participants would consider this influence to have a positive or negative impact (Fig. 30.3). The results of having a positive, negative or sense of scepticism vary amongst the different groups. The age factor highly classifies the results of having or not having installations in historic ruins. All respondents aged 18–65 years answered that having digital installations in ruins would influence their engagement. Around three quarter of them considered that this influence would be positive. They mentioned that having installations would be engaging for both adults and children. It would help to envisage the site and to narrate part of its history. It would also enhance the appreciation of what is already available on the site by implicating a sense of immersion or realism. However, 14.63% claimed that digital installations in ruins would negatively influence their engagement. Their claims referred to the need of keeping historical sites and modern technologies and creations apart. The participants of such views said that they would like to imagine the site themselves rather than being directed to look at it through modern creations. Additionally, they thought that installations can be distracting for some visitors or can decrease the attention paid to the ruins.

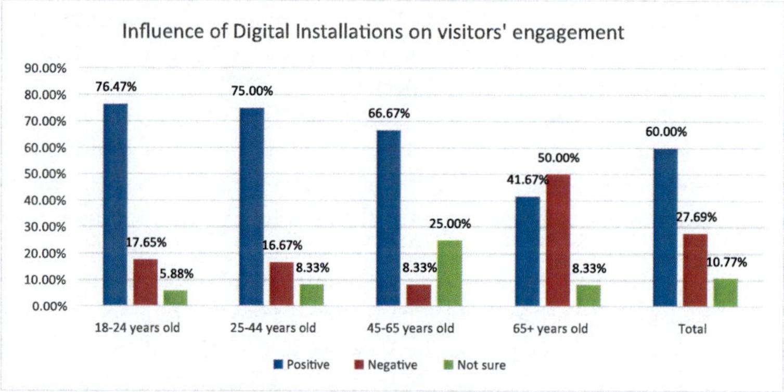

Fig. 30.3 Influence of digital installations on visitors' engagement

The responses to this question varied and was significantly different in the senior age group. The willingness of having digital interventions in ruins varied from positive to negative points of view. Also, some respondents expressed a sense of scepticism towards having digital installations on the site. As an example, 41.67% of the senior age participants said that digital interventions would have a positive impact on their engagement. However, the greater majority of them expressed negative views on having digital interventions installed on the site. The positive attributes towards having digital interventions included that such installations, in different forms, would be a further reference to the artefacts existing in display. However, the negative views suggested that encountering installations would spoil the experience or give too much direct information, which hinders independent thinking process and imagination. Additionally, 12.20% of the 18–65 years old and 8.33% of the senior respondents were not so sure about how digital installations can influence their engagement. A sense of scepticism was expressed about the installations before seeing them. A few have mentioned that they would have to encounter the installations first to be able to evaluate the positive or negative influence on their experiences. Furthermore, they claimed that not knowing the form and setting of the digital interventions in the ruins also caused a sense of doubt about their influence.

It is evident from the statements made by the study population that incorporating technology in the heritage context can cause a dichotomy. It may raise concerns that affect the visitation experience in terms of appreciating the ruins or, on the contrary, it may be of aid in better envisioning the site as a whole and have a greater understanding of the remaining relics. In this regard, Augmented Reality (AR) offers a solution which whilst meeting the interest of the younger generations in experiencing heritage in digitised modes, also takes into account concerns around preservation and reconstruction of cultural heritage. On the one hand, digital installations in AR provide visitors with visualisation enriching their visitation experiences by offering visualisation of artefacts that are not available for observation. AR broadens the chances of learning and appreciating cultural heritage whilst intervening least in the site. Additionally, AR can be coined in many modes such as wearable or portable devices, temporary and permanent installations. It, therefore, expands the possibilities of integrating and benefiting from heritage visualisation in many ways (Fig. 30.4).

30.3.2 Defined and Perceived Authenticity

The subject of authenticity has historically been interlinked with conservation and reconstruction of cultural heritage artefacts. Authenticity can be discussed from two points of view: on the one hand, from a theoretical point of view which defines authenticity according to philosophical discussions and the "Operational Guidelines for the Implementation of the World Heritage Convention" and on other hand, from the perceived authenticity in relation to the objects and domains interwoven between them, people and places. Authenticity, in "Operational Guidelines for the Implementation of the World Heritage Convention" was initially defined based on four degrees

Fig. 30.4 Augmented reality projection of tile patterns at St Augustine's Abbey (author)

of workmanship, design, materials and settings (Feilden and Jokilehto 1998). In 1994, upon the realisation of the Nara Document, the definition of authenticity was expanded to acknowledge "social and cultural values of all societies" (ICOMOS 1994). Jones (2009) refers to the dichotomy between the two definitions of authenticity as "materialist" and "constructivist" approaches. The constructivist approach to authenticity, derived from the Nara Document, looks into objects beyond their objective and materialistic features. However, the constructivist approach is culturally erected and can vary depending on by whom and in what context the object is being perceived.

In this study which evaluates visitors' experiences at St Augustine's, participants were asked to present their views on how authentic they perceive digital reconstructions of the Abbey artefacts to be. Additionally, what attributes could distinguish authenticity and inauthenticity of reconstructed imageries in digital format? The responses to this question varied from a definite yes, acknowledging the authenticity of reconstructed imageries to relatively uncertain views and definite disagreements. Indeed, 53.85% of participants expressed that reconstructed imageries are authentic. Their reasoning mainly falls into three themes. Firstly, reconstructed imagery is authentic if it is based on in-depth research, and there is sufficient knowledge backing up the creation. In that case the outcome would accurately reflect the references, and also if a specialist (i.e. a historian) approves of the quality of the content that the reconstruction aims to demonstrate. Secondly, the execution of the reconstruction has to be of professional standard. The equipment enabling the reconstructions has to have high-tech qualities, and the imageries need to be clear and have detailed visualisations. Thirdly, in the absence of the original object, the reconstructed imageries are considered reminiscent of the original, even though they are presumptions nearing the real.

A total of 27.69% of the respondents said that reconstructed digital imageries might be authentic depending on their qualities and contexts. For example, participants mentioned that digitally reconstructed imageries might be, to some degree, authentic because it is not possible to absolutely achieve the authenticity levels of a historical element, or that the imageries can be an outcome of subjective reconstructions. In terms of the context, respondents said that digital reconstruction could be considered authentic if they help the visitors to visualise or interpret the objects or provide them with a concept about the artefacts. Accordingly, the aesthetic and technical qualities of the reconstructions define the authenticity levels of the imageries. On the contrary, 13.85% of the respondents said that digital reconstructions could not be authentic due to three reasons. Firstly, reconstructed imagery is not the original and cannot replace the original and is therefore not authentic. Such justifications relate to views that correlate authenticity with originality and genuineness (Jones 2009; Pye 2001). Secondly, reconstructed imagery which have contemporary technological qualities cannot accurately represent the age of a historical element. Thirdly, authenticity is associated with keeping historical artefacts untouched.

30.3.3 Realism and Hyperrealism

In "Simulacra and Simulation" (1994), Jean Baudrillard defines the real and the hyperreal and what differentiates the two notions. He reflects on hyperreal as simulation which "is the generation by models of a real without origin or reality" (Baudrillard 1994, p. 1). In hyperrealism, as Baudrillard claims, the reality can be produced an indefinite number of times without the need of being rational as it no longer compared to an ideal. Simulation feigns what one does not have. It is not a matter of imitating or duplicating but alternating the signs of the real for the real. Also, it is not an act of pretending because it relies on the difference between the "real" and the "imaginary" not the "true" and the "false" (Baudrillard 1994, p. 3).

In the responses associated with identifying projections as real, the vast majority of survey respondents (69.23%) said that projections are not real. They mentioned that projections are hyperreal, not real and that there is a distinction between them. Hyperreality and thus projections only implicate an illusionary vision of reality which can be comparable but not equivalent. Their justifications were based on the fact that projections are representational tools for delivering visual contents. In the case of projections within the cultural heritage context, participants of such views said that projections could not be real because they are to some extent based on guesswork and are not as authentic as the original content. The concept of "representation" according to Baudrillard (1994) initiates from the sameness of the sign and the real even if from a utopian point of view. A total of 16.92% of participants were uncertain of projections being real. However, they said that they would need to encounter one first to judge the realism depending on the content simulated and its conjuncture with the surrounding area.

A total of 13.85% of participants claimed that projections are indeed real. Their explanations are related to the characteristics of the projections and the implications that projections make on humans. The reasoning as to why projections are real includes that they are light rays that create physical and tangible experiences. Viewers can see projections, interact with them and be influenced by them as well as being influenced by the content the projections display. Projections can immerse viewers and portray photo-realistic scenes. Even though projections are real, the content they visualise is not. The content can only relate to evidence in reality. However, Arva (2008) claims that visual imageries are more than a representation of the reality, they are evidence of and hence the reality.

In terms of the content of the visual information available on the site, participants' views on the VR which is built upon the sixteenth century of the Abbey was measured. In total, 70.77% of the participants used and rated the VR headsets. This population is composed of 56.52% first time visitors and 43.48% returning visitors. On the other hand, 54.35% attended a guided tour, and 32.81% were on a self-guided visit. The users rated the immersion of the VR on a scale of 1–9, with 1 being not immersive at all and 9 being very immersive ratings. The average rating for VR was 5.97. From this, 10.87% participants rated the VR as low as 1–3 ratings on the scale. However, 43.48% ranked it moderately between 4 and 6 and another 45.65% ranked it high from 7 to 9. Some participants (6.52%) rated the VR the lowest of all with 2 points on the scale. On the contrary, 13.04% marked it highest on the scale (Fig. 30.5).

Participants' opinions on the quality of the information that the VR delivers varied from positive to contrary views. In both age groups (18–65 and 65+), more than 50% of the surveyed population expressed that the VR does, to some extent, provide decent quality information. The positive comments on the VR concentrated on the fact that the VR reconstruction gives a good overall sense of the Abbey's scale and primarily

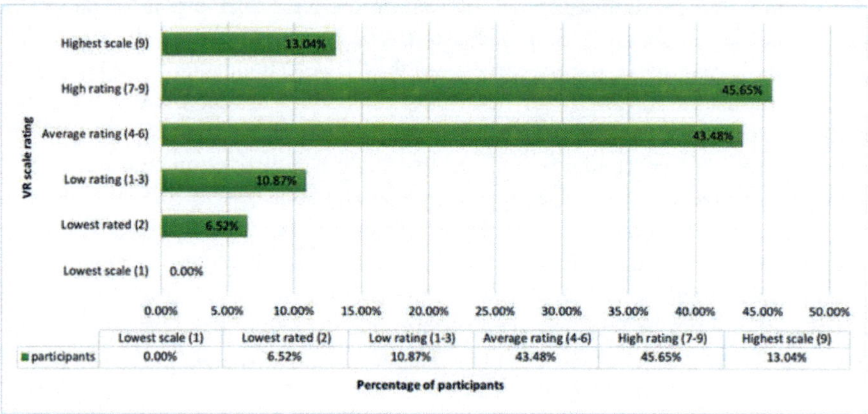

Fig. 30.5 Scale rating of VR immersion

its interior and exterior architecture. On the contrary, the negative views on the VR said that there is no proving evidence that the interface has been created based on historical facts. Often users claimed that the VR is not of good quality as it lacks details. Therefore, it is not authentic enough to be believable. Alternatively, some users said that VR is not the best way of representing the Abbey church as it does not reflect on the activities which took place there. Suggestion for a more pleasing VR reconstruction included that it should be livelier with perhaps some simulations or have human characters in it. Also, in terms of navigating in the virtual space, it would be helpful for the VR to guide the user from one environment into another and clarify what is represented in each space. Some users suggested audio tracks to introduce and explain the virtual environments.

30.4 Workflows for Digital Reconstruction of Non-existent Artefacts

Reflecting on the previous sections, the results of this study on visitors' perceived experience of digital reconstruction and virtual heritage can be linked with the procedure through which a digital reconstruction produced. This section therefore concentrates on the reconstruction of historical and archaeological sites by reintegrating three-dimensional data according to historical evidence. Different methodologies have been introduced over the last two decades. One approach to digital reconstruction is to incorporate the 3D documentation of the site in its existing state which then has to be followed by negotiations between archaeologists, historians and digital modellers in order to develop a reconstructed proposition. The first step involved in such procedures would be to generate a reality-based model. In a methodological proposition towards the reconstruction of archaeological sites, Guidi et al. (2013) introduce three procedures that when interwoven lead to an integrated procedure of reconstructing archaeological artefacts.

The first action is to obtain a thorough and accurate 3D acquisition of the artefact supposing that it is available in some format, for example, ruins or modified substances. The 3D acquisition should be based on the shape and the colours of the artefact in order to assist with generating a reality-based model which correlates with the scale, textures and other specific characteristics. Such data is usually gathered through photogrammetry and laser scanning procedures and is then processed into 3D textured polygons or 3D cloud points that highly resemble the real objects.

The second action initiates from a humanities viewpoint that is contrary to the first technological oriented actions. It relies on gathering historical and archaeological information from written, topographical, contextual and iconographic sources from the past and present of the artefact (Guidi et al. 2013). In circumstances where the object is no longer extant or was never realised, historical imageries are among the few sources to which we can refer to for reconstruction. In such situations, the comparison of different historical images is very helpful, given that not all the required information is directly provided or measurable (Münster 2013).

The third action is fulfilled through adjoining the first and second steps. It is through the process of reconstruction that "tangible no longer extant objects or intangible historic issues" become tangible and conceivable (Münster 2013, p. 201). It involves synthesising both the reality-based model and the archival data into a novel digital model. The input of all the mentioned procedures enhances the communication between the technical specialists and historians (Guidi et al. 2013).

In this approach to workflow, images are crucial as they can ease the process of transferring and comparing information between interdisciplinary teams (Münster 2013). Image-based processes of digital reconstruction has been advancing in attempts to optimise the workflow process. An example is MySon site in Vietnam, a UNESCO World Heritage site including a complex of Hindu temples which was heavily damaged by the bombings of the US troops during the Vietnam War. To document the damage levels and prepare for the restoration of the site an accurate reality-based 3D modelling of the site was first required. It was done through several procedures from data acquisition through images, point cloud creation and 3D surface modelling (Guidiet al. 2013; Patay-Horváth 2014). This approach provides a basis for negotiations between archaeologists, historians and digital modellers in order to develop a reconstructed proposition in a way to respond to multiple criteria which in turn contributes to visitors' experiences of the site upon the implementation of digital technologies and reconstruction strategies.

30.5 Conclusion

Evaluating visitors' experiences at St Augustine's, in general, survey participants said that they would like to see digital reconstructed imageries in St Augustine's Abbey. Most commonly, it was expressed that reconstructed imageries of the artefacts would positively influence visitors' engagement with the site and enhance their understanding and appreciation of the site. However, a higher number of positive views on this matter come from the younger participants and decreases in the older age groups. According to the respondents, the VR experience in the Abbey provides a sense of scale and a visual representation of a long-gone Abbey building. However, clearer and more detailed imageries that reflect the ageing of the Abbey would have been a more appropriate representation. In general, reconstructed imageries representing or conveying a sense of architecture and design of the artefacts seemed more pleasing to the surveyed visitors.

In terms of authenticity, participants acknowledged digitally reconstructed imageries as authentic based on intrinsic creation features being referential to the object they represent. However, the claims which suggest the inauthenticity of digitally reconstructed imageries refer to the perception of differentiating the original and the reconstructed and the fact that the original cannot be replaced. Concerning realism and hyperrealism, respondents most commonly said that projection cannot be real. They may represent or simulate the reality but are not the reality. The claims which refer to projections being real mainly focus on the environment that a projection creates in a physical setting.

References

Arva EL (2008) Writing the vanishing real: hyperreality and magical realism. J Narrative Theory 38(1):60–85

Baudrillard J (1994) Simulation and simulacra. University of Michigan Press

Feilden FM, Jokilehto J (1998) Treatments and authenticity. In: Rockwell C, Lawrence (eds) Management guidelines for world cultural heritage sites, 2nd edn. ICCROM, Rome, pp 59–75

Guidi G, Russo M, Angheleddu D (2013) Digital reconstruction of an archaeological site based on the integration of 3D data and historical sources. Int Arch Photogram Remote Sens Spatial Inf Sci 5:W1

ICOMOS (1988) Advisory body evaluation. 496-ICOMOS-578-en

ICOMOS (1994) The Nara document on authenticity. ICOMOS

Jones S (2009) Experiencing authenticity at heritage sites: some implications for heritage management and conservation. Conserv Manage Archaeol Sites 11(2):133–147

Morris W (1877) The principles of the society. The society for the protection of ancient buildings

Münster S (2013) Workflows and the role of images for a virtual 3D reconstruction of no longer extant historic objects. ISPRS Ann Photogram Remote Sens Spatial Inf Sci 5:W1

Ordnance Survey (2018) St Augustine's abbey, 1:10000. Historic England. Available from https://mapservices.historicengland.org.uk/printwebservicehle/StatutoryPrint.svc/21712/HLE_A4L_NoGrade|HLE_A3L_NoGrade.pdf. Accessed 27 Oct 2020

Patay-Horváth A (2014) The virtual 3D reconstruction of the east pediment of the temple of Zeus at Olympia an old puzzle of classical archaeology in the light of recent technologies. Digital Appl Archaeol Cultural Heritage 1(1):12–22

Philippot P (1972) Historic preservation: philosophy, criteria, guidelines. In: Preservation and conservation: principles and practices: proceedings of the North American international regional conference, Williamsburg, Virginia, and Philadelphia, Pennsylvania. Preservation Press, Washington, D.C, 1976, pp 367–382

Pye E (2001) Caring for the past. Issues in conservation for archaeology and Museums. James & James, London

Riegl A (1982) The modern cult of monuments: its character and its origin (Trans. by Forster KW, Ghirardo D). Oppositions J Ideas Criticism Arch (25):20–51

Roebuck J (1997) St Augustine's abbey. English Heritage, London, pp 67–84

Ruskin J (1849) The lamp of memory. The seven lamps of architecture. Smith, Elder, London, pp 221–247

Stone R, Ojika T (2000) Virtual heritage: what next? IEEE Multimedia 7(2):73–74

UNESCO (2017) The World Heritage List. Operational guidelines for the implementation of the world heritage convention. UNESCO World Heritage Centre, Paris, pp 18–32

Viollet-le-Duc EE (1990) The foundations of architecture: selections from the Dictionnaire Raisonné. (trans.: Whitehead KD, 1854). George Braziller, New York

Chapter 31
Seismic Microzonation: A Preventive Measure for the Conservation of the Built Heritage

M. Cristina García-Nieto, Marcos A. Martínez-Segura, Manuel Navarro, and Patricia Alarcón

Abstract Several earthquakes around the world verified the vulnerability of monumental buildings and the urgent action needed to protect them. This chapter assesses the necessity of performing detailed seismic studies in historical buildings (small-scale), due to the importance of this types of structures that deserve to be protected and conserved or, on the contrary, if the seismic microzonation in the city (large-scale) is enough. The case study is Murcia city and one of its most important historical buildings, the Cathedral of Santa Maria. The Murcia Region, located in southeast Spain, is classified as a seismically active zone. Multichannel Analysis of Surface Waves (MASW) method was used providing a characterisation of the materials in terms of shear-wave velocity (Vs), to obtain characterisation of the subsoil structure in historical buildings. The Vs investigations were carried out at the scale of a historic building and at the seismic microzonation scale in the city. Results evidenced a clear difference in Vs values obtained under the Cathedral and in the city. The study makes the case that the analysis of the local effect due to the shallow soil conditions in historic buildings, is a fundamental point to address the preventive analysis of the building seismic response, beyond studies of seismic microzonation carried out at city scale.

Keywords Cathedral of Santa Maria · Historical heritage · MASW method · Murcia city · Seismic microzonation · Vs structure · Soil structure · Murcia · Spain

M. Cristina García-Nieto · M. A. Martínez-Segura (✉)
Universidad Politécnica de Cartagena, Cartagena, Spain
e-mail: marcos.martinez@upct.es

M. Navarro
Universidad de Almería, Almería, Spain
e-mail: mnavarro@ual.es

P. Alarcón
Universidad Michoacana de San Nicolás de Hidalgo, Morelia, Mexico

© The Author(s), under exclusive license to Springer Nature Switzerland AG 2021

F. F. Arefian et al. (eds.), *Historic Cities in the Face of Disasters*, The Urban Book Series,
https://doi.org/10.1007/978-3-030-77356-4_31

31.1 Introduction

The vulnerability of historic masonry structures and monuments under seismic loading has been confirmed during recent major earthquake events such as the ones in Axochiapan, Mexico (2017); Lorca, Spain (2011); and in L'Aquila, Italy (2009) (Maniatakis et al. 2017). The protection of monuments affected by earthquakes is an issue of critical importance for countries which experience significant seismic activity (Tassios 2010). The Mediterranean region is one of those areas of intense seismic activity (Chiarabba et al. 2015). It also has a large quantity of cultural heritage. On the coast of the Mediterranean Sea, the southeast of Spain is an area with the highest seismic hazard in the Iberian Peninsula (Vidal 1986). Located in this area is the Region of Murcia whose capital is Murcia with a wide range historic built heritage which deserves to be protected and preserved. Specifically, the iconic Cathedral of Santa Maria is symbolic to the identity of the city (Fig. 31.1) and an important building with an area of almost 5,000 m^2. It is located in the heart of downtown Murcia, and it has the maximum degree of protection according to the Spanish Historic Heritage Law of 1985 (BOE 1985). The building was declared a Property of Cultural Interest in 1931. Being situated in the most touristic area of the city, it is an area of great administrative, commercial and residential activity throughout the year. The Cathedral is the city's main scene of worship, where festivals and religious events take place. It has an intrinsic intangible value typical to the monument as a cultural heritage, beyond the tangible value, so the historic, cultural, social and identity value of the Cathedral is incalculable. Since its construction, the Cathedral has had numerous restorations, as it was damaged by several earthquakes for example, in 1716, 1732

Fig. 31.1 Cathedral of Santa Maria in Murcia, Spain, with its Baroque frontage and the Tower (authors 2020)

and again on 21 March 1829 during which the tower was damaged (Ministerio de Fomento 2002; Bernárdez 2010; Molina Gaitán 2014).

In parallel, whereas traditional technical studies of seismic hazard were done on rock, without taking into account the influence of different types of local soil, numerous studies found that ground motion during an earthquake can be amplified by local ground conditions (Aki 1993; Bard 1999). These studies in urban areas showed that the existence of deposits of unconsolidated sedimentary materials can cause an increase in local seismic hazard. This is due to the amplification of the waves for certain frequency bands, caused by the contrast in velocity between the "soft" materials and the rigid basement. This phenomenon is known as site effect, being especially relevant when these ground amplification bands coincide with the natural periods of existing buildings, causing additional soil structure resonance effects (Martínez Segura et al. 2017). Thus, the site effect is the modification of the properties of the incoming wave field (amplitude, frequency content and duration), as a result of the characteristics of the surface topography and soil deposits (Stanko et al. 2017). The seismic microzoning process aims to address the issue. It can be defined as the process that results in the division of a geographical area into homogeneous ground zones, based on the local simulated response of each zone to a reference earthquake. This is done by taking into account the interactions between the seismic waves and the local geological, topographical and geotechnical conditions that modify the basic hazard, meaning the hazard assessed in a flat, hard ground. Seismic hazard is usually expressed as the maximum acceleration in a given time interval (Crespellani 2014).

This chapter examines Seismic Microzonation, in the case study of the Cathedral of Santa Maria, as a basic knowledge tool for supporting decision-making on heritage intervention and conservation and in general urban and territory planning, emergency planning and management and post-earthquake reconstruction. Several technical terms used in this work have been defined in a glossary at the end of the chapter.

31.2 Seismic Microzonation

A seismic microzonation study is a scientific-technical process that requires different disciplines, mainly Geology, Applied Geophysics, Applied Seismology, Structural and Geotechnical Engineering (Pagliaroli 2018). These methods also have a social dimension, since knowledge of the reference earthquake response zones is a fundamental condition in decision-making for the management of a territory (Aversa and Crespellani 2018). There are three levels of microzoning studies: Level 1 is based on the collection of existing information and geological maps, with the aim of dividing the study area into homogeneous zones, from the seismic point of view, and evaluating the effects of amplification on the geological bases by means of appropriate coefficients. Level 2 consists of mapping, establishing the homogeneous zones, quantitative parameters of interest for engineering that indicate the level of local danger.

In this case, the amplification estimation can be made associated to individual lithostratigraphic situations, through simplified processes or tables to find the amplification factors on the surface. Finally, a level 3 study provides a more detailed map of seismic microzoning, with the aim of evaluating site effects (analysis of local seismic response, liquefaction, slope stability, etc.) through the use of special and advanced software with complex input data (Crespellani 2014). Recently, a few cities underwent microzoning studies of level 3, for example, Adra, Lorca and El Ejido in Spain (See Martínez-Pagán et al. 2018; Navarro et al. 2014; García-Jerez et al. 2019); and others in Italy like Monte San Martino, Arquata del Tronto, Montedinove, Capitignano and Norcia (Pagliaroli et al. 2019).

At the urban and territorial planning scale, the knowledge of the fundamental vibration period of the soil in new residential areas helps defining heights and volumes of new buildings to avoid possible resonance phenomena if an earthquake happens. The establishment of seismic hazard hierarchies also supports the planning of green areas in high-risk zones and locating strategic buildings in safer areas, and deciding on intervention priorities for buildings already exist in high-risk areas. For disaster emergency planning and management, the creation of seismic microzoning study maps helps to locate emergency road networks, meeting points and even temporary facilities, among others. Logically, with the occurrence of an earthquake, the effectiveness of the existing seismic microzoning studies is verified and reconstruction can be based on updated knowledge of the seismic response in the different zones, thus guaranteeing greater safety and establishing new territorial planning (Aversa and Crespellani 2018).

At a more micro-scale, soil impact assessment and microzoning are essential tasks to prevent damages to buildings (Macau et al. 2018). One of the most important aspects of reducing the effect of earthquakes on monuments is to reduce the amplification of the movement of the soil on the surface based on the soil and its geotechnical characteristics (Tarque et al. 2013). Recent earthquakes in southeast Spain showed the influence of the softness of the ground surface and the thickness of the surface sediments on the strength of earthquake movement (Navarro et al. 2007, 2014; Alguacil et al. 2014). Thus, local variations can give rise to differences in spatial seismic intensity, obtaining a notable influence on the level of damage to buildings and on the distribution of significant seismic damage.

Seismic wave propagation varies depending on the shear-waves velocity (Vs) through the near-surface materials. At present, the parameter Vs is considered a good quantitative indicator of the stiffness of the geological material. Besides, the international codes classify soils according to the Vs parameter.

Vs models are applied to seismic hazard assessments through ground amplification, which may cause most earthquake-related damages, and often change with shallow ground stiffness (Park 2013). Thus, Vs is a useful indicator for soil microzoning in urban areas (Martínez-Pagán et al. 2014), and studies have been carried out to obtain the Vs structure in different places, e.g. in Spain, Portugal and Turkey (Rosa-Cintas et al. 2017; Vicêncio et al. 2018; Karabulut 2018). In Europe, the European Committee for Standardisation of the 1998 (Eurocode 8) adopted the average of Vs in the upper 30 m (denoted by Vs30) as a fundamental parameter of the terrain to be considered in the structural design of buildings against seismic events (Mahajan 2009).

Additionally, the use of state-of-the-art software and new data acquisition devices have opened up new fields of application in geophysical studies (Cardarelli and Di Filippo 2009; Evangelista et al. 2017; Vásconez-Maza et al. 2019). The Multichannel Analysis of Surface Waves (MASW) (Xia et al. 1999) a non-invasive geophysical technique is one of them. Due to its versatility, low cost and accurate results, this type of technique has begun to be used in some cities and historic buildings with the aim of characterising variations in shear-wave velocities (Martínez-Pagán et al. 2018; Grassi et al. 2019).

For examining the Seismic Microzonation of the Cathedral of Santa Maria, this study aimed to: (a) obtain the Vs structure under a building declared a Property of Cultural Interest, in this case, the Tower of the Cathedral (small-scale), by using MASW method; and (b) to compare it with the Vs structure obtained in the metropolitan area of Murcia (large-scale), also through the MASW method. It ultimately analysed the influence of the scale factor on seismic microzonation studies. These results could be fundamental as a first step in preventive analysis of historical buildings, with application beyond the Murcia Region.

31.3 The Tower of Santa Maria Cathedral

The knowledge of physical, historical and cultural characteristics of cities is a basis for establishing good planning in areas of seismic risk, reducing vulnerability and improving the protection of citizens, buildings, artistic heritage and the landscape (Aversa and Crespellani 2018).

The construction of the Cathedral of Santa Maria began in the fourteenth century on the remains of an ancient Muslim mosque (Vera Boti et al. 1994). Throughout its history, numerous interventions have been carried out both in terms of conservation and configuration. The Cathedral is composed of three connected buildings which at the same time are different by their use and architectural form. They are the temple, tower and cloister building (Fig. 31.2a). The Baroque frontage of the building and the tower are the most representative elements of the Cathedral. The construction of the tower began in 1519, although it was postponed for more than 100 years because during the construction the tower began to show signs of subsidence. It was in the eighteenth century when the construction was resumed and completed in 1793 (Vera Boti 1993). It is approximately 93 metres in height and represents diverse architectural styles owing to different periods of construction, e.g. renaissance and Ionic (Molina Gaitán 2014). Nowadays, a slight degree of inclination of the Tower can be observed. The tower is a square building, 20 m on each side, composed of different bodies that decrease in size as the height increases (Fig. 31.2b). This building is raised from two thick parallel walls, having different access ramps to the different levels. Located in the first body is the Major Sacristy and as the height increases, there are several rooms with different uses, covered with vaults inside. It

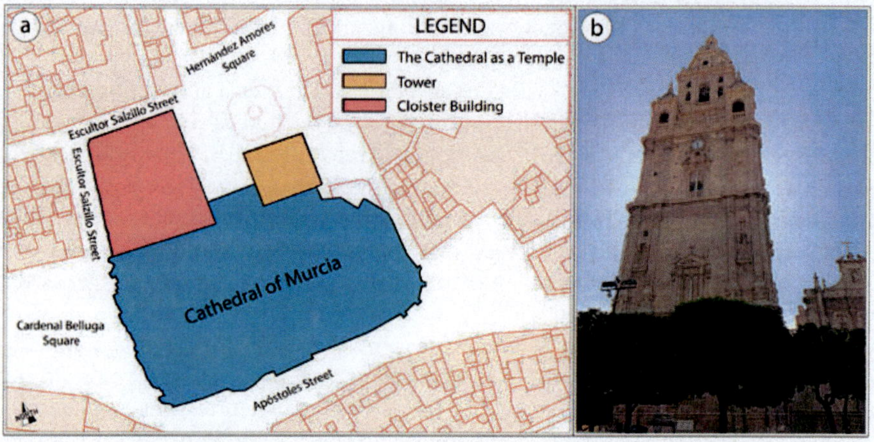

Fig. 31.2 **a** Plan view of the parts of the Cathedral; **b** North view of the Cathedral's Tower (authors 2020)

is worth mentioning the existence of four cupolas located on the roof of the third body, in each of its corners, called "Conjuratorios". In addition, the body of bells is another of the great highlights of this architectural element with twenty bells situated at three different heights.

Since the completion of the tower, the Cathedral complex with its clock-tower has acquired further prominence. Not only is it considered to be a place of worship, but it has become a symbolic and functional element. For example, since the tower was visible from all the surrounding orchards of the city of Murcia, the clock helped, among other things, agriculture workers during batch-irrigation to control the times of irrigation easily (Vera Boti 1993). Such deep interdependence to this monumental building still exists in this society. In the Murcian orchards, there is a cultural heritage inscribed in 2009 as an Intangible Cultural Heritage of Humanity called, "The Council of Wise Men of the plain of Murcia". It is a traditional court, made up of farmers, whose objective is to establish the peaceful use of water, a scarce and fundamental essential, in the Region of Murcia (UNESCO 2009). The tight relationship between tangible and intangible cultural heritage intensifies the need for the preservation of such symbolic yet functional buildings.

31.4 Large-Scale Microzonation

The city of Murcia is located in the Murcia province, which has a maximum expected Peak Ground Acceleration (PGA) value of 0.23 g for a return period of 475 years (IGN 2017). Several earthquakes have occurred in the region with moment magnitudes (Mw) of 4.7 in 1999 (Mula), 5.0 in 2002 (Bullas), 4.8 in 2005 (La Paca) and 5.2 in 2011 (Lorca), shown in Fig. 31.3. These earthquakes showed the special relevance

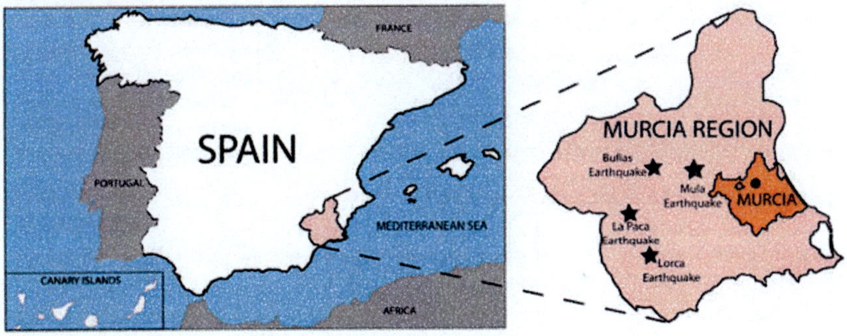

Fig. 31.3 Stars represent earthquakes with moment magnitudes (Mw) from 4.7 to 5.2 in the Murcia Region

of soil type in the degree and spatial distribution of damage observed in buildings (García-Jerez et al. 2007; Navarro et al. 2014).

The Murcia Region has a Special Plan for Civil Protection in the face of seismic risk called SISMIMUR (Dirección General de Protección Civil –DGPC 2015). This Special Plan obliges all 45 municipalities of the region to draw up an Action Plan to deal with seismic risk at the municipal level because they have a predicted forecast of earthquakes with an intensity equal to or greater than VII (EMS-98 scale). This Action Plan is called Local Scope Action Plan. Under this regulation, the City Council of Murcia promoted seismic microzonation studies as an essential tool to identify the factors that influence building-damage distributions to characterise the local site effects in Murcia city by using the Vs30 values.

Murcia Region is located in the Betic Cordillera, a set of mountain ranges (Sanz de Galdeano et al. 1995). More specifically, the city is located in the Vega Media of the Segura River. This basin is registered in an intramontane environment of tectonic subsidence whose limit coincides with important fault lines: Lorca–Alhama fault in the northwest and the Palomares fault (south) and Carrascoy fault in the southeast.

From the geological point of view, Murcia city is composed mainly of alluvial fans of different generations, made up of sands and gravels, with silts and clays (Marín Lechado et al. 2009) (Fig. 31.4).

The Multichannel Analysis of Surface Waves (MASW) method determined the velocity profiles of shear waves near the surface (Olafsdottir et al. 2018). Based on the Rayleigh wave analysis, it provided the sections of the subsoil materials as a function of their seismic characteristics. It is a non-invasive method, which does not cause damage to the roadway and pavement when it is performed. Based on previous geological data (Marín Lechado et al. 2009), five linear profiles were made with a total of 7.61 km in length. Three geological zones were studied. In the north zone, which geologically corresponds to geology of alluvial fans, profiles M1 and M5 were made. The central zone, which geologically corresponds to coluvial and undifferentiated alluvial, profiles M2, M3 and M5 were measured. And, in the south zone, which geologically corresponds to alluvial fans, profile M4 was measured.

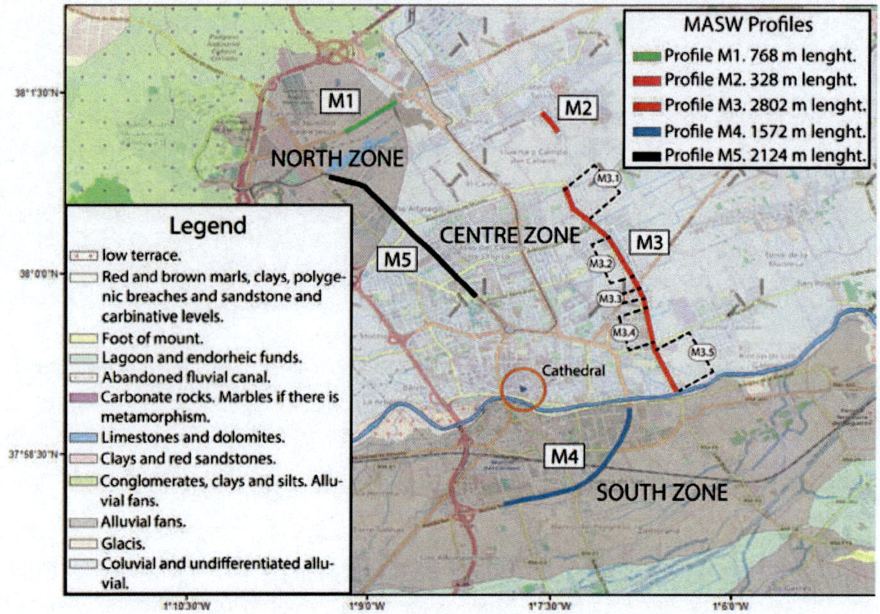

Fig. 31.4 Geological map of the urban area of Murcia city and spatial distribution of the MASW profiles. The orange circle shows the Cathedral

The analysis used a combined active and passive MASW method because it provides greater penetration depths and better resolutions of the Vs profiles (Park and Miller 2008). In the Active MASW method, the surface waves are generated from a seismic source. The Passive MASW method recorded the surface waves that are generated from the ambient noise. The equipment was displaced every 10 m after doing the active and passive measurements using an off-road vehicle. Every single geophone was georeferenced using a GPS device.

After field data acquirement, specialist software (the Surfseis package) was used to process the seismic data recorded. Through an inversion process, the software produced a dense series of dispersion curves (an example presented in Fig. 31.5) (Boiero et al. 2013). The Rayleigh-wave dispersion curve was manually picked for each shot. Then, 1D Vs models were inverted, and a cross section was interpolated.

31.4.1 Results of MASW Profiles (Large-Scale)

The results of Vs structures in the central zone of the territory are presented here. The city of Murcia has been characterised in terms of shear-wave velocity, obtaining the Vs30 values according to Eurocode 8. The focus is on the results obtained from profile M3 because it is located in the same zone of the Cathedral, near to the Segura River.

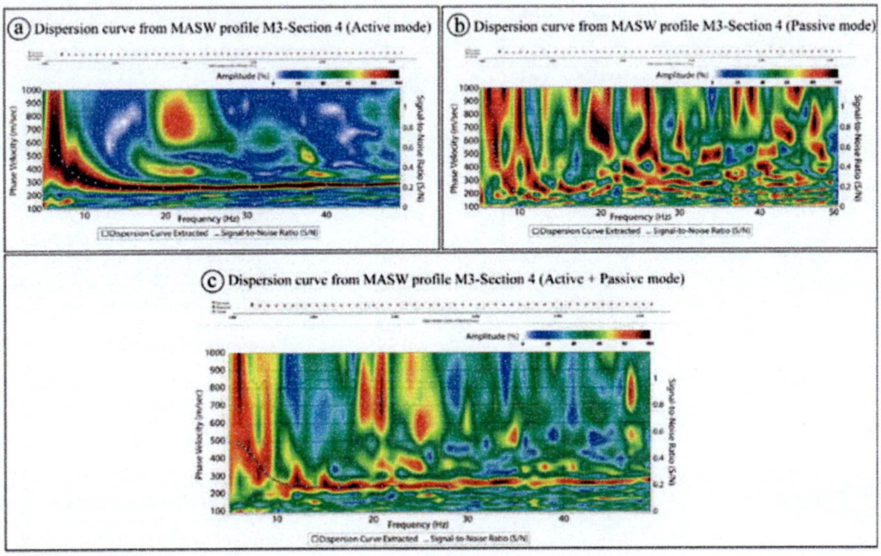

Fig. 31.5 Dispersion curve from MASW profile M3.4 (see Fig. 31.4) used for inversion in **a** Active mode; **b** Passive mode; and **c** Active + passive mode

Figure 31.6 shows the results of Vs structures in the central zone, which correspond to grounds composed of colluvial and undifferentiated alluvial materials. M3 profiles were measured in this area, in particular five profiles or sections.

In the central zone, the profiles showed differences in layer arrangement and depth. In some profiles there were two soil types according to Vs values of the distribution of depth, while others had three and even four soil types. M3.1 and M3.2 profiles, located in the upper area within this central zone, were the only ones that reached velocities greater than 800 m/s. On the other hand, the profiles M3.3, M3.4 and M3.5, despite having different dispositions of the materials, reached depths of between 33.75 and 37.50 m with velocities between 610 and 730 m/s.

31.5 Small-Scale Microzonation

The Cathedral of Santa Maria was selected as the case study because of the possible peculiar local ground conditions that can affect seismic microzonation. Given the size of this building, the tower of the Cathedral was selected as the first phase of study. It is a building located near the Segura River, in a landfill zone (Fig. 31.7a). The MASW method was used to obtain the surface shear-wave velocity structure under the tower. Two linear profiles were made on the outside of the building, profile C1 and profile C2, with a total of 55 m in length (Fig. 31.7b).

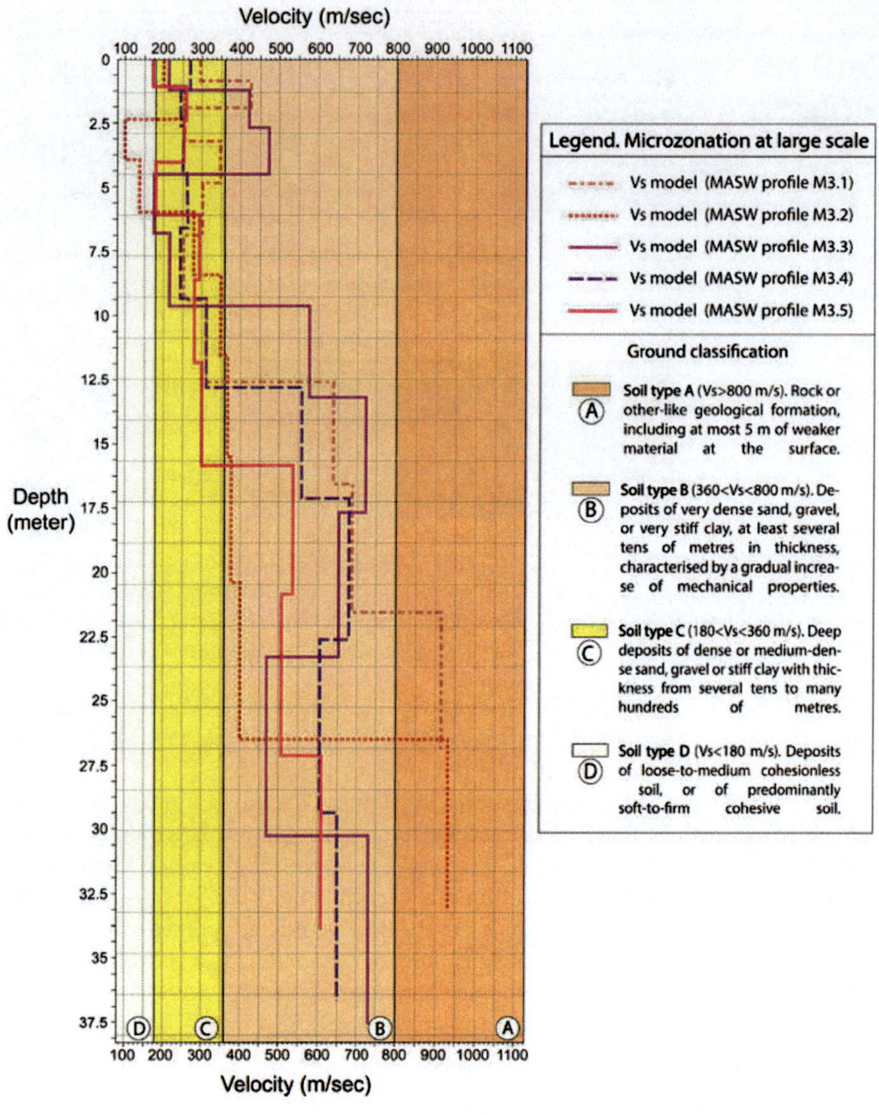

Fig. 31.6 1D Vs model obtained from the inversion of the dispersion curve in active + passive mode. Profile M3

Similar to the large-scale examination, the Surfseis package software was used to process the seismic data recorded. The software produced a dense series of Rayleigh-wave dispersion curves according to a logical sequence of seismic data processing. From these results, phase–frequency velocity spectral diagrams were obtained (Fig. 31.8). The Rayleigh-wave dispersion curve was manually picked for

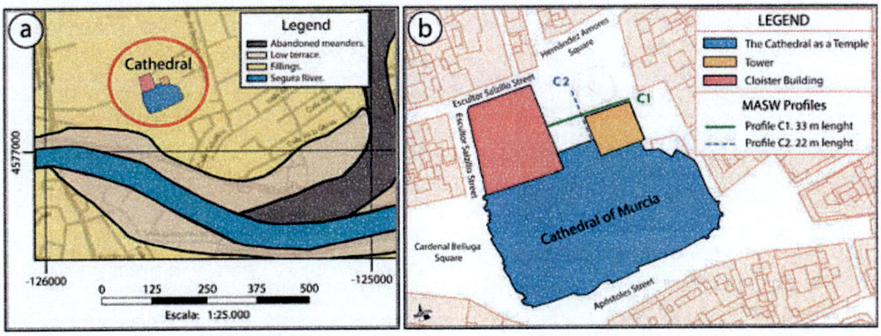

Fig. 31.7 **a** Geological map of the urban area near the Cathedral; **b** Spatial distribution of the measurements with MASW profiles

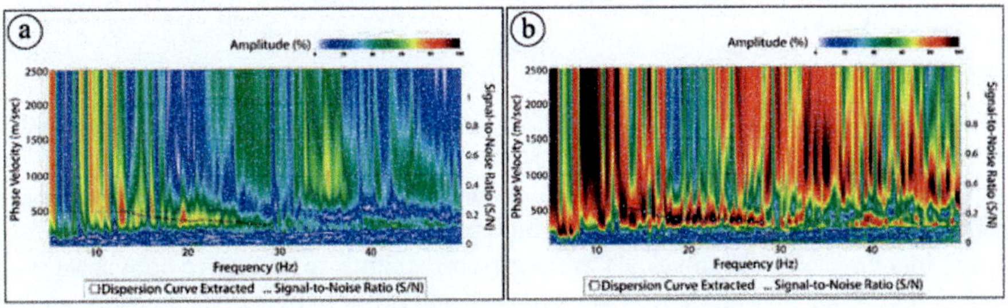

Fig. 31.8 Dispersion curve from MASW profile C2 used for inversion in **a** Active mode; **b** Active + passive mode. Dots represent the pitting of the maximum amplitude waves

each station. The 1D Vs models were inverted for the line centres, and a cross section was interpolated.

31.5.1 *Results of MASW Profiles Under the Cathedral (Small-Scale)*

A combination of passive and active MASW method was carried out to reach greater depths. Two 1D Vs profiles with depths of up to 18.5 were calculated. The values of the shear-wave velocity did not reach the depth of 30 m, so it was not possible to calculate the Vs30 parameter established according to Eurocode 8. Due to the fact that the Cathedral is a protected area, the seismic source was used with less strength than in the large-scale.

The measurements were made in the Cathedral, in a landfill zone (Fig. 31.7a) and showed different values of Vs between the linear profiles C1 and C2. The C1 profile reached a depth of 16.9 metres, while the C2 profile reached 18.5 m. Figure 31.9 shows results of 1D Vs structure in profile C1 and C2. Profile C1 is the one which

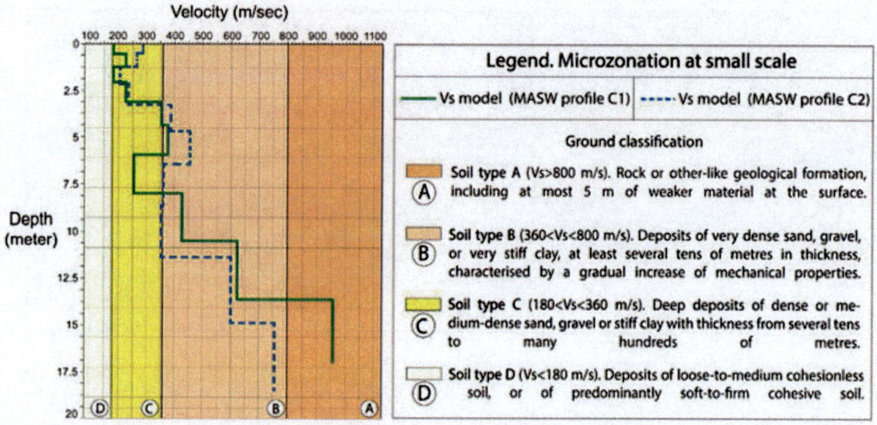

Fig. 31.9 1D Vs model obtained from the inversion of the dispersion curve in active + passive mode. Profile C1 and C2

were closest to the building. Up to 4.40 m deep, Vs values were between 180 and 360 m/s. From 4.40 to 13.60 m deep, the soil had shear-wave velocities between 360 and 800 m/s. Only between 5.9 and 8 m deep, the ground returned with Vs values less than 360 m/s. Finally, at a depth of 13.60 m, ground had velocities between 623 and 955 m/s. In profile C2, up to 3.40 m deep, corresponds to velocities between 180 and 360 m/s. From 3.40 m, there was a soil with velocities between 360 and 760 m/s, reaching depths of 18.5 m.

31.6 Comparison of Vs Values from Large-Scale and Small-Scale

In order to make a comparison of the results, the values of Vs obtained at small and large-scales have been established according to the limits of Eurocode 8 as reference levels. Therefore, four soil types are established. Soil type A with Vs>800 m/s, soil type B with Vs between 360 and 800 m/s, soil type C with Vs between 180 and 360 m/s and soil type D with Vs<180 m/s. The distance between the large-scale profile (M3.5) and the small-scale profiles (at Cathedral place) is approximately of 1.7 km.

One stretch of the seismic profile M3.5 (see location in Fig. 31.4) has been selected to compare the results obtained at the large-scale with those at the small-scale due to it being the closest to the Cathedral building, both of which are located on grounds with the same typology (Quaternary alluvial fan). For more comparison detail, the results of Vs values are shown at every 5 m of depth. The soil has been classified thus establishing the types of soils in each case, as a function of shear-wave velocities. Figure 31.10 shows the results obtained at each 5 m depth. In the first 5 m depth, on

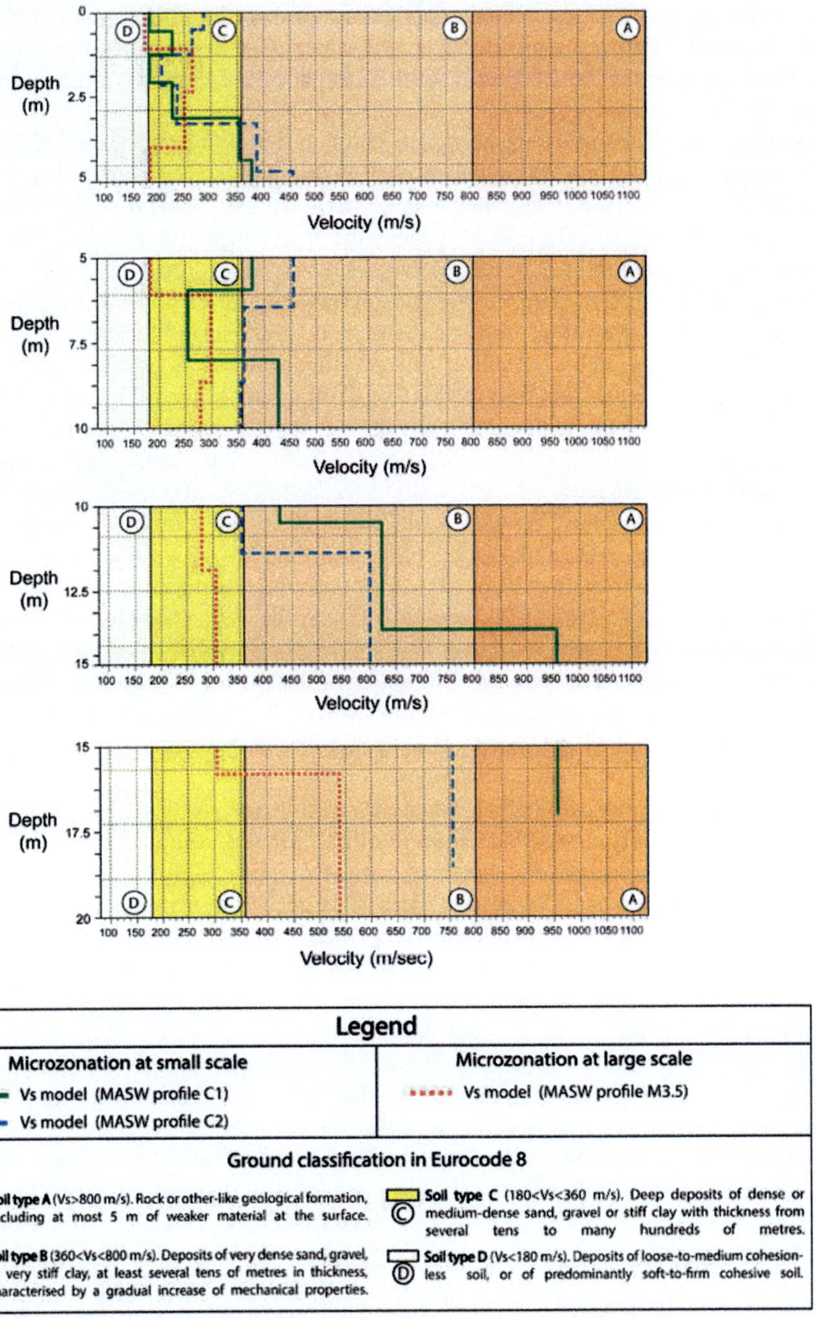

Fig. 31.10 Ten-layer velocity model each 5 metres in depth: Comparison of 1D Vs values obtained at the large-scale (red line) and at the small-scale (green and blue lines)

both large and small-scales, the materials mostly correspond with Vs values between 180 and 360 m/s. However, profile C2 (small-scale) at 3.4 m depth begins to increase the Vs values, greater than 360 m/s. Profile C1 begins to increase its Vs values greater than 360 m/s at a depth of 4.4 m.

Between 5 and 10 m deep, the results show that profile M3.5 (large-scale) has velocities among 180 and 300 m/s. However, in the small-scale microzoning, the Vs velocities increase and decrease between 250 and 455 m/s, depending on the depth. At depth of 5.9 m, the C1 profile reduces the Vs velocity from 374 to 256 m/s. However, from a depth of 8 m, the velocity increases to 426 m/s.

Between 10 and 15 m deep, large-scale microzoning shows Vs values among 280 and 310 m/s. However, the results obtained from small-scale microzoning show a progressive increase in Vs velocities. If we compare the three profiles exactly at a depth of 15 m, we can see how in large-scale microzoning a velocity of 310 m/s is shown. However, in small-scale microzoning, the C2 profile has a Vs value of 600 m/s and the C1 profile at this depth shows velocities of 955 m/s. Therefore, there are significant differences.

The last comparison of Vs values was at 15–20 m depth intervals. Large-scale microzoning begins to have higher velocities, reaching 537.5 m/s at a depth of 20 m. Small-scale studies show that, in the case of the C2 profile, a study depth of 18.5 m was reached with Vs velocities of 754 m/s. The C1 profile, the closest to the Cathedral building, reached a depth of 16.9 m and a Vs velocity value of 955 m/s. Hence, significant contrasts have been found in the Vs values from 13.6 m deep. Although the study on a large-scale classifies that area as type B, the study under the building classifies the same area as type A. This means that it changes the stiffness of the material and therefore its seismic response.

It is important to know that when an earthquake occurs, the interactions between the seismic waves that move through different soils can generate resonance phenomena, dissipation, increase in water pressure in the pores, decrease of the stiffness and resistance of the soil, being this behaviour very closely in relation with the parameter Vs (Aversa and Crespellani 2018).

The Vs values obtained in this study contribute to the characterisation of the ground and to the identification of the places prone to amplify the earth tremor during an earthquake.

Eventually, one of the most important factors in predicting the distribution of damage in an earthquake is the relationship between the fundamental frequency of a building and the fundamental frequency of the soil foundation (resonance). The knowledge of the Vs value plays an important role in the site effect and in the possible presence of resonance phenomena (Panzera et al. 2018; Stanko et al. 2017).

31.7 Conclusion

Seismic microzonation presented in this chapter examined the seismic characteristics of the city of Murcia (large-scale) and the tower of the historic Santa Maria Cathedral (small-scale) a symbolic and functional heritage in the city. To do so, the same technique of MASW method combining active and passive measurements was used in both cases.

Next to the historic building, two MASW linear profiles were made to obtain precise 1D Vs models. The results were used to compare them with the nearest profile (large-scale), in the same ground typology (Quaternary alluvial fan) and near the Segura River. Despite the close proximity of 1.7 km, the results showed different Vs structure models for the large-scale and small-scale tests, leading to different soil type classification.

In this case, the Vs values under the Cathedral were obtained at different depths. The velocities were significantly different from depths of 13.6 m. Large-scale microzoning showed Vs values from 310 to 537.5 m/s, between 13.6 and 20 m depth. On the small-scale microzoning study (profile C1), the Vs values were among 623 and 955 m/s between 13.6 and 16.9 m depth.

As the above shows, the shear-waves velocity increases, and then, the type of soil changes from a type of soil classified as B to the one as A, modifying its geotechnical properties. Materials with low Vs indicate less cohesive soils, and they are more prone to the amplification of seismic waves in the event of an earthquake.

Due to these significant differences in the Vs values, it is suggested that the increase in the Vs at the Cathedral site could be due to its foundations. For this reason, it is advisable to contrast these results with other geotechnical studies that examine this hypothesis.

Significant lateral/vertical contrasts in shear-wave velocity and soil stiffness in the near-surface volume can result in variable local amplifications of soil movement, which can be significantly different from those inferred from microzoning seismic maps. This may induce professional engineers involved in building designs to place excessive reliance on microzoning results, rather than performing on-site investigations and terrain response analyses, specifically performed at the building scale. Therefore, given the peculiarities of each local condition, seismic microzoning should be carried out on the scale of the historic building as a preventive measure (Monaco and Amoroso 2016).

The protection of historic monuments in seismic risk areas requires the study of subsurface ground conditions. To this end, the application of MASW method has been proven to be adequate to obtain a rapid, complete and non-invasive characterisation of the subsoil structure underneath artistic-historical and monumental heritage.

The seismic protection of a historic building requires advanced model analysis, based on knowledge of the building, the foundation and the subsoil, in order to evaluate the expected earth movement in the foundation of the monument, defining the dynamic interaction between the soil and the structure as realistically as possible (Giocoli et al. 2019). For this purpose, the appropriate choice of the scale factor in the analysis of the local effect due to the shallow soil conditions is a fundamental point to address the preventive analysis of the historic building seismic response.

Finally, obtaining Vs plays an important role in heritage intervention and conservation and, in general, in urban and territorial planning, emergency planning and management and post-earthquake reconstruction.

Acknowledgements "Project financed by the Autonomous Community of the Region of Murcia through the call for aid to projects for the development of scientific and technical research by competitive groups, included in the Regional Programme for the Promotion of Research (Plan of Action 2019) of the Seneca Foundation, Science and Technology Agency of the Region of Murcia".

Glossary of Technical Terms

Eurocode 8 European regulation, which establishes a set of technical rules applicable to the design and construction of buildings and civil engineering works in seismic regions, with the aim of ensuring that, when an earthquake occurs, human lives are protected, damage is limited, and structures important for civil protection remain operation.

Ground acceleration Time rate of ground velocity change, commonly measured in gravity acceleration ($g \approx 9.81$ m/s^2).

Intensity Measurement of the effects of an earthquake at a particular place. The intensity depends not only upon the strength of the earthquake (earthquake magnitude) but also upon the distance from the epicentre, the local geology at the point, the nature of the surface materials, the construction of buildings, etc.

Liquefaction Soil and sand behaving as a liquid upon agitation (such as by an earthquake) that disrupts grain-to-grain contacts.

Microzonation Identification of geographical areas having homogeneous similar behaviour under seismic action, including effects beyond the traditional seismic action parameters such as landslide and liquefaction.

Moment Magnitude (Mw) Earthquake size parameter derived from the scalar seismic momentum, which is defined as the product of the rupture surface in the fault plane, the net displacement on the fault or dislocation and the stiffness coefficient. It is the size parameter that best correlates to the energy released by the earthquake.

Peak Ground Acceleration (PGA) Maximum value the acceleration obtained from the accelerogram.

Resonance Phenomenon that occurs when a structural system is subjected to a cyclical load with a frequency equal or close to natural frequency of the system. The result is a build-up of amplitude of movement of the system due to external action.

Site effects Changes in the ground motion propagating near the site, in amplitude, frequency content and duration, relatively to the motion at a "hard" lower layer, due to the geometric characteristics and mechanical properties of surface layers surrounding the site, especially for cases of soft layer and complex geometries.

Stiffness Qualitative measure of the resistance to elastic deformation produced by a material.

Vs Shear-wave propagation velocity. In an isotropic medium, the velocity of shear waves Vs is given by Vs $= (G/\rho)1/2 = \{E/[2\rho(1 + \sigma)]\}1/2$, where G is the shear modulus, ρ is the density, E is Young's modulus and σ is Poisson's ratio

Vs30 Average shear-wave velocity of the upper 30 m of the subsurface used in several building codes for seismic response site classification.

Vulnerability Degree of loss of a system (engineering structure, network, social group, etc.) under a certain level of seismic action.

References

Aki K (1993) Local site effects on weak and strong ground motion. Tectonophysics 218:93–111. https://doi.org/10.1016/0040-1951(93)90262-I

Alguacil G, Vidal F, Navarro M, García-Jerez A, Pérez-Muelas J (2014) Characterization of earthquake shaking severity at different sites of Lorca town for the May 11th, 2011 event. Bull Earthq Eng 12(5):1889–1908. https://doi.org/10.1007/s10518-013-9475-y

Aversa S, Crespellani T (2018) Seismic microzonation: an essential tool for urban planning in seismic areas. J Urban Plann Landscape Environ Des 1(1):121–152

Bard PY (1999) Microtremors measurements: a tool for site effect estimation? The effects of surface geology on seismic motion, vol 3. Balkema, Rotterdam, pp 1251–1279

Bernárdez RM (2010) Desastres naturales en la Región de Murcia 1800–1930, vol 122, pp 141–165

BOE (1985) Ley 16/1985, de 25 de junio, del Patrimonio Histórico Español. BOE-A-1985-12534

Boiero D, Socco LV, Stocco S, Wisén R (2013) Bedrock mapping in shallow environments using surface-wave analysis. Lead Edge 32:664–672. https://doi.org/10.1190/tle32060664.1

Cardarelli E, Di Filippo G (2009) Integrated geophysical methods for the characterisation of an archaeological site (Massenzio Basilica—Roman forum, Rome, Italy). J Appl Geophys 68:508–521. https://doi.org/10.1016/j.jappgeo.2009.02.009

Chiarabba C, De Gori P, Mele FM (2015) Recent seismicity of Italy: active tectonics of the central Mediterranean region and seismicity rate changes after the Mw 6.3 L'Aquila earthquake. Tectonophysics 638:82–93. https://doi.org/10.1016/j.tecto.2014.10.016

Crespellani T (2014) Seismic microzoning in italy: a brief history and recent experiences. Ing. Sismica 31:3–31

Dirección General de Protección Civil (DGPC) (2015) Plan Especial de Protección Civil ante el Riesgo Sísmico en la Región de Murcia (SISMIMUR)

Evangelista L, de Silva F, d'Onofrio A, Di Fiore V, Silvestri F, Scotto di Santolo A, Cavuoto G, Punzo M, Tarallo D (2017) Application of ERT and GPR geophysical testing to the subsoil characterization of cultural heritage sites in Napoli (Italy). Meas J Int Meas Confed 104:326–335. https://doi.org/10.1016/j.measurement.2016.07.042

García-Jerez A, Navarro M, Alcalá FJ, Luzón F, Pérez-Ruiz JA, Enomoto T, Vidal F, Ocaña E (2007) Shallow velocity structure using joint inversion of array and h/v spectral ratio of ambient noise: the case of Mula town (SE of Spain). Soil Dyn Earthq Eng 27:907–919. https://doi.org/10.1016/J.SOILDYN.2007.03.001

García-Jerez A, Seivane H, Navarro M, Martínez-Segura M, Piña-Flores J (2019) Joint analysis of Rayleigh-wave dispersion curves and diffuse-field HVSR for site characterization: the case of El Ejido town (SE Spain). Soil Dyn Earthq Eng 121:102–120. https://doi.org/10.1016/j.soildyn.2019.02.023

Giocoli A, Hailemikael S, Bellanova J, Calamita G, Perrone A, Piscitelli S (2019) Site and building characterization of the Orvieto Cathedral (Umbria, Central Italy) by electrical resistivity tomography and single-station ambient vibration measurements. Eng Geol 260: https://doi.org/10.1016/j.enggeo.2019.105195

Grassi S, Imposa S, Patti G, Boso D, Lombardo G, Panzera F (2019) Geophysical surveys for the dynamic characterization of a cultural heritage building and its subsoil: the S. Michele Arcangelo Church (Acireale, eastern Sicily). J Cult Herit 36:72–84. https://doi.org/10.1016/j.culher.2018.09.015

IGN (2017) Actualización de mapas de peligrosidad sísmica de España 2012. Catálogo de publicaciones de la Administración General del Estado, Spain. https://doi.org/10.7419/162.05.2017

Karabulut S (2018) Soil classification for seismic site effect using MASW and ReMi methods: a case study from western Anatolia (Dikili -İzmir). J Appl Geophys 150:254–266. https://doi.org/10.1016/j.jappgeo.2018.01.011

Macau A, Benjumea B, Gabàs A, Bellmunt F, Figueras S (2018) Geophysical measurements for site effects characterisation in the urban area of Girona, Spain. Near Surf Geophys 16:340–355. https://doi.org/10.3997/1873-0604.2018004

Mahajan AK (2009) NEHRP soil classification and estimation of 1-D site effect of Dehradun fan deposits using shear wave velocity. Eng Geol. https://doi.org/10.1016/j.enggeo.2008.10.013

Maniatakis CA, Spyrakos CC, Kiriakopoulos PD, Tsellos KP (2017) Seismic response of a historic church considering pounding phenomena. Bull Earthq Eng 1–29. https://doi.org/10.1007/s10518-017-0293-5

Marín Lechado C, Roldán García FJ, Pineda Velasco A, Martínez Zubieta P, Rodero Pérez J, Díaz Pinto G (2009) Mapa Geológico Digital Continuo E. 1:50.000, Zonas internas de las Cordilleras Béticas. (Zona-2100). In: GEODE. Mapa Geológico Digital continuo de España

Martínez Segura MA, Navarro Bernal M, Martínez Pagan P (2017) Riesgos sísmicos en la Región de Murcia y métodos de evaluación de la peligrosidad sísmica a escala urbana. In: editum (ed) LOS RIESGOS AMBIENTALES EN LA REGIÓN DE MURCIA. Ediciones de la Universidad de Murcia, Murcia, pp 35–57

Martínez-Pagán P, Navarro M, Pérez-Cuevas J, Alcalá FJ, García-Jerez A, Sandoval-Castaño S (2014) Shear-wave velocity based seismic microzonation of Lorca city (SE Spain) from MASW analysis. Near Surf Geophys 12:739–749. https://doi.org/10.3997/1873-0604.2014032

Martínez-Pagán P, Navarro M, Pérez-Cuevas J, Alcalá FJ, García-Jerez A, Vidal F (2018) Shear-wave velocity structure from MASW and SPAC methods: the case of Adra town, SE Spain. Near Surf Geophys 16:356–371. https://doi.org/10.3997/1873-0604.2018012

Ministerio de Fomento (2002) Catálogo sísmico de la Península Ibérica (880 a.C.-1900). Subdirección General de Producción Cartográfica, Madrid

Molina Gaitán JC (2014) Historia de la restauración de la catedral de Murcia : intervenciones desde 1928 a 2010. Tesis Doctoral. Universidad Politécnica de Cartagena. Vol.I

Monaco P, Amoroso S (2016) Site effects from the building scale to the seismic microzonation scale: examples from the experience of L'Aquila. Proc Eng. 158:517–522. https://doi.org/10.1016/j.proeng.2016.08.482

Navarro M, Vidal F, Enomoto T, Alcalá FJ, García-Jerez A, Sánchez FJ, Abeki N (2007) Analysis of the weightiness of site effects on reinforced concrete (RC) building seismic behaviour: the Adra town example (SE Spain). Earthq Eng Struct Dyn 36:1363–1383. https://doi.org/10.1002/eqe.685

Navarro M, García-Jerez A, Alcalá FJ, Vidal F, Enomoto T (2014) Local site effect microzonation of Lorca town (SE Spain). Bull Earthq Eng 12:1933–1959. https://doi.org/10.1007/s10518-013-9491-y

Olafsdottir EA, Bessason B, Erlingsson S (2018) Combination of dispersion curves from MASW measurements. https://doi.org/10.1016/j.soildyn.2018.05.025

Pagliaroli A (2018) Key issues in seismic microzonation studies: lessons from recent experiences in Italy. Riv Ital di Geotec 52:5–48. https://doi.org/10.19199/2018.1.0557-1405.05

Pagliaroli A, Pergalani F, Ciancimino A, Chiaradonna A, Compagnoni M, de Silva F, Foti S, Giallini S, Lanzo G, Lombardi F, Luzi L, Macerola L, Nocentini M, Pizzi A, Tallini M, Teramo C (2019) Site response analyses for complex geological and morphological conditions: relevant case-histories from 3rd level seismic microzonation in Central Italy, Bull Earthq Eng. Springer, Netherlands. https://doi.org/10.1007/s10518-019-00610-7

Panzera F, Lombardo G, Imposa S, Grassi S, Gresta S, Catalano S, Romagnoli G, Tortorici G, Patti F, Di Maio E (2018) Correlation between earthquake damage and seismic site effects: the study case of Lentini and Carlentini, Italy. Eng Geol 240:149–162. https://doi.org/10.1016/j.enggeo.2018.04.014

Park C (2013) MASW for geotechnical site investigation. Lead Edge 32:656–662. https://doi.org/10.1190/tle32060656.1

Park CB, Miller RD (2008) Roadside passive multichannel analysis of surface waves (MASW). J Environ Eng Geophys

Rosa-Cintas S, Clavero D, Delgado J, López-Casado C, Galiana-Merino JJ, Garrido J (2017) Characterization of the shear wave velocity in the metropolitan area of Málaga (S Spain) using the H/V technique. Soil Dyn Earthq Eng 92:433–442. https://doi.org/10.1016/j.soildyn.2016.10.016

Sanz de Galdeano C, López Casado C, Delgado J, Peinado MA (1995) Shallow seismicity and active faults in the Betic Cordillera. A preliminary approach to seismic sources associated with specific faults. Tectonophysics 248:293–302. https://doi.org/10.1016/0040-1951(94)00279-I

Stanko D, Markušić S, Strelec S, Gazdek M (2017) HVSR analysis of seismic site effects and soil-structure resonance in Varaždin city (North Croatia). Soil Dyn Earthq Eng 92:666–677. https://doi.org/10.1016/j.soildyn.2016.10.022

Tarque N, Lai CG, Bozzoni F, Miccadei E, Piacentini T, Camata G, Spacone E (2013) Expected ground motion at the historical site of Poggio Picenze, Central Italy, with reference to current Italian building code. Eng Geol 166:100–115. https://doi.org/10.1016/J.ENGGEO.2013.09.003

Tassios TP (2010) Seismic engineering of monuments: The first Prof. Nicholas Ambraseys distinguished lecture. Bull Earthq Eng 1231–1265. https://doi.org/10.1007/s10518-010-9219-1

UNESCO (2009) Irrigators tribunals of the Spanish Mediterranean coast: the Council of Wise Men of the plain of Murcia and the Water Tribunal of the plain of Valencia

Vásconez-Maza MD, Martínez-Segura MA, Bueso MC, Faz Á, García-Nieto MC, Gabarrón M, Acosta JA (2019) Predicting spatial distribution of heavy metals in an abandoned phosphogypsum pond combining geochemistry, electrical resistivity tomography and statistical methods. J Hazard Mater 374:392–400. https://doi.org/10.1016/j.jhazmat.2019.04.045

Vera Boti A (1993) La Torre de la Catedral de Murcia: de la teoría a los resultados. Real Academia Alfonso X El Sabio, Murcia

Vera Boti A, Sánchez-Rojas Fenoll MC, De la Peña Velasco C, Pascual Martínez L, Esbert Alemany RM (1994) La Catedral de Murcia y su plan director. Colegio Oficial de Arquitectos de Murcia, Murcia

Vicêncio H, Teves-Costa P, Caetano PS (2018) Site characterisation in Barreiro urban area (Portugal) using H/V and ReMi techniques. Near Surf Geophys 16:298–312. https://doi.org/10.3997/1873-0604.2018006

Vidal F (1986) Seismotectonics of the region Betics-Alboran Sea. University of Granada (Spain)

Xia J, Miller RD, Park CB (1999) Estimation of near-surface shear-wave velocity by inversion of Rayleigh waves. Geophysics

Chapter 32
The Representation of a Resilient City: The Case of Amatrice's Reconstruction

Giuseppe Amoruso and Polina Mironenko

Abstract This chapter presents methodologies of documentation to be used for historical urban landscapes, according to the 2011 UNESCO *Recommendation on Historic Urban Landscapes*. This study proposes a protocol for the typological and parametric representation of townscapes, also introducing the application of a Building Information Modelling (BIM) tool for the planned reconstruction. This process aims to promote in-depth knowledge of the site condition: open data collection, typological survey and visualisation of the city's evolution and conformation prior to the earthquake. A further step is the application of information technology (IT) models to combine adequate technological retrofitting and resilience measures while taking into account the character of the places being reconstructed. As a consequence, it is important to make simulations to visualise typological, aggregative and functional solutions with respect to the regeneration of social capital in the places where it is expected by the local community. The strategy presented here proposes that in decision-making processes, project information modelling through Heritage BIM and the simulation of different reconstruction scenarios is included. Toolkits will be used to visualise and disseminate the recommended building solutions based on architectural repertoires such as pattern books, included in the masterplan. The heritage information modelling system allows simulations that integrate environmental modelling, semantic representation and typological documentation of each settlement. The main outcome is the typological and parametric representation of identity characteristics based on algorithmic simulations and the application of Heritage BIM architectural libraries and typological models to the reconstruction design. The project also supports the regeneration of Amatrice's local identity through documentation of its urban landscape, its material expression as an impulse for the regeneration of memory and the inclusion of the community within the process.

G. Amoruso (✉) · P. Mironenko
Politecnico di Milano, Milan, Italy
e-mail: giuseppe.amoruso@polimi.it

P. Mironenko
e-mail: polina.mironenko@polimi.it

© The Author(s), under exclusive license to Springer Nature Switzerland AG 2021
F. F. Arefian et al. (eds.), *Historic Cities in the Face of Disasters*, The Urban Book Series,
https://doi.org/10.1007/978-3-030-77356-4_32

Keywords Historic urban landscape (HUL) · Heritage BIM (H-BIM) · Urban reconstruction protocols · Amatrice · Italy

32.1 Introduction

Cities are fregile organisms according to their organic patterns based on hierarchy, functions and density; manufactured by organic creatures, they are inhabited by people who are themselves fragile. Cities follow life cycles: it takes an incredible amount of time, energy, culture and resources to build a city. Furthermore, to make it beautiful, a continuous process of architectural tuning within the natural landscape that can make a settlement the best place to live in is required. Unfortunately, with an earthquake, for example, it only takes a few seconds to destroy it. Despite these fragile conditions and its embodied resilience, cities have been continuously destroyed and rebuilt. Organic reconstruction was the natural answer to the problem according to several lessons we have learned from history.

When a devastating earthquake struck the eastern part of Sicily in 1693, the Val di Noto, the Spanish government immediately appointed the Duke of Camastra as head of reconstruction work. The Sicilian cities of Noto, Catania, Siracusa, Ragusa e Ibla—today well known as significant historical cities and therefore as cultural attractions—were rebuilt in a short time.

In addition, some new cities, such as Avola and Grammichele, were founded on a hexagonal city model according to theoretical, geometric plans inspired by Early Modern treatises. At Avola, two months after the earthquake and a century after the construction of Palmanova (1593), the Jesuit architect Angelo Italia and the master builder Antonio Vella chose the new site by tracing the image of the new Avola. The hexagonal plan allowed effective and flexible defence, a fortified, functional and compact urban system characterised by five squares, one central and the others placed on the cardinal points of the compass in correspondence to the four city gates. The perimeter of the hexagonal walls made a valid defence possible, and, in the event of an earthquake, the squares ensured easily reachable shelter inside the city.

Historically, of particular relevance is the case of the reconstruction of Ragusa which, post-reconstruction became Ragusa and Ibla, two separate administrative cities: one rebuilt respecting that part of the population that asked to return to the original city and the other built *ex novo* to favour the economic interests of the landowners. Two parishes, San Giorgio and San Giovanni, were established, and two different ways of making reconstruction resilient were deployed, privileging in both cases the requirement of security rather than that of identity, while also nourishing that sense of belonging in the absence of which the sense of community would never have been re-established (Gangi 1982).

The case of Noto is extraordinary also because it can now be recognised as a precursor of contemporary participatory processes in place-making. Due to the need to move the city to a less steep location in order to avoid any new "domino effect" destruction, survivors were able to influence decisions on reconstruction. New neighbourhoods were built according to an Islamic grid, despite the city dating back to the early eighteenth century: citizens accepted being moved to another place, but they accepted the agreement they could recreate, in the rebuilt town of Noto, urban-environmental conditions similar to those experienced in the city of origin (Trigilia 1994).

The organic reconstruction of Noto is listed as one of the best practices of efficient place-making after the earthquake, and since 2002, the townscape is included in the United Nations Educational, Scientific and Cultural Organisation (UNESCO) World Heritage list as the *Late Baroque Towns of the Val di Noto* site: "They represent a considerable collective undertaking, successfully carried out at a high level of architectural and artistic achievement. Keeping within the late Baroque style of the day, they also depict distinctive innovations in town planning and urban building" (UNESCO 2002).

This short list of examples of resilient reconstruction could be extended to include the rebuilding of Baixa after the violent earthquake of 1755 that affected Lisbon (the earthquake caused between 60,000 and 90,000 deaths, out of an estimated population of 275,000). The new quarter combined innovation by deploying for the first time a special wooden reinforcement net with traditional, local materials and a regional colour palette. Looking across history, examples show that each time natural hazards interrupt the organic evolution of a place and create a cultural gap; consequently, this phenomenon creates a vulnerability in settlement infrastructure and the population, both of which are inevitably exposed to risk, especially in a territorial context already characterised by high seismic risk or reduced resilience.

This chapter examines the reconstruction processes of Italian centres affected by the 2016 earthquake and sets out typological and parametric representations of local character based on algorithmic simulations. A further outcome is the application of Heritage BIM architectural libraries and typological templates to the reconstruction design process. This preamble introduces the real challenge of reconstruction: to maintain Amatrice's authentic character while rebuilding not only its places but also a conscious community as outlined in the principal resilience improvement goals (Fig. 32.1).

32.2 Framework Conventions for the Value of Cultural Heritage in Resilient Communities

Currently, the main cultural framework is provided by the *Faro Convention* (2005) that underlines the important issue of heritage as it is entrenched within human rights and democracy. It promotes a wider understanding of heritage and its correct relationship to communities and society. The *Convention* encourages recognition of the fact that objects and places are not, solely in themselves, what is important about cultural heritage. They are important also because of the meanings and uses that

Fig. 32.1 Plan of the historic centre of Amatrice before the earthquake in Central Italy in 2016 (authors)

people attach to them and the values they represent. The *Convention* introduced two innovative concepts that are relevant to how the concept of the resilient city and the role of local communities can be exploited:

(a) *cultural heritage* is a group of resources inherited from the past which people identify, independently of ownership, as a reflection and expression of their constantly evolving values, beliefs, knowledge and traditions. It includes all aspects of the environment resulting from the interaction between people and places through time;
(b) *heritage community* consists of people who value specific aspects of cultural heritage which they wish, within the framework of public action, to sustain and transmit to future generations (Faro Convention 2005).

The active role of communities as living heritage is underlined also by the *European Landscape Convention* that defines "Landscape" as an area perceived by people whose character is the result of the action and interaction of natural and/or human factors. The territory is a whole covering natural, rural, urban and suburban areas, landscapes that may be considered outstanding as well as every day and degraded landscapes.

"The identification, description, and assessment of landscapes constitute the preliminary phase of any landscape policy", the Convention states. "This involves an analysis of morphological, archaeological, historical, cultural and natural characteristics and their interrelations, as well as an analysis of changes" (European Landscape Convention 2000).

To achieve this target, integral knowledge of the values and meanings of a specific territory are required, because these form the cornerstone for informing every decision-making process aimed at repairing or rebuilding cultural heritage. The assessment made during the post-emergency phase suggests that it is crucial to deal not only with reconstructing the architectural character of the place but also the intangible substance through which the community is constituted.

In Italy, populations live on fault-lines and therefore historically coexist with the risk of earthquakes or seismic swarms; it is paradoxical how awareness of minimum precautions related to the maintenance and adaptation of traditional construction technologies to minimise or avoid serious damage to buildings exists, yet other things and people have been ignored and lost.

The conventional approach to post-disaster recovery has mostly been based on post-impact reaction rather than risk reduction policies: prevention, preparedness, mitigation, because cultural resilience has to be embodied in the place-making process.

According to UNESCO's *Recommendations on the Historic Urban Landscapes* (2011), the historic urban landscape (HUL) is any urban area understood as the result of a historic layering of cultural and natural values and attributes, extending the "historic centre" or city "ensemble" concept to include the broader urban context and its geographical setting. This wider context notably includes the site's topography, geomorphology, hydrology and natural features, its built environment, both historic and contemporary, its infrastructures above and below ground, its open spaces and gardens, its land use patterns and spatial organisation, perceptions and visual relationships, as well as all other elements of the urban structure. It also includes social and cultural practices and values, economic processes and the intangible dimensions of heritage as related to diversity and identity (Bandarin and van Oers 2012).

The historical urban landscape of Amatrice should be considered as a cultural landscape (Salerno 2017), because it is a material expression of adaptation templates to different sites and environmental resilience, in which each element can be identified and documented; this landscape can be described graphically through a catalogue with multiple levels of information developing appropriate semantic models (Amoruso 2017). Amatrice, a symbol of the 2016 Central Italy earthquake, faces a challenge of considerable economic and social commitment, and it will only be able to overcome these if the appropriate tools are adopted. Documentation of local identity after the emergency phase ought to deal with issues that seem inevitable and no longer postponed: living traditions, people and cultural processes.

32.3 Research Methodology: Coding Local Identity

The research methodology adopted begins with the statement that landscape is an expression of local character and culture of place. In article IV of the UNESCO *Recommendation on Historic Urban Landscape*, a series of innovative tools were presented:

> "The approach based on the historic urban landscape implies the application of a range of traditional and innovative tools adapted to local contexts. Some of these tools, which need to be developed as part of the process involving the different stakeholders, might include:
>
> (a) Civic engagement tools should involve a diverse cross-section of stakeholders and empower them to identify key values in their urban areas, develop visions that reflect their diversity, set goals and agree on actions to safeguard their heritage and promote sustainable development.
>
> (b) Knowledge and planning tools should help protect the integrity and authenticity of the attributes of urban heritage. They should also allow for the recognition of cultural significance, diversity and provide for the monitoring and management of change to improve the quality of life and of urban space. These tools would include documentation and mapping of cultural and natural characteristics. Heritage, social and environmental impact assessments should be used to support and facilitate decision-making processes within a framework of sustainable development (UNESCO 2011).

According to these principles, three different work packages are proposed: architectural lexicon analysis and pattern book release; Heritage BIM modelling for each block of the town centre to design the seismic aggregate retrofitting; and territorial and urban simulations according to the different reconstruction scenarios as proposed by the masterplan.

The research aimed to adopt, in decision-making processes, a strong support from urban visualisations based on repertoires, high-iconic databases and predictive simulations and a range of visual and descriptive images, illustrations and 3D model to enhance the decisional power. The maintenance of local identity by design allows the testing of local character documentation toolkits that could be operated by community activists as well: analysis of urban patterns, construction techniques, materials and environmental tonal characteristic for each landscape unit to describe a sequence of clusters forming the urban habitat. The workflow promotes the in-depth knowledge of Amatrice and its historic city: of its pre-earthquake conditions and visualisation of its typological character; and the application of information models that describe resilient solutions also according to the maintenance of place identity. It is therefore important to simulate and visualise typological, aggregative and functional solutions in respect to the regeneration of places and their social value as expected by the community.

For this purpose, a pattern book, the *Codice di Amatrice*, will be released: it is a collection of technical drawings and graphic illustrations of local traditional architecture, working drawings and typological solution and a guide to best building practices. The pattern book is elaborated block by block starting from the architectural lexicon and construction technologies analysis, putting *genius loci* and tradition into practice (Figs. 32.2 and 32.5).

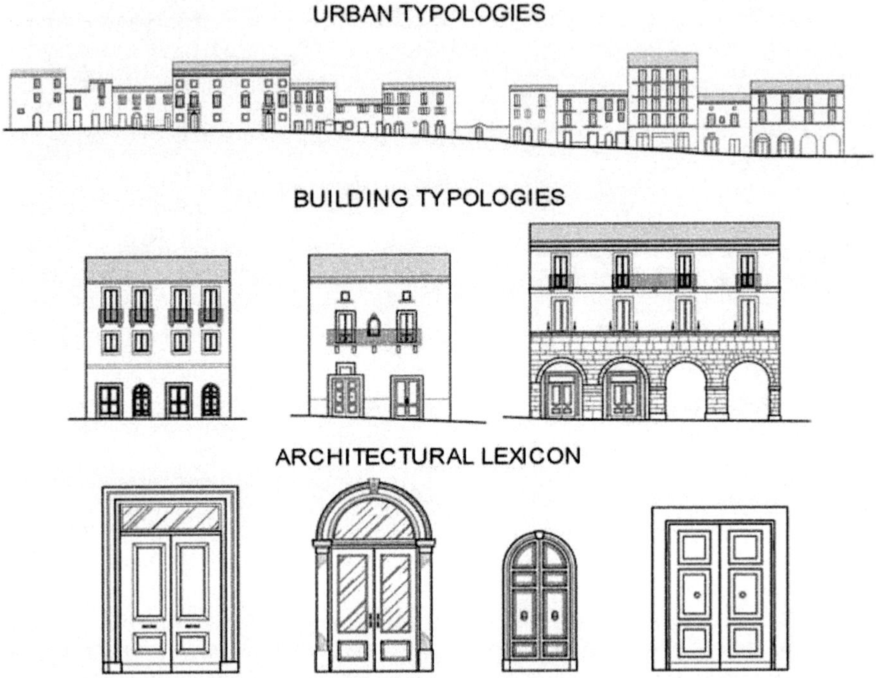

Fig. 32.2 Prototype of pattern book was elaborated for the central block of Corso Umberto I (authors)

The *Codice* will inform the creation of the design regulations and building codes to be embedded in rehabilitation plans integrating all the information on the ground, landscape and architectural typologies in the same tools. The idea for the *Codice* is inspired by a centuries-long tradition of Italian recovery guides "*manuali di recupero*" to inform place-making and rebuild communities.

Additional examples can be found in several handbooks and treatises addressing good practice for seismic reinforcement: a 1570 treatise on earthquakes by Pirro Ligorio, who served under Duke Alfonso II d'Este of Ferrara, describing plans for an earthquake-resistant home; and the *Manuale per la Costruzione Antisismica*, the first European handbook on traditional engineering and building techniques produced by the Bourbon Government after the 1783 earthquake in Calabria. The decision to reconstruct the entire region using traditional anti-seismic techniques and building craft skills known as the *baraccata* system, presented in the handbook, as in Ligorio's work, shows that the holistic practice of traditional building methods is not limited to the design itself, but also encompasses the engineering and integrity of structures according to the availability of local materials.

The main operational reference for the *Codice di Amatrice* is the *Pattern Book for Newquay* an urban design project by the Duchy of Cornwall to put in practice

Fig. 32.3 The 3D visualisation of the H-BIM model of the sample block in the centre of Amatrice as the result of the research process application, from the urban typologies to building typologies to patterns and architectural lexicon (authors)

the making of a resilient community (Duchy of Cornwall 2005). The book, rich in graphic illustrations and technical drawings, is not prescriptive, but it provides town planners, architects and builders with a useful resource of typologies and details for the design of urban space and buildings. It is intended to guide new development in a way that will provide continuity throughout the urban fabric and strengthen the built character of the town, respecting the masterplan's building code.

In Amatrice, urban landscape documentation and the graphic transcription of its semantic expressions provide a cognitive framework but also an operative vision to regenerate places and building according to local traditions, with the aim of generating new knowledge so that it can be transformed into skills, seismic culture and social resilience (Figs. 32.3 and 32.4).

32.4 Generative Algorithmic Simulation of Reconstruction Scenarios

Predictive simulations based on algorithms of "instant modelling" for drafting territorial and urban models (Calvano and Guadagnoli 2016) forecast reconstructions, and they are useful to estimate economic management issues according to town development and demographic scenarios or construction technologies (Fig. 32.5).

The research here tested several generative algorithmic simulations to process ground and terrestrial available data and 3D models of urban blocks; the outcome was a series of digital elevation models of Amatrice's townscape and 3D representation of terrain's surface (Figs. 32.6 and 32.7).

Fig. 32.4 Visual pattern book based on Google 360° immersive video developed as equirectangular-panoramic models and elaborated through the *StreetView Grabber* software (authors)

In Amatrice it was not possible to carry out the field documentation phase from the record of existing or surviving buildings because the earthquake had particularly disastrous effects and the site of the old city after its ruins were demolished remains totally unbuilt.

This research presents an innovative protocol for territorial and urban simulation starting from GIS data collection available online: open-source dataset, public authority data, numerical and vector data, while territorial images were collected from public body databases, Google Earth/Street View and Open Street Map. The simulation algorithm also included high-quality Shuttle Radar Topography Mission data (SRTM), available online as grey scale GeoTiff images: a collection of elevation models, arranged into tiles, each covering one degree of latitude and one degree of longitude. Images contain also metadata of geographical references with which to numerically reconstruct the earth's surface.

The protocol could be also integrated with GIS tools like the Arches platform: a collaboration project between the Getty Conservation Institute (GCI) and the World Monuments Fund (WMF) to develop an open-source, web-based and geospatially based information system that is purpose-built to inventory and manage immovable cultural heritage.

Following data and information collection, the territorial settlement of Amatrice was reconstructed with the 3D computer graphics and computer-aided design (CAD) tool, *Rhinoceros* (typically abbreviated *Rhino*). The software geometry is based on the Non-Uniform Rational Basis-Splines (NURBS) mathematical model, which focuses on producing a mathematically precise representation of curves and freeform surfaces

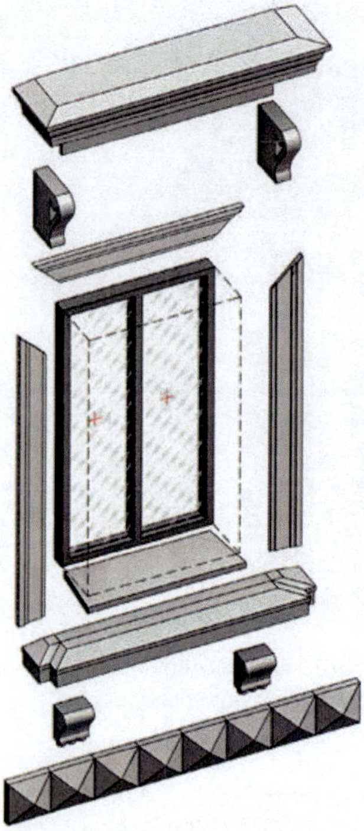

ABACO ELEMENTO DECORATIVO Dec-3		
Codice	Descrizione	Quantità
3.1	Infisso 1	1
3.2	Mensola	2
3.3	Cornice 1	1
3.4	Cornice 2	1
3.5	Cornice 2	2
3.6	Elemento di ceme	1

ABACO ELEMENTO DECORATIVI			
Codice	Descrizione	Immagine	Quantità
Dec-3	Infissi di finestra con cornice decorativa ed elementi di cemento		4

Fig. 32.5 Construction technology sampling and 3D modelling of the architectural lexicon (authors)

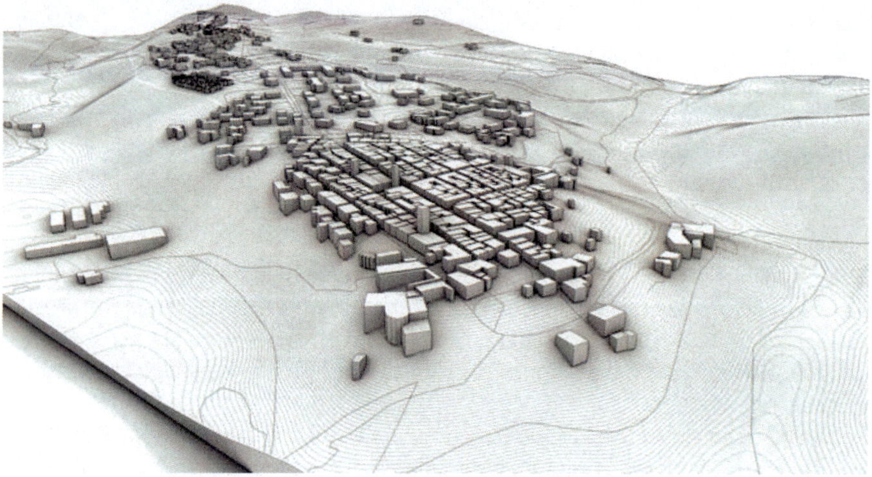

Fig. 32.6 Digital Ground Model of Amatrice as a result of the generative process based on open data from *OpenStreetMap* project and from the Shuttle Radar Topography Mission (authors)

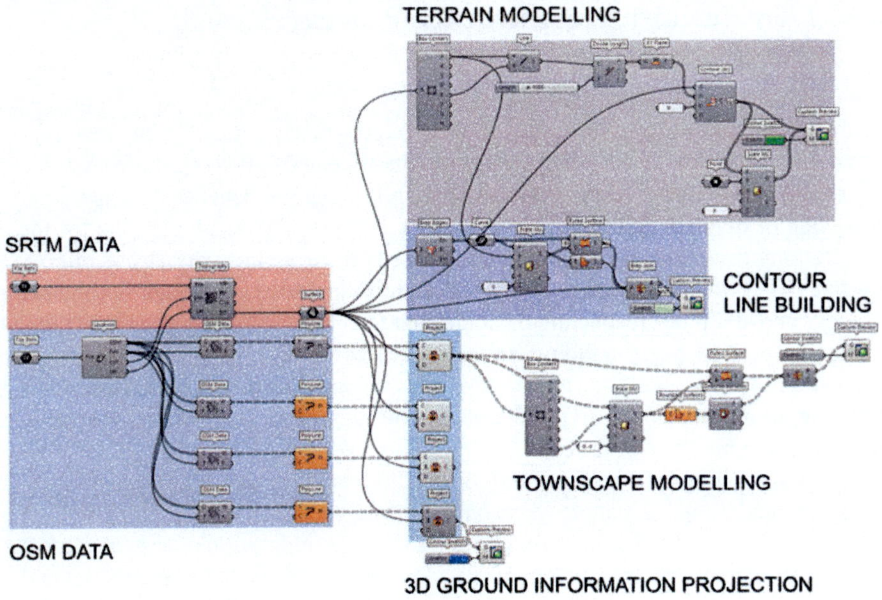

Fig. 32.7 Algorithmic modelling with Grasshopper and Elk to generate the Digital Ground Model of Amatrice and a series of visualisation and scenarios from open- source dataset

in computer graphics (as opposed to polygon mesh-based applications). *Rhino* was associated with a visual programming language and environment, *Grasshopper*: algorithms are created by dragging components onto a canvas and making visible the computational layout from data importing to 3D modelling outcome. The algorithm is designed according to a logical sequence of modules where outputs to each component are then connected to the inputs of subsequent components. Grasshopper is primarily used to build generative algorithms and many of its components create 3D geometry, but tools may also contain other types of algorithms including numeric, textual, audio-visual and haptic applications. The algorithmic modelling with *Grasshopper* included a module that, through the plug-in tool *Elk*, generates maps and topographical surfaces using *OpenStreetMap* data and USGS. *OpenStretMap.org* is an open/crowd-sourced website of mapping data that allows to export XML formatted data of a selected area. *Grasshopper* and *Elk* made possible to build collections of points and tag data first and then to create curves and other Rhino/Grasshopper geometry. Through this set of integrated tools and also data on building features, the Digital Ground Model of Amatrice was generated as well as a series of visualisations and scenarios. Following this workflow, the research tested generative processes from data collection to the 3D environmental simulation to also inform public authorities and decision-makers during the planning process and the reconstruction of the town.

32.5 Place-Identity Reconstruction Process Through Heritage BIM

The final outcome is to regenerate the town local identity according to the meetings that were organised with the municipality of Amatrice and following the visit of the Prince of Wales in 2017. Among the tangible and intangible benefits were a greater knowledge of diverse urban landscapes and local expression as an impulse for the regeneration of memory because the earthquake had particularly disastrous effects. How to manage the reconstruction process of entire historic centres, approximately 30 blocks and 400 plots, respecting the local character and rebuilding a resilient city?

It is strategic to begin building an open archive of data, urban visualisations, information and 3D models that can be developed through digital and physical models of the single minimum and independent seismic aggregate that could reasonably coincide with a block or a part of it.

Heritage Building Information Modelling allows simulations that integrate environmental modelling, semantic representation and typological documentation of Amatrice. It is the right location to experiment with the integration of a management and design tool to document or assess place-identity and also visualise recommended solutions based on the single-plot architectural design chosen from the repertoires published in the pattern book (Arayici et al. 2017).

The concept of Heritage BIM (H-BIM) pursues the modelling of architectural elements, according to artistic, historical and constructive typologies. In addition, H-BIM is considered to be an emerging technology that enables researchers to understand, document and virtually reconstruct built heritage (Lopez et al. 2018).

Several BIM environments are used by specialists in the field of modelling, virtual simulation and management of comprehensive and incremental knowledge of architectural heritage. However, it is important to highlight that tools, resources and libraries of the BIM softwares are focused on the design of new buildings including conventional construction techniques and standardised materials.

To overcome the constraints of BIM systems and the need for detailed reconstruction of cultural–historical heritage, the research released the pattern book first and started the modelling of the main street of Amatrice in order to make available historical parametric object libraries to designers.

The tested methodology for historic urban landscape documentation included analytical data collection, representation and interpretation of architectural composition, materials, patterns, textures and pigments and chromatic dominants within buildings and blocks.

The H-BIM process adopted integrated methodology of 3D modelling to produce elevation drawings, colour palettes, rectified images, 3D models and database records. The site was studied recording each building, master models, and drawing the whole block elevation, derived models, in order to enhance the understanding of chromatic landscapes and the variability of tones and building techniques. The documentation process was applied to the main street landscape where it is usual to analyse and classify a more complex variety of artefacts from a typological point of view: age, size, geometry, façade technology, lexicon, textures, colour gradient and dominants façade semantics.

The workflow follows the 2011 UNESCO *Recommendations* that "research should target the complex layering of urban settlements, in order to identify values, understand their meaning for the communities and present them to visitors in a comprehensive manner. Hence, it is essential to document the state of urban areas and their evolution, to facilitate the evaluation of proposals for change and to improve protective and managerial skills and procedures". For this purpose, the methodology proposes the definition of interpretative models for the analysis and representation of historical, cultural, aesthetic and environmental cultural heritage, its historical landscapes, their architectures and their tangible and intangible issues.

Rebuilding Amatrice means modelling the experimental Heritage BIM of the town centre, generated from the recorded pattern book and including additional information gathered during research and the field tests. The digital BIM will allow for the virtual testing and simulation of different solutions for the same plot or block as well as the comparison and checking of the impact of new buildings, blocks and interventions, thus helping to ensure the preservation of Amatrice's unique sense of identity and place. Once the parametric objects are modelled using the architectural historical documentation, the regional restoration handbooks and other technical thesaurus or data, the library of the modelled elements should be created and encapsulated within the BIM. The heritage and traditional architecture libraries could be 3D-printed and exhibited during technical sessions and community consultations. The H-BIM will be an interactive tool providing information, visual references and supporting evidence to aid the reconstruction process and associated decision-making.

In this way, data representation and visualisation target and frame knowledge not only as technology-oriented *status quo* recording process but as the expression of assets and their quality through actions of revelation and legitimation. Digital contents can be also used to put in practice processes, methods and tools of representation and narration of material values and intangible assets (identity, community, representations, etc.) for the well-being of people who live segregated in their temporary barrack or dwellings. In the research context, it was useful to assess the potentialities and difficulties of H-BIM application dealing with methodologies of parametric modelling of cultural heritage and where it is not possible to import a point cloud or survey data (Figs. 32.8 and 32.9).

32.6 Conclusion

The act of seeking technology for creating the sense of place through the representation of its identity is a prerequisite to integrate it into the strategic design process; the critical documentation and the several forms of representation address the perceptual and conceptual mapping process (Amoruso 2015). The research framework presented

Fig. 32.8 The H-BIM model of the sample block in the centre of Amatrice. Elevation with the 3D modelling of the façade pattern and architectural lexicon for each plot (authors)

Fig. 32.9 The H-BIM model of the sample block in the centre of Amatrice and pattern book (Authors)

in this chapter contributes to the knowledge of places, creating the design process and pointing out their identity, character and distinctiveness. The contribution also aims to set a methodological model of reading, interpretation and narrative of the cultural heritage qualities; a visual catalogue that becomes meta-narrative on which to build subsequent pathways of activation and enhancement. This process can be aided by technologies, but it has to include also the personal body experience to shape abstract experiences because it may be assumed the concept that much of our information is non-literal and allegorical: the knowledge of intangible world comes from the concepts with which we have learned to know the reality of which it was possible to have a sensory experience.

Heritage is of the main assets that communities wish to rebuild according to the local way of life and traditions after a disaster. The accurate knowledge of such community heritage will support the process and the steps to plan the reconstruction as fundamental key to envision a more resilient city. Cultural environment cannot be separated from the natural one and that ours are urban cultures that are linked to a social and aesthetic experience that cannot be distinguished from technical design (Mallgrave 2015). Reconstructing the interior space, almost intimate of Amatrice, means giving citizens back that lost sense of security, a strong public internal space capable of influencing people's culture, social behaviour and sense of identity. In this perspective, they are the addresses from two international networks, *Conscious Cities* (The Centre for Conscious Design 2015) and *Resilient Cities* (ICLEI 2019). The quality of the spaces can be relaunched by recognising the social nature of human creative expression, the real common good in the reconstruction of Amatrice urban interiors with historical memories, formulating proposals that redesign living according to models of high liveability and implementing an idea of health intended as "dynamic state of complete physical, mental, social and spiritual well-being, not merely the absence of disease" (WHO 1948, 1998).

The study introduced a process of integrated knowledge providing science- and evidence-based guidelines and digital models to local authorities for carrying out sustainable reconstruction within a participatory- and community-based context. Resilience is described as the overall goal of decreasing human vulnerability through the achievement of awareness, participation to planning process and well-being.

Acknowledgements This work is part of an international research project that started in 2017 entitled: *Regenerating Amatrice. The resilience of local identity* funded by the *Prince's Charitable Trust* and promoted by the International Network for Traditional Building, Architecture and Urbanism that works under the patronage of its founder HRH the Prince of Wales.

References

Amoruso G (2015) The image of historic urban landscapes. Representation codes. In: Brusaporci S (ed) Handbook of research on emerging digital tools for architectural surveying, modeling, and representation. IGI Global, Hershey, PA, pp 566–595

Amoruso G (2017) Developing semantic models for the historic urban landscape regeneration. In: Amoruso G (ed) Putting tradition into practice: heritage, place and design. Springer International Publishing, Cham, pp 769–777

Arayici Y, Counsell J, Mahdjoubi L, Nagy GA, Hawas S, Dweidar K (2017) Heritage building information modelling. Routledge, London

Bandarin F, van Oers R (2012) The historic urban landscape: managing heritage in an urban century. Wiley, New York, NY

Calvano M, Guadagnoli F (2016) 'Ricostruzione 3D della città di Amatrice. Una operazione di 'instant modeling'. DisegnareCon 9(17) (online). Available at: http://disegnarecon.univaq.it/. Accessed 15 Nov 2020

Council of Europe (2000) European landscape convention. Firenze

Council of Europe (2005) Convention on the value of cultural heritage for society. Faro

Duchy of Cornwall (2005) A pattern book for Newquay. The Prince's Foundation for the Built Environment, London

Gangi G (1982) 'Ragusa barocca'. Sellerio Editore, Palermo

ICLEI (2019) Resilient cities, thriving cities: the evolution of urban resilience, Bonn, Germany

Lopez F, Lerones P, Llamas J, Gómez-García-Bermejo J, Zalama E (2018) A review of heritage building information modeling (H-BIM). Multimodal Technol Inter 2(21) (online). Available at: https://www.mdpi.com/journal/mti. Accessed 15 Nov 2020

Mallgrave HF (2015) 'L'empatia degli spazi. Architettura e neuroscienze'. Raffaello Cortina, Milano

Salerno R (2017) Enhancing not-outstanding cultural landscapes in a European perspective: a challenge for digitization. In: Amoruso G (ed) Putting tradition into practice: heritage, place and design. Springer International Publishing, Cham, pp 3–8

The Centre for Conscious Design (2015) A manifesto for conscious cities (online). Available at: https://theccd.org/conscious-cities. Accessed 6 July 2019

Trigilia L (1994) Ricostruzione del val di Noto dopo il terremoto del 1693. Gangemi, Roma

UNESCO (2002) 'Late Baroque Towns of the Val di Noto (South-Eastern Sicily)' (online). Available at: http://whc.unesco.org/en/list/1024. Accessed 7 July 2019

UNESCO (2011) Recommendation on the historic urban landscape, Paris

WHO (1948) Constitution of the World Health Organization. United Nations, Geneva

WHO (1998) The world health report 1998: life in the 21st century A vision for all. United Nations, Geneva

Chapter 33
Evacuation Simulation Considering Tourists' Attempts to Return Home: A Case of the Kiyomizu-dera Temple Area, Japan

Kohei Sakai and Hidehiko Kanegae

Abstract Evacuation of large disparate groups of people during the immediate post-disaster event is one of the most critical issues. Particularly crucial is the safe evacuation of non-residents of the area such as domestic and foreign tourists. Well-managed and considered evacuation immediately after post-disaster enables for the efficient deployment of human resources for rescuing cultural property and initial firefighting efforts, leading to greater resilience. However, evacuation research does not consider the varying needs of domestic or international tourists including their attempts to return home. This chapter explores an evacuation simulation to examine the attempts of tourists to return home using a method combining sampling survey, scene imagination method and agent simulation. Initially, a survey was conducted with 274 domestic tourists and 107 foreign tourists in Japan. They provided their behavioural intention under various disaster scene scenarios such as guidance, road width, other tourist behaviour and route walking experiences. Based on the survey results using decision tree analysis, tourists' evacuation behaviour model was built considering Japanese and foreign tourists. Next, the evacuation agent simulation and proposed the evacuation method applied in the case of the Kiyomizu-dera Temple area, a popular tourist destination in Kyoto, Japan. As a result, the study found many tourists would flow out of the Kiyomizu-dera Temple area after disasters because of the location of evacuation sites and placement of guides. As mentioned above, this study proposed an evacuation simulation from the viewpoint of tourist's attempts at returning home.

Keywords Emergency evacuation · Agent simulation · Attempt to return home · Kiyomizu-dera Temple · Tourism · Japan

---

K. Sakai (✉) · H. Kanegae
Ritsumeikan University, Kyoto, Japan
e-mail: sakaik@fc.ritsumei.ac.jp

© The Author(s), under exclusive license to Springer Nature Switzerland AG 2021

F. F. Arefian et al. (eds.), *Historic Cities in the Face of Disasters*, The Urban Book Series,
https://doi.org/10.1007/978-3-030-77356-4_33

33.1 Introduction

Many tourists are travelling in Japan today. According to Japan Tourism Agency (2019), in 2018, approximately 270 million Japanese travelled on a domestic day trip, 291 million Japanese travelled on overnight tours and 31 million international tourists visited Japan. Over 55 million people visit Kyoto city annually, which is the focus area of this study. However, an increase in the number of tourists raises the risk of encountering a disaster, just as humankind faces more disasters due to an increase in its population densities.

Unlike local residents, tourists do not have a good sense of the locality or region. Often, they are unacquainted with local information and lack connections such as friends and relatives they can rely upon in a time of emergency. Thus, they cannot be expected to be fully familiar with evacuation procedures and other actions associated with local knowledge. Tourist sites become crowded with visitors whose numbers often exceed that of the local residents, and the duration of their stay is short, so it is impossible to raise awareness for disaster prevention among tourists. There are other issues for historical tourist sites in Japan, such as a lack of open space and the distance to shelters. Many tourists are unacquainted with local information, and for international tourists there is commonly a language barrier that makes their response to disasters more difficult. The consideration of how visitors can be prepared to respond to disasters at tourist sites is an urgent matter.

During the Great East Japan Earthquake, 5.15 million people were unable to return home due to suspension of rail services (Cabinet Office of Japan 2015). This massive movement of people attempting to return home resulted in large crowds that prevented emergency vehicles from moving freely and led to accidents. In addition to local residents, tourists in Kyoto city were faced with the possibility that they would be unable to return home, causing large crowds to be temporarily housed in public transportation terminals (Administration and Finance Bureau Disaster Prevention and Crisis Management Office 2018b). For this reason, there is a need to provide appropriate guidance on evacuation and limit the strain on public services.

Kyoto is a historic city that dates from 794 CE, when the capital was transferred to Heian-kyo. The city is now a tourist destination with cultural heritage at its core. However, Kyoto sits atop multiple fault lines, including Hanaore Fault, which raises the risk of an earthquake (Administration and Finance Bureau Disaster Prevention and Crisis Management Office 2018a).

If a disaster occurs, tourists should attempt to return home as soon as it is safe to do so, as their early return can mitigate the burden that a disaster-struck tourist site bears. By implementing appropriate responses, tourists can return home more efficiently, and more human resources can be allocated to reducing damage to historical sites, as well as their recovery. Hayashi (2016) defines resilience as prevention and recovery and notes that resilience can be improved by resolving the triangle formed by the damage and the time needed for restoration presented in Fig. 33.1. In other words, smooth evacuation of tourists during a disaster enables the use of human resources for rescuing cultural property and initial firefighting efforts, leading to greater resilience.

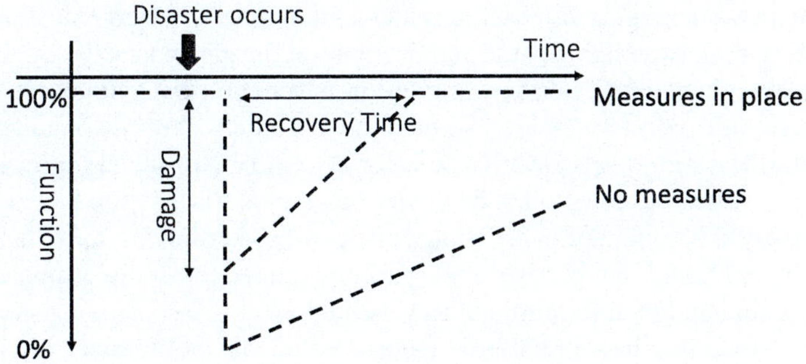

Fig. 33.1 Definition of resilience in a disaster (Hayashi 2016)

Two major contexts of how tourists should be supported in the immediate aftermath of a disaster are as follows. One setting is related to evacuations that occur immediately after a disaster, while the other focuses on temporary support for refugees and transportation for tourists stranded by a suspension of mass transit. The setting for this chapter is first one, evacuation, and proposes an agent simulation method for investigating an approach to a smooth transition for tourists to emergency shelters, while also controlling the crowds returning home via the mass transit system. Specifically, this study proposes a simulation design for tourist evacuation that extracts evacuation behaviour rules gleaned during sampling surveys conducted with tourists. This is seen as essential for the conservation of historical cities, since appropriate guidance for evacuation and an efficient evacuation process enables the effective use of human resources for firefighting activities and the rescue of cultural property.

33.2 Background: Evacuation Behaviour and Simulation

Studies have identified evacuation behaviour and decision-making of tourists during the immediate aftermath of a disaster. These studies have shown that tourists do not always evacuate to a shelter immediately, revealing behaviour in which tourists seek safety without knowing the location of a shelter or a clear destination in mind. Matyas et al. (2011) studied how factors among tourists would affect decision-making in a hurricane evacuation and found that tourists in large groups, those with children and those visiting a site for the first time, are more likely to evacuate. Arce et al. (2017) conducted interviews and questionnaire surveys with tourists in Kamakura, Japan, and revealed a relatively high level of risk awareness, as well as a willingness to evacuate after tsunamis. However, the study also found a lack of awareness regarding the direction of evacuation and the location of shelters. Nakaya et al. (2018) conducted a survey among the survivors of the Great East Japan Earthquake and found that the rate of evacuation was higher among those who had received tsunami disaster training than those who had no training. Cui et al. (2013) conducted a questionnaire survey that assumed a disastrous situation with tourists in Kyoto and identified factors such as route characteristics and signage that tourists emphasise when choosing their evacuation route and destination during an evacuation. Nishino et al. (2016) conducted an interview survey on evacuation behaviours by tourists during an earthquake disaster, asking whether they would follow the local guidance for evacuation. The survey revealed that both personal attributes such as age and travel method affect tourists' evacuation behaviours. Sakai et al. (2014) conducted a questionnaire survey that assumed evacuation during a disaster among tourists visiting Himeji castle and found that tourists tend to follow guidance but also emphasise the width of the route and tend to mimic others' behaviour.

Although the studies mentioned above have revealed that the evacuation behaviour of tourists and the evacuation destinations are not the same, these points have not been taken into consideration in evacuation simulation studies. Sugiyama et al. (2015) proposed an evacuation guide plan for Kiyomizu-dera Temple in Kyoto city and implemented agents that followed the proposed guide, which indicated that evacuation time is improved by closing some of the routes and opening routes that are normally unused. The study by Sugiyama et al. (2015) assumed that agents would follow guidance staff during evacuation, which may not occur in a real-world emergency.

On the other hand, recently, some simulation studies are conducting the implementation of realistic tourist agents with more complex rules. Takabatake et al. (2017) performed simulations using local resident agents, who moved to the closest shelter, and tourist agents, who either followed others or moved to higher elevation in the city of Kamakura. The simulation found that tourist evacuation behaviour had a significant impact on the location of bottlenecks and the number of casualties. Sonohara et al. (2018) conducted an assumed survey for evacuation behaviour and extracted behavioural rules for starting an evacuation based on the status of public transportation, availability of smartphones and environmental factors and proposed a method of designing an evacuation simulation at Himeji castle, a World Heritage Site. Behaviours in which agents lack a destination as they follow others in Takabatake et al. (2017) and decisions by agents to stay where they are as evacuation begins in Sonohara et al. (2018) are simulations that have taken this possibility into account.

Therefore, studies on evacuation behaviour and simulations require researchers to consider diverse behaviours influenced by various factors, rather than assuming uniform evacuation behaviour. This study is unique in its attempt to consider evacuation from the control of crowded mass transit systems. The study first divided evacuation behaviours into start of evacuation and route selection and then examined the attempts by evacuees to move to a shelter as well as to public transportation terminals to express the diversity of destinations and extract evacuation behaviour rules. In this study, the intention of returning home is defined as the behaviour of returning to the terminal station. This takes into account the following: personal attributes of tourists, surrounding situations at the time of disaster and whether the tourists are Japanese or foreign.

33.3 Understanding Tourist Behaviour During Evacuation

This study has two steps. First, it extracted tourists' evacuation behaviour rules on a sampling survey for tourists. Second, the study developed evacuation simulation using the extracted rules and conducted simulation experiments.

This study conducted a sampling survey on tourist evacuation behaviours during a disaster. Domestic tourists at Himeji castle, a World Heritage Site, and foreign tourists in Kyoto, a historical city, were asked about the behaviours of evacuation starting and route selection they would take if a disaster occurred in an assumed situation. Overall, 382 samples were collected (Table 33.1). A simulation is constructed using the behaviour rules obtained by the decision tree analysis as the agent rules.

33.3.1 Tourists' Evacuation Behaviour

To understand the issues at the commencement of an evacuation,s respondents were asked to imagine the situation immediately after a disaster had occurred and to consider the behaviour they would choose in a situation. This was based on three factors shown in Table 33.2 choosing whether they would "head towards a public transportation terminal", "head towards a shelter" or "stay put on site". While there were 12 cases based on these three factors, each respondent was asked about their behaviour only for a randomly selected pair of cases, since responding to all 12 cases in a face-to-face questionnaire survey would be difficult due to time constraints.

Obtained data were analysed using a decision tree analysis (used rpart from statistics software R). Figure 33.2 shows a diagram on the start of evacuation obtained from the decision tree analysis. Behaviours branch into four separate behaviours based on influence from the following factors: "presence of guidance at the start of

Table 33.1 Overview of sampling survey with tourists

Date of survey	20–22 October and 3–4 November 2018 (Japanese tourists) 20 December 2018–10 January 2019 (Foreign tourists)
Scope of the survey	Japanese tourists who visited Himeji castle, Hyogo prefectures Foreign tourists who use a kimono rental shop in Shimogyo ward, Kyoto city
The number of valid responses	Japanese tourists: 274 Foreign tourists: 107 Hong Kong: 25, Chinese Taipei: 20, Korea: 15, Thailand: 12, China: 7, Zealand: 5, USA: 5, Other: 13
Survey method	Questionnaire survey (conducted in English for foreign tourists)
Survey themes	1. Behaviour at the beginning of evacuation 2. Route selection to an evacuation area 3. Route selection to a public transportation terminal

Table 33.2 Three factors that constitute the situation at the start of an evacuation, used in the assumed survey

Factor	Detail	Level
Announcements on evacuation during a tour	An announcement made during a tour that instructs people to "evacuate to a shelter instead of heading to public transportation terminals in case of a disaster". The announcement is made via a loudspeaker installed at tourist destinations	0. No announcements 1. Announcements available
Announcements on evacuation after a disaster has occurred	An announcement made in a disaster that urges people to evacuate or refrain from moving around, which is announced aloud via a loudspeaker installed at tourist destinations	0. No announcements 1. Announcements made to encourage evacuation 2. Announcements made to encourage standby
Behaviour of other tourists	Behaviour of other tourists in the surrounding area	0. Many people in the surrounding area are waiting 1. Many people in the surrounding area are moving

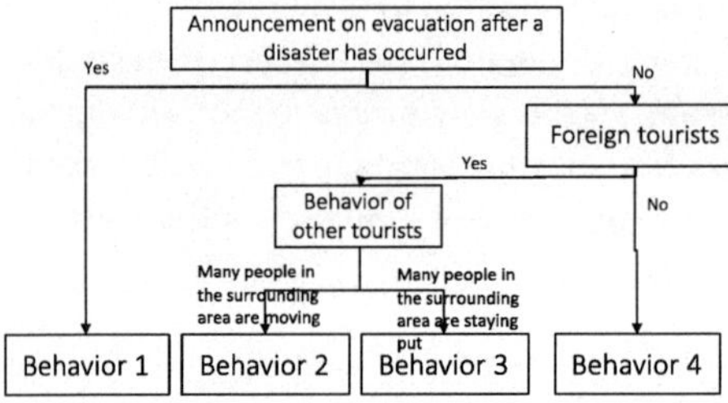

	Head toward a public transportation terminal	Head toward a shelter	Stay put
Behavior 1	0.06	0.61	0.34
Behavior 2	0.13	0.61	0.25
Behavior 3	0.07	0.35	0.58
Behavior 4	0.08	0.29	0.63

Fig. 33.2 Rules on starting evacuation by decision tree analysis

an evacuation", "whether the agent is a foreign tourist" and "behaviours of other evacuees". Each behavioural outcome was predicted as a probability. For example, if the situation was such that "there are no announcements to start evacuation", "the tourists are foreign" and "more people started to move", the agent would transition to behaviour 2 and attempt to move to a public transportation terminal at a probability of 0.13, choose to move to a shelter at a probability of 0.61 or stay where the agent is at a probability of 0.25.

From behaviour 1, tourists were more likely to start evacuation when there was an announcement that directs people to evacuate. On the other hand, tourists tended

Table 33.3 Four factors that affect route selection used in the assumption survey

Factor	Level
Width of the path	0. Road width is narrower 1. Road width is wider
Path selection by others	0. Fewer people are choosing the path 1. More people are choosing the path
Presence of a path guide	0. No path guide present 1. Path guide is leading to that path 2. Path guide is leading to a different path
Experience with the path (whether the individual has used the path before)	0. Has never used the path 1. Has used the path before

to stay where they were if there were no announcements asking them to evacuate, or if there was an announcement that orders them to stay where they are. The results also revealed that foreign tourists were affected more by behaviours of other tourists around them.

33.4 Analysis on the Selection of Evacuation Routes

Following the start of an evacuation, an evacuation route was selected. This study assumed a case in which tourists "head towards a public transportation terminal" and a case in which tourists "head towards a shelter". Respondents were asked to choose either route A or route B based on a situation consisting of five factors shown in Table 33.3. A total of 24 patterns existed for routes consisting of four factors, and the situations presented to respondents were designed such that they were always the opposite of each other. For example, if "route A is wider than route B, many people are choosing route A, and the respondent has never taken route A", then route B would be "narrower than route A, fewer people are choosing the route, and the respondent has taken the route before". However, three patterns existed with respect to the presence of a guide: a case in which the route guide is available for route A alone, a case in which the route guide is available for route B alone and a case in which both routes lack a guide. Considering there is also a limit to the ability of a single respondent to respond to all 24 cases in a face-to-face survey, respondents were asked about their behaviour for two randomly selected cases.

33.5 Route Selection Rules on the Way to a Public Transportation Terminal

Figure 33.3 shows route selection rules for evacuees on their way to a public transportation terminal. Four factors affect the outcome: "presence of route guides", "route

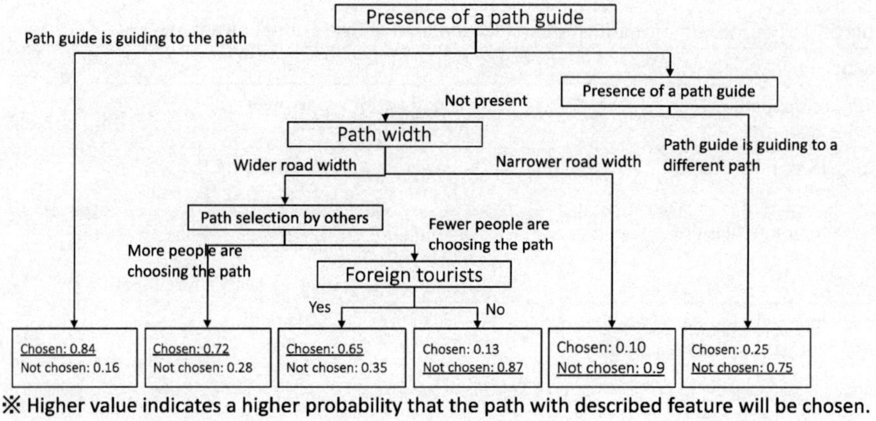

Fig. 33.3 Route selection rules on the way to a public transportation terminal by the decision tree analysis

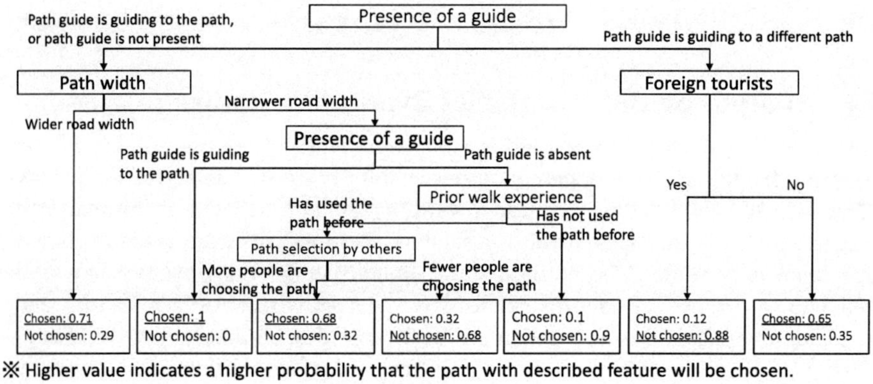

Fig. 33.4 Route selection decision-making on the way to a shelter using the decision tree analysis

width", "route selection by others" and "foreign/domestic tourist", and behaviours were split into six different behaviours. For example, if there was a route guide, the respondent may choose the route at a probability of 0.84, indicating a high probability of choosing that route.

While the presence of route guides was found to be crucial in moving towards a public transportation terminal, the results also showed that route width and the behaviour of other tourists affect route selection in the absence of route guides. Foreign tourists were more likely to choose a wider route without guidance that other tourists tended to avoid. These results indicate a difference in how Japanese and foreign tourists choose their evacuation route.

Figure 33.4 shows route selection rules for evacuees on their way to a shelter. Outcomes were divided into seven different behaviours based on five factors: "presence of a route guide", "route width", "foreign/domestic tourist", "prior experience in walking the route" and "route selection by other tourists". For example, the probability of choosing a route was only 0.12 for a case in which "a route guide is leading the crowd to another route" and the agent was "a foreign tourist", and the probability of not choosing that route was 0.84.

33.6 Route Selection on the Way to Shelter

As with the case in which tourists headed towards public transportation, tourists tended to choose a route with a guide, which reflects the tendency of tourists who do not have a sense of locality. However, tourists chose a route based on the width of the route, prior experience walking the route and the behaviours of other tourists in a case without any guidance. When the route guide attempted to guide tourists to another route, foreign tourists tended not to choose that route, but the opposite trend was observed among Japanese tourists.

33.7 Evacuation Simulation: The Case of Kiyomizu-dera Temple Area

33.7.1 Overview of Kiyomizu-dera Temple

Approximately 55 million tourists visit Kiyomizu-dera Temple annually, which is also a UNESCO World Heritage Site. According to a survey conducted by the city of Kyoto, 56.4% of foreign tourists and 65.2% of Japanese tourists who visit Kyoto visit the Kiyomizu-dera Temple (Industry and Tourism Bureau of Kyoto City 2017). However, the Hanaore fault line is found beneath these two locales. Furthermore, the district has a predominance of wooden structures and narrow routes that preserve the historical character of the cityscape (Fig. 33.5). Thus, its potential for structural and fire damage during a disaster is high, which has been brought to the attention of authorities.

33.7.2 Design of the Evacuation Simulation

The simulations were performed using spot-oriented agent role simulator (SOARS) (Tanuma and Deguchi 2007). Of the routes in the Kiyomizu-dera Temple area, Kiyomizuzaka, Matsubaradori, Gojozaka, Chawanzaka and the route inside the

Fig. 33.5 Kiyomizu-dera Temple area. (top left) the stage at Kiyomizu-dera Temple; (top right and bottom left) The Kiyomizu-dera Temple area full of tourists; (bottom right) A typical historic street in Kiyomizu-dera Temple area (authors)

Kiyomizu-dera Temple were decided upon as an evacuation route network (Fig. 33.4). Tourists entered the Kiyomizu-dera Temple area at a casual pace during the tourist period, and agents moved groups by choosing one of the four predefined tourist routes. When a disaster occurred, approximately 5,000 tourists remained within the Kiyomizu-dera Temple area. Shelters in this area are in two parking lots as shown in Fig. 33.6. Table 33.4 compares the number of people counted in the Kiyomizu-dera Temple area in Sugiyama et al. (2015) and the number of tourists at the time of a disaster in this simulation, which does not show a large difference.

The inflow of people was limited due to route width, which was defined as 0.5 person/m s in normal times and 1.5 people/m s during evacuations. Walking speed was defined as 0.5 m/s during normal times, 1.3 m/s on downwards sloping evacuation routess and 0.78 m/s on upwards sloping evacuation routes (Ministry of Construction of Japan 2000). Next, evacuation behaviour rules were implemented as behavioural algorithms of agents (Fig. 33.7). Agents repeated evacuation behaviours as shown in Fig. 33.7 from the moment an earthquake occurred until the completion of evacuations. The ratio of foreign and domestic tourists in this simulation was determined by using the ratio in the number of tourists staying overnight in Kyoto, as shown in Table 33.5.

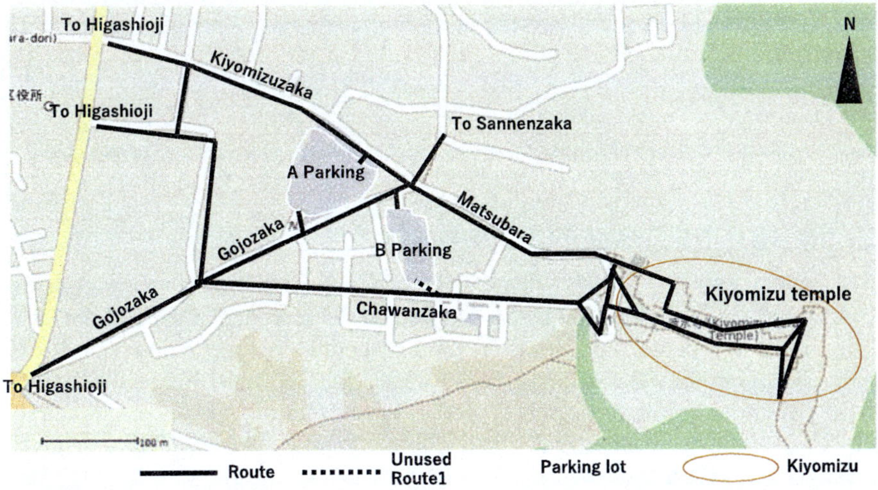

Fig. 33.6 Route network around Kiyomizu-dera Temple area in the simulation (Illustration by authors)

Table 33.4 Comparison of the count of people by Sugiyama et al. (2015) and by this simulation

Path name	# People (average of five measurements)	Simulations (average of five simulations)
Gojozaka	253 tourists	255.6 tourists
Kiyomizuzaka	561 tourists	513.4 tourists
Matsubara	1122 tourists	834.2 tourists
Chawan	298 tourists	291.2 tourists
All	3500 tourists	3981.2 tourists

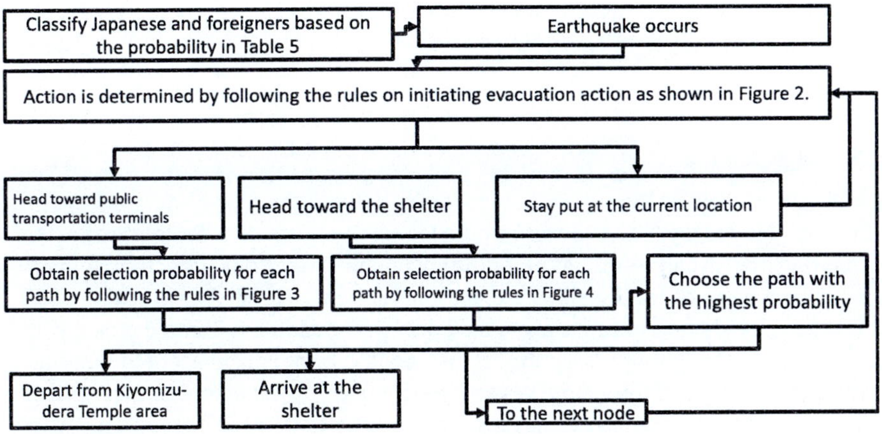

Fig. 33.7 Behavioural algorithm of agents in the simulation

Table 33.5 Ratio of Japanese and foreign tourists (Industry and Tourism Bureau of Kyoto City 2017)

Number		Ratio	
Foreign tourists	3.53 million tourists	Foreign tourists	22.6%
Japanese tourists	12.04 million	Japanese tourists	77.3%

33.8 Case Settings for the Simulation

A simulation was performed for six cases based on the presence of announcements for the start of evacuation, which was the most influential factor in starting an evacuation, and the presence of a route guide, which was the most influential factor in route selection. For announcements, both a Japanese-only announcement and a multilingual announcement were assumed. Simulation time was set to 40 min after the disaster had occurred, and each case was simulated five times. Case settings for the simulation are presented in Table 33.6.

Simulation results are shown in Table 33.7 and Figs. 33.8 and 33.9. Table 33.7 shows the number of evacuees and the number of people who left the area 40 min after the earthquake. Figure 33.8 shows changes in the number of people who left Kiyomizu-dera Temple area by case, and Fig. 33.9 shows an example of changes in evacuees by case. The greatest number of tourists moving out of the Kiyomizu-dera Temple area was observed in Case 4, in which announcements for evacuation were made in Japanese and a route guide was absent. Many tourists also left Kiyomizu-dera Temple area in Case 2 and Case 6, since both cases did not have a designated route guide, as in Case 4. On the other hand, more tourists were led to a shelter in Cases 1, 3and 5, where route guides were designated. These results mean tourists who did not have a sense of locality and were not aware of the location of the shelters, and this leads tourist to leave Kiyomizu-dera Temple area in the absence of a route guide and move to the terminal station en masse.

33.9 Discussion

The results showed that many tourists would leave Kiyomizu-dera Temple area in the evacuation simulation performed in this study due to the following factors: tourists do not have a sense of locality, they are unaware of where the shelter is, and they do not always seek shelter. For tourists to have left Kiyomizu-dera Temple area meant that they were moving towards a public transportation terminal, adjacent tourist sites or residential areas, and this study was able to simulate the behaviour of tourists that leave tourist sites after a disaster has occurred. In other words, if appropriate evacuation guidance is not provided, many tourists would leave Kiyomizu-dera Temple, and their behaviour may cause crowding at mass transportation terminals. Presence of a route guide also showed its effectiveness in preventing tourists from abandoning the Kiyomizu-dera Temple area.

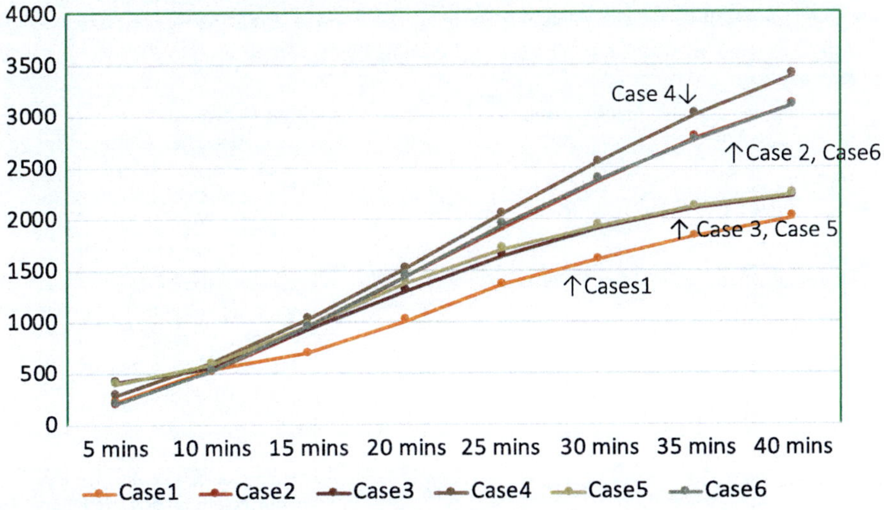

Fig. 33.8 Changes in the number of people who left Kiyomizu-dera Temple area by case

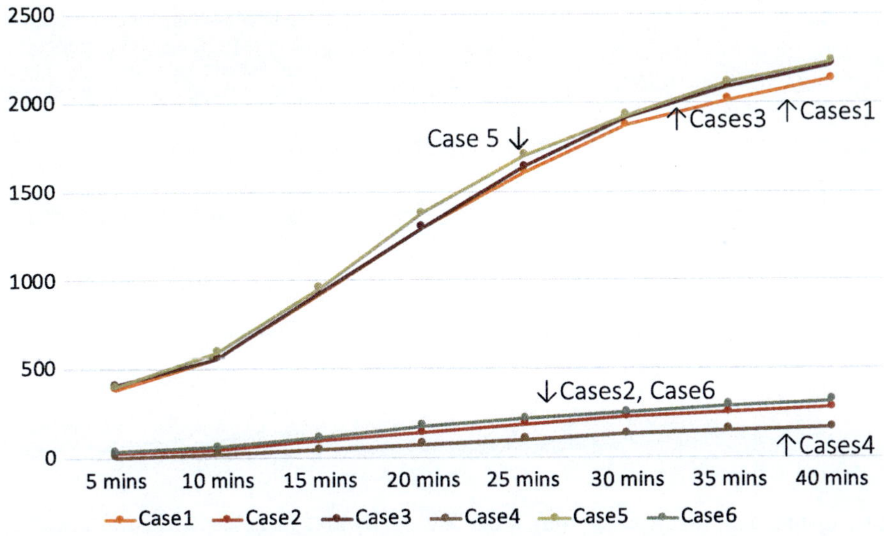

Fig. 33.9 Changes in evacuees by case

Guides are indispensable for the evacuation of tourists. In this simulation, guides were installed only in front of the parking lot, which is an evacuation site. Since tourists place importance on the width of the road when selecting a route, many tourists flowed out of the Kiyomizu-dera area from the route in the direction of Chawan, which has no guides and is wide. Thus, it is necessary to devise ways to guide more tourists to evacuation sites by installing guides on routes that are tourist

Table 33.6 Case settings for the simulation

	Announcement on evacuation after a disaster has occurred	Presence of a path guide	Details
1	Multilingual announcements	Yes	Announcements for evacuation after a disaster has occurred are made in multiple languages, and path guides are present in front of each shelter to lead people to a shelter
2	Multilingual announcements	No	Announcements for evacuation after a disaster has occurred are made in multiple languages, but path guides for leading people to a shelter are absent
3	In Japanese only	Yes	Announcements for evacuation after a disaster has occurred are made in Japanese only, and path guides are present in front of each shelter to lead people to a shelter
4	In Japanese only	No	Announcements for evacuation after a disaster has occurred are made in Japanese only, but path guides for leading people to a shelter are absent
5	None	Yes	There are no announcements for evacuation after a disaster has occurred, but path guides are present in front of each shelter to lead people to a shelter
6	None	No	There are no announcements for evacuation after a disaster has occurred, and path guides for leading people to a shelter are absent

routes that many tourists pass through. In addition, by installing the tourist route so as to pass through a place that can be an evacuation site, or by combining the evacuation route with a wide road, it is possible to make it easier for tourists to gather at the evacuation site in a disaster.

Moreover, it is impossible to prevent tourists from leaving the tourist area. It is necessary to take measures to manage tourists' evacuation and their simultaneous return home in cooperation with the tourist spots such as the Kiyomizu-dera area, the adjacent tourist spots and residential areas and terminal stations.

The effect of announcements could not be shown in this study. From the sampling survey, it was confirmed that tourist follows the behaviour of the people around at

Table 33.7 Number of evacuees and the number of people who left the area 40 min after an earthquake at Kiyomizu-dera Temple area

Case	Parking lot A	Parking lot B	Number of people who left to Kiyomizu-dera Temple area	Number of people who remained on paths in Kiyomizu-dera Temple area
1(n = 5)	1269.2	862.2	2005.6	685.6
2(n = 5)	265.4	16	3103.2	1278
3(n = 5)	1338.4	875.2	1902.6	725.6
4(n = 5)	153.6	13.8	3394.8	1098.2
5(n = 5)	1370.2	859.4	1985.4	595
6(n = 5)	297	23	3097.2	320.2

the start of evacuation. Thus, tourists' starting movement will be affected by the behaviour of the people around them and the announcement is a limited effect. However, according to the sampling survey, overseas tourists tend to aim at the terminal station instead of the evacuation site if there is no multilingual announcement. By making multilingual announcements it is expected that more tourists to move to and use evacuation sites.

In light of the aforementioned points, it shows that tourists leaving a tourist site to return home en masse without the presence of appropriate evacuation guidance could require the local area to address this issue, preventing the allocation of human resources to firefighting activities and the rescue of cultural property. This suggests that such behaviour could prevent efficient recovery of the tourist site. These findings can be utilised in other urban tourist areas such as Kiyomizu-dera Temple, but it is necessary to conduct a simulation experiment for each case regarding specific issues such as the location of the guide.

Future challenges include an analysis of the results of applying a limit to the number of tourist routes that can be accessed out of four routes used in this simulation, a survey including the relationship between public transportation information and the behaviour for starting the evacuation, and an analysis with a different ratio of foreign tourists, in light of the fact that the number of foreign tourists is expected to increase in the future.

In addition, since this agent modelling is aimed at the evacuation behaviour of tourists in a historical urban area, it is possible to conduct simulation experiments in other urban areas by constructing a route network for each area. Further, not only the area around Kiyomizu-dera, but also simulation experiments in other cases are required to further consider the evacuation of tourists in historical tourist destinations.

33.10 Conclusion

This chapter proposed an evacuation simulation capable of considering the evacuation guidance needed to control tourists returning home en masse, from the viewpoint of achieving smooth support for tourists during a disaster. Furthermore, the efficient recovery of tourist sites with historical significance or cultural value was considered. By using this simulation, more human resources can be allocated to the conservation of historical streets and rescue of cultural property, instead of deploying them to address panicked tourists and crowded mass transit systems, leading to a swift recovery of damaged historical cities and tourist sites.

References

Arce RSC, Onuki M, Esteban M, Shibayama T (2017) Risk awareness and intended tsunami evacuation behaviour of international tourists in Kamakura City, Japan. Int J Disaster Risk Red 23:178–192

Administration and Finance Bureau Disaster Prevention and Crisis Management Office, Kyoto city (2018a) Kyoto city: stranded commuters measure. Available at: https://www-city-kyoto-lg-jp.j-server.com/LUCKYOTOC/ns/tl.cgi/https://www.city.kyoto.lg.jp/gyozai/page/0000076886.html?SLANG=ja&TLANG=en&XMODE=0&XCHARSET=utf-8&XJSID=0. Accessed 15 June 2020

Administration and Finance Bureau Disaster Prevention and Crisis Management Office Kyoto City (2018b) Kyoto city: If an earthquake occurs by Hanaori fault. Available at: https://www-city-kyoto-lg-jp.j-server.com/LUCKYOTOC/ns/tl.cgi/https://www.city.kyoto.lg.jp/gyozai/page/0000015490.html?SLANG=ja&TLANG=en&XMODE=0&XCHARSET=utf-8&XJSID=0. Accessed 15 June 2020

Cabinet Office of Japan (2015) 大規模地震の発生に伴う帰宅困難者対策のガイドライン (Guidelines for people who have difficulty returning home due to a large-scale earthquake). http://www.bousai.go.jp/jishin/kitakukonnan/pdf/kitakukonnan_guideline.pdf. Accessed 21 Nov 2020

Cui Q, Yusuke T, Taniguchi H et al (2013) A study on decision making process of evacuation after earthquake: a case study of the vicinity of Kiyomizudera-Temple in Kyoto. J Disaster Mitig Hist Cities 7:23–30

Hayashi H (2016) Science and technology for disaster resilience. Disaster Preven Res Inst Ann A 59:34–45

Industry and Tourism Bureau of Kyoto City (2017) Kyoto city tourism survey. Kyoto City. https://www.city.kyoto.lg.jp/sankan/cmsfiles/contents/0000240/240130/kyosa29saishu.pdf. Accessed 7 Aug 2019

Japan Tourism Agency (2019) White paper on tourism in Japan, 2019 (Summary). https://www.mlit.go.jp/kankocho/en/siryou/content/001312296.pdf. Accessed 21 Nov 2020

Matyas C, Srinivasan S, Cahyanto I et al (2011) Risk perception and evacuation decisions of Florida tourists under hurricane threats: a stated preference analysis. Nat Hazards 59(2):871–890

Ministry of Construction of Japan (2000) Determining the calculation method for the floor evacuation safety verification method. In: Ministry of Construction Notification No. 1440. http://www.mlit.go.jp/notice/noticedata/pdf/201703/00006493.pdf. Accessed 7 Aug 2019

Nakaya N, Nemoto H, Yi C et al (2018) Effect of tsunami drill experience on evacuation behavior after the onset of the Great East Japan Earthquake. Int J Disaster Risk Red 28:206–213

Nishino T, Ohashi K, Hokugo A (2016) Tourist's behavioral tendency expected in post-earthquake fire evacuation. J Environ Eng (transactions of AIJ) 81:1–8

Sakai K, Honda A, Monkonkerd S et al (2014) A study on evacuation behavior pattern of tourists after earthquake in Himeji Castle: a survey of consciousness based on multicriteria decision making. J Disaster Mitig Hist Cities 8:189–194

Sugiyama T, Okubo T, Kim D, Hayashi M (2015) Study of the evacuation plan for the concerning stranded tourists on arounded area of Kiyomizu Temple, Kyoto. J Disaster Mitig Hist Cities 9:127–134

Sonohara M, Sakai K, Takahshi M, Kaneda T (2018) A study on agent modeling of tourist evacuation behaviors in an earthquake: a case study of an evacuation simulation of Himeji Castle. In: Innovative approaches in agent-based modelling and business intelligence. Springer Singapore, Singapore, pp 189–204

Takabatake T, Shibayama T, Esteban M et al (2017) Simulated tsunami evacuation behavior of local residents and visitors in Kamakura, Japan. Int J Disaster Risk Red 23:1–14

Tanuma H, Deguchi H (2007) Development of agent-based social simulation language: SOARS. IEICE Trans Inf Syst 90:2415–2422

Chapter 34
Public Administration Versus Social Media in Emergency Situations

Vincenzo Mini

Abstract The chapter presents a research examined whether the use of social networks, in particular Twitter, has led to improvements in the management of emergencies and post-emergency situations within public institutions, the Public Administration (*Pubblica amministrazione*, PA). Putting aside any analyses aimed exclusively at the political use of Twitter tool, and instead focusing on the possibility of building up a relationship of trust, that is, when this tool can inform citizens directly, especially so in an emergency context. At the outset, it was found that there is no legislation that requires and regulates this type of use even in normal situations. The PA regulates the use of tools that use networks such as websites and e-mails, but not the use of social network sites. Furthermore, it should be noted that the freedom of social networks does not fit well with the rigidity of the PA. At this point, the effort was twofold: to alleviate the rigidity of the PA, while remaining within permitted limits, and to exploit, but not excessively, the anarchy of social network sites. Many aspects were investigated: reference standards, guidelines, policies; recognised sources of message production; the so-called intruders in the communication process (citizens and citizen journalism, agitators and hate speech; regulation of responses (feedback and hashtags). The detection of the use of social network sites defines elements of advantage and opportunity, both from the point of view of the PA and towards citizens who choose to use the Twitter tool. Yet everything is entrusted to good will and to the sensitivity of the people on the field. There is no systemic view as had been hoped for. Twitter, in the specific PA examined, did not create adequate citizen involvement. Not all the potentialities inherent in an optimal approach to the world of social network sites, including Twitter, were deployed.

Keywords Public Administration (PA) · Social network site (SNS) · Emergency management · Social media · Abruzzo · L'Aquila · Italy

V. Mini (✉)
University of Rome, Rome, Italy
e-mail: mini@lettere.uniroma2.it

© The Author(s), under exclusive license to Springer Nature Switzerland AG 2021

F. F. Arefian et al. (eds.), *Historic Cities in the Face of Disasters*, The Urban Book Series,
https://doi.org/10.1007/978-3-030-77356-4_34

34.1 Introduction

Social network sites can be as web-based services that allow individuals to (a) construct a public or semi-public profile within a bounded system, (b) articulate a list of other users with whom they share a connection and (c) view and traverse their list of connections and those made by others within the system. The nature and nomenclature of these connections may vary from site to site (Boyd and Ellison 2007, p. 211).

After the L'Aquila earthquake on 6 April 2009, the world of social network sites (SNS) was, if not with rare exceptions, in an embryonic phase both in terms of geographical penetration and as a product used. This made it difficult to understand if these SNS created new connections, and, on the contrary, the impression was that they tended to cement the existing ones. To establish new connections, many obstacles needed to be overcome. Those which can be overcome can be analysed starting from Granovetter's definition of the strength of an interpersonal bond. According to him, "the strength of a tie is a (probably linear) combination of the amount of time, the emotional intensity, the intimacy (mutual confiding), and the reciprocal services which characterise the tie" (Granovetter 1973, p. 1361). Therefore, by pivoting to the weak links, which act as a bridge between the network segments, it is possible to create new relationships (Granovetter 1983). Further help can come from the application of Haythornthwaite's theory of latent bonds that as a basis for connectivity between unconnected people. Building on the finding that mandated media provide a substrate of connection for weakly tied pairs, the theory proposes that introducing a new medium to a group (a) creates latent ties, (b) recasts weak ties—both forging new ones and disrupting existing associations—and (c) has minimal impact on strong ties (Haythornthwaite 2005, p. 136).

A few years later, with the so-called Central Italy Earthquake, characterised by strong, temporally different tremors or seismic swarms (24 August 2016, 30 October 2016 and 18 January 2017), the situation within the digital ecosystem had profoundly changed. Even online public communication, as a characteristic activity of the PA and an ordinary mode of interaction with the citizen, had become ever more relevant since the 2000s, as part of the vast process of reform which had assigned it an increasingly greater role, aimed at greater accessibility and full transparency. These changes were further accentuated by the consolidation of Internet tools and the evolutions that are evolving from a minimal interactive environment (web) to a social tool (maximum interactivity). This has charged public communication with challenging tasks: the enhancement of skills and means, triggering training tools of civic culture that lead to citizens being aware; as well as the acquisition of credibility by web-based tools, rethinking their positioning after examining their context and consequently opening up to contamination from the outside world (Faccioli 2016). There are dozens of initiatives that experiment with the use of social media to support activities related to environmental emergencies. Although often, and especially in the initial moments, these digital flows of news and opinions arose spontaneously and in a self-organised way because of the initiative of individual citizens or local bloggers, in other cases this digital communication activism was promoted and coordinated by public bodies or administrations through their communication offices (Giglietto and Lovari 2013).

Research here has focused on Twitter: because it can be used to share facts in real time, it has the ability to give feedback, but it can also serve to convey recommendations and warnings very quickly. Also, it is not yet fully impacted by the feeling of distrust affecting other pervasive SNSs, mainly Facebook. This distrust was confirmed in recent research, carried out in the field of communication, but on the phenomenon of disinformation. What is Twitter? Twitter, through the Internet, provides a social network and microblogging service tool between friends, family and colleagues, useful for communicating and staying in touch through the exchange of rapid and frequent messages. People post tweets that may contain photos, videos, links and text. These messages are published on your profile, sent to your followers and can be found through Twitter's own search methods. Furthermore, its use is appropriate as a response to the indications previously expressed regarding the strength of weak bonds.

The research aimed to detect whether the use of social network sites, in particular Twitter, has led to improvements in emergency and post-emergency management within public institutions—the Public Administration (PA).

34.2 Premises

Taking a step back, first the word "emergency" must be defined. It must be noted that it is, indeed, adopted in Italian social research, while in American studies the word "disaster" is usually used, yet the phenomena examined are more or less the same in both cases. It should also be noted that scientific investigation in this sector is limited to relatively recent times and, until recently, conducted primarily in the USA (Quarantelli 1993). The word "disaster" was commonly associated with different meanings, even very distant from each other: (a) natural catastrophe; (b) a physical damage indicator; (c) a social damage indicator; (d) a negative evaluation indicator. The first definition, natural catastrophe, only had social consequences following the actions of human beings and society, and it can be said that what, in very old-fashioned and outdated terminology, were described as "natural disasters" are in fact social (Dynes 1974).

The meanings of and the close connection between disasters and their repercussions in the social context are certainly current and convincing. Over the years with continually elaboration and evolution, two models for disaster management have prevailed. The first, temporally dated, is the military model that focuses on social chaos and the consequent need for social control. The best models, however, highlight the attitude of the populations interested in the phenomenon of building communities starting from the fact that they are not completely annihilated and also have resources to make available. Coordination and cooperation should be the goal of disaster management through the possibility of using these resources by local communities also in ways already tested that would lead to effective problem-solving and with realistic solutions, as reported in Table 34.1 reported by (Dynes 2000a).

The idea, which can be shared with any post-emergency city, is to involve local communities, with a priority for on-site solutions that mainly involve better preparation of the same, with immediate responses, and efficient recovery also through the mobilisation of local community resources. Attention is given not only to particular events, but also to technological risks. The holistic approach is centred on awareness

Table 34.1 Assumptions and consequences of different models of emergency planning (Dynes 2000a)

Assumptions about	Military model	Problem-solving model
Characteristics of emergency behaviour	Chaos	Continuity
Character of emergency response	Command	Coordination
Character of involvement	Control	Cooperation
	Plan for dramatic change	Plan for (and with) continuity
	Plan for reduced social capacity	Plan for unexpected problems
	Create new structures	Utilise existing structures
	Predetermine new authority	Utilise pre-emergency authority
	Create centralised decision-making	Utilise decentralised decision-making and coordinate
Consequences for planning	Anticipate loss of emergency workers	Anticipate extensive helping behaviour
	Expect problems of role abandonment	Anticipate importance of family support systems
	Emphasis on providing authoritative public announcements	Emphasis on organisational intelligence and keeping public informed
	Emphasis on agent-generated demands	Emphasis on response-generated demands as well as agent-generated demands
University of Delaware Disaster Research Center PRELIMINARY PAPER #300 GOVERNMENTAL SYSTEMS FOR DISASTER MANAGEMENT Russell R. Dynes 2000	Emphasis on standardised scenarios and operating procedures	Emphasis on improvisation based on preparedness and alternative solutions
	Emphasis on creating a paramilitary structure	Emphasis on mobilising social resources

of the emergency and fully sharing information about it, together with the development of attitudes within social communities in order to be able to cope with difficult situations (Dynes 2000b).

34.3 Emergency Communication

Several studies have been concerned with emergency communication, and one in particular, an authoritative source, pointed out what went wrong with emergency communication using social media (OECD 2013). Briefly and schematically the issues identified were

- the risk of not including people who feel affected by a disaster, who now have the means to turn to the blogosphere if they feel they are not being heard;
- the instances of information from another source other than the emergency services; and
- information overload.

Lastly, it is acceptable and central to the research line, or not, to consider the unpreparedness of organisations for the use of social media and the awareness that the latter is not a new solution that can solve all problems. It should be emphasised, as reported in Table 34.2 (OECD 2013), that each of these has different characteristics and consequent use in the scenarios that arise in emergency circumstances.

On the other hand, social media can be used to improve emergency communication because of some advantages pertinent to them as:

- they are collaborative and participatory: online comparisons can improve awareness of the emergency situation;
- they are decentralised: indications can circulate very quickly among the actors, thanks to the tweets, which are immediately available online for all the organisations involved;
- they are popular and accessible: consequently, the emergency services can use them to send information or warnings; and
- they can provide geo-referential data: from which it is possible to monitor the development, even temporal, of an emergency thanks to digital traces (OECD 2013).

Underlining these characteristics, it is clear that the PA, despite its efforts, cannot make the most of its potential. Notably, in Italy, the first official document that encouraged the PA to enhance its dialogue with social media communities and use them for communication purposes is dated on May 2017 (Presidency of the Council of Ministers—Public Function 2017) well beyond the date of the earthquakes in the Centre Italy (August, October 2016 and January 2017). The current situation, in the context of regulatory uncertainty, reveals that this new regulation, which was not linked to emergency contexts, is no different in terms of uncertainty.

Table 34.2 Different types of social media used in risk and crisis management (OECD 2013)

Type of social media	Examples	Use for risk and crisis communication
Social networking	Facebook Myspace Friendster	Enhance coordination among volunteers and emergency services, allow to share information inside a community, provide swift update on emergency situation, etc.
Content sharing	YouTube Flickr Vimeo	Enhance situational awareness in real time through exchange of pictures and videos, allow emergency services to easily launch viral campaigns about risks, can help identify missing individuals, victims, etc.
Collaborating knowledge-sharing social media	Wikis Forums Message boards Podcasts	Enhance dialogues between victims and emergency services
Blogging and microblogging	Blogger Worldpress Tumblr Twitter	Convey recommendations, warnings, share facts twitter allows to have immediate information sharing with a wide reach and feedback possibilities
Specialised crisis management platform managed by volunteer technology communities (VTCs)	Mapping collaboration OpenStreetMap Crisis mappers Google map maker Online and onsite contribution Ushahidi Crisis commons Sahana foundation Geeks without bounds Public–private people partnership Random hacks of kindness (with Google, Microsoft, YAHOO, NASA, World bank)	Mapping of emergencies, community emergency response team facilitator 

Even the current rules on emergency scenarios, revisited in 2018, do not regulate communication activities in the emergency phase and limit themselves to including both information to the population on risk scenarios and the related rules of conduct and those on civil protection planning in the activities of non-structural civil protection prevention.

34.4 Social Media Data in Italy

In Italy, there were over 30 million users on Facebook, 24 million on YouTube, 14 million on Instagram and almost 7 million on Twitter, one in two from a mobile device (monthly presence/data referring to January 2017). The number of people connected to the Internet was constantly growing, with a penetration of almost 70%, and in January 2019 this reached 92% with two out of three active users on social media from mobile devices, which greatly facilitates the possibility of communicating in emergency situations. It appears that most citizens are present daily on these social platforms, which are meeting and discussion places, where surfers can interact not only with other people, but also with companies, associations and more or less also with small groups of people and therefore also with institutions. Thus, as already stated, it is important for the PA to be there, not as a trend but as an effective presence that consolidates the best practices of these institutions. One of the things that the PA has always asked itself to be is with their citizens, supporting them and helping them in case of need, and if today they surf on Facebook, post photos on Instagram and converse on Twitter, then the PA must be among them in these environments.

34.5 Analysis of the Use of #hashtags and @mentions

The seismic events in Central Italy in 2016 and 2017 began in August 2016 with an epicentre along the Tronto Valley, between the province of Rieti and Ascoli Piceno (Monti Sibillini, Monti della Laga, Monti dell'Alto Aterno). In October of the same year, other strong seismic swarms occurred with epicentres both on the Umbrian-Marche border, in the province of Macerata (Visso, Ussita, Castelsantangelo sul Nera) and in the province of Perugia (Norcia, Preci).

On 18 January 2017, a new sequence of four strong tremors with a magnitude greater than 5 took place in the province of L'Aquila (Monterale, Capitignano, Cagnano Amiterno).

It should be noted that the Abruzzo region, from the point of view of it as an institution, in a study dated September 2015 (DeRev 2015), being so close to this telluric phenomena, was included in the regions studied with regard to their "conservative" approach to social media. These regions have set up social channels tokenly, without actually supervising or managing them according to the necessary rules of engagement for users. In addition, they took no advantage of the countless opportunities by way of the tools listed above to foster the relationship between citizens and institutions.

A consolidated analysis scheme was applied to the research and can be summarised in four areas as follows:

- popularity (the number of total followers of the institution's account at the time of the survey);

- creation (the number of total tweets of the institution's account at the time of detection, purified of retweets);
- interaction (the number of tweets containing retweets); and
- contamination (use of #hashtags and @mentions).

The popularity data of the account being surveyed were always below 1000 followers in the reference periods. As analyses showed, the data on the creation, the number of total tweets of the institution's account at the time of detection, purified from retweets, were about 1000 on the occasion of the first telluric phenomenon (August 2016), rising to over 3000 in January 2017 (noteworthy there were 1000 in response to a specific case study: the institution's interest in the problem of electricity, particularly important given the winter season, with very low temperatures and violent atmospheric phenomena including snow) and also the context of another catastrophe (the Rigopiano avalanche) that affected only the Abruzzo with 29 deaths. The data on interactions, together with the number of tweets containing retweets, revealed a minimal phenomenon, below double digits in August 2016 and below the triple digits in January 2017. Contamination data, use of #hashtags and @mentions, were almost non-existent and always below double digits (August 2016 and January 2017). It was already been noted that different types of social media can be complementary within the context of emergency management.

Setting aside the vision of the Twitter tool purely aimed at a political use, this research focused on the possibility of building a relationship of trust that is necessary when this tool informs citizens directly, even more so in an emergency context. Preliminarily, it was found that there is no legislation that obliges and regulates such use even in normal situations. PAs regulate the use of tools that use the Internet (websites, e-mails) but not the use of SNS other than simple and basic indications.

At this point, the research effort was twofold: to work out how to loosen rigidity, while remaining within the permitted actions of the PA and to harness, but not excessively, the anarchy of the SNS. These efforts must not lend themselves to compromises that distort the two components at stake. The research was faced with a heterogeneous context of web tool use developed in the absence of valid reference models for any particular institution, which often opens its own channel on one or more SNS without planning within a communication strategy of wider breath. It is therefore necessary to find the right balance, with an adequate methodology, between the characteristics of each individual institution (dimensional data, staff available) and consequent strengths and weaknesses, combining them with the operating characteristics of each SNS and the target. It should also be remembered that the strength of the SNS is to suggest often unprecedented ways in the field, which in turn provide answers capable of reorienting and perfecting the service, according to a virtuous circuit for which it is the user who marks the evolution of its own service by way of the feedback given with respect to a constant, rigorous and not one-sided, but "two-way" engagement with communication activity. In order to overcome the age-old concept of "one-way" communication of the PA, real time, immediacy and the different characteristics of each social tool create an opportunity for dialogue that is unique in its own way.

In an emergency context, a structure ready for this dialogue should be able to receive information from the site of an accident or catastrophe, considering that the smartphone and the SNS tool have become commonly used and behave accordingly. Furthermore, by interpreting and setting the right alarm thresholds, situational awareness can be achieved in real time. All this can be the result of a self-assessment process for the institutions on the uses of the SNS to channel the right communication flows. Even more important is the evaluation process for the information flows that originate from the institution itself, paying attention to some fundamental rules, such as ensuring the security of sensitive information, a coded process of validating the accuracy of the information and if false information is published in error to very quickly put in place a correction. There are many aspects to be investigated: reference standards, guidelines and policies; recognised sources of message production; potential intruders entering the communication process: citizens and citizen journalism, agitators and hate speech; there is also the question of the regulation of responses, such as feedback and hashtag.

The initial research question was the prerogative of an in-depth study of the documents already existing in the PA vis-à-vis the concern with unsatisfactory results given that, as mentioned, there is no real reference legislation, therefore being present and active in the SNS is a choice for the PA. Consequently, the elements of advantage and opportunity can be listed: costeffectiveness, monitoring the situation, listening, dialogue and trust, as well as transparency and collaboration, dictated by the active presence in the blogsphere, both for citizens and the PA in support of these choices. Constant increase over the years has been observed, and today almost all users of the Internet are used to accessing one or more SNS sites on a daily basis. Being present and engaged with them allows the institution to reach citizens with greater simplicity than it could do using traditional channels. This guarantees an increase in the effectiveness of communication actions from the PA which, thanks to these new channels, can have a particularly effective contact tool with its users.

This monitoring of citizens' engagement with SNSs is the premise for a real listening exercise and consequent dialogue. Given the nature of the tools, the very presence of the PA in the SNS sites is perceived by the citizen as openness and a sign of readiness for dialogue. This implies that the citizen, who sees the PA in an SNS, interprets this presence as a promise of dialogue and, in fact, obliges the institution to keep this promise. The risk, in the event of low participation and dialogue, is to betray the trust of one's citizens even before having won them over. On the other hand, entering into conversation with them guarantees the construction of a relationship of trust that would otherwise be impossible without a real relationship of proximity. The solid relationship between the citizen and the PA is favoured by the fact that a truly open presence on the SNS necessarily leads to greater transparency for the PA. Dialogue and mutual knowledge promote an opening of the borders between the institution and citizens, which transforms them from external and passive users of the PA services into an active and involved interlocutor in the processes. This gives a new meaning to the transparency of the PA: from a regulatory point of view, this requirement becomes an instrument of openness to users and willingness to build virtuous participation processes.

The relationship of trust generated by the PA's openness to dialogue can result, on the part of the citizen, in the willingness to collaborate with the PA and support it in the management of the services it offers. The new concept of transparency, which arises from a relational approach to data, activates collaboration processes that are expressed in the virtuous development of the relationship between PA and users. Inducing the citizen to report a failure, express a preference, express an opinion are not simplistically systems to involve them, but represent real collaboration tools, which have an impact on costs, on the quality of the relationship, on the perceived quality and on the overall effectiveness of the system.

What has been expressed makes us lean towards a conscious use of the SNS in particular of Twitter, which fully responds to the opportunities listed and confirms the choice of this line of research.

34.6 Conclusion

The analysis carried out on the PAs of the Abruzzo region evaluated in a negative way the fact that, in the emergency context, it was possible to produce effective and efficient communication channels with its citizens failed to do so. Furthermore, the judgement expressed by the research that speaks of a conservative approach to social media was confirmed. Not all the potentialities inherent in an optimal approach to the world of social network sites, including Twitter for example, were deployed with the potential of creating testimonial, meta-comment and propagation contents. Testimony, that would have allowed the information function, indicating the intensity and location of an event in order to ensure an immediate and adequate response was not used. Propagation, whereby most of the tweets, falling within the information type (where, when, what) of public utility services (activated numbers, vademecum, information on meeting places) that could be conveyed by the PAs were under-deployed. Meta-comments, framed in the so-called trend topics and threads that could have created a productive conversation to respond to the emergency context, were well below optimal. Addressing the above will be a way forward to move away from such conservative approach and to realise potentials of social media in disastrous situations in future. There are undoubtedly various approaches to be explored to the use of the tool, even within the same institution.

References

Boyd DM, Ellison NB (2007) Social network site: definition, history and scholarship. J Comput-Mediat Commun 13:210–230

DeRev (2015) Analisi della strategia di comunicazione delle Regioni italiane sui social media, viewed 23 june 2019. https://www.derevworld.com/2015/09/la-classifica-delle-regioni-piu-social-ditalia-pubblicato-il-report-sulla-comunicazione-istituzionale-sul-web/

Dynes R (1974) Organized behavior in disaster. Disaster Research Center, University of Delaware, Newark, DE

Dynes R (2000a) Assumptions and consequences of different models of emergency planning. In: Governmental systems for disaster management. Disaster Research Center, University of Delaware, Newark, DE

Dynes R (2000b) Governmental systems for disaster management. Disaster Research Center, University of Delaware, Newark, DE

Faccioli F (2016) Comunicazione pubblica e media digitali: la prospettiva del public engagement. Problemi Dell'informazione 1:13–35

Giglietto F, Lovari A (2013) Amministrazioni pubbliche e gestione degli eventi critici attraverso i social media: il caso di #firenzeneve. Mediascapes J 1:99–118

Granovetter MS (1973) The strength of weak ties. Am J Sociol 78:1360–1380

Granovetter MS (1983) The strength of weak ties: a network theory revisited. Sociol Theory 1:201–233

Haythornthwaite C (2005) Social networks and Internet connectivity effects. Inf Commun Soc 8(2):125–147

OECD (2013) The use of social media in risk and crisis communication. Report of the high level risk forum. OECD Conference Centre, Paris, December 13–14

Presidency of the Council of Ministers—Public Function (2017) Attuazione delle norme sull'accesso civico generalizzato (c.d. FOIA), wiewed 04 june 2020. Available from http://www.funzionepubblica.gov.it/sites/funzionepubblica.gov.it/files/CIR_FOIA_REVe.pdf

Quarantelli EL (1993) Disastri. In: Treccani (ed) Enciclopedia delle scienze sociali, viewed 01 June 2019. http://www.treccani.it/enciclopedia/disastri_%28Enciclopedia-delle-scienze-sociali%29/

Chapter 35
Social Media and Disaster Management in Iran: Lorestan Floods as Case Study

Vahide Ebrahimnia and Somayeh Zandieh

Abstract Social media can reach millions within seconds and is therefore one of the most frequently used communication and trust-building tools in pre-, during and post-disaster activities. In the case study presented in this chapter on the Lorestan floods in March 2019, public sector decision-makers and ordinary citizens used a wide variety of mobile-based and web-based technologies (SMS and blogs) and different social media platforms for help, mapping and sending status reports, fund raising, donations and mobilising volunteers for assistance. In this context, the research analysed data to examine the role of social media in disaster response, focusing on their ability of strengthening public willingness to donate and to support relief organisations during and after the 2019 floods in the province of Lorestan, Iran. This chapter explores whether collective action via social media could bring about any improvement to post-disaster management activities, and how social media operate in such contexts. The floods also caused damages in some of the outstanding historical sites of Lorestan. Findings from the manual monitoring of social media posts by major disaster-related actors in Iran reveal that the most common functions of social media in this case have been to inform about relief operations during the crisis, to mobilise the public to help during the crisis and finally to organise human resources for relief operations. Moreover, the study highlights three key themes, namely (a) the distrust of state organisations towards social media, (b) political sensitivities surrounding the application or boycott of social media and (c) lack of sufficient media literacy and the need felt by the state for a centralised control over social media as two main challenges associated with the use of social media by state organisations.

Keywords Disaster management · Social media · Flood management · Lorestan floods · Iran

V. Ebrahimnia (✉) · S. Zandieh
Shahid Beheshti University (SBU), Tehran, Iran
e-mail: v_ebrahimnia@sbu.ac.ir

© The Author(s), under exclusive license to Springer Nature Switzerland AG 2021
F. F. Arefian et al. (eds.), *Historic Cities in the Face of Disasters*, The Urban Book Series,
https://doi.org/10.1007/978-3-030-77356-4_35

Acronyms

ACECR	Academic Centre for Education Culture and Research
BF	Barakat Foundation
HFLP	Housing Foundation of Lorestan Province
IAS	Imam Ali Society
IRCS	The Iranian Red Crescent Society
IRIA	Islamic Republic of Iran Army
LEDC	Lorestan Electricity Distribution Company
LRWSC	Lorestan Rural Water and Sewer Company
MF	Meteorology Organisation
NAO	Nomads' Affairs Organisation
NOCS	National Organisation for Crisis Management
PG	Province Governorship
RCH	Reconstruction Coordination Headquarters
RCHLF	Reconstruction Coordination Headquarters of Lorestan Floods
RSA	Rescue Society of Anjirestan
RWCL	Regional Water Company of Lorestan
SCNCM	The Supreme Council for National Crisis Management

35.1 Introduction

Iran was hit by three major waves of rain and flooding over the course of two weeks from mid-March to April 2019, affecting large parts of the country, most severely in Golestan, Fars, Khuzestan, Lorestan, and 17 other provinces out of its 31 provinces (Fig. 35.1). The impact of the floods was heightened due to the Nowruz holiday for Persian New Year, since many Iranians were travelling to historic cities and many deaths occurred due to flash flooding on roads and highways.

As a part of Ancient Near East, Lorestan history goes back to the 3rd and 4th millennium BC, when migrant tribes settled down in the mountainous area of the Zagros Mountains. Many paintings, carvings, caves and bronze objects remain from this era with more than 100 of them listed in the National Monuments. In 2017, archaeological discoveries related to the Achaemenid era were made in Lorestan for the first time. They include the ancient remains of the Achaemenid and the Sassanid eras, Gavmishan, Kashkan, Kalhor, Poldokhtar, Gap and other bridges, Shapouri, Chamshk Caravansera and many others. Amongst them is also the Falak-ol-Aflak Castle situated on the top of a large hill with the same name, within the city of Khorramabad, the regional capital of Lorestan province covering an area of 5,300 m^2. Several mosques and shrines from the post-Islamic period also show the continuity of civilisation in this region (Lorestan Province Introduction, nd).

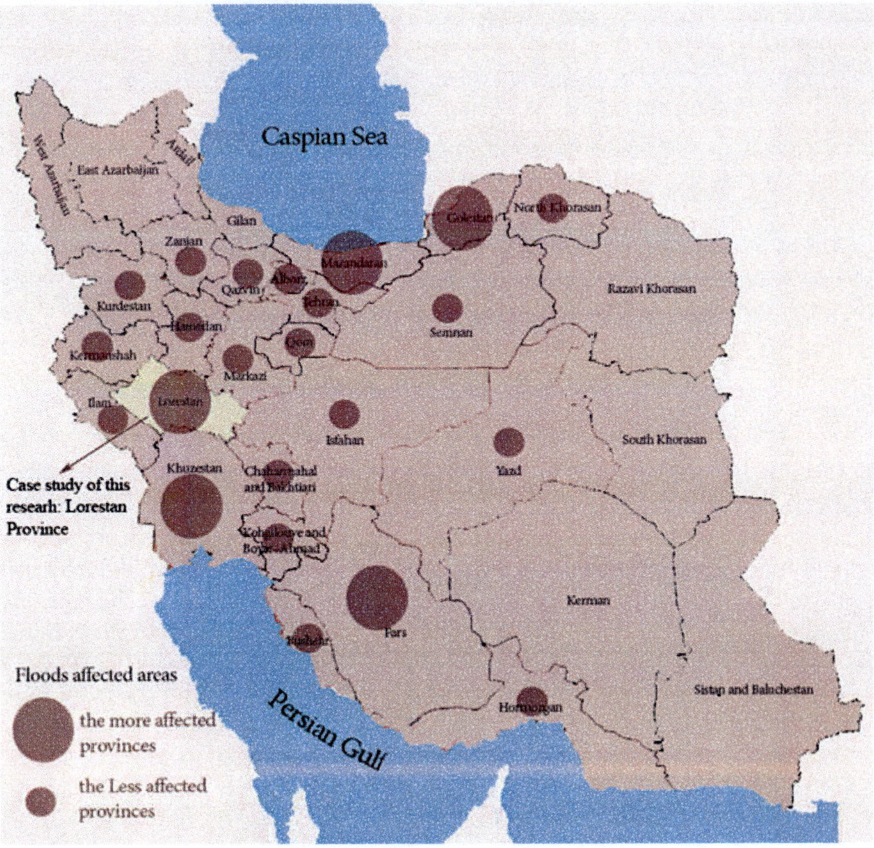

Fig. 35.1 Affected provinces by floods of Iran, March–April 2019 (authors)

The Lorestan flood damages have been estimated at over 1.1 billion US dollars, with 15 people killed and 256 injured. According to the deputy for rural reconstruction and housing of the Housing Foundation of the Islamic Revolution, more than 72,000 houses were demolished and dozens of cars and homes remaining sunken beneath thick layers of mud. In addition, the floods have severely damaged some of the outstanding historical sites in Lorestan including Kashkan, Ephrin, Kalhor and Poldokhtar Bridges and the historical Gap Bridge. For instance, the hill of the Falak al-Aflak Castle thrusted to cause the destruction of a part of the western side of the hill where the castle was built (Mahdian 2019).

Under these conditions, many Iranians including politicians took to social media platforms like Instagram, Facebook, WhatsApp and Twitter[1] to criticise the government's handling of the crisis and to help with mapping and sending status reports, fundraising, donations and mobilising volunteers for relief operations. Our research interest emerged from observing the trust-related challenges of collective action in Iran. Therefore, the main goal of the present research is to explore whether collective

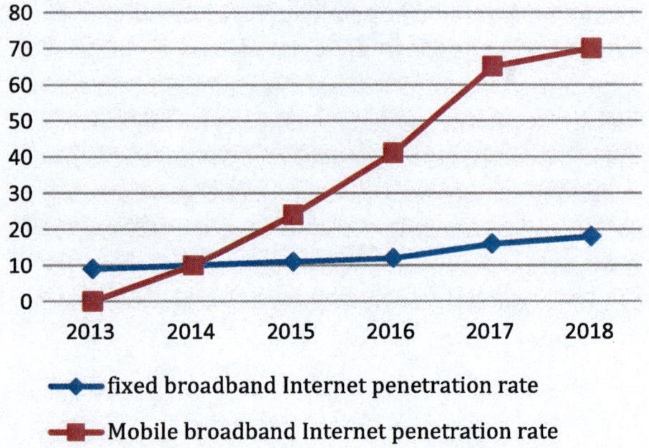

Fig. 35.2 Different Internet users in Iran (2013–2018). Adapted from Internet World Stats (2019)

action via social media has led to the development of aid communities in the wake of the Lorestan floods, and how these communities operate. Our interest in the issue stems from the fact that the communication infrastructures required to use social media in disaster management is firmly in place in Iran. According to the Internet World Stats website (2019), Iran is the world's 16th country in terms of the number of Internet users in 2019. The number of online users has exceeded 70 million from 2013 to 2018, of which more than 60 million are mobile Internet users (Fig. 35.2).

35.1.1 Collective Action and Social Media for Improving Disaster Management?

This chapter explores whether collective action via social media could lead to any improvement in post-disaster management activities and how they operate. The approach is to identify the current state and purposes of the main disaster management actors for using social media during the response phase in the case of the Lorestan floods in Iran. To this end, data were collected through (a) an online survey and (b) an online questionnaire. The online survey was conducted to understand the current state of the application of social media platforms (Facebook, Twitter, Instagram), and the questionnaire was designed to find the challenges and opportunities associated with using social media in disaster management activities.

The online survey was conducted by counting social media posts shared by disaster management authorities between April and May 2019. Then through a purposive sampling, those national and regional organisations, whose activities and responsibilities were somehow related to disaster relief, were invited to participate in the online questionnaire between June and August 2019. The representative of each organisation in charge of its social media platforms was invited to fill out the questionnaire. The questionnaire included questions about each organisation's uses of social media platforms, needs, expectations and challenges associated with the use of social media for disaster-related work. The response rate was 69.2% (nine out of 13); that is, amongst the total of 13 organisations which, according to their agenda, are responsible for disaster management in Iran, nine organisations have participated in the survey and responded to the questionnaire. Of these 13 organisations, only four had social media pages on Instagram or Twitter.

35.1.2 Background

Although the use of social media in disaster and emergency management has been investigated elsewhere in recent years using different examples and approaches worldwide (Sakaki et al. 2013; Busa et al. 2015; Shivarkar and Wei 2018; Kelly and Ahmad 2014; Kibanov et al. 2017; Shekhar and Setty 2015; Goel et al. 2019; Ahmed 2011; Alexander 2014; Kavanaugh et al. 2011; Keim and Noji 2011), only a limited number of studies have examined the role of social media in critical situations in Iran. The most notable of these studies is introduced in Table 35.1. As shown in this table, these studies have mostly focused on the negative role of social media.

Table 35.1 Studies focusing on the role of social media in times of crisis in Iran (authors 2020)

Authors	Study	Results
Akbari (2010) (in Persian)	The role of social media in creating and controlling disasters	• Social networking could be used by foreigners for stealing information from Iranian users • The skills and financial resources needed to use social media are directly related to specific groups of the society, so the impact and effectiveness of their content is not generalisable to the whole society
Afzali et al. (2014) (in Persian)	Challenges and strategies of using social media in crisis management	• Reducing the power of the official flow of information, the validity of information and infrastructure deficiencies are the major challenges facing managers in using social media • Continuous flow of information, accountability and comprehensive and targeted informing is its major advantages in disaster management
Yavari et al. (2017) (in Persian)	The role of social media in creating social disasters	• Using social networks to deliver messages may lead to instilling unwanted thoughts to users, and also turning an ordinary situation into a disaster by exaggerating it

35.2 Conceptual Framework

35.2.1 Social Media

Social media is defined as "a form of multiple-interaction communication that has an internet-based platform" (Williams et al. 2018: 539) which empowers individuals to communicate with multiple people simultaneously and in real time, across geographical and administrative borders. Blogs, micro-blogs, social book-marking, social networking, forums, collaborative creation of documents and the sharing of audio, photographic and video files are different types of social media (Alexander 2014). The definition by Obar and Wildman (2015) of social media in the context of governance challenges indicates the user-generated content as the most necessary element of social media, which enables individuals and groups to connect.

Compared with centralised information management, social media certainly offer advantages in dynamic situations. Social media make it easier to re-circulate information within teams and minimise reliance on formal liaison structures between different teams. The collective nature of social media serves to connect people across geographical boundaries and time zones via shared platforms to promote the growth of online communities with similar interests. Additionally, in contrast with most other forms of media or means of communication, the content shared via social media platforms enjoys a high degree of visibility: by participating the participants are getting aware of activities and content posted by others (Velev and Zlateva 2012). In this context, social media offer two-way communication between formal and informal institutions and the public, which exceeds the benefits of one-way communication, usually from the formal organisation to the public (Yan and Pedraza-Martinez, 2019). However, organisations can control the levels of involvements by participants by defining who can comment on social media networks (Velev and Zlateva 2012). Some studies suggest that personal characteristics may also impact on the engagement of people in social media. For example, social media are more attractive to people who are younger, more educated and more extrovert (Correa et al. 2010).

35.2.2 The Relationship Between the Disaster Management Cycle and the Use of Social Media

Scholars have studied the relationship between disaster management and the use of social media as part of a tendency to explore the functioning of social interaction using the Internet and mobile devices (Alexander 2014). The disaster management cycle includes four phases: mitigation, preparation, response and recovery (Khan et al. 2008). This research adopts an approach that recognises the functions of social media in each phase of the disaster management cycle (RPO 2013; Ahmed 2011; Mauroner and Heudorfer 2016):

- **Mitigation**: This phase focuses on preventing a natural hazard from becoming a disaster, or minimising its consequences. Corresponding communication structures in this phase implies communication amongst relevant actors and the involvement of communities. Information on how to deal with natural hazards and disasters must be provided by, e.g. the state, public actors and educational institutions.
- **Preparation**: In this phase, people, who could be affected by a disaster, are equipped with the necessary tools to survive. Early warning that comes via different channels, via both traditional mass media and social media is crucial in this phase.
- **Response**: As soon as a disaster begins, measures must be taken to minimise its effects, i.e. to prevent further suffering and financial losses. The state or private aid organisations must be able to reach and inform those affected. Media companies report on the situation to a broad public, increasingly supplemented by reports from the population itself as for
  - sharing updates and spreading awareness of the condition;
  - creating communities and volunteers for relief operation;
  - fund raising; and
  - monitoring and providing insights into the overall situation.
- **Recovery**: This phase is concerned with restoring the lives of those affected back to their normal state, which means regeneration and sustainable renewal. Long-term information campaigns, e.g. on preventative building or sustainable economic activity, may be part of the communication strategy. Therefore, the recovery phase loops back to the mitigation phase.

35.2.3 The Relationship Between Social Media and Post-disaster Management Activities

Social media can inform people in charge of critical decision-making about situational information in post-disaster phases (Kavota et al. 2020; Alexander 2014). Social media also have a critical role in crowd-sourcing and creating social cohesion (Alexander 2014). One of the main factors contributing to the citizen's willingness to support government funding for disaster relief is public trust in government and non-governmental organisations (NGOs) (Fard and Rostamy 2007); people donate when their perceptions about an organisation's efficiency are positive (Weng et al. 2015). On the other hand, Song et al. (2015) suggest that citizens are more likely to donate when they know that government organisations do not have enough resources for disaster management. In this context, social media is an information source for the public, which not only can inform people of the essential needs of disaster management activities during and after the crisis, but it is also a tool to ensure the efficient allocation of resources in disaster relief activities. Moreover, responsible authorities can use social media to monitor situations and cover emergency response and management and even enhance research to improve their future disaster responses. The function of social media as a platform for listening to the people's voice enables the authorities to know about people's behaviours and reactions after the disasters (Alexander 2014).

35.3 Results

The findings obtained from the online survey data address the main objective of the present chapter through identifying the current state and purposes of using social media by the major disaster management actors in the phase of response towards the Lorestan floods in Iran.

35.3.1 Organisations Involved in Flood Management and Their Responsibilities

The organisations which are involved in flood management could be divided into three categories (Table 35.2): governmental organisations, public organisations and

Table 35.2 Organisations involved in flood management in Iran (authors)

Organisations			**Duties**					
			1	2	3	4	5	6
Governmental organisation	National level	The Supreme Council for National Crisis Management[a] [SCNCM]		✓				
		National Organisation for Crisis Management[b] [NOCS]			✓			
		Meteorology Organisation[c] [MF]	✓					
		Barakat Foundation[d] [BF]						✓
		Nomads' Affairs Organisation[e] [NAO]				✓	✓	
		Reconstruction Coordination Headquarters[f] [RCH]						✓
		Islamic Republic of Iran Army[g] [IRIA]					✓	
	Regional level	Province Governorship[h] [PG]				✓		✓
		Housing Foundation of Lorestan Province[i] [HFLP]				✓		✓
		Reconstruction Coordination Headquarters of Lorestan Floods[j] [RCHLF]						✓

(continued)

Table 35.2 (continued)

Organisations			**Duties**					
			1	2	3	4	5	6
		Regional Water Company of Lorestan[k] [RWCL]				✓		✓
		Lorestan Electricity Distribution Company[l] [LEDC]				✓		✓
		Lorestan Rural Water and Sewer Company[m] [LRWSC]						
Public organisation		The Iranian Red Crescent Society[n] [IRCS]				✓	✓	✓
Non-governmental organisations		Imam Ali Society[o] [IAS]					✓	
		Rescue Society of Anjirestan[p] [RSA]					✓	
		Academic Centre for Education, Culture and Research[q] [ACECR]					✓	

Duties 1. Environment scanning before the crisis; 2. Division of labour and coordination with relevant organisations; 3. Issuing flood warnings and holding training events; 4. Reporting casualties and damages; 5. Rescue and fast relief; 6. Post-disaster rehabilitation and reconstructing

[a]Shora-ye Aali-ye Modiriat-e Bohran-e Keshvar [SCNCM]
[b]Sazman-e Modiriat-e Bohran-e Keshvar
[c]Sazman-e Havashenas
[d]Bonyade-e Barakat
[e]Sazman-e Omour-e Ashayer
[f]Setad-e Hamahangi-ye Bazsazi
[g]Artesh-e Homhoori-ye Eslami
[h]Ostandari
[i]Bonyad – e Maskan-e Ostan-e Lorestan
[j]Setad-e Hamahangi-ye Bazsazi-ye Manategh-e Seilzade-ye Lorestan
[k]Sherkat-e Ab-e Mantagheiee-ye Lorestan
[l]Sherkat-e Tozi-e Bargh-e Ostan-e Lorestan
[m]Sherkat-e Ab va Fazelab-e Roostaie-ye Lorestan
[n]Jamiat-e Helal-e Ahmar-e Iran
[o]Jamiat-e Imam Ali
[p]Anjomen-e Nejat-e Anijirestan
[q]Jahad-e Daneshgahi

non-governmental organisations. These organisations have various roles and responsibilities before, during and after a crisis, divided into six categories. While governmental organisations have responsibilities in all the six categories, non-governmental and public organisations are mostly concerned with post-crisis activities.

35.3.2 The Relevant Organisations and Their Use of Social Media

To investigate how the relevant organisations have used social media in disaster management in the case of the Lorestan flood, it is necessary to identify the most popular social media platforms amongst Iranian governmental, public and non-governmental organisations. They are:

- Instagram, Twitter and Facebook are the most popular social platforms amongst Iranian governmental, public and non-governmental organisations.
- LinkedIn is rarely used by these organisations.
- Flicker, YouTube, Pinterest, Google Plus and a number of other platforms are never used by these organisations.

Given the above, the research focused on the first category since it includes the most widely used social media platforms amongst Iranians. As Table 35.3 shows, in terms of frequency, Instagram dominates the list as the most widely used platform, followed by Twitter. Facebook is used less often across the board. It should be noted that both Twitter and Facebook are banned in Iran and used through proxies.

Table 35.3 Followers of Iranian disaster management organisations in social media as of June 2019 (authors)

Organisations			The number of followers in social media		
			Instagram	Facebook	Twitter
Governmental organisation	National level	SCNCM	–	–	–
		NOCS	–	–	–
		MF	–	–	✓78
		BF	✓12k	–	–
		NAO	✓196	–	–
		RCH	299	–	–
		IRIA	–	–	–
	Regional level	PG	–	–	–
		HFLP	–	–	–
		RCHLF	–	–	–
		RWCL	–	–	–
		LEDC	–	–	–
		LRWSC	–	–	–
Public organisation		IRCS	✓24k	✓2k	✓11k
Non-governmental organisations		IAS	✓57k	✓86k	✓12k
		RSA	–	–	–
		ACECR	–	–	–

Table 35.4 Frequency of social media usage by disaster management organisations in Iran, April 2019–May 2019 (authors)

Organisations			Frequency		
			Instagram	Facebook	Twitter
Governmental organisation	National level	SCNCM	–	–	–
		NOCS	–	–	–
		MF	0	–	0
		BF	Daily	–	–
		NAO	Rarely	–	–
		RCH	Weekly	–	–
		IRIA	–	–	–
	Regional level	PG	–	–	–
		HFLP	–	–	–
		RCHLF	–	–	–
		RWCL	–	–	–
		LEDC	–	–	–
		LRWSC	–	–	–
Public organisation		IRCS	Daily	Rarely	Daily
Non-governmental organisations		IAS	Daily	Daily	Daily
		RSA	–	–	–
		ACECR	–	–	–

35.3.3 *Frequency of Social Media Usage*

Reviewing the three most popular social media platforms and whether they are used by the relevant organisations shows that only four out of the 13 organisations are using social media at all. It also shows that the POs and NGOs are using a wider array of social media platforms in a more efficient manner. In addition, as the following charts show, out of a total of 167,700 hashtags about the Lorestan floods on Instagram and Twitter, only 811 belong to the organisations involved in flood management (Table 35.4, Fig. 35.3).

35.3.4 *The Quality of Using Social Media in the Lorestan Floods, 2019*

The content analysis of the posts shared by the organisations involved in the Lorestan floods (Table 35.5) shows that the government organisations have never used any relevant hashtags. The Iranian Red Crescent society has been the only public organisation to have used the relevant hashtags 44 times between 25 March and 25 April 2019 on

Fig. 35.3 Percentage of the posts about Lorestan floods in the accounts of the relevant organisations in social media (authors)

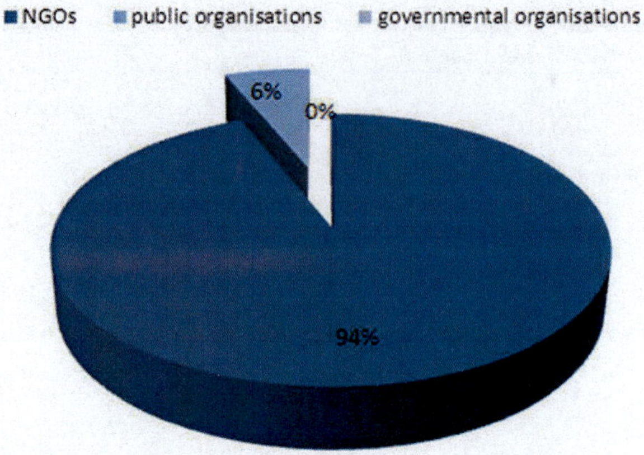

Table 35.5 Quality of using social media in the Lorestan floods, 2019 (authors)

Social media	Organisations	1	2	3	4	5	6	The period of posting
Instagram	Governmental organisation	0	0	0	0	0	0	–
	Public organisation	0	0	0	0	0	0	–
	Non-governmental organisations	9	3	43	10	158	17	March, 25th till August, 15th
Facebook	Governmental organisation	0	0	0	0	0	0	–
	Public organisation	0	0	0	0	0	0	–
	Non-governmental organisations	9	3	43	10	158	17	March, 25th till August, 15th
Twitter	Governmental organisation	0	0	0	0	0	0	–
	Public organisation	3	7	0	0	41	0	March, 25th till April, 25th
	Non-governmental organisations	9	3	43	10	158	17	March, 25th till August, 15th

Six categories of content 1. Warning and training, 2. Mobilising the public for relief operations (human resources), 3. Organising the public to provide assistance during the crisis, 4. Organising the public to provide assistance after the crisis, 5. Informing about the relief operations during the crisis and 6. informing about the relief operations after the crisis

Twitter. While the NGOs have used the relevant hashtags 760 times on Instagram, Facebook and Twitter, they have repeated the same posts in their accounts on these three platforms. They are still posting on the relevant hashtags. It should also be noted that the content of the posts could be divided into six categories, namely warning

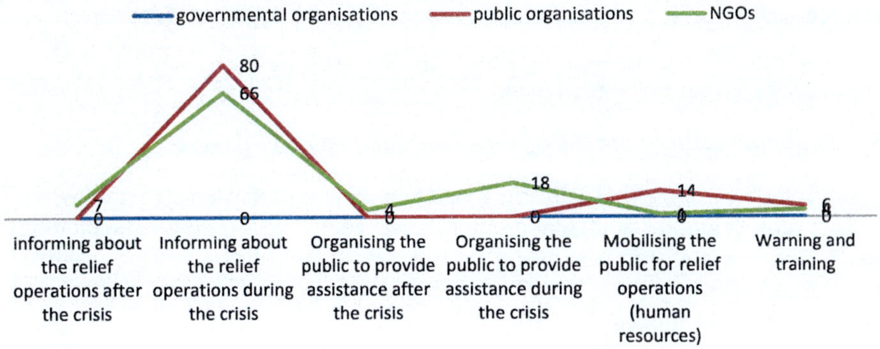

Fig. 35.4 Content of the social media posts by different organisations (%)

and training, mobilising the public for relief operations (human resources), organising the public to provide assistance both during and after the crisis and providing information about the relief operation both during and after the crisis.

The following diagram (Fig. 35.4) shows the results obtained from the above table. According to these results, most of the posts shared by the public organisations (PO) have been designed to provide information about the relief operations and organising the public to provide assistance during a crisis. The first priority for the NGOs is the same as the POs, but their second priority has been to organise the human resources for relief operations.

35.3.5 *Findings from the Interviews: Challenges for the Government's Use of Social Media*

The analysis of the data obtained from interviews identified three key themes, namely (a) distrust of governmental organisations towards social media, (b) the use or boycott of social media as a politicised issue and (c) lack of sufficient media literacy and justification of the government's centralised control over social media. Below, we are elaborating on each of the above themes.

(a) ***Distrust of governmental organisations towards social media***

The surveyed organisations claimed that Facebook and Twitter were tools aimed to discredit the Iranian government. They argued that the Iranian opposition uses these platforms to highlight the government's shortcomings. As one interviewee pointed out:

> In the case of the Lorestan floods, most social media pages were trying to undermine the sense of empathy amongst the public through spreading rumours and fake news, undermining the Iranian national identity and suggesting in-fighting amongst various disaster management organisations. [...] These pages are being managed by the country's enemies both from inside and from outside the country.

Additionally, social media is usually assumed to be the source of fake news:

> Unfortunately, most social media platforms are constantly publishing fake news for propaganda purposes and to attract a larger audience.

(b) ***The application or boycott of social media: a politicised issue***

Political parties in Iran have differing views on the role of the social media. While some are fully cognisant of the important role played by social media in facilitating disaster management activities, others assume that some decision-makers' support for social media is largely motivated by their desire to garner support for elections. One interviewee mentioned:

> [...] By supporting Telegram and Instagram, they only care about their party interests.

(c) ***Lack of sufficient media literacy and justification of the government's centralised control over social media***

The governmental organisations surveyed in this research argued that since media literacy is low amongst the Iranian public, they cannot recognise fake news in social media, and thus, the government has to control their content centrally through a systematic media policy which is at odds with the nature of such platforms. As one interviewee pointed out:

> The Government has control over the Islamic Republic of Iran Broadcasting (IRIB), and the IRIB mobilised its resources for informing people about the floods in real time without any of the risks associated with fake news and rumours.

35.4 Conclusion

This chapter is about the function of the social media in disaster management in Iran. We have attempted to examine whether collective action via social media could bring about any improvement in post-disaster management activities, and how social media operate in such contexts. This was based on an online survey and an online questionnaire designed to explore the current state of social media use in disaster management activities in Iran and the challenges for their use by governmental organisations involved in such activities. Findings obtained from social media analysis tools revealed that the common function of social media in this particular case was to promote community engagement and volunteer mobilisation, to communicate with the public regarding disaster recovery during and after the crisis and to raise funds following the disaster. The results reflected a general trend towards enhanced public participation in public communication. They also demonstrated that in their disaster response authorities need to make greater efforts aimed at enhancing public trust to ensure that critical information is efficiently disseminated and accessed by people over social media when disaster strikes. Finally, the outcomes of this research suggest the following.

- Studies in various parts of the world indicate that the large number of users following government-affiliated social media accounts demonstrates the leading role of public authorities. However, the lack of social media presence by Iranian officials has created a situation in which the country is unable to play a dominant role in this area.
- This lack of presence is rooted in the public sector's pessimism about the role of virtual networks on the one hand and its lack of awareness of social media's impact on the whole society on the other hand.
- Accordingly, the government has not been able to effectively use social media to publicise its efforts and achievements and to build public trust and is continuing to rely on formal and traditional data transmission networks instead.
- In this case, by comparing the follower base of the three categories of disaster management organisations, it becomes clear that people turn to information sources that they trust, and that the application of social media as an instrument for improving disaster management activities requires more than just providing information. Therefore, it is essential for governmental organisations in Iran to pay sufficient attention to the level of public trust enjoyed by government-affiliated social media accounts.

Findings from the questionnaires also highlight four key themes, namely (a) a distrust of governmental organisations towards social media, (b) the application or boycott of social media as a politicised issue and (c) lack of sufficient media literacy and (d) the justification of the government's centralised control over social media as the challenges to the use of social media by governmental organisations. An overarching theme to explain these four issues is a "centralised government in which there is no political willpower to legitimate any sources of information outside government control". This theme is closely related to the long history of public sector centralisation and the accumulation of power in the public sector in Iran, characterised by, and in turn reflected in the lack of political will to delegate powers and responsibilities.

Notes

i. Twitter and Facebook were banned by the government after they were used by protesters after the 2009 presidential election to inform and coordinate protest activities. However, many Iranians still use these networks, and even some officials use them to communicate with domestic or international audiences. Instagram is now the most common social network in Iran which is mostly used as a platform for small online business and entertainments, as well as a special unit of the Police which traces user's activities in this network.

References

Afzali M, Arizi SF, Khojasteh H (2014) Challenges and strategies of using social media in crisis management. The National Conference on Crisis Management and HSE. Available at: https://www.civilica.com/Paper-DMHSE02-DMHSE02_234.html. Accessed 10 May 2019 (in Persian)

Ahmed A (2011) Use of social media in disaster management (research in progress)
Akbari A (2010) The role of social media in making and controlling of crisis. Rahavard-e-No 30:32–39 In Persian
Alexander DE (2014) Social media in disaster risk reduction and crisis management. Sci Eng Ethics 20(3):717–733
Busa MG, Musacchio MT, Finan S, Fennell C (2015) Trust-building through social media communications in disaster management. In: Proceedings of the 24th international conference on world wide web. ACM, pp 1179–1184
Correa T, Hinsley AW, De Zuniga HG (2010) Who interacts on the Web?: the intersection of users' personality and social media use. Comput Hum Behav 26(2):247–253
Fard HD, Rostamy AAA (2007) Promoting public trust in public organisations: explaining the role of public accountability. Public Organisation Rev 7(4):331–344.5
Goel A, Chakraborty M, Biswas SK (2019) The role of social media in crisis situation management: a survey. In: Emerging technologies in data mining and information security. Springer, Singapore, pp 439–448
Internet World Stats (2019) Internet top 20 Countries—internet users 2020, Available at: https://www.internetworldstats.com/top20.htm. Accessed 17 July 2019
Kavanaugh AL, Yang S, Sheetz SD, Li LT, Fox EA (2011) Between a rock and a cell phone: social media use during mass protests in Iran, Tunisia and Egypt. Department of Computer Science, Virginia Polytechnic Institute & State University
Kavota JK, Kamdjoug JRK, Wamba SF (2020) Social media and disaster management: case of the north and south Kivu regions in the Democratic Republic of the Congo. Int J Inf Manage 52:
Keim ME, Noji E (2011) Emergent use of social media: a new age of opportunity for disaster resilience. Am J Disaster Med 6(1):47–54
Kelly S, Ahmad K (2014) Determining levels of urgency and anxiety during a natural disaster: noise, affect, and news in social media. In: DIMPLE: disaster management and principled large-scale information extraction workshop programme, vol 70
Khan H, Vasilescu LG, Khan A (2008) Disaster management cycle-a theoretical approach. J Manage Market 6(1):43–50
Kibanov M, Stumme G, Amin I, Lee JG (2017) Mining social media to inform peatland fire and haze disaster management. Soc Netw Anal Mining 7(1):30
Lorestan Province Introduction (nd) General Administration of the Ministry of Road and Urban Development in Lorestan. Available at: https://lorestan.mrud.ir/%D9%85%D8%B9%D8%B1%D9%81%DB%8C-%D8%A7%D8%B3%D8%AA%D8%A7%D9%86-%D9%84%D8%B1%D8%B3%D8%AA%D8%A7%D9%86. Accessed 17 May 2019 (in Persian)
Mahdian A (2019) Damage to 174,000 housing units from the flood. Interviewed for Islamic Republic of Iran Broadcasting News Agency, 29 Apr 2019. Available at: http://www.bonyadmaskan.ir/Lists/List/DispForm.aspx?ID=6817. Accessed 20 May 2019 (in Persian)
Mauroner O, Heudorfer A (2016) Social media in disaster management: how social media impact the work of volunteer groups and aid organisations in disaster preparation and response. Int J Emergency Manage 12(2):196–217
Obar JA, Wildman SS (2015) Social media definition and the governance challenge-an introduction to the special issue. Telecommun Policy 39(9):745–750
Sakaki T, Matsuo Y, Kurihara S, Toriumi F, Shinoda K, Noda I, Kazama K (2013) The possibility of social media analysis for disaster management. In: 2013 IEEE region 10 humanitarian technology conference. IEEE, pp. 238–243
Shekhar H, Setty S (2015) Disaster analysis through tweets. In: 2015 international conference on advances in computing, communications and informatics (ICACCI). IEEE, pp 1719–1723
Shivarkar P, Wei W (2018) Exploit social relations in sentiment analysis of social media content for disaster management, AMCIS 2018 Proceedings, New Orleans, LA, Augest 16–18.
Song M, Kim JW, Kim Y, Jung K (2015) Does the provision of emergency information on social media facilitate citizen participation during a disaster? Int J Emergency Manage 11(3):224–239

Velev D, Zlateva P (2012) Use of social media in natural disaster management. Int Proc Econ Dev Res 39:41–45
Weng WW, Woo CK, Cheng YS, Ho T, Horowitz I (2015) Public trust and corruption perception: disaster relief. Appl Econ 47(46):4967–4981
Williams BD, Valero JN, Kim K (2018) Social media, trust, and disaster: Does trust in public and nonprofit organisations explain social media use during a disaster? Qual Quant 52(2):537–550
Yan L, Pedraza-Martinez AJ (2019) Social media for disaster management: Operational value of the social conversation. Prod Oper Manage 28(10):2514–2532
Yavari AH, Mohammadi A, Javanmard M (2017) The role of social media in making the socail crisis. Res Danesh-e-Entezami Q 19:1–23 In Persian

Chapter 36
Environmental Issues and Energy Potentials in Post-earthquake Reconstruction

Marianna Rotilio and Valeria Annibaldi

Abstract Directive (EU) 2018/844 amending Directive 2010/31/EU on the energy performance of buildings and Directive 2012/27/EU on energy efficiency were to be implemented by European Union member states by March 2020. In particular, an explicit reference to existing historical buildings and contexts appeared for the first time in paragraph 18 of the cited Directive 844. This requires member states to encourage research into and the testing of new solutions to improve the energy performance of historical buildings and sites, while also safeguarding and preserving cultural heritage. At last, at European level, there is a sense of awareness of the importance of historical buildings and contexts that architectural, technical–construction and environmental values must be protected and preserved. To date, in fact, the current legislation has revealed deep issues over time, which has become extremely evident during the post-earthquake reconstruction. In this context, issues and practices that focus on sustainable process should be embraced. This theoretical position supports the introduction of principles that will allow the use of what technology offers and all the scientific and technical knowledge now widely available. The implementation of sustainability policies must be based on local knowledge and conditions that allow the highlighting of environmental issues and energy-efficient possibilities. This chapter presents a methodology applied to the city of L'Aquila that allows for the identification of homogeneous areas with regard to environmental behaviour, in which specific strategies could be implemented. These strategies take into consideration the analysis of the values, the transformable elements, the new users' needs, the level of damage as well as the latest technological solutions in the field of environmental sustainability. In this way, compatible design solutions will be identified and developed based on a sensitive case-by-case approach to the repair and reconstruction of historic buildings damaged during a disaster event.

M. Rotilio (✉) · V. Annibaldi
University of L'Aquila, DICEAA, L'Aquila, Italy
e-mail: marianna.rotilio@univaq.it

© The Author(s), under exclusive license to Springer Nature Switzerland AG 2021

F. F. Arefian et al. (eds.), *Historic Cities in the Face of Disasters*, The Urban Book Series,
https://doi.org/10.1007/978-3-030-77356-4_36

Keywords Post-earthquake reconstruction · Sustainability · Energy efficiency · Historic buildings · L'Aquila · Italy

36.1 Introduction

In consequence to the pressure of the great urban and territorial transformations of the industrial revolution years, around the middle of the nineteenth century, the interest for the ancient city and for the landscape developed. They were understood as elements to be safeguarded because they were a testimony of documentary and environmental values, that were stratified in its interior. This is the "urban recovery" discipline, in relation to which theorists and scholars have developed many theories supported by different approaches and hypotheses. There were several phases in the thought evolution, and it was necessary to arrive to few years ago to include themes as sustainability, environmental comfort, energy efficiency in the context of the historical heritage recovery. Between them, one of the most interesting is the one conducted by Martínez-Molina et al. (2016), which highlights how a research acceleration on the link between energy and historical buildings occurred around 2011. It also underlines that researchers normally develop case studies according to the context in which they work. However, more often than not, it also happens that nations which are rich in cultural heritage, such as Italy, are able to attract foreign researchers. Their study precisely emphasises how Italy is the country that has produced the largest number of scientific articles on the subject of energy efficiency and thermal comfort in historical buildings. It is followed by Great Britain, Spain and China. Moreover, the international collaborations between Italy and Great Britain are very frequent. Focusing on the developed countries, the authors explain this phenomenon mainly with the fact that Italy, followed by Spain and Great Britain, is one of the nations with the greatest cultural heritage (Fabbri et al. 2012, 2014). Overall, the study carried out in the international research scene has shown that from 2011 to 2014, the topic of interest was analysed from an economic point of view, through life cycle tests and profitability studies (Yung et al. 2012, Arumagi and Kalamees 2014). Between 2013 and the first half of 2014, the issue of energy efficiency in historical contexts began to become deeper through the development of "sector" research, where different methodologies were applied to study specific case studies as rural buildings (Alev et al. 2014; Li et al. 2013), academics ones (Buvik et al. 2014; de Santoli et al. 2014), religious ones (Woroniak and Piotrowska—Woroniak 2014; Martins and Carlos 2014) and urban ones (Moschella et al. 2013; Cardinale et al. 2013; De Berardinis et al. 2014). In more recent times, the research on energy efficiency in buildings and historical contexts has been directed towards the identification of specific techniques and methodologies, adopting the concepts related to the modelling (Lopez et al. 2018), the monitoring, the insertion of systems suitable for the exploitation of renewable energies (Polo Lopez and Frontini 2014).

The analysis of the state of the art has revealed how environmental and sustainability issues are a long and widely debated topic in the approach to the refurbishment and energy rehabilitation interventions of historical buildings. Nevertheless, they were very often considered incompatible with the protection of the building organism. Moreover, from a regulatory point of view, the existence of an important "void" also internationally recognised by some authors (Mazzarella 2015; Cem Dogan et al. 2015). In fact, in most cases, the attention to environmental issues is left to "ethics or conscience" of the actors involved in the building process because there are still evident shortcomings due to the determination of rules and principles that are designed mostly for new buildings and are difficult to apply in historical contexts. After all, the ancient buildings were based on self-sufficiency principles and, as they lacked systems, they were able to exploit local resources and to "dialogue" with the climate. These considerations are even more true in those contexts damaged by natural hazards.

Therefore, this chapter is aimed at illustrating an intervention in the post-earthquake reconstruction context by introducing new elements that intend to exploit energy-efficient measures and mitigate environmental impacts, promoting appropriate interventions aimed at environmental sustainability in accordance with regulatory requirements. Figure 36.1 presents an intervention case, a damaged historic building in L'Aquila. The damage exposes the presence of traditional construction techniques, namely masonry in limestone ashlars, wooden roof structure, vaulted horizontal surfaces and roof bent tiles.

Fig. 36.1 A historic building located in a side street of Via Sallustio, L'Aquila presents traditional construction techniques (Marianna Rotilio, 2019)

36.2 Context

The European Parliament and Council on energy efficiency issued the European Directive 2002/91/CE on 16 December, 2002 known as the Energy Performance Building Directive (EPBD). This was to address the problem of unsustainable energy consumption and depletion of non-renewable resources in the construction sector. The EPBD was the product of two oil crises. The first one, which occurred during the 1970s, was not the result of environmental awareness but simply the fear of oil depletion. The second oil crisis occurred during the 1990s and based on a greater awareness of the environmental issues and finally those producers who, until then, had been the main suppliers began to be seen as depleting finite resources. Thus, the Directive was issued for multiple reasons, including environmental, economic and political.

The EPBD recognised that Member States could decide to waive the application of the minimum requirements in the case of existing buildings and monuments protected as designated heritage sites or for their architectural or historical value "…in cases where compliance with provisions would imply an unacceptable alteration of their nature or appearance …" (art. 4 co. 3). Thus, the directive gave each Member State the power to decide according to its own historical and architectural heritage the application of restrictive energy efficiency requirements. Consequently, this position has been interpreted as a relaxation of the rule, potentially to avoid dealing with a complex problem, rather than as an opportunity to improve upon the status quo (Lucchi 2014).

The contents of article 4, paragraph 3 of the EPBD are also referred to in article 4 paragraph 2 of the European Directive 2010/31/EU on energy performance in buildings, issued by the European Parliament and Council. This directive went beyond this provision in Directive 2002/91/CE by addressing the issue of existing buildings. In particular, since the first article, Directive 2010/31/EU clarified that the provisions concerned the application of minimum requirements to the energy performance of (article. 1, co. 2, lett. C) stating that:

> … existing buildings, real estate units and building elements subject to major renovations; building elements that make up the building envelope and have a significant impact on the energy performance of the building envelope when they are renewed or replaced; as well as the technical building systems if installed, replaced or improved.

Therefore, according to the provisions of Directive 2010/31/EU, historical buildings, like other existing buildings, had to comply with the minimum energy efficiency requirements, without any reference to the architectural, historical, construction, environmental values which are present in it and must be preserved. These minimum requirements are often very restrictive and difficult to achieve in historical buildings.

The latest European regulatory update on the topic of energy is Directive 2018/844/EU, published in the Official Journal of the European Union of 19 June 2018, which amended Directive 2010/31/EU on the energy performance of buildings (EPBD) and 2012/27/EU on energy efficiency. This update was necessary to achieve the ambitious goal that European nations set themselves to reduce greenhouse gas emissions by at least 40% by 2030 compared to 1990 levels. To meet this goal, Member States are required to increase the proportion of renewable energy and energy efficiency measures and to decarbonise the building stock, which is responsible for approximately 36% of all CO_2 emissions in the European Union (Costanzo 2018).

Importantly, in paragraph 18 of Directive 2018/844/EU an explicit reference to historical buildings and contexts appeared for the first time. It states that:

> Member States should encourage research and test new solutions to improve the energy performance of historical buildings and sites, while safeguarding and preserving cultural heritage

Finally, at European level there is a sense of awareness of the importance of historical buildings, contexts and values must be protected and preserved (Zordan et al. 2009). Member States, including Italy, had to implement the Directive by March 2020. In Italy under the current legislation there are two approaches that are allowable. One refers to monumental buildings for which derogation is permissible (Legislative Decree. No 42.2004). The second concerns historical buildings for which compliance with the legislative prescriptions is required. Furthermore, these regulations are often very restrictive and difficult to adopt (Legislative Decree No. 192.2005 et al.). These approaches have highlighted profound conflicts over time, which became a serious issue during the post-earthquake reconstruction. In fact, the use of derogation for monumental buildings has not allowed their technological and energy update reinforcing their obsolescence. Conversely, to ensure compliance with the rules, historic buildings have often undergone interventions that do not respect their values.

The issue of improving energy efficiency in historic buildings in post-earthquake reconstruction has been addressed by many scholars (Bonomo and De Berardinis 2014; Cannistraro et al. 2013; De Berardinis et al. 2014), highlighting the importance of the need to address energy efficiency in post-earthquake reconstruction. However, the theme of energy always revolves around the focus on "building systems". This is often without taking into account the fact that the issues of sustainability inherent in the building process are many and concern the entire "lifecycle" of the building (Annibaldi et al. 2019), from its conception through design, construction to its reuse or demolition. The entire building process must aim at sustainability, paying attention to all the issues involved in the construction activity, such as the site where the work is to be carried out, the water system, materials and resources, as well as the aspects of environmental and energy quality including a specific focus on the "cost–benefit" (Annibaldi et al. 2020). Nevertheless, these concerns are a component of the complete picture in a historical context as considerable effort is also given to the commitments for the achievement of the *Sustainable Development Goals*, SDGs, 2030 (United Nations 2018a, b). Furthermore, they become a serious issue if they concern a complex and problematic situation such as that of post-earthquake reconstruction.

In addition, there is further local legislation applicable to the reconstruction of historic centres. Specifically, the Abruzzo Decree n. 39.2009, converted into Law in 2009, reference n. 77.2009, states that

> … the town mayors establish in agreement with the President of the Abruzzo Region—the Commissioner delegate for the reconstruction…and in agreement with the President of the Province for the matters of his jurisdiction, the "Piani di Ricostruzione" (Reconstruction Plans-PdR) of the city centre…

This decree states that the PdR has to ensure the socio-economic recovery of the aforementioned territory. Additionally, it must promote the reconstruction of houses and facilitate the return of people who had been living in temporary housing as a consequence of the seismic event. Furthermore, the PdR must identify appropriate interventions to guarantee better safety for new buildings and survey the status of current and other sites. In all of the issues to be included in the plan, there is no reference to the introduction of sustainability and energy principles. Furthermore, there is a lack of references to the application of these principles to build fabric having historical, environmental, landscape, architectural and technical values.

Finally, this discussion has highlighted that when considering the issues around legislation, historic buildings and city centres and the inclusion of sustainability principles after a seismic event the following concepts must be taken in consideration during the reconstruction process:

- adoption of issues and practices that focus on sustainable process;
- principles that allow the use of all available technology and introduction of all the scientific and technical knowledge now widely available;
- the implementation of sustainability policies must be based on the territory knowledge that allows to highlight environmental criticalities and energy potentials.

The awareness of these aspects would direct reconstruction towards a greater level of urban resilience where the community is able to design innovative social, economic and environmental responses to withstand the stresses of environment and history into the future. Certainly, therefore, among the many variables involved, one of the aspects that should be taken into account in the reconstruction work is the environmental question. Ultimately a concept inevitably dictated by contemporary events and needs as well as by the fact that sustainable cities and territories are places of value capable of determining economic, social inclusiveness and resilience.

36.3 Methodology

The methodological approach illustrated in this section is the result of a wider research (Rotilio et al. 2013; Rotilio and De Berardinis 2017; Rotilio 2019), where it was conceived and applied to the minor historical centres typical of the Abruzzo territory. Based on that earlier research, the methodology was finessed to be applied to the city of L'Aquila where the theme of rehabilitation/reconstruction raises complex questions and different solutions because it requires different design strategies.

Starting from a broad examination of the current European and national legislation on the energy efficiency of buildings and more specifically the requirements

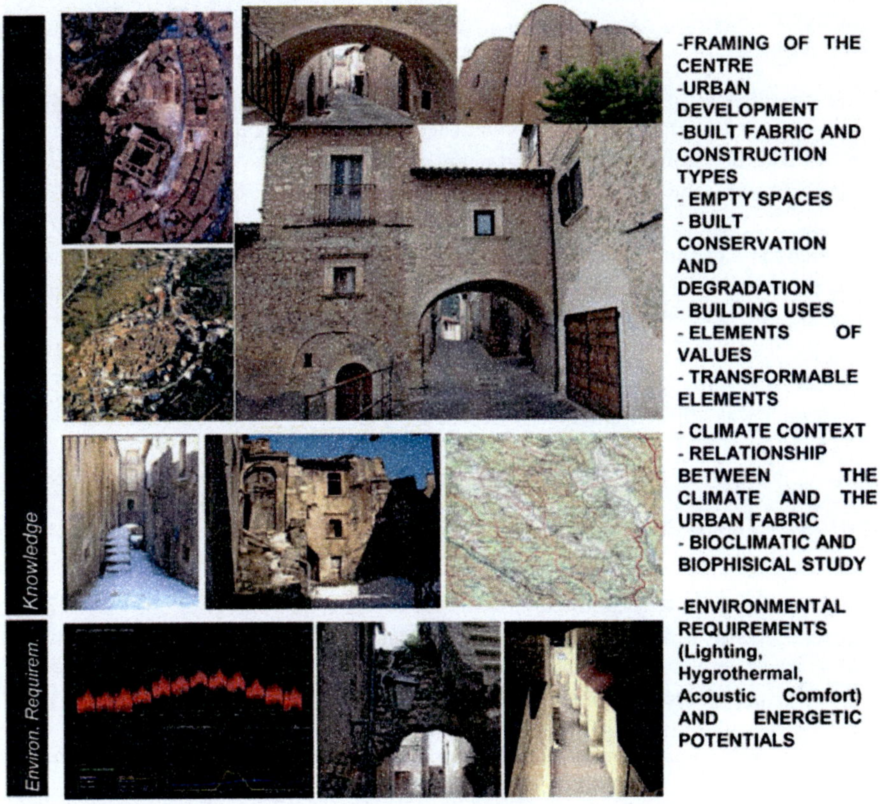

Fig. 36.2 Two types of methodological analysis: knowledge of the context and the new environmental requirements to be fulfilled (Marianna Rotilio)

concerning post-earthquake reconstruction, a methodology of two stages was developed to consider environmental issues and user needs (Fig. 36.2). The first is a cognitive study that aims to analyse the characteristics and values of the historic centre and investigate the climatic context. The synthesis of this analysis allows to draw up different types of maps to identify the transformable elements, i.e. all the elements that we could modify, the values and also homogeneous zones. The purpose of the latter map is to identify homogeneous zones in relation to the bioclimatic, biophysical behaviour but also the environmental critical aspects and potential ones. It is very useful to identify the climate risk zones and the potential of the historic centre environment. Once each homogeneous zone was identified, a score was assigned for the zones for each of the following issues:

- elevation and density of the urban fabric;
- level of sunshine (facades);
- level of sunshine (roofs);

- ventilation;
- soil moisture.

These issues encapsulate the principles that underpin key aspects of energy efficiency, namely around passive approaches to the heating and cooling of historic buildings and the reconstruction of city centres.

The second stage aims to analyse new users' needs considering the environmental requirements. For this purpose, strategies were defined for each homogeneous zone to improve the comfort conditions, through project solutions aimed at implementing the hygrothermal, visual, lighting and acoustic comfort, adopting the potential of the historical centre environment.

Finally, the process of compatibility of all these analyses led to a compatibility design through congruent solutions and according to a "case-by-case" logic. This methodology was applied to a case study located in the city of L'Aquila.

36.4 The Case Study: The City of L'Aquila

The city of L'Aquila was established with the consent of Pope Gregory IX (1170–1241) recognising a harmonious project that had no precedents in the history of medieval urban architecture and subsequently destroyed by Manfredi, in 1266. Later it was "re-founded" by Carlo I D'Angiò according to an orthogonal grid layout. The city was divided into four "quarters" that are still recognisable today and developed with a typical castrum layout, embellished with squares, palaces, churches, civic towers and bell towers, with these pieces of architecture located at the intersections of the axes. The original plan has never been lost over the centuries (Colapietra 2019), despite the numerous devastations due mainly to earthquakes that have substantially impacted the economy and the development of the city.

L'Aquila is characterised by different types of values (Figs. 36.3 and 36.4), and many scholars have devoted years of their lives to study this city at different levels: historical, architectural, urbanistic and constructive (Centofanti 1984, 1992; Vittorini 1994; Clementi 1998; Colapietra 2011; Andreassi 2012).

In such a context, it is very important to carry out a study and survey campaign according to the macro-themes identified in the methodological path illustrated in the previous section, then to elaborate specific projects for each zone. These documents act as support for the design activity. Together with the maps of values and transformability (Zordan et al. 2009), the environmental analyses were carried out. They consisted of a bioclimatic study of ventilation, level of sunshine and any relationships between the urban fabric and the climate, while the biophysical survey included analyses of the soil, vegetation and water basins. For example, Fig. 36.5 is the map indicating the different surface finishes of open spaces in the historical centre of L'Aquila in the area of Piazza Duomo. The colour coding highlights the various types of flooring: dark blue is used for cobblestones set in sand (permeable surface), cyan is used for porphyry paving slabs installed on mortar (waterproof surface), grey

Fig. 36.3 (*top*) The Church of Anime Sante in Piazza Duomo and the damage; (*bottom*) A typical multistory building and building safety intervention (Marianna Rotilio)

for asphalt (impermeable surface), bluish for the calcareous cobblestones set in sand (permeable surface).

This study is useful for identifying areas with environmental problems but also energy potential. The cases in which there are waterproof surfaces and low slopes characterised by lack of sunshine can be affected by winter problems of stagnation of humidity, mould formation and degradation. The same applies to summer. In fact, dark surfaces without suitable shading could be affected by overheating in open spaces with consequent discomfort. At the same time, however, these zones, precisely because they are not subject to shading, are suitable for the installation of systems and technologies designed to exploit solar energy.

For this reason, the planning strategies to be implemented should aim to eliminate the critical issues and exploit the energy potentials. For example, in the case illustrated, the project could include the introduction of appropriate rainwater collection and disposal systems in areas at risk of stagnation. Another option could be the replacement of dark flooring in favour of a light colour flooring that reduces the absorption of summer sunlight. The design of water fountains to cool and the installation of technological systems that exploit solar energy, such as solar trees, could be another example.

Fig. 36.4 Examples of values: (*top left*) Lowered vault, cornices decorated with stucco, decorated ceilings–a building in Via Roma; (*top right*) Limestone masonry, cross vault–Palazzo Carli; (*bottom left and right*) Corner solutions, openings in facades and balconies in wrought iron-Palazzo Carli and a building in via Paganica (Marianna Rotilio)

The synthesis of the analyses previously illustrated and listed in Fig. 36.2 led to the elaboration of the design guidelines based on the "Homogeneous Zones Map" (Fig. 36.6). This map contains the summary of all the environmental analyses, and its main purpose is to identify homogeneous zones in relation to the bioclimatic and biophysical behaviour but also the critical and potential aspects. In particular, twelve different areas were identified in L'Aquila. Accordingly, a score from 0 to 4 was assigned for each homogeneous zone in relation to:

- elevation and the density of the urban fabric;
- level of sunshine (facades);
- level of sunshine (roof);
- ventilation;
- soil moisture.

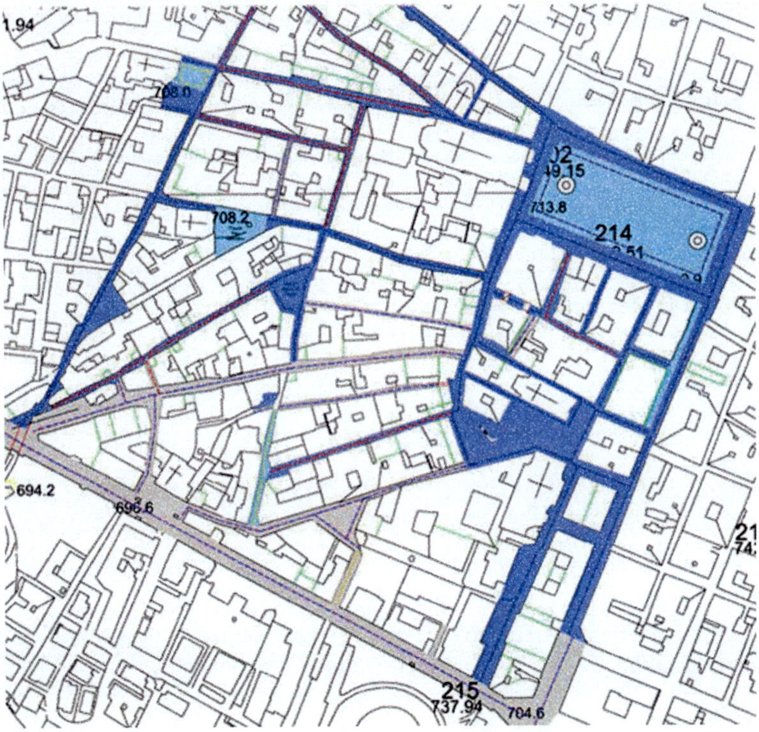

Fig. 36.5 Different colours present different types of surface finishes of the open spaces in the historic centre of L'Aquila, in the area of Piazza Duomo (Marianna Rotilio)

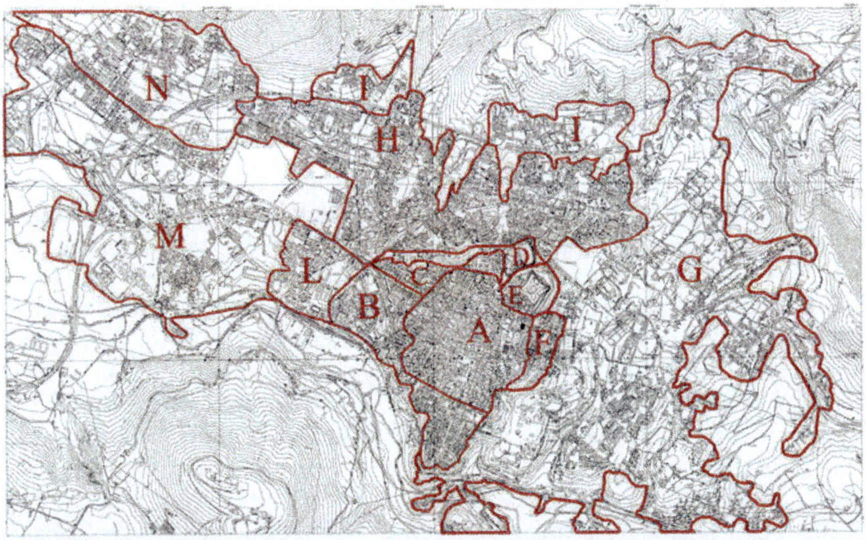

Fig. 36.6 Map of the homogeneous zones of the city of L'Aquila (Marianna Rotilio)

Table 36.1 Summary table of the main characteristics of each of the twelve zones (Marianna Rotilio)

	Elevation of the urban fabric	Density of the urban fabric	Level of sunshine (facades)		Level of sunshine (roofs)		Ventilation		Soil moisture	
Zone			Winter	Summer	Winter	Summer	Winter	Summer	Winter	Summer
A	4	4	1	2	3	4	1	1	3	0
B	3	3	3	4	3	4	1	2	2	0
C	3	3	0	0	0	1	2	1	2	0
D	2	2	0	0	0	1	2	1	2	0
E	3	2	1	2	3	4	2	2	3	0
F	3	2	2	1	1	2	3	1	2	0
G	2	1	3	3	2	3	4	4	0	0
H	3	2	4	4	3	4	3	3	1	0
I	4	1	4	4	3	4	4	4	1	0
L	1	2	3	4	3	4	3	2	4	0
M	0	1	3	4	3	4	4	4	4	0
N	0	1	4	4	3	4	4	4	1	0

and the information pertaining to each aspect was collected in a table (Table 36.1). The map of homogeneous zones is a useful tool to identify:

- the climate risk zones;
- the potential of the historical centre environment.

The classification of the areas allows to identify specific design strategies in order to achieve:

- the improvement of comfort conditions in the climate risk zones, through project solutions aimed at implementing the hygrothermal, visual and lighting and acoustic comfort;
- the exploitation of the potential of the historical centre environment.

The design strategies are defined for each homogeneous area and organised according to the main requirements to be met (Rotilio and De Berardinis 2017; Rotilio 2019). Namely hygrothermal comfort, visual, lighting and acoustic comfort, use of renewable energy, rational use of energy and, finally, energy awareness campaign for users. Every design solution is classified according to the efficiency level offered in case of greater performance and compatibility. The compatibility of the project intervention expresses the "tolerability" of the chosen solution in function of the specific context in which it is being implemented. For example, with reference to the hygrothermal comfort category, in the zones A and B of the city there are many buildings at the risk of overheating on the upper levels in summer because their roofs are continuously exposed to the sun during the day. For each critical issue, the guidelines contain an assessment of the degree of severity, the complying solutions, the performances offered by the proposed solutions and finally the compatibility search. The same applies to the "use of renewable energy" category, given that the continuous exposure to the sun makes the roofing suitable for the exploitation of solar energy.

36.5 Discussion: The Way Forward

The study of the current European and Italian legislation on the energy efficiency of buildings and of the one concerning the post-earthquake reconstruction has highlighted the gap when the two themes are coupled. For this reason, this chapter has illustrated a research that is still ongoing and whose main purpose is to understand how to carry out a conscious intervention in the post-earthquake reconstruction context, through actions that aim at environmental sustainability. In fact, resilient reconstruction can also be achieved through the implementation of sustainability policies and, as demonstrated, must be based on a methodology founded on local knowledge. This knowledge allows the highlighting of environmental issues and energy potential of the historical centre, but also its architectural, historical, technical and construction values, as well as the additions and transformable elements. It is an essential element since the application of innovative and contemporary strategies in historical contexts with values could lead to unsuitable interventions in the perspective of conservation. Therefore, these strategies must be filtered by analysing the values found and the transformable elements. The synthesis process generated by the analysis of these values and transformable elements, the new users' needs, the level of damage as well as the latest technological solutions in the field of environmental sustainability lead to the identification of compatible design solutions, developed through a sensitive case-by-case approach.

An approach of this type could potentially allow a real sustainable development of the territory through a synergy of actions: the widespread introduction of renewable energies in the territory and in the historic centre through delicate, non-invasive and fully reversible integrated systems; through the rediscovery of local materials; through the replacement of the waterproof flooring with draining type flooring along with the renovation of the infrastructure networks; through to the introduction of widespread wiring systems and the creation of rainwater storage and reuse systems; through the energy requalification of buildings but also the urban redevelopment of neglected or abandoned areas; through the reuse of demolition materials, with an accurate selection; through the use of cogeneration systems; through the use of the latest technological solutions in the field of environmental sustainability. Anyway, these strategies, although effective, can only materialise following a good design activity carried out by all the actors involved in the reconstruction process.

The opportunity to give our descendants a *better city* with a strong, recovered economy, capital of research, culture, risk prevention and the promotion of environmental sustainability. The fact that after every tragic event L'Aquila was rebuilt and enriched according to the new contemporary know-how should be inspiring. Also this time, an action to enrich and improve the city is expected, rejecting any reference to the concept "where it was before-how it was before", that would only lead to a simple physical reconstruction.

36.6 Conclusion

The study presented in this chapter is part of the research that concerns the rehabilitation of historical centres damaged by natural hazards, in order to safeguard their heritage values and to enhance their resources but, at the same time, it aims to propose technological and functional updating, also in a sustainable and resilient way. It is based on the analysis of the current regulatory scenario and the international state of the art, the study has as its main goal the definition of a method of critical control of the transformative interventions on historical buildings. The application of this method led to the development of validated guidelines in the case study of the L'Aquila historical centre. These guidelines are aimed at identifying the issues that must be resolved or the potential to benefit from in such a delicate historical moment for L'Aquila. They suggest provide recommendations to architects some project solutions, but leaving architects the discretion to decide freely, with the only condition to respect necessarily the specific historical context in which they intervene.

Acknowledgments The Authors wish to thank Professor Pierluigi De Berardinis, University of L'Aquila, who was responsible for the scientific research from which the work described in this chapter has originated.

References

Alev U, Eskola L, Arumagi E, Jokisalo J, Donarelli A, Siren K, Brostrom T, Kalamees T (2014) Renovation alternatives to improve energy performance of historic rural houses in the Baltic Sea region. Energy Build 77:58–66

Andreassi F (2012) La città evento. L'Aquila ed il terremoto: riflessioni urbanistiche. Edizioni Aracne, Roma

Annibaldi V, Cucchiella F, De Berardinis P, Rotilio M, Stornelli V (2019) Environmental and economic benefits of optimal insulation thickness: a life-cycle cost analysis. Renew Sustain Energy Rev 116:

Annibaldi V, Cucchiella F, De Berardinis P, Gastaldi M, Rotilio M (2020) An integrated sustainable and profitable approach of energy efficiency in heritage buildings. J Clean Prod 251:

Arumagi E, Kalamees T (2014) Analysis of energy economic renovation for historic wooden apartment buildings in cold climates. Appl Energy 115:540–548

Bonomo P, De Berardinis P (2014) PV integration in minor historical centers: proposal of guide-criteria in post-earthquake reconstruction planning. In: Energy Procedia 2nd international conference on solar heating and cooling for buildings and industry, Freiburg, Germany

Buvik K, Andersen G, Tangen S (2014) Ambitious renovation of a historical school building in cold climate. Energy Proc 48:1442–1448

Cannistraro G, Cannistraro M, Restivo R (2013) Messina's historical buildings after the earthquake of 1908: energy and environmental analysis through a global screening methodology. Int J Heat Technol 31(2):155–158
Cardinale N, Rospi G, Stefanizzi P (2013) Energy and microclimatic performance of Mediterranean vernacular buildings: the Sassi district of Matera and the Trulli district of Alberobello. Build Environ 59:590–598
Cem Doğan Ş, Zeynep Durmuş A, Selim Sarp T, Tor B, Gulden Gokcen A (2015) A transdisciplinary approach on the energy efficient retrofitting of a historic building in the Aegean Region of Turkey. Energy Build 96:128–139
Centofanti M (1984) L'Aquila 1753–1983, il restauro della città. Colacchi, L'Aquila
Centofanti M (1992) La costruzione dell'immagine delle piazze. In: Centofanti M, Colapietra R, Conforti C, Properzi P, Zordan L (eds) L'aquila città di piazze. Spazi urbani e tecniche costruttive. Carsa, Pescara, pp 160–187
Clementi A (1998) Storia dell'Aquila. Dalle origini alla prima guerra mondiale. Laterza, Roma-Bari
Colapietra R (2011) Una lunga storia d'amore. Pagine scelte di storia aquilana. Carsa, Pescara
Colapietra R (2019) La città di L'Aquila attraverso i secoli. Recupero e Conservazione 153:40–42
Costanzo E (2018) La Direttiva Europea 2018/844 che modifica l'EPBD. In: Amato A, Costanzo E, Di Pietra B, Hugony F (eds) Rapporto Annuale Efficienza Energetica. Enea
De Berardinis P, Rotilio M, Marchionni C, Friedman A (2014) Improving the energy efficiency of historic masonry buildings. A case study: a minor centre in the abruzzo region, Italy. Energy Build 80c:415–423
de Santoli L, Fraticelli F, Fornari F, Calice C (2014) Energy performance assessment and a retrofit strategies in public school buildings in Rome. Energy Build 68, part A, 196–202
Fabbri K, Zuppiroli M, Keoma A (2012) Heritage buildings and energy performance: mapping with GIS tools. Energy Build 48:137–145
Fabbri K, Tronchin L, Tarabusi V (2014) Energy retrofit and economic evaluation priorities applied at an Italian case study. Energy Proc 45:379–384
Li Q, You R, Chen C, Yang X (2013) A field investigation and comparative study of indoor environmental quality in heritage Chinese rural buildings with thick rammed earth wall. Energy Build 62:286–293
Lopez FJ, Lerones PM, Llamas J, Gomez–Garcia–Bermejo J, Zalama E (2018) A review of heritage building information modeling (H–BIM). Multimodal Technol Interact 2:21
Lucchi E (2014) Riqualificazione energetica dell'involucro edilizio. Diagnostica e interventi. Dario Flaccovio Editore, Palermo
Martínez-Molina A, Tort-Ausina I, Cho S, Vivancos JL (2016) Energy efficiency and thermal comfort in historic buildings: a review. Renew Sustain Energy Rev 61:70–85
Martins AMT, Carlos JS (2014) The retrofitting of the Bernardas' convent in Lisbon. Energy Build 68(part A):396–402
Mazzarella L (2015) Energy retrofit of historic and existing buildings. The legislative and regulatory point of view. Energy Build 95:23–31
Moschella A, Salemi A, Lo Faro A, Sanfilippo G, Detommaso M, Privitera A (2013) Historic buildings in Mediterranean area and solar thermal technologies: architectural integration vs preservation criteria. Energy Procedia 42:416–425
Polo Lopez CS, Frontini F (2014) Energy efficiency and renewable solar energy integration in heritage historic buildings. Energy Proc 48:1493–1502
Rotilio M (2019) I centri storici minori. Linee guida per il recupero sostenibile. Aracne Editrice, Roma
Rotilio M, De Berardinis P (2017) Sustainability in post-earthquake reconstruction planning: the case-study of San Pio di Fontecchio (Aq). In: Amoêda R, Pinheiro C (eds) SB-LAB 2017—proceedings of the international conference on advances on sustainable cities and buildings development, Porto, Portugal, 15–17 Nov 2017
Rotilio M, De Berardinis P, Marchionni C (2013) Principles of sustainability in the reconstruction plans of the minor centres struck by the earthquake. A case study: Sant'Eusanio Forconese (Aq).

In: Proceedings of the annual international conference on architecture and civil engineering (ACE 2013), Singapore, 18–19 Mar 2013. Global Science and Technology Forum (GSTF), Singapore

United Nations (2018a) Transforming our world: the 2030 agenda for sustainable development. https://sustainabledevelopment.un.org/post2015/transformingourworld. Accessed 12 Aug 2019

United Nations (2018b) UN sustainable development goals progress report. https://www.un.org/sustainabledevelopment/progress-report/. Accessed 12 Aug 2019

Vittorini M (1994) La città borghese: nuove istituzioni e nuove centralità. In: AA. VV. L'Aquila e la Provincia aquilana - Economia, Società e Cultura dal 1859 al 1920. La Provincia aquilana e le comunicazioni nell'800 – Cassa di Risparmio della Provincia dell'Aquila

Woroniak G, Piotrowska–Woroniak J (2014) Effects of pollution reduction and energy consumption reduction in small churches in Drohiczyn community. Energy Build 72:51–61

Yung EHK, Edwin HW, Chan C (2012) Implementation challenges to the adaptive reuse of heritage buildings: towards the goals of sustainable, low carbon cities. Habitat Int 36(3):352–361

Zordan L, Bellicoso A, De Berardinis P, Di Giovanni G, Morganti R (2009) Le tradizioni del costruire della casa in pietra: materiali, tecniche, modelli e sperimentazioni, Stone house building traditions: materials, techniques, models and experimentations. Gruppo tipografico editoriale, L'Aquila

Chapter 37
A Multidisciplinary Approach to Retrofitting Historic Buildings: The Case of the Former San Salvatore Hospital, L'Aquila

T. de Rubeis, M. De Vita, L. Capannolo, E. Laurini, I. Nardi, D. Ambrosini, D. Paoletti, and P. De Berardinis

Abstract Earthquakes represent one of the most disastrous natural phenomena for adverse social and urban impacts. The consequent need to recover buildings from a catastrophic event may represent an opportunity for their structural and energy retrofit, particularly for listed structures or those representing architectural heritage. However, the various aspects of energy optimisation, seismic protection and cultural heritage preservation of the building stock may be in conflict. For this reason, a multidisciplinary design approach is needed to consider all the aspects involved in the retrofit process. This chapter discusses the development and application of an Operative Design Tool (ODT). This is a tool based on a multidisciplinary approach and sustainability principle guidelines with standards for seismic and energy retrofitting. The applicability of the proposed approach is examined through the analysis of the former San Salvatore Hospital of L'Aquila, a complex city district of buildings of different eras, severely damaged by the 2009 earthquake. The development of ODT

T. de Rubeis (✉) · D. Ambrosini · D. Paoletti
Department of Industrial and Information Engineering and Economics, University of L'Aquila, L'Aquila, Italy
e-mail: tullio.derubeis@univaq.it

D. Ambrosini
e-mail: dario.ambrosini@univaq.it

D. Paoletti
e-mail: domenica.paoletti@univaq.it

L. Capannolo · E. Laurini · P. De Berardinis
Department of Civil, Construction-Architectural and Environmental Engineering, University of L'Aquila, L'Aquila, Italy
e-mail: pierluigi.deberardinis@univaq.it

M. De Vita
National Research Counci, Rome, Italy
e-mail: mari.22@live.it

I. Nardi
Energy Efficiency Unit Department, ENEA Casaccia, S.M. Di Galeria, Rome, Italy
e-mail: iole.nardi@enea.it

© The Author(s), under exclusive license to Springer Nature Switzerland AG 2021
F. F. Arefian et al. (eds.), *Historic Cities in the Face of Disasters*, The Urban Book Series,
https://doi.org/10.1007/978-3-030-77356-4_37

identifies energy improvement best practices that simultaneously allow renovations to address the issues of seismic damage, protection and valorisation of historic buildings. Such guidelines for integrating the design of seismic improvement and energy optimisation in historical buildings can help policy makers and designers during renovation processes.

Keywords Earthquakes · Structural and energy retrofit · Multidisciplinary approach · San Salvatore Hospital · L'Aquila · Italy

37.1 Introduction

This chapter analyses the potential performance improvement of a complex of buildings from both seismic impacts and energy efficiency, highlighting the need to intervene on the building envelope through processes that partially overlap. For some years now, scientific research has been proposing the creation of combined interventions that take into account both aspects upstream of the design process, resulting in a significant reduction in overall costs, an optimisation of the building's performance and also promoting the mitigation of environmental impacts (Mannella et al. 2017; De Vita et al. 2018).

Over the last few decades, structural consolidation has achieved praiseworthy results from both regulatory and applicative perspectives, basing its foundations on pure research and on the historical-technical tradition now completely practiced by designers. The objective of this improvement is to achieve pre-established performance levels in terms of safety. The experimental approach, typical of research, has meant that the design of structural interventions is always aimed at the application of the solution "appropriate" to the specific case, adapting to the context in which it operates. This methodology is the result of an interdisciplinary dialogue between restorers, engineers and technicians which has been going on for some time.

As far as the behaviour of a building under the effect of gravitational actions is concerned, the current Italian Technical Regulations substantially equate existing and new buildings as the same. However, for behaviour due to the impact of seismic actions the Italian Technical Regulations distinguish between "adaptation" and "improvement" interventions.[1] In general, it can be said that in the absence of technical and/or economic constraints, adaptation interventions are planned. However, it should be noted that the total absence of constraints is a rare condition, and therefore, in most cases, it is not possible to adapt but only to improve the performances. All possible interventions aimed at improving the seismic performance of a building can be substantially traced back to strategies of "Increasing stiffness and resistance", "Increasing ductility" and "Reducing demand". Among all the possible intervention scenarios for structural improvement, the final choice must fall on the one that, for the same performance, best integrates with distribution, technological, formal and, where relevant, conservation aspects of the building under intervention.

---

[1] In the adaptation the level of safety required as a result of the interventions must be similar to that of a new building, while in the improvement it is accepted that a more modest level of safety is sufficient.

Based on these considerations, an interesting field of application is represented by buildings with significant historical and artistic value, not necessarily listed buildings, however, falling into the category of buildings for which there is an exception to the application of the legislation[2] on energy efficiency. In fact, in Italy, the regulations (continuously revised), do not apply, or are applied with limitations, to listed buildings and to those located in historic urban centres. The interest aimed at this category of buildings is due not only to their considerable diffusion on the Italian territory, but also to the possibility of planning interventions that can bring a significant improvement, both in terms of energy efficiency (de Rubeis et al. 2016; Smarra et al. 2018) and for the reduction of seismic vulnerability. Therefore, in contexts of artistic and cultural importance, the use of non-destructive investigation techniques is desirable, to minimise the invasiveness of the analysis (Nardi et al. 2017; De Berardinis et al. 2018). Moreover, the choice of the most suitable intervention must be made on the basis of a detailed diagnostic survey able to provide information on the actual performance of the buildings under analysis (Nardi et al. 2018, 2016). In particular, the optimisation strategy of the building's energy performance can be achieved by combining the reduction of energy requirements, determined by the envelope characteristics (passive elements), with the increase in the efficiency of the technical plants (active elements) (de Rubeis et al. 2018; Laurini et al. 2018).

More difficult, and therefore particularly challenging, is the integration of the energy improvement of buildings in the design of interventions consistent and compatible with their historical value. The objective of the improvement measures is to contain energy costs, greenhouse gas emissions and, at the same time, to increase users' comfort. This is possible by intervening on the building's envelope and/or on the technical plants also encouraging the use of renewable energy systems, technologies suitable for containing water consumption and Building Energy Management Systems (BEMSs). The energy performance increase of passive elements allows the reduction of heat losses during the winter season and the improvement of performance in summer, thus reducing the primary energy requirements of the building. At the same time, the energy optimisation of the building's active elements maximises the efficiency of the systems, reducing the primary energy consumption. Therefore, interventions aimed at optimising energy performance use active and passive control strategies to reduce energy requirements and ensure that this energy is produced as efficiently as possible.

Because of the complexity of historic buildings, characterised by a difficult knowledge of structure and materials, preservation of cultural heritage and lack of documents or project data, there is a need to define tailored methodologies capable of carrying out multi-objective analyses. During an emergency and the subsequent reconstruction phase, in many cases, the priority seems to be the recovery of the normal functionality of the building, without considering the quality of the reconstruction and refurbishment interventions carried out. In this sense, an earthquake, like other disasters, represents an opportunity for structural and energy retrofitting.

[2] The Technical Regulations are contained in Paragraph 3, article 3 of the Legislative Decree no. 192 of 19 August 2005 and subsequent amendments and additions.

Therefore, this chapter describes and applies a methodology based on a multidisciplinary approach and principles of sustainability. This is based on the knowledge of the historical value of the building under analysis and their urban context. This is in order to provide practical guidelines for designers and policy makers to define combined interventions for seismic and energy improvement to be adopted during the retrofit of historic buildings. The former San Salvatore Hospital in L'Aquila is an ideal case study for the application of the proposed methodology applying a multidisciplinary approach based on combined energy and seismic improvement in accordance with the restoration principles.

37.2 Developing an Operative Design Tool

The development of an operative tool to guide and support the design, based on the strategic principles of sustainability and respect for the characteristics of historic buildings takes the form of "Guidelines" and Operative Design Tool (ODT). This tool aims to guide, in a discretionary way, the design choices towards compliant intervention strategies, developed on the basis of a deep knowledge of the building (or block of buildings) under analysis. The core of the activity is the study of the building complex to characterise and optimise it, both from the energy and structural point of view. The methodology is based on the achievement of the following objectives:

- the recognition of homogeneous areas within the building complex;
- the development of solutions for energy and seismic improvement, compatible with the historic values of the piece of heritage;
- the creation of Guidelines and ODT.

The elaboration of a detailed knowledge of the buildings' characteristics and their factory elements, supported by visual and instrumental in situ analysis, is a fundamental phase for the identification of critical points from the energy and structural point of view, and the levels of transformability, in addition to the already investigated specific values of the various factory elements (see *Acknowledgements*).

A thorough knowledge of the architectural complex through the non-destructive diagnosis allowed the identification of some sections of the building similar in construction, technological and formal characteristics and that, consequently, are comparable for the thermo-physical and structural properties. These sections are defined as Homogeneous Zones as minimum intervention units. Homogeneous Zones allow to define a certain number of criticality and transformability levels, which are therefore specific to each zone. In this way, for each zone and for each level of criticality/transformability associated with it, it is possible to develop a range of compliant design solutions for a structural and energy renewal compatible with historical values.

In this context, the research is also aimed at improving the Indoor Environmental Quality (IEQ), described as the set of indoor environmental conditions of a building in terms of comfort. It includes air quality, daylight and visual conditions (from inside and outside), acoustics, lighting conditions and thermo-hygrometric comfort, without neglecting the significant interaction of users with the environment through the good management of these variables. A proper management of the space allows, in fact, to monitor and govern through appropriate tools the conditions described above, through a functionally efficient organisation of the environments.

Interventions aimed at increasing the energy performance of buildings cannot be separated from actions aimed at improving the behaviour of the building envelope—an energy passive element—which is mainly responsible for energy losses. In addition, the efficiency of the technological systems that equip buildings must be optimised. In terms of increasing the performance of passive components, the research aims to investigate the way materials, technologies and energy improvement solutions can be adapted to a historical context, exploring their impact on the environmental performance and on the indoor comfort of the building under analysis.

Finally, through a synthesis process, the best design solutions applicable to each Homogeneous Zone are defined, as long as they are compatible and respectful of the values and architectural identity of the building complex. These design solutions constitute the intervention scenarios identified in relation to previously defined levels of criticality/transformability, the effectiveness of which is verified thanks to the use of modelling and simulation software.

The above-mentioned levels of improvement must therefore be understood as part of a single intervention that is expressed through compliant design scenarios. For all these reasons, it is essential that the construction contractors, and even before the designers, take into account all the parameters that contribute to the good enjoyment of the architecture, rather than focusing exclusively on thermo-hygrometric comfort, air quality or static safety. Therefore, the intervention scenarios represent a synthesis, for the various Homogeneous Zones, of the analysed features: energy and seismic performance, building enhancement and cost–benefit ratio.

The definition of the intervention scenarios is therefore a fundamental step to achieve the objective of an integrated design, which is consistent with the cognitive analysis and in line with the objectives underlying the energy and structural refurbishment. The method can be summarised in three successive phases:

- **Cognitive phase**: it concerns the study of the current state and conservation of the building. In this phase, the architectural development of the building since its construction until the current date is reconstructed, possible due to the study of documents and historical evidence. In situ monitoring campaigns are foreseen, aimed at precisely identifying the envelope features, and the internal and external microclimatic conditions to which the building is subject. This phase is characterised by visual, thermographic, heat-flow meter measurements, monitoring of indoor and outdoor thermo-hygrometric conditions, seismic vulnerability analysis and data processing should be carried out. In this way, the effects of improving the energy performance of the passive components can be observed and measured. The more accurately this cognitive phase is carried out, the more it is possible to define reliable energy and structural models, based on the evidence of the processed data.

- **Operative phase**: the energy and structural models of the building complex in its current state (ante-operam) are created using dedicated software. This step also identifies the Homogeneous Zones, i.e. the minimum units of intervention, on the basis of which energy and structural interventions are defined in accordance with the requirements, suitably combined so as not to interfere with each other. Subsequently, a selection of "n" possible scenarios is made, identified according to the criteria of compatibility and technical feasibility. These scenarios are modelled through dynamic simulation tool, in order to conduct the energy and structural analysis post-operam and to assess the benefits arising from the implementation of the planned interventions.
- **Final phase:** the energy and structural results of the "n" possible intervention scenarios are elaborated in relation to different targets—reduction of CO_2 emissions, management of heating and cooling requirements, minimisation of construction costs and execution times, etc. Based on results, guidelines and an ODT are defined.

  In Fig. 37.1, the flowchart summarises the methodology employed in the development of the ODT and Guidelines.

37.3 The Case Study of the Former San Salvatore Hospital of L'Aquila

The former hospital San Salvatore in L'Aquila is a building complex located on the northern edge of the historic city centre (Fig. 37.2). From the urban point of view, the building site, historically belonging to the quarter of Santa Maria Paganica, is bordered by the historic city walls in the north-west side and by one of the main roads linking to the south side (Viale Duca degli Abruzzi). Its location, given its strong centrality and its historical importance, currently has a strategic interest in the development of the entire surrounding urban fabric.

The building is located immediately behind the ancient city walls, of which some parts, built in the *pomerium*,[3] have been erased. The health functions were moved to a new hospital on the outskirt of the city. Therefore, the building has undergone progressive degradation, aggravated by damages due to an earthquake that struck the city in 2009. This has led to the current state of disuse, obsolescence and the total loss of function. The complexity of the building is linked to a) its architecture, rich in historical, architectural and cultural values, which makes it difficult to be transformed; and b) its state of preservation is reflected in the extreme complexity of the design of retrofit intervention. The necessity for an in-depth knowledge of the building, based on an identity multidisciplinary approach, allowed for the definition of a complete cognitive framework from a historical-cultural, architectural, figurative, functional, structural, energetic and economic point of view.

[3] Understood as the boundary of the city.

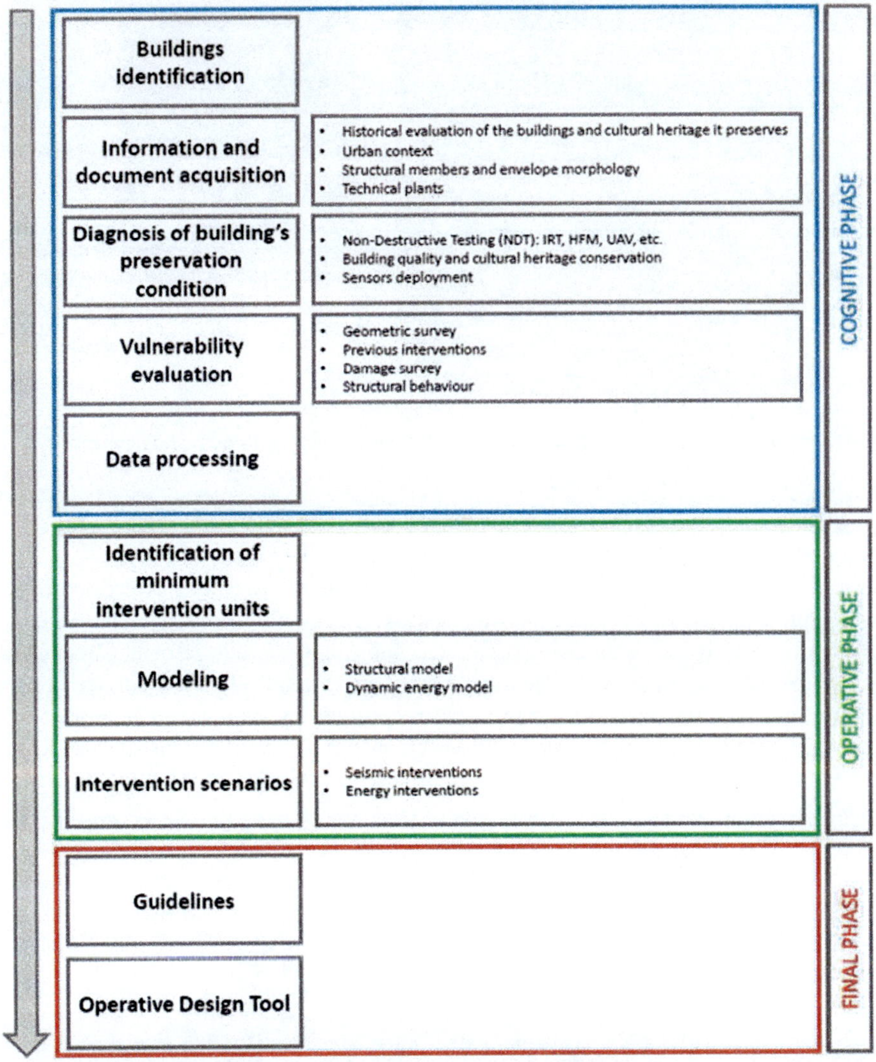

Fig. 37.1 Flowchart of the employed methodology (authors' elaboration)

The explication of individual aspects that characterise the building is fundamental to identify appropriate intervention strategies (De Berardinis et al. 2017a, b). The cognitive phase, started in previous studies (Capannolo, 2014; De Berardinis et al. 2017a, b), was then deepened by further in situ non-destructive investigations (thermography, heat-flow analysis, etc.) and mainly concerned the following aspects: metric survey of the current state, distributive-functional, historical, constructive analysis, study of historical, architectural and construction values, analysis of surface

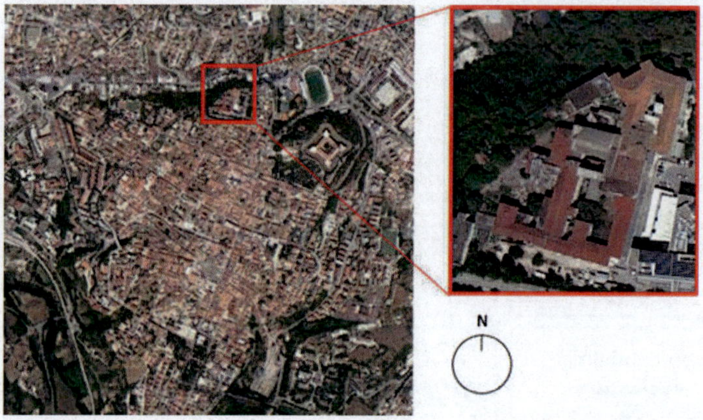

Fig. 37.2 Aerial view of the architectural complex (authors' elaboration from google maps)

degradation and structural damage and bioclimatic analysis (sunshine, radiation, ventilation).

From a volumetric-spatial point of view, the building is currently a heterogeneous set of volumes, due to the multiple transformations and additions over the years, exclusively caused by distributive, functional and hygienic requirements. Because of this, the original historical structure is difficult to read, altered by a series of buildings overlapping over time, in most cases configurable as inadequate additions or design incongruity. The complex three-dimensional metric reconstruction was done with the use of advanced survey techniques, including photogrammetric with unmanned aerial vehicle (UAV) technologies (Fig. 37.3).

From the historical-architectural point of view, the transformations of the hospital complex originate from 1875, the year of its establishment in the abandoned Santa

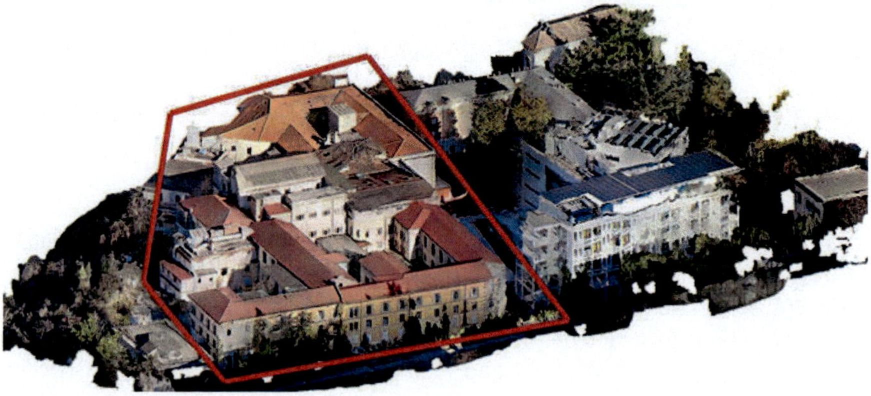

Fig. 37.3 Aerial view of the architectural complex, obtained from the UAV campaign (© Gitais S.r.l. spin off)

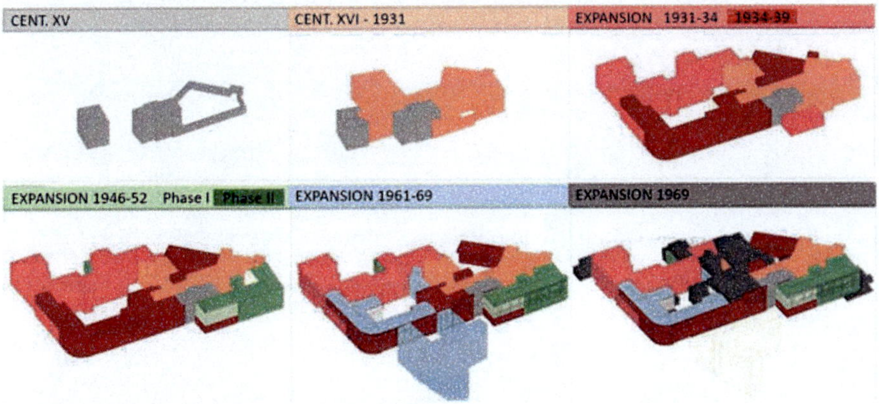

Fig. 37.4 Axonometric reconstruction drawing of the historical evolution of the former hospital complex (authors' elaboration from Capannolo 2014)

Agnese monastery, whose structures are partly recovered and fully incorporated. At present some monastic spaces including the church of Santa Agnese are still easily identifiable, while only a wall of the much later (or earlier) church of Santa Maria del Guasto remains. The subsequent interventions, including the significant extensions of the 1930s and 1950s, created new sections and reorganised existing spaces. The reason for these interventions is the need for new spaces in order to improve health services, since the hospital complex was established in an old building originally built for religious purposes (Centofanti 1984; Centofanti and Brusaporci 2011; Guarino and Vittorini 1968; Tozzi 1956). The transformations during the 1970s mainly consisted of inappropriate additions that altered the architectural and typological aspect of the ancient hospital, obstructing the external courts (Fig. 37.4).

The remarkable historical-cultural value of the complex is enhanced by the presence of numerous valuable linear elements on the facades, which are configured as further complexities to be taken into account in the elaboration of intervention strategies aimed at energy efficiency and seismic improvement.

From the technological point of view, the construction of walls is clearly visible in various points, given the obvious state of degradation of the plaster (widespread detachments). These masonry structures are heterogeneous. In the original *nucleus*, there is the presence of disordered stone masonry. Almost all of the remaining building complex has full-walled stones with a double row of brick appears every 60 cm, strengthened in some parts with reinforced concrete pillars and curbs. Additionally, the floors throughout are constructed differently, for example, in the medieval core there are vaults or mixed floors in iron and brick, such as double T beams and small vaults. While in the remaining sections of the complex there is evidence of, generally, hollow bricks and concrete floors, as shown in Fig. 37.5 (Capannolo 2014).

Despite this, the deep knowledge process was conducted adopting an interdisciplinary approach, flanking historical, bibliographic and archival research and in situ visual surveys with non-destructive investigations, including thermographic,

Fig. 37.5 Different types of walls (top) and floors (bottom) (authors)

thermo-hygrometric and thermo-flow analysis (De Berardinis et al. 2018). The non-homogeneity of the structures, in fact, entailing a nonlinear behaviour both from the energy and structural point of view, requires a thorough evaluation of the performance characteristics of the various elements (De Berardinis et al. 2017a, b). These investigations, especially the thermographic one, were also of great use for the complete definition of the historical construction phases of the building complex.

The analysis of the superficial degradation and structural damage of the portions of the building complex completed the cognitive framework to define the main homogeneous areas of intervention. The aim of such analysis is, in fact, the building devoid of all the worthless elements which, in addition to altering its appearance, are strongly compromised from the structural point of view and have contributed to a deterioration of the microclimatic conditions of the building (Fig. 37.6).

The analysis of surface deterioration shows the presence of widespread degradation phenomena on almost all of the building complex. The wall surfaces are mainly affected by chromatic alterations, detachment of internal and external plasters and efflorescence. In the interior spaces, moreover, there is humidity on the walls and floors, mainly caused by seepage from the roof. There are also widespread forms of anthropic-technological degradation, such as improper placement of technological elements and improper use of building materials. As regards the structural damage, however, the situation is different between the various portions of the building. From the oldest sections to the expansion during the 1950s, adjacent to it, are affected by more substantial damage to both the vertical structures and horizontal. In fact, there is widespread deep lesions with diagonal and horizontal directions on load-bearing walls, partial detachments and collapses of false ceilings and internal partitions, as well as partial collapses on the roof, as shown in Fig. 37.7. The remaining sections of

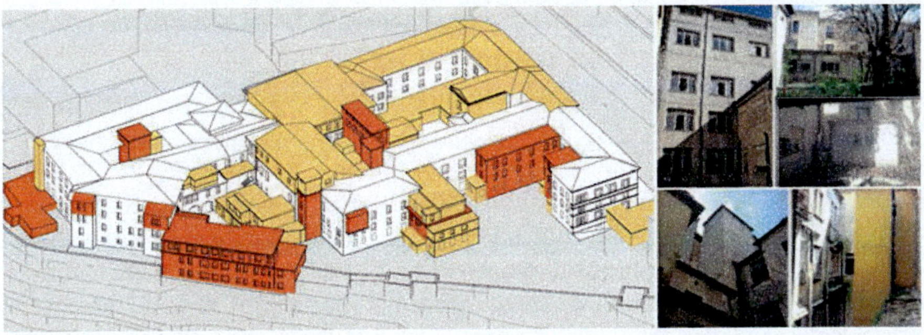

Fig. 37.6 Inappropriate additions (yellow outlines) and those of design incongruity (orange outlines) (Capannolo 2014)

Fig. 37.7 Surface degradation and structural damages (authors)

the building complex show less serious damage, but still characterised by superficial or deep lesions and collapses of internal partitions (Capannolo 2014).

Furthermore, the presence of volumes added over the years has led, following the earthquake, to the activation of some damage mechanisms, such as shearing mechanisms, partial overturning or rupture in bending of the walls and irregularities between adjacent structures.

The in-depth knowledge of this building is taken as a premise and guide first for identifying the homogeneous zones of intervention and the criticality of each from the energy and structural point of view, further structural and energy modeling and then for developing optimal intervention scenarios as proposed in the methodology.

37.4 Enhancing Architectural Heritage Through Energy and Seismic Retrofit Optimisation

The case study of a historical former hospital San Salvatore presented the integration of multiple disciplines for the application of proposed methodology for developing ODT, to achieve the objective on the recognition of homogeneous areas within the building complex. The case study highlighted some important criticalities to be considered. There was a general difficulty of having accurate measurements as a result of the dangerous conditions caused by the earthquake; furthermore, there was also a lack of knowledge of the construction phases that characterised the evolution of the building complex and of its general condition of disrepair. Finally, there was the total absence of information on the energy performance of the buildings under analysis, considering both passive and active elements. Furthermore, the main findings of the case study are summarised as follows: During the expansion phases that have taken place throughout history, the conservation of cultural heritage has not been considered, thus the expansion phase and alterations have determined a remarkable difficulty in defining the materials used. There was a lack of urban strategies during these expansion phases of the building complex that led to a substantial "isolation" of the complex itself. The knowledge necessary to have an overall view of the building complex to be reconstructed could only be obtained by the interaction of historical information, archive and in situ surveys. From this point of view, the application of methodology to develop ODT allowed to verify the construction apparatus of the building complex, highlighting materials discontinuities due to the historical evolution of the complex itself. The next development of this work will be the definition of intervention strategies and guidelines for a harmonious reconstruction of the building complex.

Developing "virtuous" and integrated strategies not only to safeguard but also for sustainable, environmental, social and economic reuse of heritage building means having a precious resource. This aspect, still little exploited today, could become crucial in urban and territorial resilience strategies developed to respond to emergencies. It is crucial to define integrated and sustainable strategies and solutions, which allow preserving and protecting the cultural heritage and, at the same time, improving the well-being and quality of life of the community. Therefore, the following results are sought:

- the conservation and protection of buildings having particular historical and architectural significance, built in the historical centre of the cities, and to date largely abandoned and in a state of decay, by employing advanced seismic improvement techniques;
- the identification and cataloguing of low-impact internal environmental control solutions;
- the development of solutions for a possible adaptation to modern environmental comfort levels (drafting of guidelines), in particular those relating to thermo-hygrometric and visual factors (lighting), for the reuse of buildings. Such aim can be pursued, with the use of non-invasive techniques and technologies and

respectful of the historical and architectural features. The approach can be a prototype to be used for other historic buildings, even of monumental value.

The large number of disused public buildings, that can be given new life through adaptive reuse, represent for society, researchers and technicians a threefold opportunity:

- the possibility of preservation of a precious asset otherwise destined for progressive decomposition and loss;
- the possibility of a useful social function again;
- the possibility of studying and preserving an important part of the Italian architectural heritage, still poorly protected, potentially undervalued and with a serious risk of loss.

In this sense, relevance comes from the guidelines for adaptive reuse and conservation that will be developed to allow similar buildings to the case study being examined and used as "nodes" in urban and regional resilience strategies and identified prototypes for architectural solutions with low environmental impact and sustainable to improve the performance of buildings, allowing the necessary adaptation to allow a new use.

The integration of the disciplines related to the study and conservation of the architectural heritage and the various phases of the construction process (from urban and territorial analysis to the definition of technological solutions for the "reactivation" of buildings) are the basis of the research project.

The ultimate vision is the "set up" of a global strategy of study and intervention on a particular type of historic building which, however, can be used, with the necessary specificities and corrections, for other types of uses, not only for historical or monumental value. As the case study presents, the methodology for developing proposed ODT will lead to opening new scenarios on the use of technologically innovative architectural solutions (activation of light tubes, solar tubes, lightweight architectural solutions, ventilation chimneys, etc.) and optimised intervention methods for the retrofit of historic buildings. The novelty relies in the formulation of low-impact solutions (environmentally and in terms of safeguarding cultural heritage) and, from an energetic and plant-related point of view, reversible and/or adaptive. Moreover, the adaptability and reversibility allow non-invasive strategies to be adopted and at the same time optimised, with the aim of achieving a good level of environmental performance.

The installation of a sensors network on the building envelope would allow monitoring of all the parameters that contribute to the IEQ, allowing to optimise the air conditioning systems, resulting in lower energy consumption and consequently improving the building energy performance.

37.5 Conclusion

In this chapter, a methodology based on multidisciplinary approach and sustainability principles, whose output is represented by Guidelines for proposing interventions for seismic and energy improvement, was presented. The proposed integrated approach was applied for analysing a specific case study harshly hit by the 2009 earthquake of L'Aquila: the former S. Salvatore Hospital, a complex city district made up of buildings of different eras.

The application of the proposed methodology for historic buildings subject to post-emergency intervention highlighted some important practical criticalities to be considered. There were difficulties, such as a lack of knowledge, available information, previous mistreatment of cultural heritage and so on. Nevertheless, the application of a multidisciplinary methodology also helps to verify the historical aspects of the built heritage and construction apparatus of the building complex, identifying architectural discontinuities due to the historical evolution of buildings and urban fabrics.

The future development of the case study work will be the definition of guidelines for a harmonious reconstruction of the former Hospital building complex. More generally, developing an Operative Design Tool for buildings undergoing post-emergency reconstruction is a useful multidisciplinary tool for policy makers and designers during the retrofit and renovation processes. Adding to this is a well-thought-out planning process.

Acknowledgments This research stem from a feasibility study for the project of a university campus in the former San Salvatore Hospital, University of L'Aquila. Scientific contributor: P. De Berardinis; work group: C. Bartolomucci (CNR-ITC), F. Giancola, L. Capannolo, M. De Vita (CNR-ITC), E. Laurini, C. Marchionni, 2017.

References

Capannolo L (2014) Piano Strategico di Sviluppo Dell'università degli Studi dell'Aquila: Riqualificazione Energetica Sostenibile dell'ex Ospedale San Salvatore a L'Aquila. Master's thesis in engineering and architecture (Supervisor Prof. P. De Berardinis). Università degli Studi dell'Aquila, L'Aquila, Italy

Centofanti M (1984) L'Aquila 1753–1983: Il Restauro Della Città. Colacchi, L'Aquila, Italy

Centofanti M, Brusaporci S (2011) Il disegno della città e le sue trasformazioni. Città e Storia 1:151–187

De Berardinis P, Capannolo L, Marchionni C (2015) Sustainable and energy efficient rehabilitation of the former hospital of San Salvatore in L'Aquila. In: Brebbia CA, Hernández S (eds) Structural studies, repairs and maintenance of heritage architecture XIV, Proceedings of STREMAH, A Coruna, Spain, 13–15 July 2015, vol 153 of WIT Transactions on the Built Environment. WIT Press, Southampton, UK; Boston, MA, USA, pp 171–182

De Berardinis P, Bartolomucci C, Giancola F, Capannolo L, De Vita M, Laurini E, Marchionni C (2017) Studio di Fattibilità per il Progetto di un Polo Universitario Nell'ex Ospedale San Salvatore. Università degli Studi dell'Aquila, L'Aquila, Italy

De Berardinis P, Rotilio M, Capannolo L (2017b) Energy and sustainable strategies in the renovation of existing buildings: an Italian case study. Sustainability 9:1472

De Berardinis P, Bartolomucci C, Capannolo L, De Vita M, Laurini E, Marchionni C (2018) Instruments for assessing historical built environments in emergency contexts: non-destructive techniques for sustainable recovery. Buildings 8(2):27

de Rubeis T, Nardi I, Paoletti D, Di Leonardo A, Ambrosini D, Poli R, Sfarra S (2016) Multi-year consumption analysis and innovative energy perspectives: the case study of Leonardo da Vinci International Airport of Rome. Energy Convers Manag 128:261–272

de Rubeis T, Nardi I, Ambrosini D, Paoletti D (2018) Is a self-sufficient building energy efficient? Lesson learned from a case study in Mediterranean climate. Appl Energy 218:131–145

De Vita M, Mannella A, Sabino A, Marchetti A (2018) Seismic retrofit measures for masonry walls of historical buildings, from an energy saving perspective. Sustainability 10(4)

Guarino M, Vittorini M (1968) Studi Preparatori per la Realizzazione di un Centro Ospedaliero a L'Aquila. Alterocca, Terni, Italy

Laurini E, De Vita M, De Berardinis P, Friedman A (2018) Passive ventilation for indoor comfort: a comparison of results from monitoring and simulation for a historical building in a temperate climate. Sustainability 10(5)

Lucchi E, Pracchi V (2013) Efficienza energetica e patrimonio costruito: La sfida del miglioramento delle prestazioni nell'edilizia storica. Maggioli Editore, Milano

Mannella A, De Vita M, Sabino A (2017) Interventi combinati di miglioramento sismico ed efficientamento energetico di edifici esistenti in muratura. In: Proceedings of XVII ANIDIS congress, L'Ingegneria sismica in Italia, Pistoia, Italy

Nardi I, de Rubeis T, Perilli S (2016) Ageing effects on the thermal performance of two different well-insulated buildings. Energy Proc 101:1050–1057

Nardi I, de Rubeis T, Taddei M, Ambrosini D, Sfarra S (2017) The energy efficiency challenge for a historical building undergone to seismic and energy refurbishment. Energy Proc 133:231–242

Nardi I, Lucchi E, de Rubeis T, Ambrosini D (2018) Quantification of heat energy losses through the building envelope: a state-of-the-art analysis with critical and comprehensive review on infrared thermography. Build Environ 146:190–205

Smarra F, Achin J, de Rubeis T, Ambrosini D, D'Innocenzo A, Mangharam R (2018) Data-driven model predictive control using random forest for building energy optimization and climate control. Appl Energy (Invited Paper) 226:1252–1272

Tozzi A (1956) L'Ospedale San Salvatore dell'Aquila nel Passato e nel Presente. Tip. F.lli Centofanti, L'Aquila, Italy

Index

© The Editor(s) (if applicable) and The Author(s), under exclusive license to Springer Nature Switzerland AG 2021

F. F. Arefian et al. (eds.), *Historic Cities in the Face of Disasters*, The Urban Book Series,
https://doi.org/10.1007/978-3-030-77356-4

B

C

D

N

Q

R

S

T

V